AN INTRODUCTION TO
THE LAW AND ECONOMICS OF
ENVIRONMENTAL POLICY:

ISSUES IN INSTITUTIONAL DESIGN

RESEARCH IN LAW AND ECONOMICS

Series Editor: Richard O. Zerbe, Jr.

RESEARCH IN LAW AND ECONOMICS   VOLUME 20

AN INTRODUCTION TO THE LAW AND ECONOMICS OF ENVIRONMENTAL POLICY:

ISSUES IN INSTITUTIONAL DESIGN

EDITED BY

TIMOTHY SWANSON

University College London, UK

JAI

An Imprint of Elsevier Science

Amsterdam – Boston – London – New York – Oxford – Paris
San Diego – San Francisco – Singapore – Sydney – Tokyo

ELSEVIER SCIENCE Ltd
The Boulevard, Langford Lane
Kidlington, Oxford OX5 1GB, UK

First edition 2002
Second impression 2002

Library of Congress Cataloging in Publication Data
A catalog record from the Library of Congress has been applied for.

British Library Cataloguing in Publication Data
A catalogue record from the British Library has been applied for.

ISBN: 0-7623-0888-5
ISSN: 0193-5895 (Series)

Transferred to digital printing 2005.

CONTENTS

SECTION A: THE LAW AND ECONOMICS OF ENVIRONMENTAL POLICY MAKING

PART 1: OPTIMAL PROCEDURE FOR POLICY MAKING

PART 2: OPTIMAL INFORMATION FOR POLICY MAKING

PART 2: OPTIMAL ENFORCEMENT

LIST OF CONTRIBUTORS

Carolyn Abbot	University of Manchester, Manchester, UK
James Boyd	Resources for the Future, Washington D.C., USA
Marcel Boyer	Université de Montreal, Montreal, Canada
Daniel H. Cole	Indiana University of School of Law, Indiana, USA
Michael Faure	Universiteit Maastricht, Maastricht, The Netherlands
Peter Z. Grossman	Butler University, Indiana, USA
Achim A. Halpaap	Yale School of Forestry & Environmental Studies, New Haven, USA
Anthony G. Heyes	Royal Holloway, University of London, Egham, UK
Andreas Kontonleon	University College London, London, UK
Richard Macrory	University College London, London, UK
Robin Mason	University of Southampton, Southampton, UK
Jonathan Remy Nash	Tulane Law School, USA
Anthony Ogus	University of Manchester, Manchester, UK
Donatella Porrini	Instituto di Scienze Economische e Statistiche, Milan, Italy

Richard L. Revesz   New York University School of Law, New York, USA

Susan Rose-Ackerman   Center for Advanced Study in the Behavioral Sciences, Stanford, USA

Richard B. Stewart   New York University School of Law, New York, USA

Timothy Swanson   University College London, London, UK

Farhana Yamin   Foundation for International Environmental Law and Development, London, UK

Richard O. Zerbe, Jr.   University of Washington, Seattle, USA

PREFACE

I am pleased to present this volume published by Elsevier Science, our new publisher. I suggested such a volume to Professor Timothy Swanson and asked if he would be willing to undertake its editorship. Professor Swanson brings to this task his own considerable reputation, an international flavor which characterizes this volume and his good judgement. He organized the conference which this volume represents and edited the results. I and our readers owe him our thanks for these useful high quality articles.

The plan for the Journal is to produce more of these specialty volumes. The next issue is one on antitrust edited by Jack Kirkwood. Those interested in undertaking editorship on special topics in law and economics are encouraged to contact me.

Richard O. Zerbe Jr.
Seattle
Zerbe@u.washington.edu

AN INTRODUCTION TO THE LAW AND ECONOMICS OF ENVIRONMENTAL POLICY: ISSUES IN INSTITUTIONAL DESIGN

Timothy Swanson

1. AN INTRODUCTION TO THE ISSUES OF INSTITUTIONAL DESIGN IN ENVIRONMENTAL POLICY

The study of the law and economics of environmental policy is interesting because it is the study of institutional design issues precisely in that area of policy where institutions are most difficult to design. Almost by definition environmental issues are those that sit at the intersection of complex natural, social and institutional phenomena. Complicated environmental media (atmospherical, hydrological, biological) render the derivation of optimal management strategies difficult and complex. Complicated social contexts (inter-personal, inter-generational, trans-national) render simple solution concepts inapplicable. And the continuing need for increasingly more complicated institutions to deal with these complexities, and those who exploit them, makes the question of institutional design a dynamic and strategic enterprise.

An Introduction to the Law and Economics of Environmental Policy: Issues in Institutional Design, Volume 20, pages 1–22.
© 2002 Published by Elsevier Science Ltd.
ISBN: 0-7623-0888-5

This volume charts the current state of the art in the analysis of the issues of institutional design in the context of these complexities. It considers the core topics of: (a) Policy Making; (b) Instrument Choice and Design; and (c) Monitoring and Enforcement. Each of the papers takes a different issue or case study as its motivation, but the sum of the papers constitutes a framework for considering institutional design issues generally. The motivating concept in this field concerns the importance of comparative institutional cost-effectiveness in the pursuit of effective policy. That is, how is it that the institutions selected to implement the chosen policy contribute to the ultimate cost-effectiveness and efficiency of that policy?

We will see in the following sections, and in the ensuing chapters, that the study of these issues in the context of environmental policy is a pursuit that is increasingly focused on the costs of complexity. As we continue to study our institutions and their weaknesses, we increasingly appreciate the large set of weaknesses inherent in simplistic approaches; however, as we attempt to address these weaknesses, we are reminded of the increasing costs of introducing complexities into our institutions. For example, the study of marketable permits once focused exclusively on the importance of incorporating heterogeneity in firm's abatement costs, while now it is seen to be equally important to incorporate other forms of heterogeneity within these systems as well (such as spatial heterogeneity in environmental impacts). Can institutional design cope with increasing complexity? This volume examines these same trade-offs again and again in various facets of institution-building.

Another common theme in the volume is the increasing recognition of the important potential role for private agents and institutions in policy design and implementation. As institutions have become more complex, we have come to recognize that there are many niches that might be inhabited by private individuals and firms rather than public agencies. The potential for privatizing many of the functions of environmental regulators is increasingly under scrutiny, as it is in many other fields of regulation. Can individuals undertake some of the monitoring of environmental performance under liability regimes? Can financial institutions perform some of the enforcement roles previously undertaken exclusively by government? What are the costs and impediments to the substitution of private for public forms of environmental regulation?

Finally, the volume also illustrates the study of the issues at the intersection between economics and the other disciplines of moral philosophy. Environmental issues raise many of the most salient problems concerning the role of the state in society: distribution, fairness, intergenerational equity and the management of uncertainty. Many of the papers discuss the nature of the limitations of the discipline in addressing these topics. Should individual preferences be the basis for deciding some or all of these important issues? Should law and economics attempt to incorporate wider values into questions of institutional choice and design? What are the core functions of the state?

We turn now to surveying the papers within a general law & economics framework. This allows us to set out the issues of institutional design, by considering the papers in an organized fashion around the important issues. This does not mean that the individual papers were constructed to meet this framework, or that the papers address only the issues mentioned here. The interested reader should turn to the individual papers to find out the full message intended by the author. The purpose of the survey below is simply to use the papers to illustrate the approach that law and economics takes to the general problem of institutional design in the case of the environmental problem.

This survey consists of three core sections on policy, instruments and compliance. In the first section (on policy making), we will survey the role of both process and preferences. In the second section (on instruments), we will survey the role of both instrument choice and instrument design. In the third section (on compliance), we will survey the role of both monitoring and penalties. In the final section, we state our conclusions about the state of the art in the law and economics of environmental policy.

2. THE LAW AND ECONOMICS OF POLICY MAKING

The optimal institutions for policy making are important for determining the frameworks within which environmental policy are made. This is crucial for determining both the procedure of policy making, and ultimately the substance of those policies. In this section we consider the choice of the optimal procedure for environmental policy making; the choice of the optimal information for environmental policy making and the ultimate role limits of the economic approach to policy making. These issues will illustrate the importance of institutional design in determining the procedures and substance of policy making.

2.1. The Optimal Procedure for Policy Making

The chapter by **Susan Rose-Ackerman et al.** opens the volume by considering the issues concerning the Aarhus Convention, and the procedural environmental rights that it would confer. The text of this international agreement has been adopted under the auspices of the UNECE in order to provide rights to information and participation within the environmental policy making process to the citizens of Europe. Professor Rose-Ackerman asks the question: Why has the convention attracted relatively little attention from western European countries, when it contains procedural rights of no greater substance than those already conferred upon the citizens of countries such as the U.S.? What is it about the systems of government that would render the procedural rights in the convention more or less interesting to the various governments concerned?

Rose-Ackerman lists three important components of procedural rights system: (a) the right to participate within the policy making process; (b) the right to be informed about the policy making process; and (c) the right to request judicial review of the outcome from the administrative process. These rights are the necessary components for enabling the engagement of individuals and groups in the policy making process, if they should desire to be so engaged. They are the core elements of the rights conferred under the terms of the Aarhus Convention, as that text was agreed by the UNECE member states in 1998. They also appear to be characteristic of the administrative procedure that is used in the U.S.

Although several eastern European states have ratified the Aarhus Convention, only two western democracies have done so to date. Why would this be the case? One reason is that the policy making process provides for engagement at other points, other than within the administrative process, and these other points might be substituted. In Germany, for example, an electoral process based on proportional representation implies the likelihood of a coalition style government capable of representing a wide range of public participation at the highest levels of government. Currently the Green party in Germany is one of the coalition partners, and it represents the interests of many environmental groups within the parliament and the executive branches. When proportional representation is the electoral mechanism, the guarantee of procedural rights in administrative proceedings is less important, as proportional representation in government is already guaranteed at the highest levels. It would then be superfluous (or repetitious) to guarantee a second round of representation. Then the issue presented by proportional representation is: Where is it most cost-effective to guarantee representation – at the executive or the administrative level? There is a trade-off between expediency and effective representation at whichever level the representation occurs.

Another reason that parliamentary forms of government are reticent in providing procedural rights is the lack of will to make long term commitments to current policies. Within a presidential form of government, the existence of checks and balances guarantees that any policy is the result of some measure of consensus. When the policy issues, there is a more broad-based will to place it into effect, and to make a commitment to its effectiveness. The granting of procedural rights to the polity represents that commitment. In a parliamentary government, however, there is less of a role for consensus and commitment; each government acts relatively unchecked during its time in office and the polity provides checks through removal and replacement of governments. In this context there is less of a role for procedural rights, as there is less of an interest in long term commitment to consensus-based policies. To convey procedural rights within a parliamentary system would be to convey a level of commitment upon policies that has never previously existed.

So, in the law and economics of environmental policy making, there are choices to be made between different types of institutions that might be used to represent individual preferences in environmental policy making. Voting mechanisms might be used to generate proportional representation, but then this might create a relatively divided (and hence less effective) executive branch. On the other hand, non-proportional representation might be used to generate undivided offices, but then representation must be achieved elsewhere to satisfy the polity. It might make little sense for states using proportional representation to duplicate this representation function in procedural guarantees within the administrative process. The optimal choice of institutions depends on what is already there.

2.2. The Optimal Information for Policy Making

The chapter by **Richard Stewart** considers the optimal level of information to be used in policy making in the context of a case study concerning the "precautionary principle". This is the substance of much policy making in the area of the environment, given the level of uncertainty that often exists about the amount and probability of potential harm from an activity or technique. When should regulation be initiated against an uncertain harm? Should the burden of proof regarding the need for regulation lie with the proponent or the regulator? Should uncertainties be dealt with differently from risks or in the same manner?

Professor Stewart analyses four different versions of the precautionary principle (PP), two weak and two strong. The weaker versions of the PP place no burden of proof on the proponent of an activity with uncertain harm potential, but they require regulators to consider regulation of all such activities and to consider implementing a "safety margin" to incorporate that uncertainty. The stronger versions of the PP place the burden of proof on the proponent of the activity with potential for uncertain harm, requiring either the best available technology to be used to manage it or (under the strongest version of the PP) the outright prohibition of any activity unless it is demonstrated to be reasonably safe. The chapter focuses on the strong versions of the PP.

What sort of information should be required for allowing an activity with potentially harmful consequences to proceed? Professor Stewart divides the information forms into three categories: Type (1) known and determinate; Type (2) known but stochastic (i.e. harm drawn from a known distribution); and Type (3) unknown (i.e. the distribution of potentially harmful consequences is itself unknown). He argues that the irony of the PP is that it attempts to treat the activities of Type 3 as more serious than those of Type 2, irrespective of the level of risk entailed by the Type 2 activity. In effect, he argues, this dichotomy must be false as it would otherwise generate inexplicably diverse treatment of similar phenomena. The approach to regulation must be fundamentally the same but with some additional treatment of the uncertainty inherent in the situation.

The optimal approach to risk management has long operated on the principle of the minimization of expected damage, where the distribution of the probability of harm is known. The same approach is generally used when the distribution is unknown. Then the usual approach is to assume that there is a family of possible distributions from which the actual distribution would be drawn, but it is unknown which of the distributions actually applies in the situation. Then the distinction between risk management (the Type 2 problem described above) and the management of uncertainty (the Type 3 problem) is simply one of compounded probabilities (the family of probability distributions from which the actual risk distribution would be drawn). In this way, uncertainty is seen to be a more difficult problem than risk management, but of the same fundamental form and substance. For this reason it deserves the same fundamental approach to management.

What is the additional factor that should be incorporated to take into consideration this problem of uncertainty? The concept of "quasi-option value" indicates that the cost of uncertainty lies in the value of waiting for further information on the distribution of costs. If there is likely to be a more well-refined understanding of the applicable distribution of potential costs by reason of allowing the passage of time, then the cost of proceeding on the current information is the cost of the lost refinement in information. It is this premium that should be added to the cost of activities with uncertain (as opposed to merely risky) outcomes.

This points to the even more fundamental problem with the strong versions of the PP. They halt the information-gathering process rather than build it into the policy making process. The absence of information should not halt the development of the activity and the policy making process in regard to it; it should rather be the basis for information-gathering activities regarding the activity. The costs of designing institutions that are precautionary is that they do not focus on generating the missing information on the costs of the activity.

2.3. *Optimal Use of Economics in Policy Making*

Two papers consider the limits of economics in resolving problems in environmental policy making. The underlying assumptions of the discipline of economics are based in the philosophy of "welfarism": the belief that the advancement of general social welfare should be the object of social choice. To this end the discipline uses (as the measure of success) the satisfaction of individual preferences, no matter what the values they represent and no matter what the aggregate outcome. This approach raises problems with both philosophers in academic journals and with laypersons and lawyers in situations of policy making and implementation. The questions raised concern: What are the limits of individual preference-based approaches to policy making? And, what refinements to preference-based approaches are necessary to make the aggregate outcomes more palatable?

The paper by **Richard Zerbe** argues the latter point. He finds that other values might be readily incorporated into the economic analysis of policy making, without violating even the rules of welfarism. His case study concerns the problem of fairness in assessing outcomes, and the importance of incorporating basic tenets of fairness alongside preference satisfaction in the analysis of policies. For example, if a society cares about the distributional effects of a policy as well as the satisfaction of preferences, then Zerbe demonstrates that the exclusion of this additional value generates outcomes that are unacceptable under welfarism. This is because individuals are able to care about others' preference satisfaction as well as their own, and to suffer if that fact is ignored. Then the failure to incorporate such concerns itself violates the fundamental principles of welfarism.

In essence the current limits of economic analysis based on narrow conceptions of individual preference satisfaction are self-imposed, and they generate problems that are difficult to resolve. For example, the problem of long term discounting is difficult to address within conventional frameworks; even a cost equivalent to the entirety of World Current GDP is acceptable given that adequate time is allowed to pass before its incurrence (witness the problems of radioactive wastes and global warming). This problem stems as much from our conceptual framework as from our current technologies. If we were allowed to express our empathy with future generations in terms of current preferences for intergenerational fairness, then the problems surrounding discounting would disappear. The exclusive reliance upon individual preferences is a narrow approach to preferences that serves us well in some situations but not so well in others.

The paper by **Andreas Kontoleon et al.** considers this framework issue further in the context of the introduction of preference-based approaches within policy making. The case study he considers is the problem of conferring legal standing on individuals to claim damages for broad environmental harms. Who should be able to claim standing to allege harms to biodiversity? Is a preference-based approach an adequate or acceptable approach to policy making concerning broad social values such as environmental services? The case study concerns the recent EU White Paper on Environmental Liability and its proposals regarding claims for damages to such things as biodiversity.

Kontoleon accepts that welfarism has long been a discredited approach to rely on exclusively in social decision making; since the time of Plato its many deficiencies as an overarching value system have been debated and discussed and agreed. There are many other criteria (fairness, human rights, etc) that are too inadequately incorporated within this system for it to be viewed as a moral or acceptable form of exclusive decision making. However, this does not imply that it should be entirely rejected, but only that its limitations should be known and incorporated into the decision making process.

In essence, individual preference-based approaches should be viewed as informational in nature rather than determinative in policy making processes. The information provided by such an analysis should advise the policy maker on how individuals perform trade-offs concerning the good under consideration and other goods and services. This is important but limited information, useful in the policy making process, and it is acceptable if seen to be advisory in nature.

But does it make sense then to utilize preference-based approaches to manage a resource such as biodiversity? For example, is it sensible to use the civil litigation system and preference-based valuation for the purpose of environmental management, if the limits of the approach are known and accepted? Kontoleon argues that it is acceptable but unlikely to be cost-effective in most cases. The problem lies in the definition of the class of injured persons in the case of broad environmental resources. Importantly, this is likely to depend on the knowledge of the individual concerned, as the absence of knowledge is likely to indicate both an absence of injury and a relative lack of interest. Except in the case of the most widely-known and understood environmental resources, this absence of knowledge is likely to be both common and costly to verify. The determination of the appropriate rate of aggregation across a society for a harm to a public resource (such as biodiversity) is likely to be costly and complicated to address on a case by case basis.

Therefore, the use of individual preference-based approaches to environmental policy making clearly have their limits. These limits need to be addressed and understood when designing institutions. There is little reason to complain about the public's unwillingness to accept economic solutions if economists are unwilling to accept the public's broader set of values and objectives. When the costs of preference-based approaches are considered, it will sometimes be the case that they will either need to be modified (as in Zerbe's argument), accepted as limited (as in the advisory nature of cost-benefit analysis) or rejected altogether. Even the law and economics approach to institutional design has its own comparative institutional costliness.

3. THE LAW AND ECONOMICS OF INSTRUMENT CHOICE AND DESIGN

The choice of the optimal instrument for implementing environmental policy is the next stage of optimal institutional design. The law and economics of instrument choice concerns the same issues of comparative costliness discussed above. The objective is to minimize the cost of the problem to society, considering all of the social costs emanating from the problem and from its resolution. The range of available approaches is also wide, ranging from command and control (CAC) based regulatory approaches to economic instruments (such as marketable permits) and ultimately to decentralized approaches such as liability regime. When will each instrument be optimal? What are the considerations in choosing between them?

The issue of instrument design is equally of interest. This raises very clearly the problem of the optimal level of complexity that is optimal in regulatory policy. As environmental policies recognize more complicated social objectives and more complex forms of interaction, the recommended forms of institutions become similarly complex. What are the costs of increasing complexity in regulation? What are the determinants of those costs, and how should they be managed? The design of instruments must take into consideration both what is theoretically desirable and what is practically achievable. This is always a lesson to be drawn from the application of law and economics to policy making issues.

3.1. Optimal Instrument Choice

The choice of the appropriate instrument for environmental policy implementation has always been seen to minimize the sum of: (a) the costs of the environmental harm; (b) the costs of abatement; and (c) the costs of the institution chosen to manage the problem. **Daniel Cole** and **Peter Grossman** argue that many proposals for the use of economic instruments in policy implementation fall down at this last hurdle. Their argument is based on the observable fact that economic instruments often require substantial ex ante institutional investment in order to give them effect. For example, there is little prospect for emissions trading in the absence of continuous emissions monitoring. The capacity to avoid emissions monitoring renders the permit for trading relatively worthless, and so the market fails to exist. Before the market can exist, the investment in the necessary monitoring technology must exist (and in the case of the U.S. SOX regime this investment amounted to several billions of dollars). The addition of this additional costliness to the objective function will render all but the most cost-effective economic instruments inefficient.

It is true that it is often the case that the proponents of economic instruments fail to recognize the inherently more costly nature of the infrastructure required for permit-based regimes. While stressing the importance of minimizing the polluters' abatement costs, they do not always consider the public costs of achieving this objective. It has long been known that the relative advantage of many CAC regimes is their relative simple and straightforward nature. Their costs often derive from the many facets of the problem that they ignore in tackling it. However, it may be necessary to await technologies, or even to first adopt policies to generate new technologies for monitoring and enforcement before it is possible to move to the next generation of instruments.

This point is especially important in the case of international environmental problems, in which the institutional and technological capabilities for monitoring vary across all of the member states. What is the purpose of adopting the latest generation of regulatory instruments if the institutional capacity to monitor and to implement is clearly unavailable in the states concerned? For this reason the objective must be to either first develop and transfer the technologies for monitoring to the member states, or to adopt an instrument compatible with national capabilities.

The general point is that optimal institutional design requires that the complexity of the problem be addressed (such as all of the facets of abatement cost and pollution cost), but also that it is equally important that the cost of introducing instruments to deal with that complexity be incorporated. The progress of institutional design is equally about both; the costs of inadequate institutions and the costs of effective institutions.

More generally, the problem of instrument choice is a problem of choosing between institutions (not just between simple and complex institutions). **Marcel Boyer** and **Donatella Porrini** provide an interesting and excellent survey on this topic, comparing the relative costliness of centralized and decentralized approaches to environmental regulation. The initial analysis of this issue is attributable to Shavell (1984), in which he laid out the basic categories of costliness as follows: (a) Information costs; (b) Limited Liability costs; (c) Transactions costs; and (d) Administration costs. The objective in choosing between institutions should be selection of that institution that minimizes the sum of these costs. Boyer and Porrini consider these categories of costliness while contemplating the choice between decentralized institutions (such as a liability regime) versus centralized ones (such as an agency-regulator).

"Information costs" concern the costs of acquiring information within a decentralized economy. When these costs are disseminated widely across an economy, then it may be easiest for the individuals themselves to gather the information. For this reason liability regimes make sense as a means of reducing monitoring costs. However, when the information is more concentrated and more complex, then it makes sense for an agency to deal with the complexity of the issue.

"Transactions costs" concern the costs of contacting similarly situated parties, e.g. grouping similar claims together, and putting together a joint action. When a harm is very diffuse, then the fixed costs of initiating the civil process may render it very difficult for any one victim to proceed. Then the political process may aggregate these claims more cost-effectively than would the civil system.

"Liability limitation costs" concern the specific problems involved in pursuing a given polluter under the terms of the civil liability system. These involve problems of judgment-proofness (i.e. firms with limited liability protection and inadequate assets) and insurance limitations (e.g. policies that limit coverage in time). In short, the civil liability system requires that liability provide some potential threat of liability in order for it to provide some incentives for internalization. Where the limited liability system negates the potential impact of civil liability, then these incentives are negated.

"Administration costs" concern the costs of making policies, monitoring for compliance and enforcement of policies. These are the primary categories of costliness for the regulatory agency. In the case of a civil liability system, monitoring is continuous and decentralized, and the development and enforcement of policy is undertaken within the judicial system on a case-by-case basis. One important issue concerns the comparative costliness of these two institutions in policy making.

Boyer and Porrini argue that there is an important and relatively overlooked sub-category of costliness under this heading. This concerns the matter of the "Costs of Capture". They are concerned about the important task of weighting and aggregation of individual preferences within the policy making institution. The political economy of the situation indicates that the administrative branch is more likely to be subject to influence by concentrated lobbying representing specific preferences, while the judicial branch is less likely to be able to be informed of the entire range of preferences. On balance, it is more likely that the problem of unrepresentative policy making is more prevalent in administrative contexts, and so this cost should be added to the centralized alternative.

These various categories indicate the sorts of situations in which centralized and decentralized alternatives are more likely to be optimal. Decentralised alternatives are likely to work best where the harm is concentrated on a few parties whose preferences are then representative of the society concerned with the problem. Centralised alternatives are probably best when the harms are diffuse and the institution must undertake the task of aggregating preferences across a wide swathe of society.

These papers on the optimal approach to instrument choice indicate that there are two important types of discussion taking place in the literature. One concerns the costs of complexity in instrument choice and design. The other concerns the relative benefits of centralized over decentralized options in instrument choice. The issues concerning decentralized instruments extend quite a bit further into the roles of insurers and financial institutions, and these are discussed further below.

3.2. Optimal Instrument Design

How should a liability system be designed in order to achieve environmental policy goals? This is a very important question to be asked now in the EU as that jurisdiction contemplates the prospect of adopting civil liability as an approach to implementing environmental policy. **Michael Faure** considers the issues of liability system design from this perspective. He considers the different ways in which environmental risks might be insured, and how liability systems might be designed to create efficient insurance systems.

In regard to the design of optimal liability systems, there are several important issues to consider: What are the obvious problems and failures of a liability system? How should one be designed for purposes of addressing problems concerning the environment? Liability systems are important potential instruments for addressing environmental problems. They operate as *ex post* pricing systems for environmental harms, assessing harms that result from given activities and charging polluters the cost of compensating that damage. As the literature has demonstrated, this is the correct cost to internalize into decision making regarding environmental use, and so the system aids in resource allocation decisions about environmental resources. It is also important that the system operates *ex post:* charging the price of the use of the resource after it has occurred. Such pricing systems enable the use of public resources (such as the environment) without the necessity of incurring the costs of transacting first with each of the rights-holders. If the completion of such transactions were necessary prior to use, then the formidable transactions costs involved would prevent the initiation of many new activities and technologies. Liability systems have built into them the opposite presumption than do systems based on the precautionary principle. So long as the costs are internalized *ex post* the timing of their internalization does not matter. Actors interested in advancing new activities will have to do so in the knowledge that they will ultimately need to pay the price of all exploited resources. The liability system is only supposed to enable this use of resources without necessity of first incurring the transaction costs.

The optimal design of liability systems requires the consideration of all the problems that might prevent the ultimate internalization of environmental costs. Some of these were intimated above, in the discussion about the relative costs of using liability systems. The first problem concerns the *diffuseness and complexity* of environmental harms, and the fixed costs of bringing civil actions. The second problem concerns the intersection between the institutions of civil liability and *limited liability*, and the potential loopholes that this intersection generates (see below). The third problem concerns the need to "get the *ex post* prices right", and the reasons why judges might or might not be capable of doing so. All of these problems raise particular design issues.

The first problem ("the diffuseness of environmental harm") implies the need for institutional solutions that are capable of minimizing the transaction costs of grouping civil actions. If these costs are not minimized, then the institution of liability does nothing but reverse the burden of proof onto the injured and away from the injuror, and then leaves the costs where they lie. Liability systems make sense (when compared to property right institutions) only to the extent that the costs of such transactions are reduced by switching to the *ex post* approach. In order to do so, diffuse costs must be managed via procedural designs that enable either: (a) group representation (e.g. class actions or interest group representation); or (b) representative actions (e.g. single plaintiff representing group harm). These design issues imply a wide range of procedural measures that must be considered, ranging from attorneys' fees rules to class actions. All such procedures must be considered as possible methods for reducing the costs of representing group-based harms in civil actions.

The second problem ("the limited liability loophole") implies the need for continuous rather than exclusively *ex post* monitoring mechanisms for certain types of agents. The capacity to exploit such loopholes in the liability system implies the need for broader forms of liability rules, such as joint and several liability. Such liability is not important so much as the guarantor of a defendant, but more as the generator of a monitor. Joint and several liability, including such forms as lender liability and successor liability, create several contemporaneous partners who have the responsibility to ensure that the potential defendant is in fact sufficiently solvent for the nature of activities it is undertaking. All such forms of liability must be considered as possible methods for creating additional monitors of potentially bankrupt actors.

The third problem ("the assessment of the correct measure of damages") implies some of the issues raised in the initial section of this introduction. The assessment of damages must include all of the costs incurred by the environmental resource, in order for the price charged for its use to be accurate. But many of the services of the environment are difficult to ascribe to any one person – so then who in particular may be said to have been harmed by their loss? These are the issues of standing and valuation that were raised in the first part of this introduction. For purposes of institutional design, a liability system needs to resolve the extent to which it will enable environmental harms to be actionable by a single individual, or group, and the system by which such harms might be valued.

Institutional design is critical for liability systems to operate effectively. These systems can generate effective decentralized approaches to environmental problems or they can simply create a strong presumption against the internalization of environmental costs, the difference lies in the design.

The problem of designing marketable permit systems is investigated by **Jonathon Nash** and **Richard Revesz** This is another example of the optimal amount of complexity required within institutions of environmental policy. Marketable permit systems were the first step toward building complexity into emissions regulation. The object of a marketable permit system is to take into consideration the heterogeneity in the costs of abatement across various users of an environmental resource, and to implement the joint solution in the most cost-effective way given this heterogeneity. Marketable permits created the incentives for all users of the resource to cooperate in reducing resource use in the most cost-effective manner. This was done by giving the aggregate amount of permits to the user group, and then allowing them to cooperate in order to generate the most efficient joint use of those permits. Such cooperation could be arrived at by means of individual trading, by permits being transferred away from those who valued them least (the "low cost" abaters) and toward those who valued them most (the "high cost" abaters). Hence, by creating a market in permits, it was possible to both create static incentives to reduce emissions at least cost and dynamic incentives to generate new methods for avoiding the use of permits in the future.

The use of such artificial markets to generate incentives for cost-effective solutions is now widely accepted, and the desire to extend and elaborate upon the approach is now widespread as well. One issue that arises is: How much complexity can such artificial markets manage? Nash and Revesz consider the case of the permit system designed to deal with two sources of heterogeneity simultaneously: abatement costs and environmental impact costs. In the case of many airborne pollutants, it is not the amount of emissions that determines the environmental damage, but rather the amount of depositions and where they fall. The authors consider whether it is possible for a marketable permit system to cope with both forms of complexity simultaneously. They argue that a somewhat centralized trading regime (operating through a central website) will be able to cope with the complexity of a deposition based trading regime.

It is important to recognize just how much more complexity the addition of this one additional facet renders the problem to be solved. A current example comes from the UNECE Second Sulphur Protocol will illustrate, as it calls for the attainment of deposition standards across Europe defined at the level of a 100 sq. km. grid. This implies that each emission of SOX within Europe must in fact meet the standards established in more than 250 different deposition districts. The solution proposed by Nash and Revesz argues that the implicit acquisition of 250 different permits would now be possible by use of a central website. In effect, this advance in communications technology negates many of the problems occasioned by the enhanced complexity of large transactions in marketable permits.

More interesting is the fact that the addition of the second dimension makes one-to-one trading a second-best approach to arranging cooperation. In a single dimension of cooperation (i.e. abatement cost heterogeneity), trading can always continue until the first-best distribution of emissions abatement is achieved. With two dimensions, however, it is possible that the first-best distribution is never attained, as prior trades (motivated by one factor) might come to block others (motivated by the other). With the addition of this one additional layer of complexity, one-to-one trading of permits is not always capable of attaining the desired solution. Some other form of intervention, i.e. greater centralization, will be required in order to cause the optimal allocation to result.

The design of increasingly complex institutions, in order to attain solutions that incorporate more of the complexity of the natural and social environments concerned, necessarily implies increased costliness. The marketable permit example demonstrates that some of this costliness can be mitigated via technological change, but some of it will also be borne in terms of increasingly costly institutions. It is important to recognize however that the failure to incorporate the important complexities that exist necessarily implies the incurrence of costs (because the social optimum cannot be attained when important dimensions of the problem are ignored). Increasing complexity of institutions is necessary consequence of working on problems, but it also implies substituting increasing costs of institutions for the formerly existing costs of imperfect solutions.

4. THE LAW AND ECONOMICS OF COMPLIANCE MECHANISMS

Optimal institutions for compliance will balance the costs of non-compliance with the costs of achieving compliance. There are two parts to the pursuit of compliance: institutions to monitor for non-compliance and institutions that penalize non-compliance. The institutions available for monitoring for non-compliance cover a wide range: individuals; special interest groups; financial and insurance institutions; agencies. One of the important issues in this area is the extent to which it is feasible and cost-effective to shift monitoring functions outside of the public sector.

Optimal penalty mechanisms have also been under review recently. The original view was that greater penalties translated into greater deterrence. More recently the literature on penalties has generated a lot of thinking about the importance of maintaining both relative costs of various forms of non-compliance, and the importance of recognizing that penalties are part of an ongoing process (not a one-shot occurrence). When the complexities of this relationship are considered fully, the form and schedule of the optimal set of penalties will be likewise much more complicated.

4.1. Optimal Monitoring Institutions

Monitoring was one of the original policing functions of the state and its agencies. With the advent of civil liability as an important institution in the implementation of environmental policy, the role of monitor was extended beyond the state to the set of individuals who were harmed through environmental injuries. As it became recognized that there were many environmental harms that occurred without immediate notice, the need for ancillary monitoring was recognized. This is the economic justification for extending liability beyond the actor itself and into the set of institutions with which it works closely. For example, an insurance company extending environmental liability insurance to a firm is accepting a premium in exchange for a potential future liability. This provides the insurer with the incentive to monitor for moral hazard on the part of the insured. Similarly, when liability is extended beyond the actor itself and toward its financial partners (lenders, asset purchasers, successor firms), this also provides these partners with the incentive to monitor for non-compliance on a more continuous basis.

When is this form of continuous monitoring important? The paper by **Timothy Swanson** and **Robin Mason** examines the conditions that determine whether a firm is more or less likely to be a risk to society. The basis for the analysis is the previously-mentioned intersection between the regimes of civil and limited liability. This intersection creates a "limited liability loophole" by means of which it is possible to push costs into the future and then liquidate the corporation prior to the arrival of the liabilities. This liquidation strategy is possible whenever there is a gap between the time of arrival of benefits and the time of the arrival of the corresponding liabilities, but when will it be profitable? The authors demonstrate that the profitability of such a strategy depends on three factors: (a) the cost of taking precaution; (b) the quantity of sunk assets (i.e. the mobility of assets); and (c) the profitability or productivity of assets in the enterprise. When a firm exists in an industry where the assets are relatively mobile but unproductive in the current enterprise, there will be the incentive to use the limited liability loophole to generate immediate benefits and then shift them out before the arrival of the corresponding liabilities. This indicates that the risky firms are those for which assets are either extremely low or highly mobile.

This indicates that monitoring under liability regimes may take one of several forms. First, monitoring may take the direct form of continuous monitoring for compliance with existing rules and for potential liabilities. This would imply the necessity of relatively intrusive activities such as continuous emissions checks, and ongoing review of internal risk assessment procedures. Essentially, this form of monitoring would consist of gathering most if not all of the information within the firm, and then "second guessing" the firm on issues of potential hazards.

A second and less intrusive form of monitoring would be to check continuously only on the financial status of the firm. So long as the firm continues to hold relatively sizeable amounts of relatively fixed assets, the risk of the firm undertaking the liquidation strategy is relatively small. This is akin to the "reserve requirement" strategy for banking institutions. All that is necessary is the maintenance of a relatively significant (i.e. relative to size of potential hazards) share of sunk assets within the industry.

The manner in which such continuous private financial monitoring might be encouraged is by means of bonding forms of institutions. **James Boyd** provides a survey on the set of bonding institutions that have been adopted in recent years, and examines their capacity to manage this problem. Bonding requirements have been applied to specific industries in the U.S., and precisely in those where they should be expected to be important (i.e. large scale producers of potentially accumulative wastes). These bonding regimes have required that these firms place some financial assets into reserve for the purpose of guaranteeing future liabilities. Once again the problems arise in dealing with the complexities of the situation, and the costs that arise with complexity.

These costs derive from the fact that bonds are contractual instruments attempting to deal with unforeseeable contingencies of all forms and sizes. Thus, to some extent bonding requirements represent attempts to substitute written contractual arrangements between the state and the regulated entity for monitoring arrangements. The costs arise from attempting to answer the questions about what the future might hold. What amount of bond should be required of the firm? For what amount of time should it provide these funds? It is straightforward to see that, whatever the answer given to these questions, any limit on the length or the amount of the bond will leave a tail of limited liability somewhere in the future (and hence the continuing availability of a limited liability loophole).

This points to one of the major problems with the bonding regimes as they currently exist. Most of them allow for the prospect of self-assurance or self-bonding, i.e. where the firm places its own assets on reserve without involving another institution as guarantor. The problem here is that this manner of bonding provides assets without providing any incentives for other firms to become involved in the monitoring process. Having a second firm act as the bonding agent provides the incentives for continuous monitoring of the other firm's financial status, and without the necessity of placing so many assets in reserve. Self-bonding either places more assets on reserve than are necessary for the purpose of compliance, or it places too few on reserve and does not ensure compliance.

Optimal institutions for monitoring for noncompliance are taking an increasingly wide range of forms, as the liability institutions in the U.S. become increasingly more complicated. These new forms should be viewed as opportunities to involve other institutions in the role of the regulator, without the intrusiveness of state reporting and monitoring. Financial institutions are able to specialize in the forms of audits required to ensure viability of the corporation for liability purposes. Insurance institutions are able to monitor and regulate for moral hazard. Regulators now have a wide range of options for generating continuous monitoring, including bonding and insurance requirements. This requires that the state participate in complicated forms of contracting and contract review in place of the need for continuous monitoring. The comparative costliness of the two approaches will determine which will be the optimal approach to monitoring.

4.2. *Optimal Institutions for Penalising Non-compliance*

It would seem that at least the problem of optimal penalties would be a relatively straightforward one. More frequent penalisation of non-compliance at higher levels of penalisation would surely generate a better record of compliance. The law and economics literature of the past decade has generated a long list of exceptions to this approach. Here it is the complexity of motivations (across the regulated public) and the complexity of long term relations (between the regulated and the state) that renders simple institutions insufficient.

The paper by **Anthony Ogus** and **Carolyn Abbot** first assesses the record of penalisation within the U.K. for noncompliance with environmental standards. They discover that the number of firms penalized is very low, and the level of the penalties is low as well. They explain this observation in part by reason of the nature of the branch of governance that handles the matter of penalisation in the U.K.. There it is the judicial branch that determines whether the deviations from legislated environmental standards constitute a criminal breach of law and, if so, the level of the criminal penalty that should be applied. Since it is the judicial branch that handles these matters, the standards of proof and rules of evidence are all derived from within the criminal system. Criminal procedures are intended to protect defendants, and so the capacity and willingness to penalize for minor breaches of environmental standards is lacking.

The authors contrast the administrative system of penalisation used in civil systems, such as in Germany. There the question of liability is a civil one, termed *objectionable behaviour* rather than *moral guilt.* The burden of proof is corresponding lightened. In addition, there is an entire ladder of potential penalties, ranging from informal warning to administrative fine. These lesser penalties, together with the lesser standard of guilt and proof, make it easier to deal with minor transgressions of the law. The authors conclude that use of a civil system of enforcement provides the necessary flexibility required within a system of environmental compliance. In short, a more complicated sort of relationship than exists in criminal law is required between the environmental governance institution and the governed.

The paper by **Anthony Heyes** lists several additional explanations for this often-observed pattern of low and infrequent penalisation within common law systems. The explanations all have their source in existence of long-term relationships between the state and the regulated. For this reason the regulated knows that it is choosing its conduct not only by reference to a single event of possible penalisation, but by reference to an entire future of potential penalisation. Once it is recognized that the future can be a very long time (and the regulated agent knows this), then it is less important to impose the maximum penalty for any single event. In fact, there is leverage to be gained from withholding penalties from single (potentially mis-detected) events and instead reserve large penalties only for repeat offenders. This allows the regulator to reserve its resources for the worst offenders, and to create a virtual wall between the "good" and the "bad/ugly". Under such a strategy it makes sense to ignore the occasional violation (and even not to expend resources to detect it) as it is as likely to be the result of inadvertence or mistake (by regulator or regulatee), and for this reason the overall record of the regulator will appear to be rife with excused violations.

In general this principle points out that penalty institutions must take into consideration the potential for complexity in the relationship between the state and the regulated. The state must be able to be forgiving not only for honest mistakes, but also for mitigating behaviour that would be forsaken as self-reporting. The regulated constituency has many options for costly behaviour other than simply non-compliance: over-compliance, failure to mitigate, pollution levels far in excess of noncompliance levels. All of these other decisions must also be regulated by the penalisation institution (and implicitly they will be). The institution must be complex enough to deal with several decisions by the regulated, other than simply the decision about non-compliance.

One final, interesting point by Heyes concerns the complexity of the motives of the regulated. Sometimes a particular decision is best motivated by means other than prospective penalisation. He makes note of the fact that the commercialization of certain activities (supply of blood, recycling) appears to dampen the supply motivated by public-spiritedness. In such contexts, the implementation of explicit incentive/disincentive schemes fails to generate the desired results, because the public's motives are more complex than the incentive scheme assumes.

In general, the research in the area of optimal institutions for penalizing non-compliance indicates the importance of taking complexity into account. The archetypal and intuitive understanding of the optimal approach of the state to non-compliance simply does not take into consideration anything more complicated than the single event of noncompliance. Once a longer view is taken, and once a broader set of motivations are considered, it is apparent that first-best institutions for penalisation must be as complicated enough to deal with the full range of possible decisions under the full range of potential motivations. Given the costliness of this complexity, it becomes apparent why more flexible forms of regulatory dealing and less inflexible forms of penalisation are in play.

5. CONCLUSION: THE STATE OF THE ART IN THE LAW AND ECONOMICS OF ENVIRONMENTAL POLICY

This volume was the result of a two day workshop at the Faculty of Law, University College London in September 2001. It was an outstanding group of scholars, representative of the wide and diverse range of perspectives operating in the field of law and economics in environmental policy. They also represented the approaches of many different jurisdictions to problems of environmental policy. The papers ranged from political economy to marketable permits to optimal penalties, and yet they covered only the tip of the iceberg in this area of research. Nevertheless, it is apparent from this volume that the set of papers did an excellent job of surveying the three major areas of investigation in environmental policy: policy making; instruments; and compliance. It is also apparent that they identified most of the most important issues under investigation in the field today. This volume has identified those issues as: (a) the costs of complexity; (b) the limits to decentralization; and (c) the limits to economics in issues of environment and of policy.

After reading this chapter and the papers in the volume, it is apparent that one of the items highest on the agenda of law and economics scholars is the inclusion of greater complexity in environmental policy. This is important because the omission of the complexities of nature and of society necessarily renders the derivative policies sub-optimal. Assessing complexity and incorporating it into policy is a necessary task in the building of better institutions. However, it is also clear that the incorporation of complexity has its costs. For this reason a secondary objective must be the development of policies and institutions that will minimize the costliness of generating new technologies and new policies that allow us to complexity. This is the lesson to be learned, for example, from the discussions about marketable permits in this volume. It is clear that increased complexity in such markets has its costs, but it is equally clear that there are investments in technology (e.g. emissions monitoring) and in institutions (e.g. web-based trading) that will help to minimize those costs. The pursuit of greater complexity in institutions is another task that must be managed and optimized.

The second theme concerns the ability of institutions to engage other actors in the tasks formerly the sole domain of the state. Initially this involved individuals in the monitoring of harms under liability regimes. Now this includes much more complicated relationships between financial institutions and insurers continuously monitoring all forms and varieties of behaviour on the part of the regulated. This area of research indicates the precise manner in which various aspects of monitoring might be devolved to various agencies, under the auspices of bonding and assurance regimes. It also indicates how these institutions fail to eliminate the state's role completely, but merely shifts its focus away from policing compliance and toward writing contracts. It is similar to the franchising movement generally in public services: the state is able to engage other agents to perform certain of its former roles but only to the extent that it writes a complete and contingent contract. Again, the increased complexity in the institution comes with its costs.

Finally, the last theme noted in the volume concerns the limits of economics itself in the field of policy and in the area of environment in particular. As with any discipline it is crucial that it knows its own boundaries. Economists who work with lawyers or in the field of law will always be brushing up against others who do not share the same conceptual or framework or professional biases. For these reasons it is important to recognize that economics sometimes has difficulties in dealing with the management of the public goods and unowned resources that lie at the core of environmental problems. The use of preference-based approaches to valuing harms to biodiversity in liability suits strikes most in the legal community as far-reaching and unwarranted. On the other hand, the failure to consider anything other than individual preferences in making policy recommendations seems far too narrow-minded to generate reasonable policy. The law and economics scholar working across many jurisdictions must be aware of the limits of the discipline in order to know what recommendations make sense.

In sum, the volume gives a nice depiction of the field and a decent understanding of the enterprise. The interested reader, still interested up to this point, will find it worthwhile to make his or her way through the entire volume.

SECTION A:
THE LAW AND ECONOMICS OF
ENVIRONMENTAL POLICY MAKING

PART 1:
OPTIMAL PROCEDURE FOR
POLICY MAKING

THE AARHUS CONVENTION
AND THE POLITICS OF PROCESS:
THE POLITICAL ECONOMY
OF PROCEDURAL
ENVIRONMENTAL RIGHTS

Susan Rose-Ackerman and Achim A. Halpaap

ABSTRACT

In 1998, UNECE Member States completed negotiation of the Aarhus Convention to enhance public participation in environmental decision-making. Three years later, only two western democracies have ratified the agreement. This paper suggests why parliamentary democracies in Western Europe have been slow to ratify the Convention. We argue that their political structure discourages strong public participation in bureaucratic policy making, in contrast to separation of powers regimes, such as the United States. To illustrate our point, we discuss examples from the U.S., selected European countries, and the European Community, which has separation-of-powers features similar to the U.S.

An Introduction to the Law and Economics of Environmental Policy: Issues in Institutional
Design, Volume 20, pages 27–64.
Copyright © 2002 by Elsevier Science Ltd.
ISBN: 0-7623-0888-5

INTRODUCTION

On 25 June 1998, thirty-one out of fifty-five Member States of the United Nations Economic Commission for Europe (UNECE) signed the Aarhus Convention, an international agreement designed to strengthen democratic environmental governance.[1] Unlike other international environmental agreements, the Convention does not address substantive environmental issues, such as ozone depletion or climate change. Instead, it establishes procedural obligations for policy making, implementation, and enforcement with the aim of enhancing public participation.

The Convention is based on the premise that "every person has the right to live in an environment adequate to his or her health and well being".[2] To achieve this goal, the Convention grants citizens the right to obtain environmental information, to participate in environmental decision-making, and to appeal to courts or non-judicial bodies. These three pillars of the Convention – collectively known as procedural environmental rights – are interdependent.[3] They assume that meaningful participation in policy making depends on access to environmental information and that access to justice guarantees individuals and organizations that their participation and information rights can be exercised. Although regional in scope, it may serve as a model for strengthening procedural environmental rights in all United Nations member states.[4]

The number of signatory nations has now grown to forty, but only seventeen have ratified the Convention.[5] Among the ratifying countries, only two are western European countries with a long-standing democratic tradition, while the great majority are post-Communist countries without either a strong democratic or an administrative law tradition. The goal of this paper is to analyze and explain the first half of this seeming anomaly. Why have France, Germany and the United Kingdom not yet ratified a convention that seems to represent widely-held democratic values? Are there explanations for the ratification delay that go beyond the complexity of fitting the Convention's provisions into existing legal structures? We argue that the parliamentary structure of the West European democracies may well have played a role in discouraging quick ratification because parliamentary regimes may see little benefit in enhanced public participation in bureaucratic processes.

Before moving to the political–economic analysis, one needs to understand the strengths and weaknesses of the Convention itself. Part I argues that the Convention, although it would require increased participation rights in most signatory countries, is a moderate document designed to accomplish marginal changes. Nevertheless, many signatories have not been in a hurry to ratify the Convention. To explain this foot dragging, Part II develops a positive political–economic analysis of a legislature's motivations to create procedural environmental rights. We build on existing work to show that parliamentary systems have little incentive to establish such rights. Part III follows up the conceptual arguments with case study material that shows how procedural environmental rights differ in practice under different political systems.[6] The cases are broadly consistent with our conceptual scheme, but they also reveal some interesting nuances. Part IV concludes with a discussion of the relationship between the Aarhus Convention and the European Union (EU). This linkage is of interest as all EU Member States have signed the Convention and the European Community itself is a signatory. EC ratification requires both changes in EC legislation that affect member states as well as modifications in the practices of EC institutions.[7] We suggest that the separation of powers characteristic of EC institutions may work to spur EC ratification and, in turn, may raise the salience of the Convention in member states with less accommodating systems of public law.

PART I. THE AARHUS CONVENTION

The Aarhus Convention is motivated by the claim that environmental protection policy requires participation from ordinary citizens as well as from scientists and other experts. Policy makers face two ways: toward public accountability and toward technical competence. The public may be uninformed about scientific and economic factors, but the technocrats may be uninformed or uninterested in the opinions of ordinary citizens.

Some political systems seek to separate environmental policy making from implementation. The ideal in such systems is for politically accountable politicians to make policy and then delegate implementation to the professional bureaucracy. The bureaucracy ought to consult with technical experts before implementing a statute, but, under this view, officials need not elicit the opinions of organized pressure groups or ordinary citizens. These actors should exercise their influence at the legislative stage through party representatives and legislative hearings. If ordinary citizens do participate at the implementation stage, it is only to complain about violations of their individual rights.

This sharp division of labor sounds sensible at first, but it ignores the realities of democratic life. The legislature is uninformed about many technical aspects of environmental issues, and in a complex world where time is scarce, this is as it should be. The result is laws that delegate many of the details of implementation to the bureaucracy. These "details" are not mere technical gaps but often determine the policy impact of environmental laws. This basic feature of environmental implementation raises the issue of public participation when the administration fills in the gaps. In practice, there is no sharp distinction between political concerns and technical matters. The latter can only be decided in light of the former.

Even case-by-case implementation can raise policy issues beyond the protection of individual rights. Local decisions on protecting wetlands, building roads, expanding airports, and licensing industrial facilities can have regional environmental impacts. Participants frequently combine an individual claim of harm with a public-spirited concern for policy. This will be especially true if the implementing agency has avoided promulgating general rules or if, in doing so, it has failed to consult with interested groups and citizens.

These arguments for open and accountable administrative procedures that permit public involvement are reflected in the Aarhus Convention. However, as we will see, worries about the costs of too much openness are also reflected in language that is sometimes vague and deferential to existing national laws. We first summarize the provisions of the three pillars of the Convention – access to information, public participation, and access to justice – and then analyze the probable effectiveness of its provisions.[8]

A. Access to Information

The definition of environmental information in Article 2 includes: (1) information on the state of elements of the environment (for example, air, water, and soil) and the interaction among these elements; (2) factors affecting or likely to affect the elements of the environment (for example, substances, energy, as well relevant activities and measures ranging from administrative measures and legislation to cost benefit and other economic analysis and assumptions used in environmental decision making); and (3) information on the state of human health and safety, conditions of human life, cultural sites, and built structures in as much as they relate to or are affected by the above environmental factors.

Article 4 of the Convention establishes the general principle that information on the environment – or relevant to the environment – held by public authorities must be accessible to any party without a need for the party to state a particular interest. This is the key feature of freedom-of-information acts of this type. Those requesting the information do so as interested citizens; they do not have to explain why they want the information. The goal is to give those outside of government better access to the information and reasoning behind the internal decisions of the executive.

Article 4 also defines procedures for information disclosure and several categories of information that are exempted from disclosure.[9] A demand may also be refused if the information is not available, if a request is manifestly unreasonable, or if the request concerns material in the course of completion. Refusals to all exceptions must be interpreted in a restrictive way, taking into account the public interest served by the disclosure. Refusals must be made in writing if the request was made in writing. Where only part of the information requested falls within one of the exempt categories, the remainder of the information must be separated out and made available.

Article 5 addresses issues of active information disclosure and dissemination. It places obligations on public authorities to develop national information systems and procedures that ensure systematic and periodic dissemination of environmental information, for example, through state-of-the-environment reports, or national pollutant inventories or registers. These systems must also provide sufficient product information to enable consumers to make informed environmental choices. All efforts should be made to provide environmental information in electronic format that is easily accessible through public telecommunications networks.

B. Public Participation

Requirements for public participation in environmental decision making are addressed through Article 6 (decisions on specific activities), Article 7 (plans, programs and polices), and Article 8 (preparation of executive regulation and legally binding normative instruments). Activities under Article 6 generally include activities subjected to the environmental impact assessment (EIA) procedure under the UNECE Espoo Convention on Environmental Impact Assessment in a Transboundary Context, as well as activities subject to the Integrated Pollution Prevention and Control (IPPC) directive of the European Community [10]

Many activities referred to under Article 6 of the Convention are likely to have a potential impact at the local level. For such activities, the Convention prescribes a fairly formal and detailed public participation process. Required elements include public notice of the proposed activity, detailed information on the proposed activity, transparent opportunities for public comment and participation, reasonable timeframes for participation, and full information disclosure on all relevant aspects of the decision-making process. Participation should take place early in the process when options are still open, and due account must be taken of the outcome of the public participation.

Participation requirements related to plans, and programs (Article 7) are not specified in similar detail. Public participation should take place in a transparent and fair framework and follow several of the principles established in Article 6, including reasonable timeframes, early participation, and due consideration of the outcome of the participation. As far as the development of policies is concerned, Article 7 merely specifies that each Party shall, to the extent appropriate, endeavor to provide opportunities for public participation, without further defining the concept.

Article 8 of the Convention addresses public participation in the preparation of executive regulations and legally binding normative instruments. It stipulates that draft rules be published or otherwise be made publicly made available, that the public should be given the opportunity to comment directly, or through representative consultative bodies, and that the results shall be taken into account as far as possible. This article is even less precise than Article 7 and hence gives considerable leeway for individual countries to interpret the provisions differently.

C. Access to Justice

The access-to-justice provisions (Article 9) of the Convention are closely linked to the first two pillars of the Convention. First, any person who considers that his or her request for information was not addressed in accordance with Article 4 of the Convention has, in accordance with national law, access to a judicial or non-judicial review procedure. Second, members of the public – and any non-governmental organization – have access to a review procedure to challenge the substantive and procedural legality of decisions or omissions related to Article 6 of the Convention (as well as for other relevant provisions of the Convention, if so provided by national law). The claimant should, however, have a sufficient interest, or maintain impairment of a right if national administrative law requires this as a precondition. Third, members of the public have access to administrative or judicial procedures to challenge acts and omissions by private persons and public authorities which contravene national law relating to the environment, subject to access criteria as they may be specified by national law. References to national law, of course, mean that the impact of this section will depend upon the way courts interpret their own national laws in light of the Convention.

D. Analysis and Discussion

The Aarhus Convention is celebrated by the international community of environmental non-governmental organizations as an important instrument which "promotes citizen involvement as key to combating environmental mismanagement" (Petkova & Veit, 2000). One Convention delegate pointed out that the Convention would require legal changes in almost all of the participating countries.[11] Questions arise, however, as to whether the various provisions of the Convention really have "teeth" and whether or not the Convention will actually trigger fundamental change in the way countries involve the public in developing, implementing and enforcing environmental decisions.

Its challenge to the status quo in Western Europe is suggested by the fact that only two established European democracies, Denmark and Italy, had ratified the Convention by October 2001 even though all EU Member States are signatories. This suggests that politicians view ratification as creating new rights and as requiring the amendment of existing statutes. Most of the countries that have ratified the Convention are in Central and Eastern Europe and a few are outside of Europe in the post-Communist countries of Central Asia. These are mostly countries with no existing set of administrative practices that are challenged by the provisions of the Aarhus Convention. However, many of them are also countries with little capacity to implement the Convention and with only a few, weak non-governmental interest groups.[12] The Convention has formally come into force but is still mainly useful as a guide to law reform in the transitional economies in the former Soviet Union and in Central and Eastern Europe. It has yet to have any significant impact on public access in Western Europe. We turn now to an analysis of the Convention's three pillars.

1. Access to Information
Despite the range of exemptions contained in the access-to-information pillar, the information and right-to-know provisions of the Convention seem extensive and comprehensive. The definition of environmental information goes beyond the European Community's directive on access to environmental information.[13] Thus, in agreeing on Articles 2, 4 and 5, Member States of the European Union indirectly acknowledged deficiencies in the legal framework of the EU in the area of environmental information disclosure.

The Convention – through Article 6, Section 9 – is also likely to have a major impact on the development of community right-to-know programs in participating countries. The Convention calls for parties to establish a nationwide and publicly accessible system of pollutant inventories through standardized reporting. This will trigger the development of national reporting systems that resemble the U.S. Toxic Release Inventory (TRI).[14]

In some signatory states that have parliamentary governments and powerful professional bureaucracies, this pillar of the Aarhus Convention is likely to be controversial. Even the more limited EC directive was very controversial in

Germany. Although Germany did eventually pass conforming legislation, the debate raised questions about whether information disclosure was consistent with German democratic principles.[15] Subsequent disputes have tried to narrow the definition of "environmental information" to exclude information collected by ministries dealing with issues such as highway construction that have environmental impacts but are not under the jurisdiction of the environmental ministry.[16] As we argue below, the constitutional structures of most West European governments have not produced actors within government with much incentive to encourage disclosure to groups and individuals who might make life difficult for them.

2. Public Participation

The sections on public participation in environmental decision making are likely to prove controversial in Western Europe, because they are written with the aim of increasing public involvement in traditionally rather secretive processes. Nevertheless, it is difficult to believe that the practical effect will be very dramatic.

Considering just the German case, a narrow reading of the Convention's provisions related to public participation in specific activities (Article 6) is largely consistent with existing German practice which provides for participation of the concerned public in a range of licensing and approval processes with environmental consequences.[17] The main break with existing requirements in Article 6 is that the "public concerned" is defined in the Convention to include "non-governmental organizations promoting environmental protections." The Convention, however, permits governments to require such groups to meet "requirements under national law," and German statutes already permit citizen groups to participate in some administrative processes.[18] Article 6 includes numerous places where requirements are to be "in accordance with national law," thus, possibly limiting its impact.[19] The provisions in Articles 7 and 8 for public participation in "plans, programs, and policies" and in "executive regulations", respectively, are much vaguer. True, it is the "public", not just the "public concerned" that can participate, but governments can decide who falls into that category. Process is left vague. For "plans and programs" in Article 7, governments shall establish "a transparent and fair framework" that establishes reasonable time frames, occurs early in the process, and is structured to take account of public comments. However, unlike the procedures for specific projects, there are no requirements for information disclosure, hearings, or reason giving. For "policies", there are no requirements at all. The Convention simply asks those states that ratify it to "endeavor" to provide participation opportunities.

The procedures in Article 8 for promulgating legally binding regulations are likely to have little impact on European practice. The Convention only says that Parties "shall strive" to promote effective public participation and "should" take various steps. There are no requirements. The Convention suggests that notice and an opportunity for comments would be desirable, but does not recommend formal statements of reasons.

The Convention has no provisions for Regulatory Negotiation in the promulgation of policies or rules.[20] The only hint of support for participatory decision-making processes is in Article 6, Section 5 where the Convention suggests that Parties "should, where appropriate, encourage prospective applicants to identify the public concerned, to enter into discussions, and to provide information regarding the objectives of their applications before applying for a permit." This provision only applies to the approval of specific projects where a private firm or public authority is applying for permission to carry out a project.

Thus, insofar as public participation in administrative processes is concerned, the Convention is likely to encourage more participation in decisions with respect to specific projects. However, it does not present a major challenge to executive-branch rule making and policy making. Signatories could ratify the Convention and do little to open up such processes. Nevertheless, parliamentary governments used to closed-door rule making, may be reluctant to put the Convention before their legislatures simply because it opens such issues for discussion. In this, they are likely to be supported by the professional bureaucracy. The Convention, in spite of vague and permissive language, clearly views increased public participation at all levels as desirable. It also represents a commitment to including organized environmental groups in the regulatory process. In spite of signing the Convention, sitting Western European politicians have little incentive to further this goal on their own. For instance, with the Green Party part of the coalition government in Germany, it is not obvious that their leaders will endorse procedural policies that require them to solicit the views of outside groups. Thus, countries with parliamentary democracies are only likely to ratify the Convention if pressured by external forces, such as public opinion or the EC.

3. Access to Justice
Judicial review of the provisions on access to information seems comprehensive and clear (Article 9.1) and give that portion of the Convention real bite. Thus, these provisions provide one more reason why some governments and bureaucracies – in particular those that feel threatened and bothered by public access to information – may not push for ratification of the Convention.

Judicial review of decisions with respect to specific projects under Article 6 is weaker (Article 9.2). Because this provision is hedged by references to national law, a Party might be able to deny review of any kind. Assuming that some review is granted, the main object of controversy is likely to be the standing requirements. These give not only individuals, but also environmental organizations the ability to challenge substantive and procedural matters so long as they assert a sufficient interest or, under other legal systems, the impairment of a right. If judicial review of executive decisions is seen as mainly a way to protect individuals against an overbearing state, the notion of giving standing to organizations runs against the grain. The idea that courts might be monitors of the democratic accountability and policy making competence of the state is threatening.[21] In a parliamentary system, the government would have no incentive to create such constraints on its own. A written constitution may build in some measure of judicial oversight, but politicians in power would not have such an interest and neither would professional bureaucrats. Organizations are only given standing if they have a "sufficient interest" or assert "impairment of a right", but the Convention asserts that organizations fulfill these conditions if they claim to be promoting environmental protection. The only escape for the government is to establish restrictive requirements that limit the number of organizations that are licensed in the first place. This provision is thus part of the tilt toward organized environmental groups that permeates the Convention, and it is likely to give pause to the party-centered states of Western Europe.

Nevertheless, the judicial review provisions fall far short of the provisions under the American Administrative Procedures Act.[22] The Convention fails to endorse the review of legally-binding rules as provided under the USAPA, and there is no review of policy making decisions. Article 9.2 simply states that if national law provides for review of such decisions, then its provisions apply. Thus, to the extent its provisions broaden standing, the Convention may deter signatories from expanding judicial review of regulations. It states that *if* review is granted, then it must be made broadly available. A Party might wish to increase judicial involvement but only if access to the courts is restricted to a limited class of plaintiffs. The Aarhus Convention makes such restrictions more difficult to impose. The Convention in Article 9, Section 3 makes an interesting stab at providing for citizen suits to force both public authorities and private parties to comply with environmental laws. The section is full of caveats and references to national law, but it does say that, subject to conditions, parties "shall" permit members of the public to challenge acts and omissions by private parties and public authorities contravening national environmental law. This section follows several American environmental statutes that permit such suits.[23] The section, however, makes no mention of the payment of attorneys' fees. In the United States, one-sided fee shifting applies. Citizen suit statutes provide that the victorious plaintiff has attorneys' fees paid by the defendant but need not pay the defendants' fees in the event of a loss. If the plaintiff is performing a public service by bringing a suit, this makes sense.[24] Unfortunately, European courts generally enforce two-sided fee shifting, a practice that if applied here, would have a sharply chilling effect on the willingness of citizens and organized groups to bring suits. Because the Convention simply requires that such suits be possible, it is unlikely to mean much in practice, but may, of course, be another reason for opposition.

In short, the references to existing national legislation recognize that countries have fundamentally different approaches regarding the role of the judiciary in environmental policy development, implementation, and enforcement. In comparison to American practice, the access-to-justice provisions seem limited and constrained. However, because courts in several European countries have limited standing rights for individuals and non-governmental organizations, the Convention would likely require some changes in practice. For example, Denmark, the host country of the Aarhus Convention and first western democracy that ratified the agreement, had significant difficulty complying with the access-to-justice provision.[25] To the extent these changes – like those for access to information and public participation – challenge established practices, they are likely to make ratification an unattractive prospect.

4. Conclusions

All three pillars to the Aarhus Convention threaten – to various degrees – some established practices in the democracies of Western Europe. They incorporate a view of democratic accountability that is at odds with accountability based solely on elections and political parties that form governing coalitions. Thus, the political will to ratify Aarhus is unlikely to come from within parliamentary governments. The Convention also challenges the authority of professional bureaucracies that are used to deciding for themselves whom to consult and what use to make of this information. Thus, career officials are also unlikely to push for ratification. Furthermore, the silence from some organized private interests suggests that they are satisfied with the privileged access that they obtain in the status quo. Pressure for change must come from members of the public and from organized groups excluded from present processes. This will be difficult because groups may have little incentive to organize at present given their marginal impact under present practices.

PART II. THE POLITICAL ECONOMY OF PROCEDURAL ENVIRONMENTAL RIGHTS

Almost all Aarhus signatories are democracies. But this category embraces many different types of parliamentary and presidential systems. In some participating countries, the democratic tradition stretches back hundreds of years, but it is very recent in others. Paradoxically, some of the countries with the longest and most well established democratic traditions are likely to find the Convention most in tension with their traditional modes of operation.

Despite broad-based international consensus, proposals to strengthen procedural environmental rights often face considerable opposition and are subject to intense debate at the national level in some countries, while widely accepted in others. What are the origins of these differences? What institutional or political factors affect the substantive and procedural aspects of environmental policy making? Does the structure of existing political institutions matter? This section presents a framework for thinking about these questions that suggests that ratification of the Aarhus Convention will be a slow process under certain political structures.

A. Government Decisiveness

George Tsebelis (1995) has developed a way to predict the potential for policy change (decisiveness) across different types of regimes, legislatures, and party systems. He demonstrates that policy stability (or, in other words, the difficulty of changing the status quo) increases with the number of veto players. For example, a system where both houses of a bi-cameral legislature and the president must approve before a bill becomes a law will be relatively stable. Similarly, a large number of political parties, control of institutions by parties with significant differences in opinion, and diverse views within veto players are all factors which make changing the status quo difficult. Conversely, countries with only one veto player and two major parties are in a better position to trigger policy change. Thus, Tsebelis's analysis predicts that a polity with many veto points – characterized by separation of powers, a federal structure, and bi-cameralism – is likely to under-supply public goods, such as environmental quality, due to its incapacity to change the status quo. Matthew Shugart and Stephan Haggard (1997) make a similar argument, pointing out that presidentialism (which is characterized by at least two veto points) tends to reduce legislators' interest in providing public goods at the national level.

The above analysis, however, does not necessarily allow one to make predictions about the procedural aspects of environmental decision making –

the focus of this essay. In order to proceed, we need to introduce some theoretical concepts developed in the field of analytical comparative politics.

B. Presidentialism vs. Parliamentarism

Presidential democracies and parliamentary democracies are two fundamentally different types of regime. In the former, the chief executive (the president) and the legislature are separately elected; in contrast, in parliamentary democracies, the head of government is appointed by and dependent on the legislature. Thus, the executive and the bureaucracy (which is supervised by the executive) are directly accountable to the parliament.

Terry Moe and Michael Caldwell (1994) predict that statutes in parliamentary regimes will place few procedural requirements and constraints on the executive and will provide the bureaucracy with more leeway than it will have in a presidential system. The major reason is that in parliamentary democracies, the majority party (or majority coalition) has, via the executive, direct control over the bureaucracy. Thus, one of the primary objectives of national civil servants is to implement the political will of their masters (that is, the parliament and their respective ministers) as closely as possible. This direct relationship between the legislature and the bureaucracy is also likely to reduce the chance that narrow interests can capture the bureaucracy. Consultation with interested parties could still take place, but the emphasis of such consultation would be on enhancing the technical capacities of the administration to develop sound regulations.

Just as procedural protections are not needed in a unitary system to implement the will of the majority, they are likely to be ineffective with a change in government. New parliamentary majorities have the power to overturn decisions made by old majorities. Therefore, the incentive to establish procedural constraints and/or strong procedural environmental rights are weak, at least from within political institutions.

Civil servants in presidential systems, like their counterparts in parliamentary democracies, have the task of implementing national legislation developed by the legislature. But, the top official of an agency may represent a different party from that of the majority party in the legislature.[26] The potential conflict between presidential appointees and the legislature may pull national bureaucrats in various directions. The legislators who draft environmental legislation may, therefore, include procedural and institutional measures in statutes in order to "lock in" their political interests and to hedge against future changes in political majorities.

Here is the connection to Tsebelis's work. Statutes are more difficult to change in presidential than in parliamentary systems because of the multiple veto points. If the political will exists at one point in time to enact a new law, that law can be structured to limit the damage done by a subsequent president who is hostile to the aims of the law. One way to do this is to give outside interests a legally protected role in implementation. This, in turn, may increase the stability of the rules that are promulgated because transparent processes may make the rules more credible and acceptable.

But, one might ask, if a coalition exists to pass a law, why not constrain the executive by including substantive detail instead of procedural constraints? First of all, this *does* happen. Second, the compromise that produces the new law may require some vagueness about how it will be implemented, and procedural protections can give something to everyone – those supporting business interests as well as environmental activists and consumers. Our discussion of chemicals control in Part III illustrates this point particularly well. Finally, delegation to the bureaucracy may be justified by the complex and fluid nature of the underlying environmental problem. The legislature does not wish to be overly precise because it wants the executive to be able to respond to changed conditions. However, when the executive does respond, the statute contains built-in procedures that require consultation with outside groups affected by any changes in the rules.

For these reasons, Roger Noll (1989) argues that statutory environmental legislation in presidential systems can be expected to contain procedural rights. These may include, for example, prescriptions to make the rule-making process public and transparent, requirements for the agency to consult certain stakeholder groups during the rule-making process to ensure that "preferred" views are taken into consideration, obligations to provide strict evidentiary criteria and use of analytical tools such as policy analysis or cost-benefit analysis, and opportunities for judicial review of regulations – including granting of liberal legal standing rights for individuals and interest groups.

C. Electoral Rules

Electoral rules and district magnitude shape the constellation of political parties and the incentives of legislators to focus on procedural issues. First, the electoral system may have an impact on whether or not a small party running on an environmental platform has a chance of entering the legislature. Second, electoral rules may create incentives (or disincentives) for legislators to run personal rather than party-based campaigns. These factors, in turn, affect the provision of broad-based public goods (such as environmental quality), patterns of interest group organization, and the level of procedural safeguards demanded by (and provided to) special interest groups from legislators.

Electoral rule and district magnitude help determine the number of political parties. Plurality-voting rules generate "winner-take-all" systems and are often coupled with a district magnitude size of one.[27] They tend to result in two parties with positions located close to the political center. In this situation, groups with relatively narrow interests are forced to join a larger party since no small party alone is in a position to obtain the majority.[28] They may seek alternative routes to influence through legislative provisions that give them access to the administrative process.

A proportional representation (PR) voting system is likely to generate several parties that can play an active role in national politics. As the district magnitude increases, so does the theoretical number of parties.[29] PR may allow a party running on an environmental platform to enter the national legislature. Most systems limit party proliferation by imposing a cutoff vote share, say 5%, below which no seats are awarded. Thus, an environmental party would need to exceed that threshold and would only be able to do so if a significant portion of the population ranks environmental issues highly.[30] In that case, environmentalists may concentrate on influencing the legislative process through "their" political party and put relatively less emphasis on implementation.[31]

However, all PR systems are not alike. Some have closed lists (CLPR), where the party determines the order of candidates, and others have open lists (OLPR), where voters rank candidates. OLPR promotes candidate-centered choices and creates incentives for individual legislators to run on narrow interests and to provide favors to particular groups of constituents.

Candidate-centered electoral systems, such as system with plurality electoral rules or OLPR, may lead to sub-government structures, such as sub-committees within the legislature, that, according to Gary Cox and Mathew McCubbins (1997), tend to favor narrow interests and under-supply public goods. Such systems also decrease the internal cohesion of parties and weaken the position of party leaders – yet another factor which may reduce provision of national public goods. John Carey and Matthew Shugart (1995) point out that in countries with open voting lists, the incentive of candidates to run personal campaigns increases with the size of the district magnitude. In countries with closed voting lists the opposite result holds: the larger the district magnitude, the larger the incentive for individual candidates to put the party position in the center of their campaign.

A country's voting system may affect the way interest groups with a stake in environmental matters organize themselves. For example, in parliamentary systems, like the U.K. or a CLPR case, with only one veto player and strong parties, lobbying of individual legislators by a relatively small interest group is unlikely to be effective. In such cases, corporatist patterns of interest group organizations, therefore, can be expected. Here, industry develops consolidated positions through so-called "peak associations" rather than lobbying individually. The weight of such groups in national policy making is likely to be significant; few policies will be accepted unless peak associations, in particular those representing economic interests, have made their positions clear and provided some type of consent.

In contrast, pluralist interest groups patterns are likely to be more effective in countries with several veto players, such as the U.S. presidential system, or in countries with weak parties, as in OLPR regimes. Cox and McCubbins (1997) point out that in such cases, interest groups may be better off influencing individual actors rather than the whole governing parties. This may be an important reason why pluralist interest group structures develop in countries with personal vote systems.

The important question for us is how legislative and interest group organization maps onto preferences for procedural protection in implementation. Our claim is that systems that encourage the development of multiple interest groups with focused agendas will be likely to draft laws that permit these groups to play a role ex post during implementation. Both the United States and OLPR parliamentary systems are in this category. This prediction overlaps with arguments about the way legislative compromise can lead to vaguely drafted statutes. If laws represent compromises between interests groups and among political actors, this may produce both vague laws and procedural guarantees that permit the groups that lobbied for the act to play a role in implementation. These groups will want that role to be formalized in legal provisions the greater the risk that they may be left out of informal, closed-door consultations. In contrast, both majoritarian parliamentary systems, such as the U.K., and CLPR systems, such as Germany, are likely to feel less pressure to include procedural provisions in environmental laws. However, majoritarian and CLPR systems will differ if the latter includes a distinct environmental party, such as the German Greens. The legislative visibility of an environmental party, even if it is not in the majority coalition, will help put environmental issues on the policy agenda. However, its members are likely to push for strong substantive statutes and be less interested in the administrative process than environmental groups in more pluralistic systems. The basic point we are making here is that the content of environmental laws, especially their provisions for public participation in executive branch policy making and implementation, may be a function of the underlying structure of the electoral system and the pattern of interest group organization that it produces.

D. Judicial Review

The courts can be an additional veto player. Judicial review mechanisms are likely to be strong in countries that have more than one veto player. Each veto player may want to have access to an independent entity within the political system to serve as a mediator in cases of disputes or differences in interpretations. Therefore, if there are two veto players, they may support enhanced judicial review to limit each other's power and, as a consequence, produce a third veto player. Politicians in parliamentary systems that are based on the U.K. model or that use CLPR will have little interest in creating independent judicial fora to review their decisions.

Countries with a strong role for the courts in monitoring legislative and administrative decisions will probably have rather liberal policies on access to information in order to permit informed challenges. Such policies are particularly important if legal standing rights are provided not only to individual citizens, but also to groups that can represent common interests, such as the protection of the environment. In addition, full access to documentation and underlying information used in decision making is an important input for credible and successful judicial review.

In short, widespread judicial review and broad public access to information are likely to go together and to be associated with systems that have multiple formal veto points and widespread organization of interest groups. Parliamentary systems, especially those on the Westminster model or ones operating by CLPR, are likely to have relatively weak provisions both for access to information and to judicial review. Only widespread public outcry, unmediated by political parties, is likely to convince politicians to move toward a regime with greater openness and oversight of administrative decisions.

PART III. PROCEDURAL ENVIRONMENTAL RIGHTS UNDER DIFFERENT POLITICAL STRUCTURES

Although a full analysis of the links between political structure and environmental policy making is beyond the scope of this essay, we provide several examples to illustrate such linkages focusing on the differences between the United States, on the one hand, and Germany and the United Kingdom, on the other. First, we argue that differences in the electoral systems between the United States and Germany appear to shape the way popular demands for environmental quality are channeled into the environmental policy making process. Second, the different regulatory cultures of the United Kingdom and the United States seem to have had an impact on freedom of information policies in each country. Third, we examine the control of industrial chemicals.

In all three cases, the United States has broader participation rights and broader access to information in line with its separation of powers constitutional structure. However, pressures for greater openness are being felt even in the established parliamentary democracies. For example, the United Kingdom recently passed a Freedom of Information Act and participation opportunities appear to be expanding in Germany. The case of chemical regulation, however, illustrates that broad participation opportunities are not sufficient to produce higher levels of protection for human health and the environment. Also important is the character of the underlying law and the identity and influence of participants. Recall that our theoretical discussion did not claim that American-style democracies necessarily will do more to further the interests of the general public, only that there are political reasons for legislators to write statutes that include participation rights for all influential interests. These cases suggest that broad framework laws such as the Aarhus Convention or the United States Administrative Procedures Act are better ways to assure evenhanded participation rights than reliance on individual substantive statutes produced through narrowly focused political bargaining.

A. Electoral Rules and Environmental Interests: The U.S. and Germany

The United States has a long history of providing freedom of information, public participation, and access to justice. As early as 1946, the Administrative Procedures Act required agencies to comply with notice and comment procedures for rulemaking.[32] In the environmental policy arena, the National Environmental Policy Act (NEPA) of 1969 provided comprehensive participation opportunities and legal standing rights to individuals and interest groups related to the planning of certain development projects which would have an impact on the environment.[33] The Freedom of Information Act was passed in 1966 and has been amended several times since then to keep up with changes in information technology.[34] The Emergency Planning and Community Right-to-Know Act (1986) expanded the freedom of information concept by providing communities (and others) with emission data from industrial facilities in their neighborhoods.[35] The Government in the Sunshine Act requires that most high-level decision-making meetings are open to the public.[36]

In contrast, Germany had no strong procedural environmental rights in its first generation of environmental statutes and has not been a driving force towards strengthening procedural environmental rights at the level of the European Union. Germany's administration carries out many of its regulatory rule-making activities under the presumption of confidentiality.[37] Germany's record of providing the public with environmental information relevant to regulatory and administrative decision making is therefore rather modest. Although several environmental statutes state that interest groups should be consulted in developing implementing rules and regulations, such generic guidance provides significant leeway for civil servants to manage and control consultative processes, including decisions about which specific groups should be involved.[38]

Yet, Germany is considered a leader in environmental performance in the European Union.[39] One explanation for this mixture of weak participation rights and apparently strong environmental commitment is the organization of the political system. However, we also argue that Germany's reputation hides a number of important weaknesses that could be mitigated by increased public participation.

Germany's electoral system of proportional representation has allowed the Green Party – a small party running on an environmental platform – to shape policies from within the parliament, and nowadays from within the government. As a consequence, the political system has channeled popular demands for environmental quality into political decision-making processes. In contrast, the pluralistic U.S. electoral system does not favor the establishment of small parties with a focus on relatively narrow issues (such as a Green Party). This creates niches and opportunities for public interest groups to shape environmental policies from outside the party system. These groups, however, could not function without strong procedural rights in place, such as access to government information, participation opportunities in decision-making processes, as well as liberal legal standing rights. The logic of our political/economic argument may explain why the introduction of procedural environmental rights into the German system of environmental governance has not organically evolved in the context of national politics.

Although a push for greater participation rights is unlikely to come from inside the political system, their creation would benefit the German policy making process. As one of us has argued previously, the high level of delegation to the executive under German law leaves the policy making and implementation process open to excessive influence from technically oriented groups with an industrial orientation because there are few formal participation opportunities.[40] True, the underlying statutes are quite stringent, but the implementation process leaves many important aspects to be decided by relatively closed-door processes.[41] In fact, German policymakers have recognized this weakness, and some recent initiatives are attempting to increase the level of procedural environmental rights. Nevertheless, the EC is not satisfied; the Commission has, for example, sent a final warning about

Germany's failure to comply fully with the EC directive on access to environmental information.[42]

B. Freedom of Information Legislation in the U.S. and the U.K.

National policies to provide access to environmental information cannot be viewed in isolation either from the overall freedom of information (FOI) policy of a particular country or from national policy-making practices. Information policies in the U.S. and U.K. seem to be deeply embedded in the regulatory structures of each country. In the case of the U.S. freedom-of-information legislation was introduced, in part, to make the notice and comment procedures of the rule-making process under the Administrative Procedures Act more effective. Enhanced transparency through enhanced information access allows Congress and the public to better monitor rule-making processes and reduce the chance that regulatory agencies will be captured by special interests.[43]

Britain, in contrast, has a parliamentary democracy with a strong alliance between the executive and the bureaucracy and a long-standing history of neutral competence of the civil service staff. The development of regulations has traditionally been undertaken through rather closed processes within the bureaucracy that has the discretion to consult with interested parties when considered appropriate. This pattern of secrecy is, according to Debra Silverman (1997), a result of the exclusive accountability of ministers to parliament and results in the control of information flows by ministers.

Over the past decade, the U.K. government has come under severe pressure from outside groups to provide greater public access to information policies. Following several unsuccessful attempts,[44] a Freedom of Information Act was passed by parliament in November 2000.[45] The Act, however, will not come into force for central government departments until summer 2002, and for other authorities only in stages afterwards.[46]

Passage of a general FOI law was facilitated by U.K. compliance with the EC directive concerning freedom of information about the environment.[47] A paper published by "The Campaign for Freedom of Information" in 1993 asserts this linkage.[48] It asks: "If ministers can accept the case for a broad and (albeit weakly) enforceable right of access to environmental information, why not a similar right for information about, say, safety, public health, consumer protection, education, the NHS, social services – and everything else?"[49] Established principles on access to environmental information – triggered by the EC directive – provided one of the benchmarks for the U.K. FOI legislation. In its review of the draft 1999 bill, Friends of the Earth (FOE) referred to the EC directive when criticizing some of the proposed exemptions[50] and also referred to the U.K.'s commitment to ratify the Aarhus Convention. Thus, international commitments provided not only driving forces to enhance transparency in environmental matters, but also contributed to a broader freedom of information reform process in the U.K.

C. Control of Toxic Chemicals in the United States and Germany

Traditionally, information on chemical hazards and the assessment of chemical risks have been considered purely scientific and technical exercises designed to generate probability estimates that indicate the potential of a particular substance to cause harm to human health or the environment under different exposure scenarios. Economic and social considerations – and related public preferences and values – were only to be considered when management and control options were evaluated. This view is under challenge. For example, a report by the United States National Research Council concluded that public participation and stakeholder involvement should become an integral aspect of both risk assessment and risk-management processes.[51] This section takes a closer look at the issue of public participation by examining procedural constraints affecting hazard testing, risk assessment, and risk management of industrial chemicals in the United States and Germany.

1. The United States
In 1976, the U.S. Congress adopted the Toxic Substances Control Act (TSCA)[52] following several years of legislative bargaining and lobbying by interest groups.[53] TSCA provides the Environmental Protection Agency (EPA) with authority to gather information about the toxicity of chemicals, to collect human and environmental exposure data, to identify chemicals which pose unreasonable risks to humans and the environment, and to take actions to control these risks. TSCA also requires EPA to review new chemicals before they are manufactured.

In the absence of clearly agreed policy goals and lack of agreement among interest groups, TSCA delegated many policy choices to the EPA. In order to protect special interests from bureaucratic discretion, legislators built a range of procedural safeguards into TSCA.[54] TSCA confirms Roger Noll's argument that statutes in a separation of power regime – with multiple veto points – are likely to include a range of procedural measures which allow interest groups to actively participate in the law implementation process.

One of the main goals of TSCA was to generate data on chemicals that have the potential to cause harm to human health and the environment. To achieve this goal, TSCA provided EPA with authority to require testing data for chemicals from industry.[55] However, reports published by major U.S. environmental groups in 1987 and 1998, revealed significant gaps in knowledge concerning the potential hazards of thousands of chemicals on the market.[56] A study issued by EPA in 1998 pointed to a similar conclusion. It stated that "no basic toxicity information, i.e. neither human health nor environmental toxicity, is publicly available for 43% of high volume chemicals manufactured in the U.S. and that a full set of basic toxicity data is available for only 7% of these chemicals".[57]

Given that the statute was passed with the goal of gathering information, the lack of data is troubling. Among the reasons for the information gaps are the procedural requirements that must be satisfied before the EPA can order industry to carry out tests under Section 4 of TSCA. EPA may only require testing *after* finding that: (1) a chemical may present an unreasonable risk of injury to human health or the environment, and/or the chemical is produced in substantial quantities that could result in significant or substantial human or environmental exposure; (2) the available data to evaluate the chemical are inadequate; and (3) testing is needed to develop the needed data. Robert Haemer (1999) points out that the rule-making process as well as judicial review and related judicial decisions have practically stifled test rules. This is an example of procedural rights favoring the regulated industry in a way that allows environmental progress to stall. They were part of the legislative deal that permitted the act to pass in 1976.

In addition to the difficulties of forcing industry to conduct tests under TSCA, banning a chemical is also difficult in the U.S.. The case of asbestos illustrates the potential burden of participatory rule-making processes. In 1979, EPA announced plans to ban all remaining uses of asbestos under Section 6 of TSCA.[58] Following ten years of preparation, a final rule was issued in 1989[59] only to be challenged in court by the asbestos industry.[60] In *Corrosion Proof Fittings v. EPA*, the U.S. Court of Appeals for the Fifth Circuit Court held that EPA did not present – as required by TSCA – sufficient evidence to justify the ban of asbestos.[61] The court also expressed its "regret that this matter must continue to take up the valuable time of the agency, parties, and, undoubtedly, future courts". As in the case of generation of data for hazard assessment, banning a chemical proved difficult because the statute's procedural provisions placed a high burden of proof on the agency – at least as interpreted by the federal courts.

Learning from these experiences, EPA's Chemical Right to Know Initiative and Existing Chemicals Program have employed more cooperative and voluntary methods as first approaches to reduce or eliminate the likelihood of harm to human health and the environment.[62] Under the Citizen Right to Know Initiative, industry may voluntarily submit relevant data to EPA. This program was developed with the help of the chemical industry and the Environmental Defense Fund, a moderate environmental group. However, it was criticized by an animal rights group that was excluded from the early stages of the process.[63]

Citizen involvement in the Existing Chemicals Program takes place via comment and consultation on topics ranging from risk assessment to pollution prevention and risk-reduction actions. The information generated is made publicly available through an "administrative record".[64] This EPA alternative relies on voluntary cooperation, but it is being carried out in the shadow of underlying statutory provisions that can be invoked by the agency if cooperation fails.

Under TSCA, EPA's rule-making responsibility is difficult to trigger because of the high burden of proof on the agency to justify moving forward to require the testing or banning of individual chemicals. The EPA sought alternatives to rule making that were based on voluntary and collaborative efforts between industry and one segment of the environmental community. But once it settled on one such alternative under the Right to Know Initiative, it was criticized by a group that was left out during early stages of the process. The agency's pragmatic effort at voluntary compliance permitted it to move forward, but its actions are less transparent and participatory than standard American rule-making processes.

The TSCA experience illustrates the way interest groups can bargain over process as well as substance at the legislative drafting stage. This can produce procedural provisions that favor particular organized interests. One conclusion to be drawn from this case is that broad framework statutes, such as the Aarhus Convention or the U.S. Administrative Procedures Act, are likely to produce more evenhanded procedural requirements than those that arise from the effort to produce statutes covering particular, contested substantive areas. Relying on voluntary, cooperative agreements may sound like an attractive alternative to legal mandates, but they raise difficult problems of fairness and access.

2. Germany

In Germany, chemical control efforts have focused on the classification and labeling of chemicals, notification and assessment of new chemicals, and risk evaluations and control measures for existing chemicals as called for by EC legislation.[65] If a risk evaluation reveals an unacceptable risk to human health or the environment, the German chemicals law allows the government to issue a regulation to ban or severely restrict the chemical, following consultation with the Bundesrat (the second chamber of the German parliament) and with affected and interested parties. Environmental groups, according to Article 17 (7) are included among those groups "to be selected" for prior consultation. This provision is of interest because requirements for consultation with environmental groups in the implementation of law are an exception rather than the rule in German environmental law.[66] Even here, however, the bureaucracy remains in control since it can select which groups to permit to participate. More important, due to the fact that existing chemicals are directly addressed through an EC regulation, German law cannot require chemical firms to generate additional testing data on existing chemicals beyond data they have generated on their own. Thus, in contrast to the United States, the problem is not biased procedural constraints but weak underlying substantive requirements.

For other matters pertaining to implementation, such as the development of risk reduction strategies and recommendations for control actions, the law allows the establishment of committees (Ausschuesse) by regulation (Article 20b). Public participation opportunities are not clearly defined.[67] The work of the committees lacks transparency and stands in sharp contrast to the U.S. Existing Chemicals Program, which is subject to public scrutiny through the administrative record.

In addition, the German government makes use of advisory committees that are not formally established by regulation to support its work in the area of chemicals management. For example, the Advisory Committee on Existing Chemicals (BUA) was established in 1992 through collaboration of the German government, the chemical industry and the scientific community, as a "neutral" scientific committee to develop initial assessments for priority chemical substances.[68] These collaborative efforts are executive branch innovations that are consistent with Germany corporatist traditions. They are not the result of legislative initiatives, and they are not broadly participatory. Representatives of environmental groups are not members of BUA.[69]

The BUA is, however, considered a success (even) by the present Minister of Environment from the Green Party, and BUA itself points out that "currently there is no committee nationally or internationally that has assessed more existing chemicals than BUA".[70] However, a mere count of the number of chemicals assessed is not a strong recommendation. Involvement of experts from environmental groups might have generated different conclusions concerning environmental or human health risks. Because chemical hazards are not country specific, one way of assessing the German procedure would be to compare its decisions with those generated by the EPA under TSCA. Such comparisons might be of limited value, however, if German and American scientists and risk management specialists share information and are influenced by each others' decisions.

German chemicals' policy is difficult to evaluate. The most that can be said is that it is constrained by EC directives and regulations and that it includes little formal public and environmental group participation, especially on technical implementation issues. Selected environmental groups must be consulted under one part of the law, but various committees with closed memberships seem central to the development of policy. Procedural issues may be relatively unimportant, however, because the underlying law – in particular for existing chemicals – seems weak. Nevertheless, the Social Democratic/Green coalition government has taken the initiative in controlling several chemical substances and has proposed that the EC take action.[71] The political coalition controlling the German government has limited legislative authority, however, because it must conform to EC chemicals' legislation. Stronger participation rights could provide a better balance in administrative deliberations, but this will be of limited value if the underlying laws continue to be relatively weak. Such rights could, however, publicize issues that might otherwise remain unexamined and might facilitate legislative changes both in Germany and at the EC level.

D. Discussion

The three cases illustrate several aspects of our politicial/economic analysis. They are broadly consistent with our speculations although they are obviously not a rigorous test of our ideas.

First, the United States separation-of-powers system has produced far more procedural rights than has Germany's system of parliamentary democracy. However, Germany has strong laws protecting the environment in spite of little public participation in rule making. This is, in part, a reflection of the Green Party's role under the country's proportional representation voting system. However, we also argue that the apparent strength of the German statutes needs critical review. Our analysis suggests that German executive branch policy making and implementation leave a good deal to be desired and appear slanted toward participation by industry.

Second, the United States has long had much stronger laws on public access to information than the United Kingdom. This may be the result of their contrasting constitutional structures. Pressure for change in the U.K. has come from the public and from some organized interests. Compliance with the EC directive on freedom-of-information about the environment as well as discussion of the Aarhus Convention informed a broader debate on freedom of information that apparently changed the political calculations.

Third, the procedural biases of TSCA produce high benchmarks for EPA to initiate action and appear to be a major reason for slow progress in U.S. chemicals' control. The transparency of the law-making process combined with public participation opportunities at the implementation stage have not been sufficient to guarantee effective control. In the U.S., the weakness of the underlying statute has been challenged by public pressure and by the publications of environmental groups. These led to innovative action by EPA and the chemical industry to address key issues outside the formal legal processes. However, these innovations create risks of their own. In Germany, legally required procedures are not elaborated in much detail and hence cannot introduce much bias. However, the weakness of the substantive law, combined with traditional, industry-centered regulatory practices, suggest that the interests of the general public and the environmental community may not be taken into account very well. One needs to know more about how requirements to consult with environmental groups work in practice and about the operation of powerful advisory committees.

Our aim in reviewing these cases is as much normative as positive. We seek to understand the incentives for establishing or failing to establish procedural rights. However, we are also interested in the role that participation of all affected interests can play in making the regulatory process more democratically legitimate. The challenge is to design participation, judicial review, and right-to-know programs so that they can contribute to creating a government that can protect the public interest and can achieve results. Determining the public interest is a difficult task. It is not simply a question of business versus the environment or management versus labor. Rather, as illustrated by our example of chemicals' regulation, there may be sharp clashes between organized groups (in this case, environment and animal protection organizations) each of which considers itself a representative of the "public interest".

We have also seen that administrative procedures are not per se desirable. Poor process can bias or unduly delay implementation. The goal of reformers concerned with democratic legitimacy should be to isolate procedures that are balanced and fair and that do not involve agencies in a procedural morass. If procedural rights place very high burdens of proof on groups with fewer resources than other powerful interests or if the bureaucracy must spend a good deal of its time defending itself in the courts, agency credibility and efficiency will suffer. Excluding the public and certain groups from policy implementation, and making deals behind closed doors, is anachronistic at a time that is characterized by free flow of information and moves towards more citizen involvement and responsibility.

PART IV. PROCEDURAL RIGHTS AND THE SEPARATION OF POWERS IN THE EUROPEAN UNION

Our positive political–economic analysis of procedural environmental rights suggests that most of the signatories of the Aarhus Convention will find some portions of the Convention in tension with existing practice. To the extent that this practice reflects underlying political and constitutional structures, ratification may prove difficult. The Convention is a challenge not just to established habits, but also to political practices that are rooted in the incentives faced by politicians and civil servants. Procedural environmental rights similar to those enacted under the United States separation-of-powers regime may not be easy to transfer to parliamentary systems.

In spite of this difficulty, more open bureaucratic processes will, we believe, benefit the policy making environment by improving the information available to officials and by increasing the legitimacy of executive branch policy making. Even in established parliamentary democracies this possibility needs to be recognized. However, such rights may conflict with the interests of politicians and challenge the prerogatives of bureaucrats. Advocates of procedural rights must, therefore, understand that although political institutions in some countries may favor (or even require) procedural rights, the opposite situation may prevail in countries with different types of regime.

What is the future of procedural environmental rights in Europe? Will the Aarhus Convention become a dead letter that serves merely as an idealized guide for law reformers in the post-socialist countries? Our political–economic analysis suggests that most signatory states in Western Europe would find ratification – in the absence of external driving forces – difficult and unattractive. It also suggests that countries with parliamentary democracies that ratify the Convention are likely to seek to limit its impact. However, these observations ignore an important actor – the European Community. In conclusion, we argue that actions at the level of the EC may serve as a catalyst for change within Member States. The EC is one of the signatories of the Convention, and the Commission has initiated the legal changes needed for accession – a move that could increase the visibility of the Convention in Member States of the EU and spur national ratification processes.[72]

The EC has, over time, developed the characteristics of a separation of powers regime. The Council and the directly-elected Parliament could be considered the upper and lower houses of a bicameral legislature, the European Commission serves as an executive with an administration attached to it, and the European Court of Justice is an independent body responsible for judicial review. Over time, the European Parliament has gained significant influence in the legislative decision-making process. Since the Amsterdam Treaty, all matters concerning the environment are subject to a co-decision procedure between the Council and the Parliament, which means that the Parliament has a veto right in all environmental matters.[73]

Yet, the EC has been severely criticized for its non-transparent decision-making procedures and its related lack of accountability. Together these are often referred to as the democratic deficit of the EC.[74] For example, risk assessment and risk management decision making for existing chemicals is undertaken through a complex web of committee and "meeting" structures. Working under the rules of the EC "Comitolgy" processes,[75] a committee, composed of representatives of Member States and chaired by a representative of the Commission, assists the Commission.[76] The committee is informed by regular meetings of the Competent Authorities (CAs) who are designated by Member States. These CAs review and approve the recommendations of Technical Meetings where individual risk assessments are discussed. Industry and environmental interests groups may informally participate as observers at all three levels in the decision-making structure.[77] Christian Hey (2000) argues that this complex process has prevented meaningful participation of environmental groups and represents, in essence, negotiations between the chemical industry and public authorities that put the burden of proof on public authorities. The "formal pluralism" established under the regulation, according to Hey, does not take into account the "imbalances in resources between industrial representatives and the representatives of public interests". Environmental groups are simply unable to afford the time and resources needed to participate in decisions concerning individual chemicals.

Thus, the role of environmental groups, at least in the regulation of chemicals, has been marginal at best and has not benefited from any strong right to be included or heard. However, theory would predict that, due to the separation of powers in the EC political system and the number of veto players, procedural environmental rights should begin to play a role in the EC context if the Parliament asserts the power it gained under recent reform of the European Union Treaties. Might participation patterns and judicial review processes in the EC resemble the level of transparency associated with the United States in the not too distant future?

At present, the EC can still be viewed as a treaty of sovereign states, not a government controlled by its citizens. The pressure for procedural rights is part of the ongoing debate about the nature of the European Union project. Recent changes in the treaties governing the EU suggest a move in the direction of more public access and participation. For example, Article 255 of the Amsterdam Treaty formally established the right of the public to have access to EU documents and a general Freedom of Information law is in preparation. A recent Commission initiative addresses this issue in a comprehensive and consistent manner for all institutions of the EC that are involved in the development of community legislation.[78] The proposal refers not only to access to documents prepared by the EC, but also to those held by EC institutions, thus expanding the scope of access to documents prepared by Member States and used in the context of EC deliberations.

Similarly, public involvement in policy making has become a major concern of the European Union. A discussion paper issued by the President of the Commission (1999) highlights shortcomings in non-governmental organizational (NGO) involvement in EC policy making in the past and outlines suggestions for strengthening the dialogue between the Commission and NGOs. The paper makes clear that the decision-making process in the EC is "first and foremost legitimized by the elected representatives of the European people".[79] However, it also points out that NGOs can make a contribution to fostering a more participatory democracy within the European Union.

In response to these developments more and more interest groups are establishing offices in Brussels.[80] At the national level in EU Member States, interest groups are usually organized through peak associations; however, at the EU level, large enterprises increasingly have their own representation. Although corporatist at the level of most EU Member States, at the EC level patterns of interest group organization resemble those associated with pluralism. These developments suggest that, in spite an absence of formal participation requirements, interest groups, including environmental groups, anticipate increasing opportunities to participate in EC decision-making processes.

These arguments are supported by recent evidence. For example, environmental groups have played a major role in the ongoing EC-wide policy dialogue on reforming the present chemicals control regime. The Chemicals Charter adopted by a consortium of environmental groups in Copenhagen[81] significantly contributed to a white paper that was recently published by the Commission.[82] Similarly, the Environmental Council, when discussing the EC chemicals control strategy in June 2001, recognized the importance of procedural environmental rights and agreed to "elaborate and implement all the relevant provisions of the new chemicals policy fully in line with the requirement laid down in the UNECE Aarhus Convention".[83]

Perhaps, the EC Aarhus ratification process can help to inform and stimulate a debate on the strengths and weakness of participation rights at the level of EC institutions. The Commission, however, is only taking cautious steps to use Aarhus to increase the openness and legitimacy of the EC's own procedures. So far, it is mainly concerned with incorporating Aarhus requirements into Directives that will have an impact on Member States, but that do little to democratize environmental policy making in the Commission.[84] Of course, the Commission may not be the right organization to push this issue forward and it has its hands tied by EU Member States that are skeptical of procedural changes, in particular in the area of public participation and access to justice.

The European Union is in the midst of a debate about the openness of its procedures. The increasing openness of EC decision making, coupled with political support from some EU Member States, seems to have triggered a reform process that may result in fundamental changes. Consistent with our earlier analysis, it is the European Parliament that has provided the leadership to democratize EC decision-making processes. In 1999, the Parliament adopted a Resolution on openness within the European Union calling for a more open administrative culture to be developed within the various EU institutions and bodies.[85] The Parliament requested, for example, that all comitology texts be placed on the Internet and that criteria be developed – based on the U.S. Government in the Sunshine Act – to open up more meetings of EU institutions to the public. Both EU Member States and the Commission have responded to the call of the European Parliament. A White Paper on European Governance that addresses a range of EU governance issues, including the democratic deficit problem, was released in 2001 as the basis for broad-based consultations.

These developments are partly a result of a strengthened European Parliament facing a European Commission with some independence from the legislature. Short of more dramatic changes in the constitutional structure of the EU, the Parliament is likely to support procedural guarantees that limit the direct control of member states and increase the influence of pan-European groups, as well as its own members and political groups. Thus, participation practices in the EC may be evolving in a direction that is more similar to United States patterns than to those in EU Member States.

Even though the EC is unlikely to ratify the Aarhus Convention in the immediate future, the Convention is "alive" and appears to have already had an impact on environmental policy-making processes within EC institutions. Although some aspects of the Convention have already been addressed in the broader context of EC governance reform (that is, freedom of information and access to EU documents), others, for example, public participation in EC decision-making processes, are being addressed through a pragmatic approach which follows the principles of the Aarhus Convention

With momentum developing at the EC level, it may well prove difficult for EU Member States significantly to prolong ratification of the Aarhus Convention and to avoid an open discussion at the national level on the strengthening of procedural environmental rights. Such discussions are likely to move beyond the environmental dimension to raise questions of open governance in general, as was the case in the U.K. Thus, in the long term, the Aarhus Convention may contribute to fundamental change at two levels. First, procedural environmental rights of citizens will be strengthened and, second, administrative decision making may become more democratic and accountable, not only in the former communist countries, but also in the established democracies of Western Europe.

NOTES

1. The text of the Convention and other supporting documents can be found at *http://www.unece.org/env/pp/*. Although Germany and Russia actively participated in the negotiation process – and according to some commentators "watered down" various provisions of the Convention – neither country signed the Convention in June 1998. In both cases, domestic constitutional obstacles were put forward as the main reason for abstaining. See McAllister (1998). In the case of Germany, however, "constitutional obstacles" were overcome, and Germany signed the Convention following change in the ruling government coalition in late 1998. The U.S. and Canada, although UNECE Member States, opted not to participate in the negotiation of the Convention. The U.S. pointed out that the principles of the Convention were already firmly embedded in its domestic laws. However, leaders of non-governmental organizations claim that citizens in the U.S. and Canada could have benefited from the active participation of both countries in the Aarhus process. See McAllister (1998).

2. Preamble, para 7. For a general introduction to the economic arguments for government intervention to regulate environmental quality, see Oates (1999).

3. For an introductory discussion on the definition and concept of procedural environmental rights, see Anderson (1996) and Mason (1999).

4. Principle 10 of the Rio Declaration 1992, adopted by more than 150 Heads of State or Government at the Earth Summit in 1992, called for such strengthening. Rio Declaration on Environment and Development, U.N. Doc. A/CONF.151/5/rev.1 (1992). Article 19 of the Aarhus Convention specifies that any Member State of the United Nations may accede to the Convention upon approval by the conference of the parties.

5. On 1 October 2001, the Convention had forty Signatories (including the European Community) and seventeen Parties. At that time, Parties included Italy, Denmark and fifteen former Communist countries. The Convention entered into force on 30 October 2001. See *http://www.unece.org/env/pp/ctreaty.htm*

6. Although the United States, a UNECE member state, decided not to participate in Aarhus Convention process, particular attention will be devoted to the U.S. as it is considered a front-runner in the area of democratic environmental governance.

7. The Aarhus Convention is the first legally binding instrument signed by the EC that also applies to European Community Institutions. See European Commission (1999).

8. A detailed guide to the Convention can be found in United Nations, Economic Commission for Europe, *The Aarhus Convention: An Implementation Guide*, New York and Geneva: United Nations. 2000.

9. Exemptions from information disclosure include: matters of confidentiality related to administrative proceedings; international relations, national defense or public security; course of justice in judicial matters; confidentiality of commercial and industrial information; intellectual property rights; personal data; and certain environmental information, such as breeding sites of rare species.

10. Council Directive 96/61/EC concerning integrated pollution prevention and control (IPPC), September 1996, Official Journal of the European Communities, L 257/26.

11. See McAllister (1998).

12. See Regional Environment Center (1998) pp. 9–14.

13. Steps have recently been taken to align Directive 90/313/EEC with Articles 2, 4, and 5 of the Aarhus Convention. See Commission of the European Communities (2000), *Proposal for a Directive of the European Parliament and of the Council on Public Access to Environmental Information,* COM (2000) 402 final, Brussels, 29.06.2000.

14. TRI was established under the U.S. Emergency Planning and Community Right-to-Know Act. In a recent development under the Aarhus Convention, steps have been undertaken by the signatories to develop a legally binding instrument to mandate the establishment of what is now internationally known as Pollutant Release and Transfer Registers (PRTRs).

15. Rose-Ackerman (1995) pp. 113–115.

16. It took Germany until 1994 to transpose directive 90/313/EEC into national law. Since then, Germany was challenged and defeated several times in front of the ECJ on matters of access to environmental information. Case C-231/96 (Wilhelm Mecklenburg v. Kreis Pinneberg der Landrat) dealt with the refusal by local civil servants to provide information to an interested citizen related to a highway construction permit. It revealed weaknesses in the transposing of the EC directive into German legislation. In Case C-217/97 (Commission of the European Communities v. Germany) the issue was the transposing law itself. The ECJ ruled that Germany must change its legislation in two areas: to ensure access to information related to administrative proceedings and to ensure that charges are not made for information refusals.

17. Rose-Ackerman (1995) pp. 82–93.

18. Rose-Ackerman (1995) pp. 86–87.

19. Article 6.1(b), 1(c), 6(f).

20. For a brief introduction to the concept of negotiated rule making in the U.S., see Percival, Miller, Schroeder, and Leape (2000) p. 173. For critical reviews, see Coglianese (1997) and Rose-Ackerman (1994).

21. Rose-Ackerman (1995) pp. 7–17, 126–131, 134–139.

22. U.S. American Administrative Procedures Act5 U.S.C. Secs 701–706.

23. Citizen suit provisions are, for example, included in: the Clean Water Act (CWA), 33 U.S.C. Sec. 1365; the Safe Drinking Water Act (SDWA), 42 U.S.C. Sec. 300j-8; the Clean Air Act (CAA)42 U.S.C. Sec. 7604; the RCRA, 42 U.S.C. Sec. 6972; and the Comprehensive Environmental Response, Compensation and Liability Act (CERCLA), 42 U.S.C. Sec. 9659. Citizen suits typically allow actions against both private violators of the act and government agencies that have failed to perform non-discretionary duties.

24. Rose-Ackerman (1995) pp. 129–130.

25. See Tuesen and Simonsen (2000)

26. In contrast, this will occur in a parliamentary system only in a coalition government where a small party is given a key cabinet post to keep it from defecting.

27. Cox and McCubbins (1997).

28. Of course, there are exceptions. The most obvious is a party with a regional base as in India or Quebec, Canada. See Cox (1997) for a full analysis.

29. For a mathematical model which allows one to calculate the theoretical number of parties as a function of district magnitude, see Cox (1997).

30. For a brief discussion of how ideology may shape voter behavior, see Hinich and Munger (1997).

31. Rose-Ackerman (1995) pp. 8–12.

32. Codified at 6 U.S. C. 551–559, 701–706. For an excellent overview of how U.S. rule-making process developed from an agency-controlled process to a more transparent process, see Strauss (1996).

33. 42 UCSA 4321–4307f.

34. 5 U.S.C 552.

35. 42 U.S.CA 11001–11050.

36. The public must be provided notice of the time, place, and subject of any meeting at least one week in advance. Only a few exceptions apply. The act only requires public observation, not participation, in the meetings. 5 U.S.C Sec 552b.

37. Brickman, R., Jasanoff, S. and T.Ilgen (1985) p. 44.

38. Rose-Ackerman (1995) p. 67.

39. Rose-Ackerman (1995) p. 18.

40. See Rose-Ackerman (1995) p. 69. Consistent with this view, the German Council of Environmental Advisors (1996) criticized the deficiency of German environmental law to provide legally guaranteed participation opportunities to environmental groups in the development of environmental standards. The Council recommended that the government amend environmental statutes to ensure such participation in order to "offset the strong influence of business interest groups and to make an open discussion of issues possible". See The German Council of Environmental Advisors, English Summary (1996).

41. Out of 154 environmental standard setting processes examined by the Council of Environmental Advisors (1996), 17% of the cases allowed participation of concerned parties, while in 6% any interested party or individual could participate. All other standard setting processes involved experts only, or no information about participation opportunities was available. For a more detailed discussion see Der Rat von Sachverstaendigen fuer Umweltfragen (1996), full report in German pp. 251–315.

42. The European Commission sent a final warning to Germany in 2001 concerning it failure to comply with Access to Environmental Information Directive 90/313/EEC. Austria has also run afoul of the Commission which has referred a case to the European Court of Justice. "Public Gains Influence under EU's New Green Law," *Environmental News Service* February 5, 2001 (LEXIS/NEXIS)

43. Strauss (1996) p. 755.

44. For a history of attempts to introduce FOI legislation in the U.K., see Lewis (1989).

45. The U.K. Freedom of Information Act 2000 is available at *http://www.legislation.hmso.gov.uk/acts/acts2000/20000036.htm*

46. See Campaign for Freedom of Information at *http://www.cfoi.org.uk*

47. The EC directive on access to environmental information was implemented in the U.K. through Environmental Information Regulations 1992, which came into force on 31 December 1992.

48. The Campaign for Freedom of Information is a non-profit organization that works towards eliminating "unnecessary official secrecy and to give people legal rights

to information which affects their lives or which they need to hold public authorities properly accountable". It is supported by some 80 national bodies including leading consumer, environmental, civil liberty and legal groups, professional bodies, civil service and other trade unions and organizations representing journalists, newspapers and authors. See http://www.cfoi.org.uk.

49. See The Campaign for Freedom of Information, *Why Britain Needs a Freedom of Information Act,* 1993, available at *http://www.cfoi.org.uk/whyfoi.html*

50. See Friends of the Earth (1999) Press Release: *Worse Than it Ever Was:"Freedom of Information Bill" Threatens Key Environmental Rights. Available at http://www.foe. co.uk/pubsinfo/infoteam/pressrel/1999/19990712000120.html*

51. NRC (1994).

52. 15 U.S.C. secs. 2601–2671.

53. See Brickman et al. (1985).

54. The procedural provisions of TSCA include, for example: judicial review of regulations; citizens suits against anyone alleged to be in violation of TSCA or against the EPA administrator for failing to perform non-discretionary duties; and citizen petitions to initiate EPA rule-making proceedings.

55. Section 2 of the Toxic Substances Control Act (TSCA) reads as follows: "It is the policy of the United States that adequate data should be developed with respect to the effect of chemical substances and mixtures on health and the environment and development of such data be the responsibility of those who manufacture and those who process such chemicals and mixtures."

56. See Conservation Foundation (1987) and EDF (1997).

57. A basic testing schedule established by the OECD known as SIDS (Screening Information Data Sets) covers six end points: acute toxicity, chronic toxicity, developmental and reproductive toxicity; mutagenicity; eco-toxicity; and environmental fate. EPA estimates that the full battery of basic SIDS screening tests costs about U.S.D 205,000. See Percival, Miller, Schroeder, and Leape (2000) pp. 376–377.

58. 44 Fed. Reg. 60,061.

59. 54 Fed Reg. 29,460.

60. The process of preparing the final rule involved 22 days of public hearings, thousands of pages of testimony, and some 13,000 pages of comments from more than 150 interested parties. See Percival, Miller, Schroeder and Leape (2000) pp. 470–471.

61. *Corrosion Proof Fittings v. EPA,* 947 F.2d 1201 (5th Cir.,1991).

62. Risk management activities in EPA's Existing Chemicals Program are divided into RM1, RM2 and Post-RM2 stages. The Risk Management 1 stage focuses on screening and selecting chemicals that appear to be of greatest concern to human health and the environment and identification of additional testing needs is a possible outcome. During Risk Management 2 options are framed for reducing or eliminating the risks posed by particular chemicals, including possible requirements for additional testing. Post-RM2 activities consist of the implementation of one or more of the risk reduction and/or testing options identified during RM2.

63. In a letter to the EPA in early 2000, the Animal Legal Defense Fund (ALDF) pointed out that the HPV chemical testing initiative was not compatible with Section 4 of TSCA which requires EPA rule making – and therefore full public transparency – prior to initiation of testing. ALDF argued that EPA had not given notice in the Federal Register about the voluntary industry testing initiatives. It also noted that the program was established in co operation with only two interested parties: EDF and the Chemical

Manufacturers Association. See letter dated January 7, 2000 addressed to EPA Office of Policy and Reinvention, available at http://www.epa.gov/reinvent/stakeholders/public/aldf.htm

64. The administrative record includes a screening dossier, summaries of major studies cited in the dossier, summaries of risk management meetings, letters of concern to industry or others and replies, and comments or correspondence from other parties outside of EPA. The screening dossier contains relevant exposure and hazard information, recommendations from the screening workgroup, and the supporting rationale for that decision.

65. Existing chemicals are regulated by 93/793/EEC. Directive 67/548/EEC (implemented through articles 13–15 of the German Chemicals Law) covers classification and labeling.

66. German environmental statutes which include participation requirements include the Nature Protection Law, the Atomic Law and the Immission Protection Law Control.

67. Rose-Ackerman (1995) pp. 65–66.

68. See http:www.gdch.de/projekte/bua.htm for an introduction to the work of BUA.

69. See http:www.gdch.de/projekte/bua2.htm for a current list of BUA Members.

70. See German Chemical Society, 1999, pp. 5–7.

71. For example, Germany recently decided to ban – without waiting for concerted action at the Community level – the chemical TBT, a persistent substance used mainly in ship painting. The action was taken because the Commission had not responded positively to an earlier German proposal to ban TBT at the Community level through an amendment of directive 76/769/EEC relating to restrictions on the marketing and use of certain dangerous substances and preparations. In this particular case, Germany is likely to face the problem that, in principle, banning or severely restricting a chemical falls clearly within the authority of Community legislation which aims at ensuring free trade among its Members States.

72. The Commission does not expect ratification to occur until 2002/2003. Report by the European Commission on Work Undertaken to Prepare for Ratification by the European Community of the Convention on Access to Information, Public Participation in Environmental Decision-Making and Access to Justice in Environmental Matters to the Second Meeting of the Signatories to the Convention, Catvat-Dubrovnik, Croatia, 3-5 July 2000.

73. Lindseth (1999) pp. 651–654.

74. For a discussion of the democratic deficit in the EC, see Douglas-Scott (1966).

75. See Council Decision 1999/468/EC of 28 June 1999 laying down the procedure for the exercise of implementing powers conferred to the Commission, Official Journal L 184, 17/07/1999 pp. 0023–0026. The decision does not spell out how the Commission should consult with the public in its deliberations. It merely instructs the Commission to inform the public about the work of the Committees.

76. For a comprehensive discussion on the EC "Comitology" process, see Egan and Wolf (1998) Bignami (1999) and Hey (2000).

77. See Hey (2000).

78. See European Commission COM (2000) 30 final/2 of 21.2 2000 and *Proposal for a Regulation of the European Parliament and Council regarding public access to European Parliament, Council and the Commission documents,* Document 500PC0030, 19.02.2001.

79. European Commission (1999), p. 4.

80. Mueller-Brandeck-Bocquet (1995).

81. EEB, 2000.

82. Commission of the European Communities, 2001. One major proposed change for chemicals control is a shift in the burden of proof. At present, public authorities are required to document that a particular chemical is unsafe; if the proposal is adopted, a greater burden will be placed on industry to prove that a chemical does not pose unacceptable risks to human health and the environment. The shift was one of the main demands of the Copenhagen Chemicals Charter.

83. See Press Release of the 2355th Council Meeting – Environment, 7 June 2001.

84. "Environment: Top EU Official Explains Delay in Ratifying Aarhus Agreement," *European Report*, October 21, 2000; "Commission Takes Further Step Towards Ratification of the Aarhus Convention," *RAPID*, January 26, 2001; "Public Gains Influence Under EU's New Green Law," *Environmental News Service*, February 5, 2001 (all articles available on LEXIS/NEXIS).

85. Parliament of the European Communities, *Resolution on Openness with the European Union*, adopted on 12/01/1999. For a detailed background on the resolution see European Parliament (1999).

ACKNOWLEDGMENTS

The comments of Bruce Ackerman and participants in the Symposium at University College London, September 5–7, 2001, are greatly appreciated.

REFERENCES

Anderson, M. R. (1996). *Human Rights Approaches to Environmental Protection: An Overview*. Clarendon Press, Oxford.

Bignami, F. E. (1999). The democratic deficit in European Community rule making: a call for notice and comment in comitology. *Harvard International Law Journal, 40*, 451–515.

Brickman, R., Jasanoff, S., & Ilgen, T. (1985). *Controlling Chemicals: The Politics of Regulation in Europe and the United States*. Ithaca: Cornell University Press.

Carey, J., & Shugart, M. S. (1995). Incentives to cultivate a personal vote: a rank order of electoral formulas. *Electoral Studies*, 417–439.

Coglianese, C. (1997). Assessing consensus: the promise and performance of negotiated rule making. *Duke Law Journal, 46*(6).

Conservation Foundation (1997). *State of the Environment: A View Toward the Nineties*. Washington, D.C.

Cox, G. (1997). *Electoral institutions, cleavage structures, and the number of parties, in: Making Votes Counts*, Chap. 11. Cambridge.

Cox, G., & McCubbins, M. (1997). Political structure and economic policy: the Institutional determinants of policy outcomes. In: S. Haggard & M. D. McCubbins (Eds), *Political Institutions and the Determinants of Public Policy: When Do Institutions Matters?* (pp. 22–96). San Diego: University of California.

Dalton, R. J. (2000). Politics in Germany. In: G. A. Almond, G. B. Bingham Powell, Jr., K. Strom & R. J. Dalton (Eds), *Comparative Politics Today* (pp. 273–322). New York: Longman.

Douglas-Scott, S. (1996). Environmental rights in the European Union: participatory democracy or democratic deficit. In: A. E. Boyle & M. R. Anderson (Eds), *Human Rights Approaches to Environmental Protection* (pp. 109–128). Oxford: Clarendon Press.

Egan, M., & Wolf, D. (1998). Regulation and comitology: the EC committee system in regulatory perspective. *Columbia Journal of European Law, 4*, 499–523.

Environmental Defense Fund (1997). Toxic Ignorance.

European Commission (2001). Strategy for a Future Chemicals Policy (White Paper). COM (2001) 88 final, Brussels.

European Commission (2000). Report from the Commission to the Council and the European Parliament on the Experience Gained in the Application of Council Directive 90/313/EEC of 7 June 1990 on Freedom of Access to Information on the Environment. COM (2000) 400 final, Brussels.

European Commission (1999). The Commission and Non-Governmental Organizations: Building a Strong Partnership, Commission Discussion Paper. Brussels.

European Environmental Bureau et al. (2000). Copenhagen Chemicals Charter: Chemicals under the Spotlight International Conference. Copenhagen.

European Parliament (1998). Committee on Institutional Affairs, Report on Openness within the European Union. A4–0476/98.

German Chemical Society, Advisory Committee on Existing Chemicals (BUA) (1999). *Assessment of Chemicals: A Contribution Towards Improving Chemical Safety.* Gesellschaft Deutscher Chemiker, Frankfurt am Main.

German Council of Environmental Advisors (1996). *Environmental Report 1996* (Summary) available at *http://www.umweltrat.de/gut96en0.htm.* See also: Der Rat von Sachverstaendigen fuer Umweltfragen.

Haemer, R. B. (1999). Reform of the toxic substances control act: achieving balances in the regulation of toxic substances. *Environmental Lawyer, 6.*

Hallo, R. (1996). *Access to Environmental Information in Europe: The Implementation and Implications of Directive 90/313/EEC.* Kluwer Law International, London, The Hague, Boston.

Hey, C. (2000). Towards Balancing Participation: A Report on Devolution, Technical Committees and the New Approach in EU Environmental Policies: The Cases of Standardization, Chemicals Control, IPPC and Clean Air Policies in a Comparative Perspective. EEB, Brussels.

Hinich, M. J., & Munger, M. C. (1997). *Analytical Politics.* Cambridge: Cambridge University Press.

Khanna, M., & Damon, L. A. (1999). EPA's voluntary 33/50 program: impact on toxic releases and economic performance of firms. *Journal of Environmental Economics and Management, 37,* 1–26.

Lewis, J. R. T. (1989). Freedom of information: developments in the U.K. *International Journal of Intelligence and Counter Intelligence, 3*(4), 465–474.

Lindseth, P. L. (1999). Democratic legitimacy and the administrative character of supra-nationalism: the example of the European Community. *Columbia Law Review, 99,* 628–737.

MacAllister, S. T. (1998). The convention on access to information, public participation in decision making, and access to justice in environmental matters. *Colorado Journal of International Environmental Law and Policy, 187.*

Mason, M. (1999). *Environmental Democracy.* New York: St. Martin's Press.

Moe, T. M., & Caldwell, M. (1994). The institutional foundation of democratic government: a comparison of presidential and parliamentary system. *Journal of Institutional and Theoretical Economics, 150/1,* 171–195.

Mueller-Brandeck-Bocquet, G. (1996). *Die Institutionelle Dimension der Umweltpolitik: Eine Vergleichende Untersuchung zu Frankreich,* Deutschland and der Europaeischen Union. Nomos Verlagsgesellschaft, Baden-Baden.

Noll, R. G. (1989). Economic perspectives on the politics of regulation. In: *Handbook of Industrial Organizations* (Chap. 22, Vol. II). Elsevier Scientific Publishers.

NRC (U.S. National Research Council) (1994). *Science and Judgement in Risk Assessment.* Washington, D.C.: National Academy Press.

Oates, W. (1999). An economic perspective on environmental and resource management. In: W. Oates (Ed.), *The RFF Reader in Environmental and Resource Management, Resources for the Future.* Washington, D.C.

Petkova, E., & Veit, P. (2000). Environmental Accounting Beyond The Nation State: The Implications of the Aarhus Convention. *Environmental Governance Notes, WRI.*

Rat von Sachverstaendigen fuer Umweltfragen (Council of Environmental Advisors) (1996). *Umweltgutachten 1996.* Metzler-Poeschel, Stuttgart.

Regional Environment Center (REC) (1998). *Doors to Democracy: A Pan-European Assessment of Current Trends and Practices in Public Participation in Environmental Matters.* Regional Environment Center, Szentendre.

Rose-Ackerman, S. (1995). *Controlling Environmental Policy: The Limits of Public Law in Germany and the United States.* New Haven and London: Yale University Press.

Rose-Ackerman, S. (1994). Consensus versus incentives: a skeptical look at regulatory negotiations. *Duke Law Journal, 43,* 1206–1220.

Shugart, M., & Haggard, S. (1997). Institutions and public policy in presidential systems. In: S. Haggard & M. D. McCubbins (Eds), *Political Institutions and the Determinants of Public Policy: When Do Institutions Matters?* (pp. 97–144). San Diego: University of California..

Silverman, D. L. (1997). Freedom of information: will Blair be able to break the walls of secrecy in Britain? *American University Journal of International Law and Policy, 13.*

Strauss, P. (1996). From expertise to politics: the transformation of American rule making. *Wake Forest Law Review, 31,* 745–777.

Swedish Committee on New Guidelines on Chemicals Policy (2000). *Non-Hazardous Products: Proposals for Implementation of New Guidelines on Chemicals Policy.* Ministry of Environment, Sweden, Stockholm.

Percival, R. V., Miller, A. S., Schroeder, C. H., & Leape, J. P. (2000). *Environmental Regulation: Law, Science, and Policy.* New York: Aspen Law & Business.

Tsebelis, G. (1995). Decision-making in political systems: veto players in presidentialism. parliamentarism, multicameralism and multipartyism. *British Journal of Political Science, 25.*

Tuesen, G., & Simonsen, J. H. (2000). Denmark & Estonia: compliance with the Aarhus Convention. *Environmental Policy and Law, 6(30),* 299–306.

Verschuuren, J., Bastmayer, K., & van Lanen, J. (2000). Complaint Procedures and Access to Justice for Citizens and NGOs in the Field of the Environment Within the European Union. *IMPEL Network,* 30–39.

COMMENTS ON PAPER BY
SUSAN ROSE-ACKERMAN
AND ACHIM A. HALPAAP

Farhana Yamin

This paper is a thorough and helpful guide to the substantive contents of the Convention, especially the political economy section linking political institutions, literature and the environmental policy field.

Aarhus has not been the focus of as much concerted European NGO attention as it might have been, given that many NGO's resources were diverted by other negotiations e.g. the Kyoto and Biosafety protocols. But those who followed it were extremely committed activists.

1. Paper's remit: to analyze and explain the anomaly between the rapid ratification of the convention by post-communist countries (Central Asian states) and lack of equivalent speedy ratification by EU and other UN/ECE countries. The paper *assumes,* rather than establishes, that these are explanations for this disparity, other than those relating to the complexity/timing of completing the ratification process.

Conventions, like other bills and regulation, compete for parliamentary/ beaurocrats' time. Aarhus seems to be in a 'holding pattern', maybe because other aspects of environmental policy, e.g. climate change and biosafety and GMOs have taken up the lion's share of beaurocrat's time and are "waiting to hand". Given that the EU Commission has not prioritized Aarhus ratification until 2002/2003 it should not be surprising that MS are waiting for the EU to take the lead.

An Introduction to the Law and Economics of Environmental Policy: Issues in Institutional Design Volume 20, pages 65–67.
© **2002 Published by Elsevier Science Ltd.**
ISBN: 0-7623-0888-5

I am not clear how the analysis in the paper handles how the executive and beaurocrats set environmental priorities. Some empirical work, e.g. interviews with legal staff in environmental ministries might have been revealing. Were they busy transposing other directives/conventions and why did these come first? The capacity of executive/beaurocrats to "churn out" legislation is limited. Also, how many NGOs set Aarhus as ratification as a priority? In most cases, NGOs set a handful of environmental priorities per year: they can't pursue everything. In short, there may be many other explanations for the seemingly long period of "delay" in ratification by EU that are not examined in the paper but merit at least further reflection.

Additionally, there is a need to examine the motives and merits of "speedy" ratification by the Central Asian states. The lack of administrative practices of these states is a fact; it is not in and of itself an explanatory factor for why these states rapidly ratified this particular Convention. (They haven't done that for other Conventions in which they equally lack administrative practice, e.g. Kyoto.) Also, is their speedy ratification to be applauded? Do their officials actually understand the implications of the Convention?

My (untested) hunch is that they do not fully appreciate what they have let themselves in for and some may have thought this was a quick and easy way to get environmental credibility points with foreign counterparts and foreign aid donors – the latter are very concerned about the "participation of NGOs" and now require extensive consultation/participants procedures as a condition of lending money. If the paper's goal is to examine and explain the anomaly between the speedy and laggard ratifiers of the convention, it would be helpful to include more explanation about the motives and factors influencing the "speedy" countries' ratification.

2. This brings me to a related point. The paper is at times a little schizophrenic about evaluating the real impact of the Convention. For example, in the introduction (p. 2) The Convention is described as "a moderate document designed to accomplish marginal changes". Yet in the same place, and in the body of the paper, the analysis makes clear that the Convention would require, I quote "increased participation rights in most signatory countries". Rights that would go beyond state practice, particularly where the exercise of such rights run *against* the grain of hundreds of years of political culture and decades of environmental policy making traditions. It seems to me that these changes cannot be described as "marginal changes " especially as the Paper itself demonstrates that some of the Convention's provisions will have, a "major impact" (e.g. p. 7, Article 6, section on the development of community 'right to know' programmes). It is precisely because these would

require additional human resources/budgets which most environmental agencies are fighting to enhance and retain in these budget-pressed times, that beaurocrats are reluctant to begin their implementation. Perhaps this is an explanatory factor rather than beaurocrat's reluctance to "open up government"!

It seems to me that the Convention is in fact groundbreaking in many respects even if the ground is fairly familiar to many countries and has already been worked upon. I think the merits of the Convention lie in the breadth and scope of its general principles. Principles which are relatively new to major parts of the European continent (and the rest of the world, except the U.S.!) It is true that the Convention is short on the specifics of how to translate such broad principles into domestic frameworks but this, I would argue, should not detract from the ground-breaking nature of the principles it establishes for the first time in a legally binding form for such a large and varied group of countries to accept. Particularly where the implementation of those principles is going to cost the environment agencies time and money to put in place (e.g. the community right to know/register positions).

3. My final comments touch on broader questions: what contribution can Conventions (on any subject) that go beyond the state practice of the vast majority of signatories make, and over what time frame? Secondly, are gereral across the board "participation rights" for NGOs and individuals – which the paper endorses as preferable – the way forward?

My own paper on NGOs in the symposium volume questions what role NGOs can actually play in international environmental policy but its insights are relevant to domestic processes. It is not clear that NGOs can or should deliver all the expectations set up for them as a result of the Aarhus Convention, given their constraints of funding and their governance structures, and issues of NGO accountability. The increased participation of NGOs and greater rights for them to oversee implementation and exact enforcement raises more fundamental questions about the role of environmental agencies. As far as policy making is concerned, increased NGO involvement raises fundamental questions about the operation of democratic societies, and the ability of democratic processes to take the interests of all concerned into account.

The capture by stealth of democratic institutions (as described by a new wave of activists e.g. Klein), the violent demonstrations at Seattle/Gottenburg/Genoa, voter apathy, all call into question our views on the health and efficacy of existing instruments to deliver environmental results acceptable to all.

PART 2:
OPTIMAL INFORMATION FOR POLICY MAKING

ENVIRONMENTAL REGULATORY DECISION MAKING UNDER UNCERTAINTY

Richard B. Stewart

ABSTRACT

Strong versions of the Precautionary Principle (PP) require regulators to prohibit or impose technology controls on activities that pose uncertain risks of possibly significant environmental harm. This decision rule is conceptually unsound and would diminish social welfare. Uncertainty as such does not justify regulatory precaution. While they should reject PP, regulators should take appropriate account of societal aversion to risks of large harm and the value of obtaining additional information before allowing environmentally risky activities to proceed.

[T]he public generally benefits from individual activity. Holmes, The Common Law.

When in doubt, don't pump it out! Greenpeace.

This article examines the principles applicable to governmental decisions to regulate activities that may result in environmental harm when there is uncertainty regarding the probability and extent of any such harm. It analyzes and evaluates the precautionary principle (PP), which, in its weakest version, asserts that uncertainty regarding the adverse environmental effects of an activity

An Introduction to the Law and Economics of Environmental Policy: Issues in Institutional Design, Volume 20, pages 71–126.
Copyright © 2002 by Elsevier Science Ltd.
All rights of reproduction in any form reserved.
ISBN: 0-7623-0888-5

should not automatically bar adoption of measures to prohibit or otherwise regulate the activity, and, in stronger versions, further asserts that uncertainty provides an affirmative justification for regulating an activity or regulating it more stringently than in the absence of uncertainty. Strong versions of PP hold that regulators should adopt "worst case" presumptions regarding the harms of activities posing an uncertain potential for significant harm; should prohibit such activities or require them to adopt best available technology measures; that regulatory costs should be disregarded or downplayed in such decisions; and that the proponents of such activities should bear the burden of establishing their safety in order to avoid such regulatory controls. The article also considers the relevance for regulatory decisions of uncertainties regarding the costs of regulating an activity as well as uncertainties regarding harms. It further considers the implications of the circumstance that regulatory decisions about a given environmental issue may be made sequentially over time and benefit from additional information developed in the interim between earlier and later decisions. The essay concludes that, while preventive regulation of uncertain risks is often appropriate and should incorporate precautionary elements where warranted by consideration of risk aversion or information acquisition, strong versions of PP do not provide a conceptually sound or socially desirable prescription for regulation.

I. INTRODUCTION:
ENVIRONMENTAL DECISION MAKING UNDER UNCERTAINTY AND THE PRECAUTIONARY PRINCIPLE

The following are examples of regulatory decisions involving uncertain risks. In each case, consider whether a regulator should permit the potentially harmful activity to commence or continue, or, alternatively, to prohibit or otherwise regulate it, and the implications of sequential regulatory decision making and the opportunity to develop additional information to reduce uncertainties regarding the risks of harms posed by an activity and/or the costs of regulation.

- Whether to prohibit the sale of meat products from cattle that have received bovine growth hormone injections.
- Whether to prohibit the construction of an astronomic observatory atop Mt. Graham, in New Mexico, the only known habitat of the Mt. Graham Red Squirrel, a subspecies of the western red squirrel located only on Mt. Graham; other subspecies of the red squirrel are abundant in the Western United States.

- Whether to prohibit field releases of crop plants that have been genetically modified using DNA technologies.
- Whether to adopt a National Ambient Air Quality Standard (NAAQS) to limit short-term (10-minute) exposures to elevated levels of sulfur dioxide (SO_2). Several laboratory studies indicate that asthmatics exposed to higher short-term SO_2 exposures experience temporary airway resistance that makes breathing more difficult.
- Whether to prohibit introduction of sucralose, an artificial beverage sweetener that has been touted as safer than saccharin or aspartame, the artificial sweeteners currently in use.
- Whether to prohibit the dumping of wastes of any sort at sea.
- Whether to develop defenses against collisions with the earth by asteroids and other near-earth objects.
- Whether to eliminate the use of chlorine to treat drinking water.
- Whether to prohibit the use of glyphosate ("Roundup"), a broad-spectrum non-selective herbicide that is harmless to animals.
- Whether to prohibit or tightly regulate the conversion of rainforest to agricultural or other uses.
- Whether to immediately initiate strict limits on greenhouse gas emissions in order to limit the potential adverse effects of climate changes attributable to such emissions.

In evaluating these and other environmental regulatory decision-making issues, the extent of available knowledge regarding environmental harm that may be caused by the activity in question can be conceptually classified in three ideal type categories:

Type 1. The harm that the activity will cause is known and determinate. If, for example, the Mt. Graham observatory is built, the Mt. Graham red squirrel population will be wiped out within 20 years.

Type 2. The harm is probabilistic in character but its probability distribution is known. For example, if the observatory is built, there is a 40% probability that the squirrel population will be wiped out within 20 years and a 60% probability that it will survive another 1000 years, at which point it will become extinct from natural "background" causes. In this situation we deal with a risk of harm, but the risk (comprising both the probability of an adverse effect occurring and the magnitude of the adverse effect if it occurs) is determinate.

Type 3. There is a risk of harm that is uncertain. Thus, the probability of harm occurring, and/or the magnitude of the harm if it occurs, is not determinate and is subject to substantial uncertainty. To take the Mt. Graham example, it may be uncertain, based on current knowledge, whether any adverse effect on the squirrels will occur. If any adverse effect does occur, its magnitude is uncertain. Thus, it may be uncertain what percentage of the population may be lost, whether any given level of loss will result in extinction of the subspecies, and when possible losses or extinction may occur. There may also be cases where the type of harm, if any, that may occur is not known.

The difference between Type 2 cases and Type 3 cases is obviously one of degree, but the distinction between the two ideal types is very useful for purposes of analysis. Very many environmental problems are Type 3 cases, characterized by uncertainty regarding risks of harm, although the nature and degree of uncertainty varies widely from case to case. These uncertainties have many potential causes, including lack of data, limitations in scientific understanding of causal relationships, medical and ecosystem complexity, and "trans scientific" gaps in the capacities of science.[1] There may also be substantial uncertainties regarding the costs of prohibiting or otherwise regulating the activity in question. In many situations, such uncertainties can be reduced by the development of additional information and knowledge as discussed further below.

The common law traditionally awards damages only *ex post* for harm that has occurred and has been shown to have been caused by another's activity. It grants injunctive relief *ex ante* only if an activity poses an imminent and substantial likelihood of serious irreparable harm. Many environmental risks of types 2 and 3 would not qualify for an award of damages or prophylactic relief under this standard. In theory, *ex post* liability for harm caused could provide the requisite incentives for actors to manage their activities so as to prevent excessive risks of harm appropriately. In practice, however, these incentives have for a variety of reasons proven inadequate to prevent excessive environmental harm from occurring.[2] Accordingly, administrative programs of preventive *ex ante* regulation have been widely adopted in the United States and other countries to regulate activities that pose substantial risks of environmental harm, even in cases where it is not certain that harm will actually occur. International agreements, such as the Vienna Convention for the Protection of the Ozone Layer and the Framework Convention on Climate Change, have also been adopted to address such risks.

Preventive regulatory programs have been adopted not only in cases where activities have been shown to cause harm, but also in cases involving risks of harm, including cases of substantial uncertainty in the risk of harm.[3] Quantitative risk analysis and cost-benefit analysis are increasingly being used in connection with the preventive approach to regulation.[4] Under many preventive regulatory programs, regulators have the burden of establishing a significant risk of harm before imposing regulatory controls,[5] although under some programs, such as U.S. FDA new drug and food additive approvals and EPA registration of pesticides, the applicant bears the burden of showing product safety.

In recent years, environmental advocates and many environmental law scholars, particularly in the field of international environmental law, have argued that environmental regulatory decisions and policies should follow a precautionary principle (PP).[6] The focus of PP is on appropriate regulatory policy in Type 3 cases where the risks of harm posed by an activity are characterized by substantial uncertainty. PP advocates argue for a precautionary approach to regulation in the face of such uncertainty. They often criticize prevailing preventive approaches to regulation on the grounds that they place the burden on regulators to show that an activity will cause serious harm or poses a high probability of serious harm before regulatory controls may be adopted. They argue that, given the lack of scientific capacities to predict which activities will cause serious or irreversible harms, this approach results in seriously inadequate environmental protection. They often also contend that existing preventive regulatory approaches give undue weight to costs in establishing controls.[7]

Various versions of PP, mostly weak ones, have been incorporated or invoked in a number of recent international environmental declarations and conventions, including the Framework Convention on Climate Change[8] and the EU Maastricht treaty.[9] These documents and the writings of PP advocates and of academics provide widely varying formulations of PP. It has been claimed that PP is already, or is becoming established as a binding principle of customary international law.[10] PP skeptics and critics, however, have contended that the heterogeneity of PP formulations, many of which are quite vague and indeterminate, demonstrates that that there is no single or determinate PP.[11] Thus, they have concluded that the precautionary principle is a "composite of several value-laden notions and loose, qualitative descriptions" and that accordingly its "operational usefulness . . . is doubtful."[12] They also deny that PP has been established as customary international law.[13]

Criticisms of PP as indeterminate and conceptually fuzzy have merit. With a very few exceptions, there is a remarkable lack of analytic care or rigor regarding the substance of, and justification for, various versions of PP by those who advocate or favor their adoption. One can, however, identify four different PP conceptions that have emerged in legal instruments, international and national governmental declarations, advocacy statements, and the academic literature that can serve as a useful basis for analysis and evaluation. These four versions of PP are as follows:

PP1. Scientific uncertainty should not automatically preclude regulation of activities that pose a potential risk of significant harm (Non-Preclusion PP).

PP2. Regulatory controls should incorporate a margin of safety; activities should be limited below the level at which no adverse effect has been observed or predicted (Margin of Safety PP).

PP3. Activities that present an uncertain potential for significant harm should be subject to best technology available requirements to minimize the risk of harm unless the proponent of the activity shows that they present no appreciable risk of harm (BAT PP).[14]

PP4. Activities that present an uncertain potential for significant harm should be prohibited unless the proponent of the activity shows that it presents no appreciable risk of harm (Prohibitory PP).

What unites these different formulations is a focus on uncertainty regarding risks as the key factor guiding regulatory decisions. Some discussions of the PP blur the distinction between known (Type 2) and uncertain risks (Type 3), but the most careful commentators make clear that the precautionary principle is addressed to uncertain risks (Type 3) as such.[15]

PP1 and PP2 are weak versions of precautionary approaches. Unlike the strong versions, PP3 and PP4, they do not mandate regulatory action and do not make uncertainty regarding risks an affirmative justification for such regulation.

Thus, PP1 is negative in character; it states that uncertainty should not preclude regulation but does not provide affirmative guidance as to when regulatory controls should be adopted or what form they should take. This is the approach that is most widely invoked in international treaties and declarations. While the exact wording may vary, this principle of non-preclusion always sets up a threshold, e.g. an uncertain risk of serious damage, and then makes the negative prescription that, once that threshold has been triggered, regulators cannot rely on this fact alone to deny regulation. For example, the Bergen Ministerial Declaration states: "Where there are threats of serious or irreversible damage, lack of full scientific certainty should not be used as a reason for postponing measures to prevent environmental degradation."[16] The Cartagena Protocol goes further by clarifying that uncertainty can not, in and of itself, justify the decision not to regulate, nor, presumably, the alternative decision to impose regulation:

> Lack of scientific certainty due to insufficient relevant scientific information and knowledge regarding the extent of the potential adverse effects of a living modified organism shall not prevent [a] Party from taking a decision, as appropriate. Lack of scientific knowledge or scientific consensus should not necessarily be interpreted as indicating a particular level of risk, an absence of risk, or an acceptable risk.[17]

This principle of non-preclusion rejects the common law position that harm must be shown to have occurred or be imminent before legal liabilities or controls may be imposed. It also rejects the position, often asserted by industry, that significant uncertainty about risks should preclude imposition of preventive regulatory controls. Of all the formulations of the PP, this approach is the most often invoked and is most likely to be recognized as a part of customary international law; it is already widely accepted that a preventive approach, under which regulatory controls are adopted to prevent or reduce risks of harm even though the magnitude or even the occurrence of harm is uncertain, is justified in at least some circumstances.[18] Yet, the very generality and lack of specific prescriptions of PP1 may preclude it from being recognized as a binding norm.[19]

PP2 likewise fails to specify when or what form of regulation should be adopted, but instructs that, whenever regulation is adopted, it should incorporate a margin of safety. Unlike PP1, PP2 is operative only after regulators have made the determination to regulate. Once this decision is made, regulators must first determine the maximum "safe" level of an activity, and then only allow the activity at some degree lower than that level (the "margin of safety"). This is a common approach in U.S. environmental law. An example is the Sustainable Fisheries Act of 1996, in which the optimum allowable yield from a fishery "is prescribed on the basis of the maximum sustainable yield from the fishery, as reduced by" relevant factors including "ecological" factors.[20] PP2 is consistent with (although it does not necessarily mandate) many commentators' views that PP requires that regulators allow "large margins for error" in risk assessments.[21] It represents one formulation of the PP premise that: "Given scientific ignorance, prudent pessimism should be favoured over hazardous optimism."[22] PP2 is not explicitly set forth in any international agreements and declarations, but its approach is implicit in some international agreements that require or provide for the adoption of precautionary measures.[23]

The weak versions of the PP are fully compatible with and are often reflected in many well-established preventative regulatory programs that have been adopted at the domestic level by many countries and by international agreement over the past 30 years. These programs often authorize prophylactic regulation of uncertain risks in appropriate circumstances even in the absence of a showing that harm will actually occur. In many cases, they explicitly require a margin of safety in setting regulatory standards.[24] Thus the weak versions of PP do not represent or justify any basic change in the preventive approach to regulation that has generally prevailed over the past 30 years. They accordingly provide no basis for arguing that existing preventive regulatory programs are not sufficiently "precautionary" and need to be fundamentally changed in order to reflect precautionary principles.

There are, however, important differences between established programs of preventive regulation and the strong versions of PP. Weak precautionary programs generally do not make the existence of uncertainty regarding risks as such a mandatory or distinct basis for imposing regulatory controls. PP3 and PP4, on the other hand, require regulators to regulate, or regulate more stringently, activities that pose risks that are more uncertain relative to risks that are less uncertain, and thus represent a significant change in regulatory concept and result.

Under PP3, when regulators determine that there is a serious but uncertain risk, they must impose BAT measures. For example, the Second International Conference on the Protection of the North Sea calls for parties to:

> [R]educ[e] polluting emissions of substances that are persistent, toxic and liable to bio-accumulate at source by the use of the best available technology and other appropriate measures. This applies especially when there is reason to assume that certain damage or harmful effects on the living resources of the sea are likely to be caused by such substances, even where there is no scientific evidence to prove a causal link between emissions and effects ("the principle of precautionary action").[25]

Such a prescription does not appear to allow regulators to decide what sort of regulation is required, including no regulation: if there is an uncertain risk of serious harm, BAT measures should be imposed. However, some flexibility may remain under PP3 since the intensity of BAT controls may vary depending on the magnitude of the potential risk relative to the costs of controls, in accordance with a principle of proportionality.[26]

PP4 imposes an even more stringent prescription upon regulators. Under this formulation, if there is an uncertain but serious risk of harm, the activity in question should not be undertaken at all until it is proven to be safe by the proponent of the activity. Thus, the Final Declaration of the First European "Seas at Risk" Conference provides that:

> The "burden of proof" is shifted from the regulator to the person or persons responsible for the potentially harmful activity, who will now have to demonstrate that their actions are not/will not cause harm to the environment. If the "worst case scenario" for a certain activity is serious enough, then even a small amount of doubt as to safety of that activity is sufficient to stop it taking place.[27]

While this version of the PP presumably allows regulators some latitude to determine how serious an uncertain risk must be to invite regulation, it requires prohibition of the activity once the relevant risk threshold is met.

The strong versions of PP, PP3 and PP4, are the focus of this essay. Accordingly, unqualified references to PP in the following discussion should be understood as referring to the strong versions of PP. Unlike the weak versions of PP and the preventive approach to regulation generally, they make the existence of uncertain risks of significant harm both a sufficient and mandatory basis for imposing regulatory controls. We may term this the "uncertainty-based potential for harm" prescription for regulation. Different PP formulations incorporating this precept vary in the criteria for determining the potential for harm threshold that triggers the requirement of regulation, including how great the probability of harm must be, its character, and its magnitude. Some formulations, for example, stress that the probability of harm must be substantial and the harm that may eventuate must be "serious and irreversible."[28] Other formulations enunciate less demanding criteria.[29] In some strong PP formulations, once the applicable risk threshold is met, regulation is mandatory; regulatory compliance costs, including the social costs involved in forgoing the benefits of activities subject to regulatory prohibition or restriction, are not included as a factor to be considered in the regulatory decision.[30] Some formulations explicitly allow for consideration of costs, but relegate them to a distinctly secondary role, while others introduce the principle of proportionality, tailoring the extent and character of the regulatory response adopted to the gravity of the risk in question.[31] Under PP3, for example, the costs of BAT controls might be taken into account in determining whether a given technology is "available." It might be concluded that very costly technology controls are not as a practical matter "available." Under PP4, in cases where potential risks are judged less serious or where the social benefits of the activity are high, prohibitory controls might be adopted for only a limited initial period subject to "sunset" provisions or reconsideration, or field trials may be permitted.[32]

Strong versions of the PP also often hold that the burden of resolving uncertainty should be borne by the proponent of an activity rather than by regulators or opponents of the activity.[33] Accordingly, in order to avoid or lift regulatory prohibitions or BAT requirements, the proponent of an activity bears the burden of demonstrating that it does not present a potential for significant harm. Proponents of regulation, however, bear some initial threshold burden of production and persuasion. They must establish that an activity poses risks (albeit uncertain) of harm, including a potential for significant harm. Once that threshold burden is satisfied, however, the burden shifts to the activity proponent to resolve the uncertainty and show that that it does not have a potential for significant harm.[34]

The normative core of the strong versions of PP, which distinguishes PP-based regulation from preventive regulation generally, is the principle that uncertainty regarding risks is an affirmative justification for adopting regulatory controls or adopting more stringent controls than would be appropriate in the case of activities posing more determinate risks. In the face of uncertainties regarding risk, PP holds that decision makers should err on the side of precaution and environmental protection and, in effect, make "worst case" presumptions about the probability and magnitude of harm that an activity poses; precisely how "worst case" is defined ("reasonable worst case," etc.) varies in different PP formulations.[35] The justifications advanced by PP proponents for adopting its prescriptions center around limitations in our ability to predict which activities will cause serious, irreversible environmental harms.[36] The predictive capacity of science is limited. For example, science has often been unable to predict, in a sufficiently timely fashion to support effective preventive action, the occurrence of serious environmental harms such as asbestosis, stratospheric ozone depletion, or the ecological harms caused by DDT. Thus, a regulatory policy that requires regulators to demonstrate that an activity causes harm or even a significant risk of harm before imposing controls will result in the occurrence of serious environmental harms. Some of these harms, such as biodiversity loss or highly disruptive changes in natural systems resulting from rapid global warming, may be irreversible and seriously harm future generations. Accordingly, decision makers should err on the side of precaution and protection of the planet by adopting PP-based regulatory controls on activities involving uncertain risks that pose a potential for significant harm.

The PP literature provides little in the way of helpful guidance on what regulators must show in order to establish a potential for harm that triggers PP.[37] While some PP proponents appear to assume that nature is inherently vulnerable and precarious rather than resilient, such a general presumption is not sufficient to show that a given activity triggers PP. The bovine growth hormone dispute suggests that a showing that a substance similar in chemical structure to the substance in question can cause harm may be sufficient.[38] The BtCorn-Monarch butterfly controversy suggests that a report of a single experimental study, albeit one quite unrepresentative of field conditions, can be enough to trigger PP controls if it is sufficiently widely publicized.[39]

Under strong versions of PP, once the risk posed by an activity satisfies the threshold that triggers a worst case presumption, regulators must then follow a set of relatively stringent regulatory prescriptions. They must prohibit or impose BAT requirements on the activity; shift the burden to the activity proponent to show that the activity is "safe" in order to avoid or lift these regulatory requirements; and disregard or downplay regulatory costs in implementing regulatory requirements. Thus, PP can be analyzed as containing two basic components: First, a worst case presumption for uncertain risks that meet a triggering threshold. Second, a set of regulatory decision rules that are mandatory once the presumption is triggered. These components can be analyzed separately.

The potential justifications for the worst case presumption are examined in the remainder of this section and in Part III. The mandatory regulatory decision rules are considered in Part IV.

This essay concludes that the strong PP prescriptions for regulation suffer from two basic deficiencies. They lack a valid justifying principle. They also lead to socially undesirable outcomes.

Uncertainty regarding risks of harm as such does not provide valid grounds for mandating regulation of an activity, or for regulating it more stringently than risks that are determinate or less uncertain. The well-developed economics literature on decision making under uncertainty, which is almost totally ignored in the legal literature on PP, focuses on two considerations that might justify a precautionary approach when regulating uncertain environmental risks. The first is with respect to the potential unavailability in the future of environmental resources that an individual might want to use or enjoy.[40] This essay concludes that this and other forms of risk aversion should play an appropriate role in regulatory decisions on activities that threaten very serious harm to society as a whole, but that, contrary to PP, risk aversion with respect to uncertainty as such is not a justifiable basis for regulation. The second consideration identified in the economic literature that might justify precaution in the face of uncertainty is the potential benefit from acquiring additional information in the future about the risks posed by an activity before allowing it to proceed, in the (usual) circumstance where an initial decision whether or not to prohibit or otherwise regulate an activity can be revisited in the future with the benefit of such information.[41] As discussed below, this consideration may sometimes justify an initial decision to prohibit an activity or allow it to proceed only under certain limitations in order to gather information that will enable regulators to revisit the decision in the future with the benefit of the additional information. However, this information-acquisition justification is applicable only in certain circumstances and under certain conditions. This information-gathering rationale for precaution accordingly does not justify strong versions of PP, which are not subject to these qualifications.

In addition to lacking a valid conceptual foundation, PP-based regulation would be undesirable because it would secure less overall environmental protection and diminish societal welfare. PP-based regulation would systematically tend to allocate relatively more in the way of regulatory resources to risks that are highly uncertain relative to risks that are less uncertain or are known to result in harm. As a result, systematic application of PP, advocated in the name of environmental protection, provides less environmental protection than prevailing preventative regulatory approaches. Further, the PP prescription is either perverse or useless as applied to the significant number of cases where regulation itself poses a potential for significant harm, creating a risk/risk tradeoff. For example, ozone air pollution near the earth's surface reduces human exposure to ultraviolet radiation. Regulatory measures to reduce such pollution in order to prevent other health effects can result in increases in skin cancers. Under PP, the regulatory agency must either disregard the environmental costs of regulation, or decide simultaneously to regulate and not regulate.[42] Also, the PP mandate that requires regulators to adopt specific types of regulatory measures – prohibitory or BAT regulations – whenever an uncertain potential for significant harm is presented, disregarding or downplaying regulatory costs, would lead to indiscriminately stringent and excessively costly regulation and to perverse strategies by regulators to avoid this result.

In addition to critiquing PP, this article outlines an alternative approach to regulation of uncertain risks. Contrary to PP, uncertain risks should be regulated under the same general decisional framework as risks that are well-characterized. In the face of uncertainty, regulators should resolve the uncertainty as best they can and treat the uncertain risk as the equivalent of a risk whose probability distribution is known. In resolving uncertainty, decision makers should make their best estimate of the relevant probability distribution of an uncertain risk, relying on the available evidence, scientific theory, expert judgments, and guidance from analogous regulatory problems and experience. They should not proceed, as the PP requires, on the basis of "worst case" presumptions.

Having resolved uncertainty by estimating a probability distribution for a risk, regulators should follow an appropriate decision rule for preventive regulation. The most appropriate general rule is to maximize expected net benefits to society, balancing the expected harm posed by the regulated activity against the expected costs (broadly understood), subject to relevant distributional and structural constraints. In the case of Type 2 and Type 3 risks that pose an appreciable risk of large harm from the viewpoint of society as a whole, risk aversion will often justify adding a risk premium in determining the expected value of harm. A risk premium may likewise be appropriate in determining the expected value of regulatory costs where there is a substantial probability of very high costs from the viewpoint of society as a whole. While uncertainty regarding risks of harm (or cost) as such does not justify risk aversion in regulatory decision making, large uncertainties may result in relatively flat probability distribution estimates that include significant probabilities of large harms (or costs). Such a probability distribution may justify adding a risk premium to the expected value of harm (or cost) and using the resulting enhanced value as the basis for regulatory decision making.

Having evaluated an uncertain risk in accordance with these procedures, regulators should adopt the most appropriate type of regulatory instrument to deal with a given activity and risk. Contrary to PP, they should not be restricted to using only the prohibitory and BAT versions of command and control regulation. They should be free to consider adoption of other regulatory instruments, including command controls based on achieving environmental quality standards, total residuals loading targets, and economic incentive systems, such as environmental taxes and tradable pollution or resource-use quotas and credits. They should select the instrument that will be most efficient and cost-effective in achieving regulatory goals for the type of activity and risk in question. The stringency of regulation should be established based on a balancing of social costs and benefits with the objective of maximizing net societal benefits, subject to relevant constraints.

In determining net benefits, regulators should carefully consider the implications of the circumstance that regulatory decisions typically are not to be made once and for all, but sequentially. Thus, a decision can be made at T1 either to regulate or not to regulate an activity. That decision can be revisited at T2, and again at T3, and so on. What considerations should be taken into account in deciding whether or not to regulate at T1 when the activity poses an uncertain risk of serious, irreversible harm and the decision maker can later reconsider its initial decision with the benefit of new information? Making a decision at T1 to allow the activity to proceed deprives the decision maker of any additional information that may be available in the future and that would permit a more informed, and therefore socially beneficial, decision about the activity. Forgoing the benefit of such information by authorizing an activity to proceed at T1 is therefore a cost that must be factored into the T1 decision. In some (but only some) cases the benefits of acquiring new information may justify precautionary decisions to impose regulatory controls on activities. In such cases, however, it may be necessary to take steps to develop the new information so as to permit timely and more informed reconsideration of the initial regulatory decision to prohibit or otherwise regulate the activity. In some cases, moreover, developing the additional information may require and justify allowing the activity to proceed, perhaps in a limited and controlled manner, in order to take advantage of "learning by doing."[43] Some strong formulations of the prohibitory PP, however, exclude the possibility of learning by doing; they accordingly cannot be justified by information acquisition arguments for precaution. Greenpeace, for example, asserts that the uncertainties associated with the risks of GMO crops requires prohibition on their use, including prohibition of limited controlled field trials that provide the best means of resolving the uncertainties in question. This "Catch 22" consequence provides an additional reason for rejecting such versions of PP.

In critiquing strong versions of PP, this essay does not argue that stringent preventive environmental regulation should never be adopted. Nor does the analytic framework that it advocates entail such an outcome. As society places a very high value on the environment and its protection, stringent preventive regulation of uncertain environmental risks is often justified and appropriate. Rejecting strong versions of PP does not mean that serious harm must be proved before regulatory controls may be adopted. Policies for preventive regulation must, however, be defined and implemented in a rational and therefore discriminating manner rather than through inflexible and unwarranted PP risk prescriptions and regulatory mandates..

II. DECISIONS TO REGULATE: THE ROLE OF UNCERTAINTY IN CHARACTERIZING AND VALUING ENVIRONMENTAL HARMS

This part of the article considers the appropriate regulatory treatment of uncertainty in the risks of environmental harm posed by an activity in order to assess the validity and desirability of strong versions of PP. The first section provides a general conceptual framework for preventive regulatory decision making. It initially considers the appropriate regulatory treatment of an activity known to cause a determinate harm. It then considers what additional considerations are presented when the activity poses a known risk of harm, i.e. it threatens harm that is probabilistic in character where the probability distribution is known. Finally, it considers what further additional considerations are presented when the risks of harm posed by an activity are uncertain. Using this framework, the second section considers the treatment of uncertain risks advocated by strong versions of PP and the justifications advanced for such treatment. The conclusion is that, contrary to strong versions of PP, uncertainty regarding risks as such does not justify greater precaution in regulation. Strong preventive regulation of activities posing uncertain risks is appropriate in certain circumstances, but for other reasons. Regulators should take into account societal risk aversion towards risks of very large harms. Regulators should also acknowledge and give due regard to the serious limitations in our ability to assess many environmental risks and predict which will cause serious harms. Neither of these considerations, however, justifies PP; these are more appropriately considered through the general framework for preventive regulatory decision making. The section concludes that PP can not be justified by aversion to uncertainty, as such, and that the PP prescription for regulating uncertain risks leads to socially undesirable regulatory outcomes.

A. A Structure for Environmental Regulatory Decision Making

In order to provide a conceptual framework for analyzing and evaluating PP, this subsection presents a basic structure for preventive environmental regulatory decision making. This structure, I believe, provides the appropriate standard or default approach for regulators to follow and captures the essential features of much regulatory practice, although I will not attempt to justify these propositions here. The structure has three components: a base case for regulatory decisions for known harms, a second stage for dealing with known risks of harm, and a third stage for dealing with uncertain risks. It involves the use, to the extent practicable, of quantitative and comparative risk assessment, and aims to maximize net social benefits, subject to relevant distributional and other constraints. As discussed below, this structure accommodates many of the important concerns of policy and principle that inform PP. At the same time, it highlights the character and operation of PP's distinctive prescriptions for regulatory decision making and the reasons why those prescriptions are unsound and undesirable.

1. Known Determinate Harms: The Base Case

Assume that the environmental harms that will be caused by an activity are known and determinate (Type 1). The most persuasive general rule of regulatory decision for such activities is to decide so as to maximize net benefits to society, which requires a consideration of the benefits of regulation (in terms of harm avoided) and its costs. The mode and intensity of regulation should be selected so as to maximize the excess of social benefits over social costs. These costs include adverse environmental effects that will or may occur as a result of regulatory measures. Type 1 regulatory decisions may be presumptively regarded as the base case for regulatory decision making. Decisions regarding activities that present risks of harm that are known and determinate and of activities that present uncertain risks of harms can, in most instances, appropriately be reduced to the Type 1 base case decision framework. As discussed below, however, other approaches to regulating activities that pose a risk of harm may be appropriate in certain circumstances.

The decisional criterion of maximizing net societal benefits does not necessarily require that all benefits and costs be measured in a common metric such as money. A multi-attribute decisional analysis can potentially encompass a variety of different types of values, including market costs and benefits, non-market health benefits and costs, non-market ecological benefits and costs, the general social values associated with innovation and the growth of knowledge, and so on. These different types of values may have different weights. The prevention of certain types of environmental harms – such as wide-scale losses of biodiversity or catastrophic consequences resulting from rapid global warming – are justifiably regarded as enormously important. These harms should accordingly be given extremely strong weight. Further, the weighting of harms would reflect qualitative judgments by people of the character of different harms, giving greater weight to those harms viewed as unfamiliar, involuntary, or "dread."

For purposes of the analysis that follows, it is assumed that there exists a social decision process that assigns determinate weights to these various benefits and costs as bases for regulatory decisions. It is also assumed that there exists a social decision process to compare costs and benefits over different time periods. Determining the regulatory alternative (including no regulation) that will maximize net social benefit is, of course, very far from a mechanical exercise. Limitations in data and methodological conundrums require decision makers to exercise a large degree of judgment in implementing this criterion. Yet, as Cass Sunstein rightly points out in the context of regulatory standard setting for arsenic concentrations in drinking water, "cost-benefit analysis, even with wide ranges, provides an important improvement over the 'intuitive toxicology' of ordinary people."[44]

Considerations of equity and justice also play a significant role in environmental regulatory decisions.[45] They include, for example, considerations of distributional justice, including justice to future generations, and, potentially, duties to other species and to nature. Structural norms, such as those based on the importance of diverse environments to human self-development and flourishing, are also relevant.[46] These considerations may be treated as additional values to be included and considered in a multi-attribute benefits maximization decision, or as factors to be applied in adjusting the weights assigned to such values, or they may operate as absolute or weighted side constraints on such decisions. The analysis here assumes that these various considerations can and are appropriately incorporated in the base case decisional framework.

Regulators' decisions must take into account the limitations imposed by the political process on their budgets and authority and on other regulatory resources (notably, the amount of societal resources that society is willing to devote to regulatory compliance), which in turn constrains the number of risks that they can regulate and the intensity of such regulation. These constraints require prioritization of risks in connection with application of the net benefits maximization principle. Much criticism of risk regulation focuses on wide disparities in the cost effectiveness of existing regulations: an example is the wide disparity among regulations in costs per life saved.[47] Comparative risk analyses should be performed by regulators not only when deciding the relative stringency of regulatory controls for different risks, but also in connection with the threshold decision whether or not to regulate a given risk. Given resource constraints, a decision to regulate a given risk is necessarily also a decision not to regulate some other risks, which may or may not be more significant. These opportunity costs must be considered by regulators in making the benefit-benefit tradeoffs involved in setting regulatory priorities. One consequence of regulatory resource limitations is that regulators are likely to follow a general policy or practice of not regulating activities that cause *de minimis* harms.

2. Known Probabilistic Risks of Harm: The Second Case

Now assume that the harm caused by an activity is not determinate but instead characterized by a known probability distribution determined, for example, by quantitative risk assessment (Type 2). Thus, the regulatory decision maker may know that the probability is 30% that an activity will cause a given harm and 70% that it will cause no harm at all. Or, it may be characterized by more complex probability distributions. How should the probabilistic character of the harm be handled in regulatory decision making?

The most appropriate general approach in deciding whether and how to regulate a Type 2 risk is to take the expected value of the probability distribution and proceed with regulatory decision making as if the harm caused by an activity were determinate and equal to the expected value, using the base case decisional framework. Thus, if an activity poses a 30% probability of causing a harm valued at 100 and a 70% probability of no harm, the regulator should decide to regulate the activity on the assumption that the activity will cause a harm of 30.

It is often asserted, especially by PP proponents, that society should be risk averse in dealing with significant environmental harms.[48] Accordingly, regulators should proceed as if the harm that the activity would cause is greater than 30 – perhaps, at the extreme, close to 100. Risk aversion, however, is generally considered in the context of individual decision making in relation to maximization of individual subjective utility. On the premises of declining marginal utility of wealth and starting point orientation, individuals place a greater disutility on a small risk of a large loss than a certain probability of a small loss with the same expected value. They will pay a risk premium to avoid the former risks (or demand such a premium for bearing them): risk averse individuals will accordingly purchase insurance against small probabilities of large losses even though the expected value of the risk of loss is less than the cost of insurance.

Just because individuals are risk averse in their individual decisions, it does not necessarily follow that society should be risk averse in its decisions, including environmental regulatory decisions. For one thing, societies have much greater ability to self-insure and engage in internal risk pooling than do individuals. To the extent that societal risk aversion is appropriate, regulatory decision makers should presumably be risk averse with respect only to those losses that are large relative to the well-being of society as a whole. Most environmental regulatory decisions – for example, those regarding most pollutants and wastes and the ecological effects of most development activities – do not involve risks that are large in this sense. Thus, the appropriate decision rule in such cases is presumptively that discussed at the outset of this subsection: regulate each activity that poses a known probability of harm as if it will cause harm equal to its expected value. If the expected value of an activity is less than the *de minimis* threshold, it should not be regulated at all.

There may, however, be some activities that pose smallish probabilities of harms that society places a very high value on avoiding. Examples of such harms include fundamental alternation of the climate system that could cause catastrophic harms or the destruction of an especially rare, beautiful and revered natural resource that can be regarded as "unique," irreplaceable, and consti-tuting an appreciable fraction of society's ecological wealth. Societies also tend to place a higher value on losses that are "lumpy" and involve concentrated harms.[49] In such cases, it is plausible that regulators should be risk averse in valuing the risk for decisional purposes.[50] In such cases, regulators should add a risk premium to the expected value of harm generated by the probability distribution of the risk in question. Neither science nor economics, however, can answer the question of when regulators should be risk averse and the extent of risk premiums. That answer can only be provided, at a normative level, by social values, and, at a positive level, by politics.

To the extent that risk aversion is appropriate in valuing the risks of harms posed by activities and thereby determining the expected benefits of regulatory activities, the same principles of risk aversion should be applied to the costs of regulation. The appropriateness of this approach is apparent in risk/risk tradeoff situations, where the adoption of regulatory measures themselves pose a risk of environmental harm. Thus, if regulation creates a risk of potentially large health or ecological losses, regulators should follow the same principles of risk aversion in valuing such risks as they do with respect to risks caused by the activities subject to regulation. Thus, the framework for decision offered here can readily accommodate risk/risk tradeoffs. As previously discussed, strong versions of PP, at least as presently formulated, are incapable of dealing with such tradeoffs satisfactorily.

The same general principles of risk aversion should also be applied to large regulatory costs, such as widespread unemployment, that do not consist (at least directly) of environmental harms. Regulation, especially prohibitions on new products and activities, may create a risk of large social costs in terms of the market and non-market benefits forgone from innovation and new investment. Such risks raise the issue whether society should be as risk averse to such harms as to harms to health and the environment. A potential argument in favor of being more risk averse with regard to large environmental losses is that such losses (e.g. the destruction of an ecologically rare wilderness) are likely to be "lumpier" than large non-environmental losses, which tend to be more disaggregated (e.g. a general rise in oil prices resulting from a decision to ban drilling in a wilderness area).[51] However, some social costs of regulation – for example, unemployment among Appalachian coal miners due to greenhouse gas regulation – may be lumpy, whereas many environmental harms, such as the adverse effects of widespread air pollution, may not be. Resolving this issue would require further inquiry into the appropriate grounds for societal risk aversion in connection with environmental regulatory decisions that is beyond the scope of this article.

In sum, the general approach proposed for the preventive approach to regulating environmental risks of known probability is that regulators should determine the expected value of the risks of harm that would be prevented by regulation and the expected value of regulatory costs, adjust the expected values by any applicable risk premiums, and decide in accordance with the base case criteria. There may, however be cases where an alternative approach to decisionmaking is appropriate.

For example, one potential alternative approach for regulatory decision making is PP2, which calls for regulation that incorporates a margin of safety below observed or predicted no-significant-effect levels. This approach has been widely followed in regulation of pollution and chemicals. In such circumstances, there are often major uncertainties regarding the shape of the damage function for human exposure to such substances and their potency; it will accordingly be difficult to estimate a probability function for dose-response relations (in moving from step 3 to step 2) and also difficult to determine the appropriate risk premium, if any, for such an estimation (in moving from step 2 to step 1). Because of these difficulties, PP2 may serve as a useful shortcut to decison making and be used in appropriate cases as a substitute for the general decision framework. This approach can take into account relevant social costs and benefits in determining what level of effect counts as significant and in determining the margin of safety. Decision rules to prohibit activities that present a sufficiently large risk (say, one in a million risk of 10,000 deaths from a nuclear plant accident), without engaging in cost/benefit balancing, represent another form of shortcut.[52] Of course, there is the danger that such rules-of-thumb will assume a life of their own and result in excessively conservative (or excessively risky) regulatory decisions.

Another type of alternative decision rule for Type 2 decisions may be especially appropriate for dealing with cases where a regulatory decision maker resolves uncertainty regarding a risk (Type 3) by estimating a probability distribution for purposes of Type 2 decision that is multi-peaked – for example, a distribution with a significant probability of low harm and a significant probability of substantial harm. Such an estimation may be especially likely to result in cases in which the views of experts are sharply divided, for example on issues such as the ecological effects of widespread application of GMO crop plants or potential non-linear effects of global warming on ocean currents or Antarctic ice sheets.[53] In such circumstances it may be appropriate for the decision maker to conduct a sensitivity analysis, first examining the implications for base case decision making if the risk is not significant, and, secondly, examining the implications if the risk is significant, as well as examining the implications of regulation based on expected value. The decision maker may appropriately conclude that relying solely on expected value in deciding such cases would not be responsive to the structure of the risk, and conclude, based on the result of the sensitivity analysis, that a maximin strategy is appropriate for regulating this type of risk. If so, the decision maker would first evaluate the net benefits of a regulatory decision based on the assumption that risk will not be significant (the left-hand peak). In that case, the criterion of maximizing net benefits may dictate not imposing any regulatory controls at all, or only modest controls. There is, however, the danger that society will suffer large losses if the harm turns out to be significant (right-hand peak) and could have been prevented by more stringent regulation. The decision maker would then evaluate the net benefits of a regulatory decision based on the assumption that harm will be high. In that case net benefits would be maximized by more stringent regulation. But, there is the risk that the higher costs of stringent regulation will have been incurred unnecessarily if harm turns out to be low. Under maximin, the decision maker chooses the alternative that will maximize the minimum net gain. This approach might lead to the adoption of more stringent regulation than under the standard framework.[54]

3. Uncertain Risks of Harm: The Third Case
The third, and perhaps most common, category of environmental regulatory cases involves activities that pose uncertain risks of harm; they threaten

environmental harms that are characterized by a probability distribution that is uncertain (Type 3).[55] The degree of uncertainty of course varies, has a variety of causes, and can be characterized in many different ways. PP is directed at this category of harms. This subsection addresses the principles that should govern regulatory decisions of regulation of uncertain risks of harm, disregarding the possibility that such a decision does not have to be made once and for all time but can be revisited in the future when more information will be available. The additional issues presented by sequential regulatory decision making are discussed in Part IV.

The fundamental question in third case decisions is what approach the regulatory decision maker should follow in characterizing an uncertain risk of harm when uncertainties are sufficiently great that probability estimations based upon long run frequencies in data and models validated by such data are of only limited utility. Bournelli's principle of the insufficiency of reason holds that, in the face of irreducible uncertainty (i.e. no additional information can be obtained to reduce the relevant uncertainty before a decision must be made), all risk probabilities should be deemed equal, which implies a perfectly flat probability distribution. The prevailing welfare economic approach, however, follows a Bayesian approach. Under this approach, individuals deciding a course of action in the face of significant uncertainty about outcomes should make their best judgment in estimating the relevant probability distributions. They should then act, based on that estimation, so as to maximize their expected subjective utility. When faced with the same or similar decisions in the future, they should update and adjust their prior estimations giving appropriate weight to new information.[56] This approach to decision making emphasizes the relevance and importance of sequential learning, taking into account multiple lines of evidence, analysis, and experience.[57] Accordingly, it is well adapted to sequential regulatory decision making, as developed in Part IV.

What approach should regulators take with respect to uncertain risks? It would be infeasible and undesirable for regulators to attempt to aggregate the probability estimations and expected subjective utilities of all of the individuals in society as the basis for regulatory decisions in such cases. As a practical matter, individuals lack the information necessary to evaluate the myriad of environmental risks that are potentially subject to regulation and would not want to spend their lives acquiring and processing it. Regulators should therefore act as representatives of society as a whole. In order to promote accountability and transparency, regulators should make their best estimation of the probability distribution for the uncertain risk in question through a process that invites public and expert input and makes explicit and public the relevant uncertainties and the bases for the regulators' determinations. They should first seek to reduce uncertainty in the probability distribution to the extent that available information and science, including quantitative risk assessment, permit.[58] Because of limitations in data and scientific understanding, very substantial uncertainties are likely to remain in many cases. They should then exercise their best judgment in resolving the remaining uncertainty, estimating a probability distribution that they should then treat as if it were a known distribution for purposes of the later stages in the decisional process. In this way, third case decision problems are assimilated to second case decision structures. Next, regulators should apply a recognized second stage decisional procedure to the estimation. The general procedure would be to determine the expected value of the harm, adjusting for any relevant risk premiums, and to make a regulatory decision in accordance with the base case rule of net benefit maximization, subject to relevant side constraints. In this way, the third case regulatory decision is ultimately reduced to a base case decision. Or, regulators may in appropriate cases follow an alternative second case procedure and decision rule such as those discussed in the previous subsection.

These decisions by regulators will, of course, be institutional rather than purely personal ones, although personal perspectives and judgments are inescapable. In estimating uncertain risks, empirical "hunches" and the attitudes and values of the decision makers, which, it is to be hoped, are in some way representative of the attitudes and values of the larger society, will necessarily play a large role. All of the problems of decison making under uncertainty, including the influence of heuristics, framing effects, and other sources of potential bias and error, must be addressed.[59] Regulators' decisions will be complicated by the circumstance that, in many regulatory proceedings, experts other than the ultimate decision makers may hold sharply opposed views as to the nature and extent of the uncertainty and how it should be resolved. In a highly adversarial regulatory process, such as that found in the U.S., industry experts and environmental advocacy experts often assert such sharply opposed views. Even in less adversarial settings, experts from different disciplines and backgrounds often hold markedly opposed views. It must be emphasized, moreover, that there is no "correct" scientific or welfare economic way of making such estimations. Experience has, however, shown that practicable and relatively robust methods for addressing uncertainty in the context of specific types of regulatory and other decisions can be developed through careful and systematic effort.[60] The application of best institutional judgment to resolve uncertainty, through an explicit, open process of decision with opportunity for a wide input of views from professionals and the public, is better than any other means for resolving the problem.

B. Regulatory Treatment of Risk and Uncertainty Under PP

PP proponents advocate a different approach. They hold that, in the case of uncertain risks that exceed a threshold of appreciable potential for serious harm, the decision maker should presume the worst and proceed as if the activity will in fact cause harms that approach the maximum severity that might potentially occur ("worst case"). In one PP formulation:

> Precautionary action requires reduction and prevention of environmental impacts irrespective of the existence of risks . . . The crucial point is that environmental impacts are reduced or prevented even before the threshold of risks is reached. This means that precautionary action must be taken . . . even if risks are not yet certain but even probable, or, even less, not excluded.[61]

As previously noted, there are significant questions that must be resolved in defining and applying the triggering PP threshold and the resulting presumption of harm. How great must the probability and severity of the risk of harm be in order to trigger the PP threshold? What sort of evidence can be used to establish that the threshold is met? How should the resulting presumption be defined? Do we assume the "absolute worst case"? The "reasonable worst case"? How are these concepts to be defined? But these problems need not concern us for purposes of the present discussion, which addresses the basic principle, not the details of its application.

If the PP consisted solely of requiring worst case presumptions in the face of uncertainty and did not include the mandatory PP regulatory rules discussed in Part III, it could be incorporated within the structure of decision making set forth above. In the case 3 analysis of an uncertain risk, the uncertainty would be reduced, by virtue of the worst case presumption, to an estimate characterized by very high probability of very high damage values. Under the general decisional rule for stage 2, that presumed distribution would be reduced to its expected value. Under the PP worst case presumption, the assumed probability distribution would be highly peaked and located at or near the upper end of the range of damage values. Such a distribution would allow little scope for the introduction of risk aversion and the need to add any risk premium to the already high expected value of the harm in determining the value of harm (which would be quite high) used in the net social benefits maximization decision in the base case. Alternatively, if worst case is defined as a single outcome, the decision maker could directly use the value of the harm associated with that outcome in the base case decision.

The worst case presumption component of PP has two fundamental flaws. It lacks a sound theoretical foundation, and leads to bad regulatory outcomes.

PP proponents have failed to advance a persuasive justification for using a worst case approach to resolving uncertainty rather than having regulators make their best judgment of the probability distribution. PP advocates generally advance two related reasons for a worst case presumption.[62] First, they contend that it is very important to prevent certain types of serious environmental harms. Second, they contend that, due to serious limitations in data and scientific understanding, our ability to predict whether an activity will cause serious environmental harms is quite imperfect. PP advocates point to past instances where new technologies or other activities have been initiated without stringent regulatory controls because there was no contemporary evidence that would prove that they would cause serious harms, and where it was later discovered that they have caused and are causing serious harm. Many PP proponents also appear to assume that nature is inherently vulnerable rather than resilient. Accordingly, PP advocates argue, we should err on the side of caution and adopt a worst case presumption in order to prevent the occurrence of serious and perhaps irreversible harms.

While the concerns adduced by PP proponents are legitimate, they do not require adoption of a worst case presumption, and can be appropriately accommodated within the decisional structure presented in the previous subsections. In presuming the worst, PP introduces deliberate and serious error into the assessment of risk in order to generate more conservative risk management decisions.[63] This tactic conflates risk assessment and risk management, and prevents a sound and open approach both to the determination and to the appropriate regulatory treatment of uncertain risks. The framework for preventative risk regulation provides a far better, more transparent method for addressing these concerns. If avoidance of certain types of environmental harms is very important to society, as PP proponents generally believe, then such harms will be weighted quite heavily in the base case. If it is appropriate that society be risk averse towards small possibilities of very serious environmental harms, that circumstance will appropriately be taken into account in the standard decisional process by adoption of a significant risk premium. And, if science has a poor track record in predicting environmental risks, or if nature is appropriately believed to be vulnerable rather than resilient, decision makers should take those circumstances into account in making their best judgment resolution of uncertainty in the third case analysis. To the extent that skepticism regarding our predictive capacities, a healthy respect for the potential for nasty surprises, and concerns about nature's vulnerability are warranted in particular types of cases, regulators will estimate a relatively flat probability distribution for uncertain risks. Such a distribution will produce a relatively high expected value of harm. Such a distribution will also include an appreciable probability of very large harm, which in appropriate cases will generate a significant risk premium that will be added on to an already high expected value. The high expected value plus risk premium will lead to a high harm value for the base case analysis, resulting in highly protective regulation. Thus, the decisional structure at each of these stages allows ample and appropriate scope for giving due weight to important environmental values, to risk aversion, to limitations in knowledge, and to concerns about ecosystem fragility, and does so in a more open, structured, and accountable fashion.

Under the framework for preventative risk regulation presented in this article, not all cases of uncertain risk would or should be resolved so as to yield a very high expected value. Some uncertain risks can be better characterized. In many cases, the best estimate will be that it is highly probable that any harm that is caused will be insubstantial or modest; the possibility of more serious harm may not be entirely excluded, but the probabilities in many cases will be extremely small. Even assuming that some risk premium is appropriate in such cases, based on a small possibility of more serious harm, the total value of harm in the base case will be rather low. In other cases, the best estimate of an uncertain risk will have a different shape and will yield a higher value of harm. It is all a matter of context, of relevant evidence and knowledge, and of judgment. These variations, which are largely disregarded by strong versions of PP, are appropriately dealt with through the standard decisional framework presented here.

These conclusions are consistent with and supported by the economic learning regarding the concept of option value for environmental resources that are "irreplaceable" in the sense that they have no close substitutes.[64] It concludes that, under conditions of uncertainty, individuals will in some circumstances be willing, because of risk aversion, to pay a premium (option value) to preserve the opportunity to use a unique and highly valuable resource in the future, over and above the expected value (consumer surplus) to them of such use. Such a premium can be understood as a measure of the value of precaution. To the extent that individuals are willing to pay such a premium, regulators applying the criterion of net benefits maximization should take it into account in making decisions whether and how stringently to regulate activities that might destroy or harm such a resource and preclude its future use or enjoyment. This is an alternative means of conceptualizing the risk premium that may, in appropriate cases, be introduced in moving from the second case to the base case under the decisional framework presented in this article.

The discussion of this issue in the economic literature first focused on what might be called demand-side option value. It dealt with situations where an individual is uncertain whether she will use a resource in the future due to uncertainties about her future preferences and/or income.[65] The initial conclusion was that such an individual would normally be risk averse about losing the benefits that an irreplaceable resource would provide if it turned out that she wished to use the resource in the future, and that, accordingly, she would pay a premium over and above the expected value of future use in order to preserve the option of such use.[66] Later, however, it was shown that this option value might not be positive but rather might be negative (the individual would be willing to pay less than the expected use value for the option) or zero, depending on the individual's utility function and wealth.[67] While an individual could be risk averse with respect to the possibility that the resource would not be available if it turned out that she wished to use it in the future, but she could at the same time be risk averse with regard to the opposite circumstance – that she would pay for an option to preserve the resource and later find that she didn't wish to use it.[68]

Subsequent analysis turned to the supply side aspect of the issue. It addressed the situation where an individual knows that he wants to use or preserve a resource in the future, but it is uncertain whether or not an activity subject to potential regulation will destroy or seriously harm the resource and thereby deprive the individual of the use. Here the analysis has concluded that the option value is always positive, and that accordingly regulators should add an appropriate premium (likely to be small in most cases) to expected value in applying a criterion of net benefits maximization.[69] This analysis, however, did not consider uncertainties in regulatory costs and the possibility that they might generate some countervailing risk aversion. Notwithstanding this possibility, it seems quite likely that in many cases risk aversion towards the loss of an irreplaceable environmental resource that one wishes to use or enjoy in the future would generally justify the addition of some risk premium, over and above the expected value, in regulatory decisions addressed to risks of potential harm. This "precaution premium" for unique or irreplaceable environmental resources and services however, can readily be accounted for in the standard framework for regulatory decison making set forth above, through its treatment of risk aversion. Option value thus does not provide any justification for adopting strong version of PP, which would artificially inflate that value through inappropriate worst case presumptions.

Accordingly, uncertainty as such does not justify regulation, or more stringent regulation, of an uncertain risk just because it is uncertain. Only in specific circumstances, involving specific type of risks, is additional precaution justified, and then on grounds of risk aversion to large expected losses, not uncertainty as such. Pervasive, stringent regulation of all uncertain risks could be justified under the standard framework for regulatory decision making, but only if one assumed an essentially flat distribution of risk in all cases of uncertain harm; such an assumption would require a quite radical, across-the-board epistemological skepticism that is wholly unwarranted. PP, however, goes even further than Bournelli. It assumes that an uncertain risk should always be treated as one posing a very high probability of very large harm. Why does the mere fact of uncertainty regarding the risks of activity justify the automatic, across-the-board introduction of an additional and extreme degree of risk aversion in decision making? I do not believe that the PP literature provides any coherent or sound answer to this question.[70]

In addition to lacking a sound justification in theory, the PP worst case presumption leads to bad regulatory outcomes that diminish societal well-being and lead to less, rather than more, overall environmental protection. By virtue of its worst case presumption, PP would produce these bad outcomes even if the mandatory regulatory decision rules that form the second component of PP were disregarded. There are at least three reasons why PP leads to socially undesirable regulatory results.

First, and most obviously, the unrealistic worst case presumption leads to unnecessarily stringent and costly regulation in many cases – the overwhelming majority – where the worst case presumption is not justified and activities posing uncertain risks are unlikely to cause serious harm.[71]

Second, PP leads to a disproportionate allocation of limited regulatory resources to those activities posing relatively more uncertainty, because the worst case assumption inflates their harm value relative to risks that are better characterized. Adoption of the PP leads to a sort of regulatory Gresham's Law, in which the regulation of more uncertain risks that may never eventuate in significant harm, tends to drive out regulation of risks which are better known and more likely (on average) to cause harm. In short, PP prevents rational environmental regulatory priority-setting. In so doing, PP leads to a quite perverse regulatory paradox: the adoption of more precaution in the face of risks that are more uncertain produces less environmental protection.[72]

Third, PP is incapable of dealing with risk-risk tradeoffs and setting intelligent regulatory priorities. It provides no guidance in dealing with the important class of regulatory problems where regulatory measures to reduce risks (target risks) themselves create risks (non-target risks). Should regulators simply disregard non-target risks because precaution demands strong measures to reduce or eliminate target risks? Or does PP prohibit the adoption of regulatory measures that would create a potential for serious harm? PP provides no answer to this dilemma.[73] PP also fails to provide any basis for setting regulatory priorities among different risks that meet the PP threshold, or as between uncertain risks and more determinate, serious risks.

Fourth, PP is likely to lead to perverse efforts by regulators to avoid the draconian impacts of the PP prescription for regulation. Uncertain risks that meet the triggering PP threshold are subject to worst case presumptions and stringent regulatory controls that, in many instances, will be unjustified and disproportionately costly. Regulators, for a variety of reasons, will in many circumstances seek to avoid imposing such controls. In order to do so, they will seek to avoid making threshold determinations of potential risks that trigger PP, either by postponing decisions or applying the decisional criteria inconsistently. This is the lesson of experience under Section 111 of the U.S. Clean Air Act before its amendment in 1990. As widely interpreted, the section mandated elimination of any risk from toxic air pollutants that were non-threshold in their effects, a requirement that would require widespread shutdowns of thousands of industrial facilities. In order to avoid this result, EPA dragged its feet in making decisions under Section 111 or otherwise sought to avoid its mandate. The lesson is that inflexible, draconian regulatory requirements that are not justified by the standard regulatory decision making framework are likely to provoke evasive tactics on the part of regulators that undermine the transparency and integrity of the regulatory process.[74] Strong versions of PP threaten precisely this result.

III. MANDATORY PRECAUTIONARY PRINCIPLE REGULATORY RULES

In addition to requiring worst case assumptions about harms, strong versions of PP also require regulators to follow specified regulatory rules for controlling activities posing uncertain risks that trigger the PP threshold of potential for serious environmental harm. These mandatory rules include prohibiting the activity in question or requiring adoption of BAT measures; disregarding or downplaying regulatory compliance costs in imposing controls; and shifting the burden of proof to the proponent of the activity to show that the activity is safe in order to avoid or lift these controls.[75] This section briefly discusses these rules, finding them unjustified as a uniform, mandatory prescription for regulation of uncertain risks that pose some potential for significant harm.

Given the premises of PP, these mandatory regulatory rules can be regarded as logical complements to the worst case presumption. PP assumes that regulators applying the PP trigger threshold criteria are capable of distinguishing those uncertain risks that pose a potential for serious harm (PP risks) from those that do not (non-PP risks). PP further assumes, however, that in many and perhaps most cases regulators lack the capacity to distinguish those PP risks that are more serious and more likely to cause serious harm from those that are less serious. Because PP is highly risk averse with regard to uncertainty, it concludes that the worst case presumption should be applied to all PP risks. Accordingly, regulators should act as if all activities posing such risks will in fact cause serious and often irreversible harm to public health and the ecosystem. Given that assumption, it would be logical to regulate all PP risks quite stringently, as the mandatory PP regulatory rules require.

The most fundamental defect in the PP regulatory rules is the implausibility of the underlying epistemological assumptions. PP assumes a quite odd and highly implausible discontinuity in regulators' capacity to assess uncertain risks. Regulators are capable of applying the PP threshold criteria to distinguish PP from non-PP risks. But, thereafter, regulators' risk-assessment capacity largely disappears. They are consigned to ignorance regarding which PP risks are more serious and which less serious. Experience, however, fails to establish that risk assessment capacities have the sharply discontinuous function assumed by PP. Instead, the ability to assess risk seems to display a more continuous variation, progressing by increments from a high degree of confidence to very considerable uncertainty, depending on the particular activity and type of risk in question. These circumstances call for a more flexible approach to regulation than PP, an approach that applies judgment in balancing the available evidence and knowledge of risk against the uncertainties, makes appropriate estimates of the likely probability and magnitude of harm, and tailors regulatory controls in accordance with those estimates. The framework for regulatory decision making set forth in Part II.A, which operationalizes the principle of proportionality, provides just such an approach.

As shown in Part II, the PP worst case presumption, if applied to the standard risk regulatory decision making framework, would result in excessively stringent and costly regulation even without the mandatory PP regulatory rules. PP, however, also requires regulators to impose a set of stringent regulatory controls for all risks that exceed a triggering threshold in the continuum of uncertain risks.[76] Any selection of such a threshold will be arbitrary. Further, the mandatory package of controls triggered by the threshold will be inappropriately stringent for many of the risks that exceed that threshold unless the threshold is set at a very high level of risk. PP, however, is highly skeptical of regulators' ability to discriminate between more serious and less serious uncertain risks; accordingly, the logic of PP requires that the threshold be set at a substantially lower level. The result will be to require imposition of stringent regulatory controls on many risks for which they are inappropriate. Also, for reasons previously explained, the PP regulatory prescriptions will result in less overall environmental protection because they will allocate a disproportionate share of limited regulatory resources to those risks that are relatively uncertain relative to those known to cause harm. In order to avoid such outcomes, regulators will, as discussed in Part II.B, adopt various stratagems to avoid the PP threshold trigger.

Alternatively, PP's stringent regulatory prescriptions may be rooted in institutional rather than epistemological skepticism. The assumption may be that, even if regulators do have some ability to discriminate among uncertain risks of varying magnitude, they will fail to impose adequate regulatory requirements on activities that pose significant risks of harm as a result of inadequate institutional capacities and incentives, political pressures, or established conceptions of property rights.[77] PP's cure for distrust of administrative discretion is to mandate stringent controls for all uncertain risks that meet a relatively low risk threshold. Further, PP proponents may assume that, for similar institutional and political reasons, there will be an enormous degree of slippage in the actual implementation of the regulatory controls that are adopted. This assumption may be especially strong in the context of international environmental law.[78] PP's cure for these apparently inevitable regulatory failures is to mandate stringent controls at the outset. Experience indicates, however, that a strategy of imposing draconian mandates and deadlines is generally a poor cure for perceived institutional failures, often producing perverse results.[79] If institutions are performing poorly, they will have to be improved. Imposing arbitrary mandates is likely to impede rather than advance the needed improvements.

The remainder of this Part considers the three basic elements in the strong PP package of regulatory controls.

Prohibitions or BAT Controls:
Strong versions of the PP require regulators to prohibit activities (PP4) or impose BAT controls (PP3) on activities posing PP risks. Adoption of these categorical rules for all activities posing some uncertain potential for significant harm would lead to excessively stringent and costly regulation. By requiring regulators to either prohibit such activities or require them to adopt BAT controls, such rules would prevent regulators from using other available regulatory instruments that are often much more appropriate and cost-effective in dealing with many types of activities and risks. Further, PP requires regulators to disregard or mute considerations of regulatory costs in applying this narrow menu of regulatory instuments.[80] Regulators should be free to follow a more discriminating approach, tailoring the choice of regulatory instrument and its stringency to the particular type of activity and risk in question.

Under the regulatory framework presented in this article, categorical prohibitions of certain types of activities with uncertain risks of harm may be justified based on their likelihood and the potential for serious cumulative and/or irreversible harm, and also taking account of the social costs of the prohibition. The ban on disposal of radioactive wastes at sea is an example of such a strategy. In other instances, the potential for serious harm, the difficulties in applying pollutant-by-pollutant risk judgments, and scale economies may justify categorical BAT requirements for a class of risks. The BAT controls for toxic air pollutants mandated by Section 112 of the U.S. Clean Air Act as amended in 1990 are an example of this strategy.[81] Any such categorical regulatory requirements, however, must be devised and justified on a case-by-case basis for particular activities and risks, rather than mandated across the board for all uncertain risks with some potential for serious harm. Furthermore, there are many types of regulatory instruments other than prohibitions or BAT controls that regulators should be free to select. For example, some activities, including pollution and waste generation, may be addressed through controls based on environmental quality standards.[82] Where damage functions are approximately linear or are uncertain, measures to limit or gradually reduce total waste or pollutant loadings are often appropriate.[83] In many cases, the most efficient and cost-effective measures for accomplishing these objectives are economic incentive systems, such as tradable quota or credit systems and environmental taxes.[84] Furthermore, different types of economic incentive systems are often far better suited for addressing environmental problems, such as climate change, involving significant uncertainties regarding either harms or regulatory costs than are traditional command instruments.[85] The application of these different instruments should appropriately take into account the costs and benefits of different levels of regulatory stringency for different pollutants and activities.

Further, as discussed in Part IV, regulatory requirements should be adjusted in response to the development of new information regarding relevant harms and control costs. Under PP's mandatory regulatory rules, such adjustments are impossible (in the case of prohibitions) or quite constrained (in the case of BAT controls). Other instruments, including case-by-case screening under an unreasonable risk standard and economic incentive systems, allow for much more flexibility.

Disregard of Regulatory Compliance Costs:
To the extent that the PP prescription for regulation disregards or slights regulatory compliance costs in imposing prohibitory or BAT controls on all PP risks, it will produce unduly stringent and excessively costly regulation.[86] PP regulation will also result in increases in other environmental risks that are not

the direct target of the controls.[87] Sound regulatory policy, exemplified by the decision framework set forth in Part II.A, requires that the costs of regulation be explicitly taken into account and appropriately weighed against the expected harm posed by an activity, with the objective of tailoring regulation of different activities and risks in order to maximize net social benefits, subject to any applicable structural, distributional, or other side constraints. Because of resource limitations and the many environmental risks that we must deal with, we can not devote all or even a large part of society's resources to reducing any given risk, however serious. Much less can we afford PP's injunction to disregard costs in regulating all PP risks. Efforts to pretend otherwise by prohibiting consideration of compliance costs in regulatory decision making are inevitably evaded by one means or another. Candor is required in order to promote regulatory accountability. This does not, however, mean that we should do nothing in cases where the costs of reducing significantly serious risks of harm posed by current activities are large. Such risks can be addressed through the use of regulatory instruments designed to stimulate the development of innovative, less costly means of reducing such risks. Economic instruments, including environmental taxes, tradable permit systems, deposit/refund systems, and market-based information systems, are in most cases much more likely than PP's command prohibitions and BAT requirements to stimulate the necessary technological and institutional changes that will enable us to enhance overall environmental protection.[88]

Burden shifting rules:
A third component of the mandatory PP regulatory rules is shifting the burden of presenting evidence and the burden of persuasion regarding the uncertainties in the risks posed by an activity from the regulator to the regulated. In many, but not all, environmental regulatory programs, the burden is on the regulator to show that an activity poses a significant risk of harm before imposing regulatory requirements.[89] Under PP, once the regulator shows that the risk posed by an activity triggers the PP threshold, the burdens shift to the proponent of the activity to show that it does not present significant risk of harm in order to avoid prohibitory or BAT controls.[90] Thus, the ultimate burdens of resolving uncertainty regarding risk are placed on the proponent of the activity rather than the regulator.

PP's shift in the burdens of proof on risk has sometimes been attacked by PP critics on the ground that it requires the activity proponent to prove a negative – an absence of any appreciable risk – which by its very nature is virtually impossible to carry.[91] This criticism is overbroad. If, as PP presumes, we were almost totally incapable of assessing the relative severity of different

PP risks, then the burdens of showing that an activity that triggers the PP threshold did not pose a significant risk would indeed be impossible or extremely difficult to carry; PP regulatory controls would in most cases remain applicable in perpetuity. As discussed above, however, such skepticism is unwarranted. The difficulty in carrying the burdens of showing that an activity is "safe" depends on how safety, its proof and the relevant persuasion burden are defined. There are regulatory programs, such as the U.S. FDA process for approval of new drugs or food additives and for EPA registration of new pesticides, that impose the ultimate burdens of establishing safety on the manufacturer of the drug or pesticide and define the burdens in a workable fashion that allows the burden to be carried in many cases.

Regardless of which party bears the ultimate proof burdens regarding safety, the standards of proof must be defined in a reasonable and workable fashion in order for a regulatory program to function successfully. Under some strong PP formulations, an activity may not proceed unless it is shown that any potential for significant harm is included.[92] Under such a decision rule, it will be extraordinarily difficult for the activity to proceed, no matter which party bears the burdens of proving the absence of any such risk.[93]

Assuming a sound decision rule, the most important question in allocating the burden of producing evidence is which party, the regulator or the regulated, is better able to generate and marshal the relevant data and analysis. As the new drug, food additive, and pesticide regulatory programs illustrate, in some instances the party regulated is better able to develop the relevant evidence. But in other instances, including those involving individuals or small-scale enterprises, the regulator often has the superior capacity. Thus, the burden shifting requirement of PP shares the same basic defect as the other PP regulatory requirements: it is overboard.[94] Once again, sound regulatory policy requires a more flexible and discriminating approach.

IV. SEQUENTIAL DECISION MAKING, UNCERTAINTY, AND THE VALUE OF INFORMATION

Decisions whether or not to regulate seldom have to be made once and for all time. A decision maker can decide at T1 to regulate and then decide at T2 whether to continue the regulation, taking into account new information that has developed and other changes that may have occurred in the interim. Or, the decision maker can decide at T1 not to regulate, and then at T2 revisit the decision whether or not to initiate regulation, taking into account new information and developments. How should uncertainty regarding harms or costs affect a regulatory decision maker's decision at T1 whether or not to regulate, given the possibility of revisiting the decision later in the light of new information? Does such uncertainty justify a precautionary approach to regulation at T1, especially where the decision not to regulate an activity poses some uncertain risk of serious irreversible harm?

The economics literature, discussed further below, provides valuable insight in addressing this question. It shows that acquisition of additional information that will improve decisions by reducing uncertainty is often of considerable value. Assume that the regulator is now faced with a decision (T1) to approve or prohibit (or otherwise regulate) an activity that poses an uncertain risk of potentially serious harms that cannot be subsequently remedied, such as a risk of species extention. If the regulator approves the activity at T1, it will not have the benefit of the additional information that may be available in the future and that will allow the risk to be better characterized. Serious, irreversible harm may occur as a consequence of the activity. This circumstance will eliminate or greatly reduce the decisional value of any new information regarding the risk that is subsequently developed. Accordingly, overall welfare may be enhanced if the regulator decides to prohibit the activity at T1 and reconsiders the issue at T2 when the additional information is available. The value of additional information that will improve regulatory decisions regarding uncertain risks is, thus, a potential justification for a measure of precaution in making the decision at T1. This justification, however, is applicable only in certain circumstances, and only if the requisite steps are taken to develop the additional information in the interim. Strong versions of PP generally fail to respect these conditions, and impose controls far more broadly than would be justified by value-of-information considerations.

In order to provide an analytical framework for exploring these questions, this part of the essay first considers sequential regulatory decision making in cases where harms and costs are determinate and known or are characterized by a known probability distribution. It then considers the case of uncertain risks.

A. Harms and Costs That Are Determinate and Known or Characterized by a Known Probability Distribution

When both harms and control costs are determinate and known, or characterized by a known probability distribution, and where the harm caused by future activities will be greater than the harm caused by present activities and/or the costs of control will decrease in the future, it may be appropriate not to initiate regulation at T1 (because the costs of control outweigh the benefits) but to defer regulation until T2 (when the benefits of control outweigh the costs). Of course the harm that will be caused by deferring regulation must be considered in the analysis, including latent harms that are caused by activities conducted between T1 and T2, but that are not realized until after T2, and irreversible harms, such as the extinction of a species or latent harms that have a "long tail" and that can not be prevented by the subsequent adoption of regulation.[95] Savings in regulatory costs, including compliance costs that would be incurred in the interim between T1 and T2 if controls are not adopted at T1, must also be considered.

Climate change is an example of an environmental problem where postponing regulation on such grounds is an important option. Significant reductions in greenhouse gas emissions will require major changes in the capital stock, especially in the energy and transportation sectors. Immediate scrapping and replacement of the capital stock will be much more costly than replacing capital stock with low-GHG technologies at the end of its useful life. Further, R&D and innovation may result in significant reductions in the costs of reducing GHG emissions in the future. Further, because GHGs are a stock pollutant, and annual flows are small relative to the stock, the harms associated with postponing regulation may be low relative to the savings achieved by postponing regulation.[96] There is, however, the counter-argument that initiating regulation now will stimulate innovation that will reduce the costs of control, and that the harms caused by interim uncontrolled emissions may be substantial and long-tailed.

The analysis must also take into account lags in the implementation of regulation. A decision to initiate regulation at T1 rarely means that controls will become effective at or shortly after T1. The lag will be especially significant in the case of entirely new regulatory programs, and even greater in the case of a new international regulatory program such as that contained in the Kyoto Protocol. Entirely new international institutions must be developed, requirements adopted at the international level, and those requirements must subsequently be implemented domestically.

A further practical consideration is the impact of administrative and political inertia. Experience indicates that once a regulatory program is undertaken, it becomes very difficult later to eliminate or relax the program even though it is no longer justified. Regulatory programs, like other government programs, create constituencies, both within and outside government, who benefit from their continuation and will strongly oppose their abolition. This problem can, however, be addressed to some degree by "sunset" provisions. On the other hand, it is possible that a decision not to initiate regulation at T1, thereby allowing a harm-causing activity to continue or even expand, will strengthen constituencies opposed to regulation, making it more difficult to adopt regulation in the future. This problem can to some degree be addressed by agenda-forcing provisions similar to the action-forcing provisions contained in the citizen suit provisions in the major U.S. federal environmental regulatory statutes.

B. Uncertain Harms and Costs

Where the risk of harms and/or regulatory costs are uncertain, decisions and societal welfare can be improved, sometimes dramatically, by reducing uncertainty through the development of new information. The additional information can help ensure that regulatory decisions are made on the basis of more accurate estimations of costs and benefits, which will in turn make it more likely that such decisions will enhance social welfare.[97]

In order to promote better regulatory decision making, governments should have a systematic process for gathering relevant information prior to a decision being made. Examples of such processes in the U.S. include notice and comment rulemaking, environmental impact statements, and procedures for the submission of evidence in licensing and other adjudicatory decisions. Depending on the underlying legal rule defining the status quo in the absence of a decision, the resulting lag in the timing of the regulatory decision can either impose regulatory costs (if an activity may not proceed without affirmative government approval) or result in environmental harms (if the activity may proceed unless the government adapts and enforces regulatory measures). Thus, regulatory laws often provide that a new plant or development project cannot be undertaken or a new drug or pesticide marketed until after government regulatory approval is granted following an information-gathering process. This delay imposes regulatory costs by postponing realization of the benefits that the activity would produce had it been allowed to begin earlier. On the other hand, an existing activity generally cannot be subjected to new or additional controls until after completion of a regulatory decision making process in which information is gathered. In this case, delay may involve interim harm resulting from unregulated activity.[98] There are also hybrid procedures, as exemplified by the U.S. Toxic Substances Control Act, where manufacturers must submit basic information on new chemical products but may introduce them commercially within a short period following this submission unless EPA affirmatively takes action to require additional testing and study of potential environmental risks posed by such products.

Now consider the case where a regulatory decision involving uncertain risks of harm and/or uncertainty in regulatory costs need not be made once and for all, but can revisited one or more times in the future. In this situation, the decision maker must consider the value and the cost of information that could be gathered in the future, the possibility of an improved future decision based on that information, and the implications of such an improvement for the current regulatory decision.

There are three categories of additional information that could be developed in the future: information generated by investment in research, information that results from learning-by-doing, and autonomous information that is produced by experience generally. Research includes theoretical and laboratory studies, simulations and small scale tests, and other direct investments in the generation of knowledge. Learning-by-doing consists of information developed as a result of carrying out an activity or imposing regulatory controls.[99]

Research and information collection can be undertaken or commissioned by the regulatory agency or other government authorities. In making these decisions, the government will have to consider the expected value and cost of the information that would be developed. Incentives can also be provided to the regulated industry and other non-governmental actors to obtain and develop new information. The regulated industry's interest in obtaining favorable regulatory decisions is a powerful incentive to obtain and develop such information. The force of that incentive may be enhanced by imposing on the regulated industry the burden of production, the burden of persuasion, or both, in regulatory proceedings. Further, regulators may be empowered to require the regulated industry to gather information, conduct research, and report the results.

Learning-by-doing with respect to harms can occur when an activity posing uncertain risks of harm is allowed to proceed and monitoring of evidence of harm is conducted. For example, field releases of bioengineered crop plants presenting a potential risk of ecologically harmful transgenic transfers may be monitored for evidence of such harm. Learning-by-doing regarding regulatory costs can occur when regulatory controls are imposed and compliance costs and regulatory performance are monitored. For example, imposition of performance-based BAT controls on air emissions will lead the regulated industry and its suppliers to develop and install control technologies whose costs and performance can be monitored.

The value of the information produced through learning-by-doing will vary depending on the circumstances. If, for example, an activity has already been conducted for a substantial period of time and its effects have been monitored, the value of the additional information regarding harm that may be obtained as a result of continuing the activity may be far less than the information regarding harm obtained by allowing a new activity, such as field release of new bioengineered crops, to occur. The value of learning-by-doing regarding harms also depends on the extent to which adverse effects from an activity can be accurately detected. Thus, the health effects of very low level pollution exposures generated by certain industrial activities may be very difficult to detect through epidemiological studies; if so, the value of learning-by-doing with regard to harms by allowing these activities to continue may be low. The informational value of learning-by-doing regarding regulatory costs will also vary depending on the circumstances. If, for example, industry believes that regulatory controls adopted at T1 may be eliminated or relaxed at T2, or alternately if it believes that such controls will be substantially tightened at T2, its response to regulatory requirements during the period T1–T2 is unlikely to be representative of its response if the controls were expected to be imposed on a long-term basis.

Against the background of these considerations, how should the value of additional information be taken into account in an initial regulatory decision? The discussion that follows assumes that the choice is between not regulating at all or imposing regulation of a given level of intensity. The same basic analysis can, however, be extended to cases where the choice is between more stringent and less stringent regulation.

Regulate:
The regulator can decide at T1 to regulate the activity, and to re-evaluate that decision at T2 on the basis of any new information developed in the interim, including the information about regulatory costs obtained through learning-by-doing. In doing so, the regulator forgoes the additional information regarding harm that might be obtained by allowing an activity to go forward without regulation.[100] The regulator also runs the error risk of imposing on society control costs that later prove to have been unwarranted because new information shows that the activity in question does not cause any significant harm.

Defer:
Alternatively, the regulator can decide not to regulate at T1 and to reconsider that decision at T2, taking into account the additional information obtained in the interim, including information about harm obtained by learning by doing by allowing the activity to go forward and monitoring its effects. In doing so, the regulator forgoes the information regarding regulatory costs that might be obtained by imposing regulatory controls. The regulator also runs the error risk of allowing the conduct of activities that cause harms that, in the light of new information, turn out to be both significant and unwarranted because they could have been prevented at reasonable cost by the adoption of regulatory controls at T1.

The decision maker will have to consider the nature and extent of these two different types of error risks, and the costs associated with them, in its decision whether or not to regulate at T1. These risks will be characterized by significant uncertainties, which must be addressed using the decision analytical framework previously discussed. Because the relevant considerations will vary so widely from case to case, it seems very difficult to make any generalizations about which approach to follow.

The most stringent version of PP, however, asserts that the decision maker should always decide to regulate an activity and, further, that the regulation should take the form of a BAT requirement or a prohibition whenever the risks posed by the activity are sufficiently uncertain that a possibility of appreciable harm can not be excluded. The previous discussion has established a number of reasons why this decision rule is unsound. The contribution to improved regulatory decision making of new information potentially generated by learning-by-doing provides an additional consideration against adopting a prohibitory PP requirement that new activities posing uncertain risks of potentially serious harm should be barred.[101] Such a rule precludes learning-by-doing regarding the risks posed by such an activity. At the same time, prohibiting a new activity is unlikely to generate useful information about regulatory costs. In this circumstance, regulatory costs do not consist solely of compliance outlays, such as the costs of installing and operating pollution control equipment, but the opportunity costs of investments and innovations forgone as a result of the prohibition. Such costs are notoriously difficult to determine.[102] Thus, the prohibitory PP as applied to new activities deprives decision makers of any significant information benefits from learning-by-doing. Allowing the activity to proceed while monitoring it closely, however, can produce significant learning-by-doing information benefits. These benefits are likely to be especially pronounced when considered not in the isolated case of a single activity but new activities generally. Allowing various new activities to proceed while monitoring their effects can promote societal learning about which types of activities do not pose significant risks and which do. The accumulation of such knowledge, based on experience with many different types of activities and risks, may advance more general scientific and societal understanding of the characteristics of risks and how best to respond to them, thereby enhancing societal resilience to risk.[103]

This is by no means to argue for an across-the-board rule allowing all new activities posing uncertain risks of harm to proceed, regardless of the circumstances. In some cases, the risk of serious harm may be sufficiently substantial to justify prohibiting the activity and concentrating on research rather than learning-by-doing to develop new information about risks of harm. Considerations based on the information value of learning-by-doing are also less weighty in the case of a prohibitory PP as applied to existing activities, where learning-by-doing regarding harm may have already occurred and where the benefits of the activity can be estimated as a basis for projecting the costs of prohibiting its continuation. Learning-by-doing considerations may also be less important in the case of a BAT PP, whether applied to existing or new activities.

These conclusions are consistent with and supported by the insights of economic analysis. In 1974, Kenneth J. Arrow and Anthony C. Fisher first analyzed the importance of information value in connection with decisions regarding regulatory and management decisions for environmental resources involving uncertain risks of serious, irreversible consequences.[104] They pointed out that the future availability of additional information to reduce the uncertainty regarding this risk would permit a more informed decision to be made in the future. Such information accordingly has a positive economic value. If the decision is made to allow the development and it has irreversible consequences, the value of this information will be lost. This loss and of the associated loss of opportunity for an improved decision in the future should accordingly be counted as a cost in the decision whether or not to allow development. In close cases where the benefits and the other costs of development are fairly closely balanced, the inclusion of this information cost in the analysis could result in a decision not to develop. In other words, considerations of information value tend to favor flexibility in the timing of development decisions, and should lead decision makers to err on the side of not allowing development when the effects may be irreversible. The applications of this analysis to environmental regulatory decisions are apparent.

The analytic framework pioneered by Arrow and Fisher has since been extended and applied to a range of decisional problems, including decisions with respect to climate change.[105] This framework strongly supports a Bayesian approach to decision making, in which initial judgments about uncertainties are continually revised in the light of new information.[106] It also provides an analytical structure for examining the benefits and costs of learning-by-doing strategies for reducing uncertainty within the context of a Bayesian approach.[107] For example, in the context of climate policy, one study has concluded that the informational benefits of learning-by-doing, with respect to the potential extent of warming, support a decision not to impose stringent GHG controls in order to obtain valuable information on the warming effects of GHG emissions.[108]

Subsequent analysis has shown that the Arrow-Fisher framework is applicable to activities that threaten serious but remediable losses, as well as to those that have irreversible consequences.[109] The subsequent literature also confirms that the value of information does not mean that a decision to undertake or allow activities that may cause serious or irreversible harms should always be postponed in order to obtain the decisional benefit of additional information. The value of additional information depends on the circumstances and is only one factor to be balanced among other relevant costs and benefits, which in some cases justify an initial decision to allow a risky activity to proceed.[110] For example, one analysis has concluded that the probabilities of significant irreversibilities associated with global warming are too small to justify a decision to impose stringent controls on GHG emissions pending the development of further information on warming, especially where such controls would require large capital investments that could not be reversed if new information failed to show that they were justified.[111]

The value of additional information and the consequent desirability of flexibility in regulatory decision making can readily be incorporated into the standard framework for preventive regulatory decision making developed in Part II.A. The loss of information value involved in decisions to allow activities with potentially serious, irreversible consequences to proceed should be accounted a cost in decisions whether or not to regulate the activity, and that cost should be incorporated in the base case decision. Under this decisional framework, the benefit of acquiring new information may in some circumstances justify precautionary decisions to prohibit, or otherwise regulate, certain activities, provided that affirmative steps are taken to develop the new information so as to permit timely reconsideration of the regulatory decision with the benefit of such information. In other cases, however, developing the additional information may require and justify that the activity be allowed to proceed, perhaps in a limited and controlled manner, in order to take advantage of "learning-by-doing."[112] And, in still other cases, the social benefits of allowing an activity to proceed may so far outweigh the risks of harm that including considerations of information value would not affect a decision to allow the activity to proceed.

Strong versions of PP impose regulatory requirements on activities regardless of whether such requirements are justified by value-of-information considerations. Moreover, strong formulations of the prohibitory PP prevent learning-by-doing with respect to uncertain harms.[113] Greenpeace, for example, asserts that the uncertainties associated with the risks of GMO crops requires prohibition on their use, including prohibition of limited controlled field trials that provide the best means of resolving the uncertainties in question. This consequence provides an additional reason for rejecting such versions of PP.

Perhaps due to a perceived conflict of interests or goals between environmentalists and economists,[114] the value of additional information in the context of sequential regulatory decision making is generally not discussed as a justification for a precautionary approach in the PP literature.[115] Yet, the underlying concept is sometimes acknowledged. The EC Communication on the Precautionary Principle, for example, provides that: "Measures based on the precautionary principle shall be re-examined and if necessary modified depending on the results of the scientific research and the follow up of their impact."[116] This provision implicitly acknowledges the value and importance of continuing to gather and examine information throughout the regulatory process and take new information into account in subsequent regulatory decisions. Some international declarations contemplate broad adaptive approaches to managing environmental risks; such approaches could include the deliberate development of new information on risks and consideration of its implications for sequential regulatory decisions.[117]

Furthermore, writers on PP have proposed a "risk-oriented" or "procedural precautionary approach" to encourage the development of and then use valuable information obtained during a process of sequential regulatory decision making.[118] This "procedural" approach contemplates an adaptive process designed to encourage the development of new information, while continually re-evaluating and appropriately modifying previous regulatory decisions. These procedures function within an edict of precaution. Yet, the approach recognizes the drawbacks of rigid PP prescriptions and seeks to introduce flexibility. Flexibility can also be introduced by applying the principle of proportionality to precautionary regulation. For example, the EU Communication on the Precautionary Principle states that: "The measures adopted presuppose examination of the benefits and costs of action and lack of action."[119] In various ways, these approaches are similar to attempts by scholars to bridge the perceived gap between PP and cost/benefit analysis.[120] They have the potential to create a far more flexible and workable conception of PP that may not in substance be very different from the standard framework for preventive regulatory decision making set forth in Part II.A.

V. CONCLUSION

This essay has examined two basic versions of the PP, weak and strong. The weak versions – PP1 and PP2 – enunciate principles that have been widely followed over the past thirty years in environmental regulatory programs in most industrialized nations, which generally follow a preventive strategy that includes regulation of uncertain risks. Many environmental regulatory programs in international agreements follow the same approach. Accordingly, the weak versions of PP propose nothing new. On the other hand, the strong versions of PP – PP3 and PP4 – would, if adopted, make important changes in the prevailing concept and substance of environmental regulatory law and practice at both the domestic and international levels. This essay concludes that the strong versions of PP lack a sound foundation in regulatory theory and would lead to undesirable regulatory outcomes, diminishing societal welfare and reducing overall environmental protection. They should accordingly be rejected. Instead, the valuable insights of PP, clarified through careful economic and legal analysis, should be integrated into the existing paradigm of preventive regulation. This essay offers a framework for accomplishing this integration.

PP proponents may complain that my characterization of strong versions of PP is too rigid and extreme, and that both the spirit and practice of PP would be more flexible and accommodating than my analysis has supposed. Any suggestion that this essay attacks a "straw man" version of PP would, however, be unwarranted. Of course one could, as some commenters have done,[121] interpret PP in a very "soft" fashion as a collection of general admonitions for regulators such as the following: environmental protection is very important, especially if long term or irreversible harms are threatened. In assessing uncertain environmental risks, pay close attention to the limits of our ability to assess uncertain risks and the longer run and cumulative consequences of activities. Be protective where very serious harms might occur, even if their probability is relatively small. Such admonitions, if followed with judgment and attention to context, are entirely appropriate. As discussed above, these admonitions of precaution, appropriately clarified and refined, can and should be incorporated within the net-benefits-maximization approach to preventive regulation.[122]

Proponents of strong versions of PP, however, advocate something quite different and more radical. They condemn the dominant approach to environmental regulation as seriously deficient and propose an entirely new paradigm targeted on risk uncertainty. PP proponents assert that a basic shift towards a far more precautionary approach in environmental regulation based on this paradigm is needed, requiring fundamental changes in the preventive approach to environmental regulation that has generally prevailed over the past 30 years.[123] The basic elements of PP3 and PP4 – worst case presumptions, prohibitory or BAT regulatory controls, disregard or downplaying of compliance costs, and burden shifting – would accomplish those changes. It is not unfair, indeed it would be disrespectful, not to take these arguments and proposals at face value and evaluate them as such. Of course, the strong versions of PP will require judgment and a degree of flexibility in application to particular cases. But the strong PP prescriptions are regulatory rules that are to be applied across the board to all PP risks, which would include a potentially quite large array of human activities, including both existing, well-entrenched activities and activities involving new product and process technologies. It would be inconsistent with the basic premises of PP to suppose that the adverse

consequences of such rules could be avoided by applying them selectively to
the cases where they are most justified and carefully balancing regulatory burden
against risk of harm. Selective application and balancing would require some
means for distinguishing those uncertain risks most likely to pose serious harm
from those that are less likely to do so. But, that would require epistemological and institutional capacities for resolving uncertainty that PP denies. It would
also require significant discretion in regulatory decision making, discretion that
PP proponents seek to minimize. If there are indeed criteria, consistent with PP
premises, to guide selective application of the PP regulatory prescriptions and
a balancing approach so as to avoid unduly rigid and costly regulation, those
criteria need to be identified and justified. Thus far, proponents of PP have
failed to provide them.

NOTES

1. For discussion of the various sources of uncertainty regarding environmental
risks, see Cameron, J., Jordan, A., & O'Riordan, T., 2001. The Evolution of the
Precautionary Principle. In: Cameron, J., Jordan, A. & O'Riordan, T. (Eds), *Reinterpreting the Precautionary Principle*. Cameron May, London, pp. 23–24 (finding
sources of uncertainty in the complexity and interdependency of natural phenomenon
and the chaotic and discontinuous nature of many natural processes) [hereinafter
Reinterpreting the Precautionary Principle].

2. These reasons include the circumstance that private actors may not have sufficient assets to pay for harms that they cause if the harms are large; the circumstance
that private decision makers may have a higher rate of time discount or be less risk
averse than is judged appropriate from a pubic policy perspective; the difficulties
presented in resolving the issue of causal responsibility when a given harm may have
a number of different causes; problems caused when harm is the product of activities
by many actors; the circumstance that some private decision makers, including individuals and small business, may not have the knowledge or capacity to adjust their behavior
so as to appropriately reduce risks of harm; and general public distrust of the reliability
of ex post incentive systems to prevent risks of serious harm.

3. See, e.g. Ethyl Corp. v. EPA, 541 F.2d 1 (D. C. Cir. 1976), *cert. denied*, 426
U.S. 941 (1976).

4. See Pildes, R., & Sunstein, C. (1995). Re-inventing The Regulatory State.
University of Chicago Law Review, 62(1); Stewart, R. B., (2001). The Importance of
Law and Economics for European Environmental Law. *Yearbook of European
Environmental Law*, 2.

5. See, e.g. Industrial Union Department, AFL-CIO v. American Petroleum
Institute, 448 U.S. 607 (1980).

6. See, e.g. Wingspread Statement on the Precautionary Principle, Jan. 25, 1998,
available at http://www.gdrc.org/u-gov/precaution-3.html; Re-interpreting the
Precautionary Principle, *supra* note 2; Harding, R., & Fisher, E. (Eds), (1999).
Perspectives on the Precautionary Principle. Federation Press, Sydney.; De Fontaubert,

A. C., et al. (1998). Biodiversity in the Seas: Implementing the Convention on Biological Diversity in Marine and Coastal Habitats. *Georgetown International Environmental Law Review, 10*. The PP derives its intellectual heritage for the principle of Vorsorgenprinzip, adopted in Germany's domestic environmental law. See Cameron, Jordan, & O'Riordan *supra* note 2 at 11.

7. See, e.g. De Oliveira Souza, H. F., (2000). Genetically Modified Plants: A Need for International Regulation. *Annual Survey of International & Comparative Law, 6*, 173; Smith, K. L., (1999). Highly Migratory Fish Species: Can International and Domestic Law Save the North Atlantic Swordfish? *Western New England Law Rev., 21*, 35–39. See also Ewald, F., (1999). The Return of the Crafty Genius: An Outline of a Philosophy of Precaution. *Connecticut Insurance Law Journal, 6*, 60–61 ("the precautionary principle does not target all risk situations, but only those marked by two principal features: a context of scientific uncertainty on one side, the possibility of serious and irreversible damage on the other").

Some PP proponents contrast between a risk-based approach to regulation in the U.S. based on a "rationalist" weighing of probabilities and the costs and benefits of various alternatives, and a "political" model in Europe under which uncertainty is a normative principle that triggers precaution. See Cameron, Jordan, & O'Riordan *supra* note 2 at 27. To the extent that this distinction exists, however, it has not led to adoption of more stringent regulation of uncertain environmental health and safety risks in Europe relative to the U.S. See Wiener, J. B., & Rogers, M. D. (forthcoming). Comparing Precaution in the United States and Europe. Journal of Policy Analysis & Management.

8. See, e.g. United Nations Conference on Environment and Development: Framework Convention on Climate Change, May 9, 1992. In: Report of the Intergovernmental Negotiating Committee for a Framework Convention on Climate Change on the Work of the Second Part of Its Fifth Session, INC/FCCC, 5th Sess., 2d Part, at Annex I, U. N. Doc. A/AC.237/18 (Part II)/Add.1, *reprinted in* 31 I. L. M. 851. Any of the relevant texts are collected in: Cameron, J., 2001. The Precautionary Principle in International Law, in: Reinterpreting the Precautionary Principle, *supra* note 2.

9. Treaty on European Union and Final Act, Feb. 7, 1992, *reprinted in* 31 I. L. M. 247 (entered into force Nov. 1, 1993).

10. See generally McIntyre, O., & Mosedale, T. (1997). The Precautionary Principle as a Norm of Customary International Law. *Journal of Environmental Law, 9*, 221. See *also* Communication on the Precautionary Principle, Communication from the Commission of the European Communities, COM (2000) 1 final (Feb. 2, 2000); Hohmann, H. (1994). Precautionary Legal Duties and Principles of Modern International Environmental Law. Graham & Trotman/Martinus Nijhoff, London.; Unger, R., 2001. Note, Brandishing the Precautionary Principle Through the Alien Tort Claims Act. *New York University Environmental Law Journal, 9*, 673–675.

11. See, e.g. Freestone, D. (1999). International Fisheries Law Since Rio: The Continued Rise of the Precautionary Principle. In: Boyle, A., & Freestone, D. (Eds), *International Law and Sustainable Development: Past Achievements and Future Challenges.* Oxford University Press, New York, pp. 135–136.; Charnovitz, S., (2000). The Supervision of Health and Biosafety Regulation by World Trade Rules. *Tulane Environmental Law Journal, 13*, 291–295.; Hickey, Jr., J. E., & Walker, V. R. (1995). Refining the Precautionary Principle in International Environmental Law. *Virginia Environmental Law Journal, 14*, 431–432.

12. Dovers, S. R., & Harndmer, J. W. (1999). Ignorance, Sustainability, and the Precautionary Principle: Towards and Analytic Framework, In: *Perspectives on the Precautionary Principle, supra* note 6 at 173.

13. See, e.g. Tinker, C. (1996). State Responsibility and the Precautionary Principle, In: Freestone, D., & Hey, E. (Eds), *The Precautionary Principle and International Law: The Challenge of Implementation.* Kluwer Law International, The Hague, The Netherlands, (pp. 3, 5). ("it is difficult to conclude that the precautionary principle is customary international law at this time.").

14. A modification of this version calls for the Best Available Technology Not Entailing Excessive Costs ("BATNEEC"). See, e.g. Agreements on the Protection of the Rivers Scheldt and Meuse, Apr. 26, 1994, *reprinted in* 34 ILM 851 (1995) ("the Contracting Parties shall strive to use the best available technology . . . under economically acceptable conditions."). It is unclear whether this involves a cost-benefit analysis of the reduction achieved through a given technology, or simply rules out technologies that are judged "too costly." Other versions of PP also invoke vague cost considerations. See, e.g. United Nations Rio Declaration on Environment and Development, June 13, 1992, principle 15, *reprinted in* 31 I. L. M. 874, 879 (". . . lack of full scientific certainty shall not be used as a reason for postponing cost-effective measures to prevent environmental degradation."). "Cost-effective measures" is a concept that makes little logical sense in this situation, since presumably the decision to regulate cannot be based upon a cost-benefit analysis. All this seems to require is that regulators not choose one measure that costs significantly more than another but achieves the same result. Most likely, it is intended to mitigate undue rigor in regulatory requirements and add some degree of flexibility to the application of PP. For additional commentary on cost-effective precautionary measures, see Fullem, G. D. (1995). Comment, The Precautionary Principle: Environmental Protection in the Face of Scientific Uncertainty. *Willamette Law Review, 31,* 505–508; Bodansky, D. (1993). The United Nations Framework Convention on Climate Change: A Commentary. *Yale Journal of International Law, 18,* 456; and Goldberg, D. M. (1993). Negotiating the Framework Convention on Climate Change. *Touro Journal of Transnational Law, 4,* 162.

15. See, e.g. Communication on the Precautionary Principle, *supra* note 10 ("[PP]is part of risk management, when scientific uncertainty precludes a full assessment of the risk"); Ewald, *supra* note 8 at 63 (1999). ("Precautionary logic does not cover risk (which is covered by prevention); it applies to what is uncertain – that is, to what one can apprehend without being able to assess.")

16. Bergen Ministerial Declaration on Sustainable Development in the ECE Region, G. A. Prepatory Committee for the United Nations Conference on Environment and Development, 44th Sess., U. N. Doc., A/CONF.151/PC/10 (1990).

17. Cartagena Protocol on Biosafety, Jan. 29, 2000, Art. 10(6) & Annex III, reprinted in 39 I. L. M. 1027, 1045. See also Convention on the Protection and Use of Transboundary Watercourses and International Lakes, Mar. 17, 1992, art. 2(5), reprinted in 31 I. L. M. 1312, 1316; Communication on the Precautionary Principle, *supra* note 10; United Nations Framework Convention on Climate Change, *supra* note 8 at 854; United Nations Rio Declaration on Environment and Development, *supra* note 14 at 879.

18. See Communication on the Precautionary Principle, *supra* note 10.

19. See Unger, *supra* note 10 at 677.

20. Sustainable Fisheries Act of 1996 § 106(b), 16 U.S.C. § 1802(28) (2001). See *also* Geistfeld, M. (2001). Reconciling Cost-Benefit Analysis with the Principle that

Safety Matters More than Money. *New York University Law Review, 76,* 114; Gray, G. M., & Graham, J. D. (1995). Regulating Pesticides. In: Graham, J. D., & Wiener, J. B. (Eds), *Risk vs. Risk: Tradeoffs in Protecting Health and the Environment.* Harvard University Press, Cambridge, Mass., p. 173.

21. Fullem, *supra* note 14 at 501.

22. Jacobs, M. (1992). *The Green Economy: Environment, Sustainable Development and the Politics of the Future.* Pluto Press, London, p. 100.

23. See UNCED Text on Protection of Oceans, U. N. GAOR, 4th Sess., UN Doc. A/CONF. 151/PC/100/Add. 21 (1991).

24. See, e.g. Clean Air Act § 109, 42 U.S.C. § 7409(b)(1).

25. Second International Conference on the Protection of the North Sea, Nov. 25, 1987, Art. XVI(1), reprinted in 27 I. L. M. 835, 840. See also Agreements on the Protection of the Rivers Meuse and Scheldt, *supra* note 14 at 855; Bamako Convention on Hazardous Wastes Within Africa, Jan. 30, 1991, Art. 4(3), reprinted in 30 I. L. M. 773, 781 ("The Parties shall cooperate with each other in taking the appropriate measures to implement the precautionary principle to pollution prevention through the application of clean production methods . . . .").

26. See discussion *supra* note 14 on BATNEEC.

27. Final Declaration of the First European "Seas at Risk" Conference, Annex I, Copenhagen, 1994, available at the Common Wadden Sea Secretariat, Virchowstraße 1, D-26382 Wilhelmshaven, Germany. See also Wingspread Statement on the Precautionary Principle, *supra* note 6; World Charter for Nature, GA Res. 37/7, Annex, para. 24, UN GAOR, 37th Sess., Supp. No. 51, at 17, UN Doc. A/37/51 (1982), reprinted in 22 I. L. M. 455, 455.

28. See, e.g. World Charter for Nature, *supra* at 455 ("Activities which are likely to cause irreversible damage to nature shall be avoided.").

29. See, e.g. Convention for the Protection of the Marine Environment of the North-East Atlantic, Sept. 22, 1992, Art. 2, reprinted in 32 I. L. M. 1069, 1076 ("preventive measures are to be taken when there are reasonable grounds for concern that substances or energy introduced, directly or indirectly, into the marine environment may . . . harm living resources and marine ecosystems."); Priess, H-J., Pitschas, C., 2000. Protection of Public Health and the Role of the Precautionary Principle Under WTO Law: A Trojan Horse Before Geneva's Walls? *Fordham International Law Journal, 24,* 523 (". . . 'there can be no question but that the requirements of the protection of public health must take precedence over economic considerations.'" (quoting Alpharma Inc. v. Council of the European Union, Case T-70/99 R, [1999] E. C. R. II-2027)). *C.f.* Taylor, P, 2000. Heads in the Sand as the Tide Rises: Environmental Ethics and the Law on Climate Change. *UCLA Journal of Environmental Law & Policy, 19,* 257 ("Economic criteria and the operation of the market are notoriously ill equipped to cater to long-term objectives, such as the interests of future generations."); Gundling, L., 1990. The Status in International Law of the Principle of Precautionary Action. *International Journal of Estuarine & Coastal Law, 5,* 26 ("[PP] requires reduction and prevention of environmental impacts irrespective of the existence of risks . . . and it also requires action even if risks are not yet certain but only probable, or, even less, not excluded.").

30. See, e.g. Convention for the Protection of the Marine Environment of the North-East Atlantic, *supra* ("preventive measures are to be taken when there are reasonable grounds for concern"); Bacon, B. L. (1999). Enforcement Mechanisms in International

Wildlife Agreements and the United States: Wading Through the Murk. *Georgetown International Environmental Law Review*, *12*, 331 ("uncertainties about the ability of a fish stock to sustain a harvest level must be resolved in favor of the fish.").

31. See Communication on the Precautionary Principle, *supra* note 10 at 18; Reinterpreting the Precautionary Principle, *supra* note 1.

32. See Communication on the Precautionary Principle, *supra* note 10; Heyvaert, V., 2000. Uncertainty and the Role of Law: Towards a Sustainable Legal Framework for Health and Environmental Protection (unpublished manuscript, on file with the author).

33. See, e.g. Final Declaration of the First European "Seas at Risk" Conference, *supra* note 27; Wingspread Statement on the Precautionary Principle, *supra* note 6 ("In this context the proponent of an activity, rather than the public bears the burden of proof."); Juda, L. (1996). *International Law and Ocean Use Management*. Routledge, London, p. 287 (". . . the burden of proof moved from those who seek to protect the environment to those who maintain that some ocean use or activity is not harmful.").

34. The burden-shifting component of the strong PPs presents a number of intriguing problems. For example, how can proponents of regulation meet the threshold require-ment of showing a potential for significant harm when the risks posed by an activity are highly uncertain and it may not be known what type of harm, if any, it might cause? Once the regulator has made an initial showing that the threshold requirement is satis-fied, can the proponent of an activity escape regulatory controls by rebutting the regulators' prima facie case, or must it in all cases affirmatively carry the burden of establishing the activity's safety? These important issues have yet to be seriously addressed in the PP literature.

35. See, e.g. Cartagena Protocol on Biosafety, *supra* note 17 at 1045 ("Lack of scientific knowledge or scientific consensus should not necessarily be interpreted as indicating a particular level of risk."). Compare, e.g. Tinker, *supra* note 13 at 779 (". . . the precautionary principle should mandate a policy of 'no action.'").

36. See, e.g. Hughes, L. (1998). Note, Limiting the Jurisdiction of Dispute Settlement Panels: The WTO Appellate Body Beef Hormone Decision. *Georgetown International Environmental Law Review*, *10*, 933. (". . . [we must] appreciate the inability of science to predict environmental and physiological interactions."); Cameron, Jordan, & O'Riordan *supra* note 1.

37. See Applegate, J. S. (2000). The Precautionary Preference: An American Perspective on the Precautionary Principle. *Human & Ecological Risk Assessment*, *6*, 413.

38. See Appellate Body Report, GATT Doc. WT/DS26/AB/R (Jan. 16, 1998).

39. See Raeburn, P. (2000). Clamor Over Genetically Modified Foods Comes to the United States. *New York University Environmental Law Journal*, *8*, 610–13.

40. See Part II.B, *infra*.

41. See Part IV.B, *infra*.

42. See Wiener, J. B. (2001). Precaution in a Multi-Risk World. In: Paustenbach, D. D. (Ed.), *The Risk Assessment of Environmental and Human Health Hazards* (2nd ed.), (forthcoming).

43. By the same token, as discussed below, learning by doing with respect to uncertain regulatory costs may in some circumstances justify initial imposition of regulatory controls.

44. Sunstein, C. (forthcoming). *The Arithmetic of Arsenic*, 2.

45. Issues of distributive justice may include what Mark Geistfeld calls a "safety principle," applicable to cases where one person imposes an unconsented risk of harm on another person and harm results. See Geistfeld, M. (2001). Reconciling Cost-Benefit Analysis With the Principle That Safety Matters More than Money. *New York University Law Review, 76,* 114. An overview of issues of risk, equity and distributive justice is provided in Frailberg, J. D., & Trebilcock, M. J. (1998). Risk Regulation: Technocratic and Democratic Tools for Regulatory Reform. *McGill Law Journal, 43,* 865–867.

46. See Stewart, R. B. (1983). Regulation in a Liberal State: The Role of Non-Commodity Values. *Yale Law Journal, 92,* 1537.

47. See, e.g. Graham, J. D. (1996). *Making Sense of Risk: An Agenda for Congress,* In: Hahn, R. W. (Ed.), *Risks, Costs, and Lives Saved: Getting Better Results from Regulation.* (p. 183), Oxford University Press, New York; and, AEI Press, Washington, D.C.

48. See Fullem, *supra* note 14 at 498 (". . . all current embodiments of the precautionary principle or approach implicitly reject risk neutrality."). *C.f.* Rachlinski, J. J. (2000). The Psychology of Global Climate Change. *University of Illinois Law Review 2000,* 299; Dernbach, J. C. (1998). Sustainable Development as a Framework for National Governance. *Case Western Reserve Law Review, 49*(1).

49. For example, societies appear to weight harms that are concentrated in a given community and that injure identified individuals in that community much more heavily than the same harm when it is diffuse and harms individuals who can not be identified as victims. For example, the general public may regard an industrial accident that kills all 500 people in a town as much worse than widespread industrial pollution that is known, on the basis of epidemiologic studies, to causes 500 excess deaths. This enhanced weighting probably cannot be explained by considerations of distributional equity or justice with regard to local communities.

50. Alternatively, such a substantial probability of harms may trigger justice-based or other structural constraints on benefits maximization. For example, undertaking activities that cause such harms may violate duties of justice to future generations.

51. The counterargument is that the appropriate relative weight to be given to environmental harms on the one hand, and to regulatory costs that do not consist of such harms on the other, should be resolved in the base case analysis, and that there is no justification for revising such weights and placing a higher value on environmental harms simply because the harm is probabilistic in character rather than determinate. Thus, if the loss of a rare wilderness is of greater weight than a large rise in oil prices, that circumstance is appropriately taken into account in the base case analysis where the harm is determinate and known. There is no justification for an additional thumb on the scale in favor of the environment at the stage of second case analysis. At that stage, the same principles of risk aversion should apply to losses valued in the base case as "very large," whatever their character.

52. See Frailberg & Trebilcock, *supra* note 45 at 877–878.

53. See Woodward, R. D., & Bishop, R. C. (1997). How to Decide When Experts Disagree: Uncertainty-Based Choice Rules in Environmental Policy. *Land Economics 73,* 492(discussing means of addressing expert disagreement in the context of climate policy). The sources of disagreements among experts are examined in Cooke, R. M., 1991. *Experts in Uncertainty: Opinions and Subjective Probability in Science.* Oxford University Press: New York.

54. For a discussion of other environmental regulatory problems for which a maximin approach might be appropriate, see O'Riordan, T. (2001). *The Precautionary*

Principle and Civic Science, in: Reinterpreting the Precautionary Principle, supra note 2 at 103.

55. The distinction between risk (Type 2) and uncertainty (Type 3), was systematically examined by Knight, F. H. (1921). *Risk, Uncertainty and Profit.* New York: Harper & Row.

56. Bayes' Theorem provides that a person should asses the implications of new information for his pre-existing estimation of uncertain probabilities by multiplying his prior probability estimation by the ratio (1) of the probability that the new information would have been observed if his prior estimation were true with (2) the probability that the information would have been observed if the prior estimation were not true. For a discussion of Bayes' Theorem and Bayesian approaches to addressing uncertain probabilities in the context of the law of evidence, see Posner, R. A. (1999). An Economic Approach to the Law of Evidence. *Stanford Law Review, 51,* 1477.; and Friedman, R. D. (2000). Presumption of Innocence, Not Even Odds. *Stanford Law Review, 52,* 873 (commenting on Posner).

57. See Montgomery, W. D., & Smith, A. (2000). Global Climate Change and the Precautionary Principle. *Human & Ecological Risk Assessment, 6,* 399.

58. I put to one side for present purposes the question of the weight, if any, that an informed regulatory decision maker should give to the much less informed risk estimations of members of the general public.

59. For a review of the relevant literature on these problems, see Machina, M. J., (1987). Choice Under Uncertainty: Problems Solved and Unsolved. *Journal of Economic Perspectives, 1,* 121.

60. See Brown, R. (1999). Discussion Paper, Using Soft Data to Make "Probabilistic Risk Assessments" Realistic. Institute of Public Policy, George Mason University.; Montgomery & Smith, *supra* note 57 at 399; Pomerol, J-C. (2001). Scenario Development and Practical Decisionmaking Under Uncertainty. *Decision Support Systems, 31,* 31; Sunstein, C. (Forthcoming), The Arithmetic of Arsenic, *supra* note 44.

61. Gundling, L. (1990). The Status in International Law of the Principle of Precautionary Action. *International Journal of Estuarine & Coastal Law, 5,* 26, quoted. In: Harding, R., & Fisher, E. (1999). Introducing the Precautionary Principle, in: Perspectives on the Precautionary Principle, *supra* note 6 at 11.

62. See, e.g. Cameron, Jordan & O'Riordan, *supra* note 1; De Fontaubert et al., *supra* note 6 at 766; Fullem, *supra* note 14 at 499.

63. See Wildavsky, A. (2000). Trial and Error Versus Trial Without Error. In: Morris, J. (Ed.), *Rethinking Risk and the Precautionary Principle* (pp. 33–34). Butterworth-Heinemann, Oxford.

64. See Weisbrod, B. A. (1964). Collective Consumption Services of Individual Consumption Goods. *Quarterly Journal of Economics, 77,* 71; Henry, C. (1974). Option Values in the Economics of Irreplaceable Assets. *Review of Economic Studies 1974, 89*; Smith, K. (1982). *Option Value: A Conceptual Overview.*; Bishop, R. C. (1982). Option Value: An Exposition and Extension. *Land Economics, 58,* 1. The analysis of option value for use values can be extended to include option value for non-use values, such as the value to an individual of preserving a resource even if he or she does not intend to ever visit or use it.

65. This analysis could be extended to deal with cases where one is considering whether to preserve a resource for the benefit of future generations whose preferences and wealth are uncertain.

66. See Kutrilla, J. V. (1967). Conservation Reconsidered. *American Economic Review, 57*, 777; Ciccetti, C. J., & Freeman, A. M. (1971). Option Demand and Consumer Surplus: Further Comment. *Quarterly Journal of Economics, 85*, 85.

67. See Schmalensee, R. (1972). Option Demand and Consumer's Surplus: Valuing Price Changes Under Uncertainty. *American Economic Review, 62*, 62.

68. See Smith, *supra* note 64.

69. See Bishop, *supra* note 64; Smith, *supra* note 64.

70. It might be thought that a potential justification could be found in Ellsberg's paradox and the notion of ambiguity aversion. Ellsberg found, based on experimental observations, that individual subjects did not regard the chances of drawing a black ball from an urn containing an unknown number of black and white balls as equivalent to the chances of drawing a black ball from an urn containing 50 black and 50 white balls. He further found that the subjects also did not regard the chances of drawing a white ball from the first urn as equivalent to the chances of drawing a white ball from the second urn; hence the paradox. See Ellsberg, D. (1961). Risk, Ambiguity, and the Savage Axioms. *Quarterly Journal of Economics, 75*, 643. The relevance for regulatory decisions of this "ambiguity" phenomenon on the part of untutored experimental subjects is at best unclear. Regulators are or can be educated about the issues presented by decision making under uncertainty. PP proponents have not developed or relied on the concept of ambiguity to justify the PP prescriptions for regulatory decisionmaking.

71. See Adler, J. H. (2000). More Sorry Than Safe: Assessing the Precautionary Principle and the Proposed International Biosafety Protocol. *Texas International Law Journal, 35*, 206.; Ewald, *supra* note 7 at 78.

72. See Cross, F. B. (1996). Paradoxical Perils of the Precautionary Principle. *Washington & Lee Law Review, 53*, 908–915.

73. See id.; Wiener, *supra* note 42.

74. See Dwyer, J. (1990). The Pathologies of Symbolic Legislation. *Ecology Law Quarterly, 17*, 233. See *also* Melnick, R. S. (1984). Pollution Deadlines and the Coalition for Failure. *Public Interest, 75*, 123 (analyzing the perverse consequences of imposing unrealistic deadlines for achieving Clean Air Act air quality standards).

75. See, e.g. Final Declaration of the First European "Seas at Risk" Conference, *supra* note 27 ("The use of the 'economic availability' reservation in the application of precautionary measures ... is inconsistent with [the requirement of precaution] and must be abandoned ... If the 'worst case scenario' for a certain activity is serious enough then even a small amount of doubt as to safety of that activity is sufficient to stop it taking place."); Second International Conference on the Protection of the North Sea, *supra* note 25 at 840 ("... reduc[e] polluting emissions of substances that are persistent, toxic and liable to bioaccumulate at source by the use of the best available technology ..."); Wingspread Statement on the Precautionary Principle, *supra* note 6 ("... the proponent of an activity, rather than the public bears the burden of proof.").

76. Some of the more sophisticated PP literature does invoke a principle of proportionality, which includes considerations of cost-effectiveness and margins of error so that the selected degree of restraint is not unduly costly. This appears to offer regulators some degree of flexibility even after threshold criteria are met. However, any cost-benefit analysis that follows is intended to be weighted to favor regulation. See Communication on the Precautionary Principle, *supra* note 10.

77. See, e.g. Taylor, *supra* note 29 at 255–258 (arguing that property rights conceptions undermine the effectiveness of current regulatory programs); Dernbach, *supra* note
48 at 61 (arguing that the interest in economic development unduly influences current
regulatory decisionmaking).

78. See generally Farber, D. A. (1999). Taking Slippage Seriously: Non-compliance
and Creative Compliance in Environmental Law. *Harvard Environmental Law Review 23*,
297. See also Yoshida, O. (1999). Soft Enforcement of Treaties: The Montreal Protocol's
Noncompliance Procedure and the Functions of Internal International Institutions.
Colorado Journal of International Environmental Law & Policy, 10, 121–127.; Viscusi,
W. K. (1996). Regulating the Regulators. *University of Chicago Law Review, 63*, 1450.

79. See Melnick, R. S. (1983). Regulation and the Courts: The Case of the Clean
Air Act. Brookings Institution, Washington, D.C.; Dwyer, *supra* note 74.

80. See Adler, *supra* note 71 at 196–199; Cross, *supra* note 72 at 859–862.

81. See Clean Air Act § 112, 42 U.S.C. § 7412(i)(6)(A) (2001).

82. PP proponents, however, generally disfavor use of such standards because such
standards assume that there is a "safe" level of pollution exposure for humans or that
the environment can safely assimilate a given level of pollution. Such standards are
often based on evidence that shows no occurrence of harm at pollution concentrations
below the standard. PP proponents, however, argue that there is generally no affirmative proof that such harm does not or will not occur, only evidence that harm has not
yet been detected. As a result, it remains uncertain whether such harm exists.
Accordingly, under the PP, regulatory requirements, such as prohibitions on pollution-
generating activity or BAT requirements, should be adopted to address the risks
associated with this uncertainty. See, e.g. Bamako Convention on Hazardous Wastes
Within Africa, *supra* note 25 at 781 (". . . appropriate measures to implement the precautionary principle to pollution prevention [are] through the application of clean production
methods, rather than the pursuit of a permissible emissions approach based on assim-
ilative capacity assumptions . . ..").

83. See Stewart, R. B. (2000). Economic Incentives for Environmental Protection:
Opportunities and Obstacles. In: Revesz, R. L., Sands, P., & Stewart, R. B. (Eds),
Environment, the Economy and Sustainable Development. Cambridge University Press,
Cambridge, U.K.

84. See Stewart, R. B. (2001). A New Generation of Environmental Regulation?
Capital University Law Review, 29, 94–127.

85. See Newall, R. G., & Pizer, W. A. (1999). Regulating Stock Externalities Under
Uncertainty. Resources for the Future Discussion Paper 99-10-REV 1999.

86. There is a degree of flexibility in BAT requirements, which can be made more
or less stringent depending on the level of cost and performance reliability invoked in
determining whether a technology is "available." It is, however, often difficult to use
this flexibility in order to achieve a more appropriate balance between regulatory costs
and benefits, because the BAT requirements are generally set for industry categories and
the environmental benefits of different levels of control often vary substantially among
different facilities (costs may vary substantially as well). But, to the extent that such
flexibility is invoked, its exercise requires a more discriminating approach to the analysis
of uncertainty and a more explicit balancing of regulatory costs and benefits than the
precautionary principle allows.

87. The inability of PP to deal satisfactorily with such tradeoffs is discussed *supra*.

88. See Stewart, *supra* note 84 at 94–95.

89. See, e.g. Industrial Union Department, AFL-CIO v. American Petroleum Institute, 448 U.S. 607, 614 (1980).

90. See, e.g. Dernbach, *supra* note 48 at 61 (". . . the precautionary principle would shift the burden of proof from those supporting natural systems to those supporting development."); Barton, C., 1998. Note, The Status of the Precautionary Principle in Australia: Its Emergence in Legislation and as a Common Law Doctrine. Harvard Environmental Law Review 22, 509 ("[pp] places the burden of proof on proponents of change to show that their actions will not cause serious or irreversible environmental harm.").

91. See, e.g. Milne, A. (June 12, 1993). The Perils of Green Pessimism. *New Scientist.*

92. See, e.g. Roht-Arriaza, N. (1996). Book Review. *Columbia Journal of Environmental Law, 21,* 196. (Hohmann, H., 1994. *Precautionary Legal Duties and Principles of Modern International Environmental Law.* Graham & Trotman/Martinus Nijhoff, London.) ("This burden shifting, according to Hohmann, runs along a spectrum, from weak obligations . . . [to a] not-yet-implemented total reversal of the burden of proof, requiring potential polluters to prove the safety of their activities before they can be permitted.").

93. Such a decision rule will have a far greater impact on the regulatory outcome than the allocations of the production and persuasion burdens because it requires proof of facts that are inherently very difficult to establish by anyone. Take the example of the prohibitory PP. The decision rule makes it very easy for the regulatory agency to prohibit an activity, and very difficult for the agency to approve it. This is true regardless of how the burdens of production and persuasion are assigned. If, for example, the burden of production is on the regulator, it will be relatively easy for the regulator to introduce evidence of sufficient uncertainty regarding risk of harm such that the possibility of appreciable harm cannot be excluded; once that production burden is satisfied, it will also be very difficult for the activity proponent to rebut the regulators' prima facie case. On the other hand (still assuming that the production burden is on the regulator) it will be very difficult for the regulator to make a prima facie case for approving an activity by introducing proof that excludes the possibility of harm from an activity where there is considerable uncertainty regarding risk. When the burden of production is on the activity proponent, it will also be very difficult for it to introduce such proof. Thus, whichever party bears the burdens of production and persuasion, it will be very difficult to make out a prima facie case for approval of the activity because of the stringency of the decisional rule for granting such approval; it will also be easy for the regulator to make out a prima facie case that it should be prohibited. For similar reasons, the allocation of the persuasion burden is also less significant than the decision rule. This is not to say that the allocation of the production burden (for example, in cases where expensive testing is required) or of the persuasion burden (where marginal uncertainty leaves the question whether the possibility of appreciable harm has been sufficiently excluded) may not have a significant impact on outcomes. But such impacts will be less significant than that of the underlying PP decision rule itself.

94. If the burdens of producing evidence and of persuasion regarding risk are imposed on the regulator because it is better able to develop the necessary evidence, regulatory controls will not be imposed in cases of indeterminacy where the relevant risk threshold issue cannot be resolved one way or another. If this consequence results in inadequate regulatory protection, there are ways to avoid that result other than by

shifting the burden to the activity proponent, even though it is generally less able to generate the relevant evidence. For example, the relevant risk threshold that the regulator is required to establish can be lowered.

95. Examples of such long tail harms include long-latency cancers caused by exposure to toxic substances, and the adverse effects of climate change caused by emissions of greenhouse gases with long residence times in the atmosphere.

96. Stock pollutants are those that can accumulate in the ecosystem and whose adverse effects are a consequence of the total stock, not the additions or subtractions from the stock at a given time. Examples include greenhouse gases and toxic substances that are emitted at relatively low levels but that bioaccumulate. Flow pollutants, such air pollutants like particulate matter or CO, do not reside in media for an extended period of time; their adverse effects are a function of the amounts emitted during a given time period. Stock pollutants have potentially distinctive implications for the decision to regulate uncertain risks now rather than deferring to later. If annual flows are small relative to the stock, postponing a decision whether to regulate will not result in large additions to the stock in the interim before the next decision and, as a result, may not pose a significant risk of harm if the damage function is approximately linear. Thus, in the face of considerable uncertainty, it may be preferable to postpone regulation until additional information about the extent of the risk, the means for addressing it and their costs are developed. If the damage function is approximately linear, then irreversible harm will not occur. In such cases, T1 regulatory "mistakes," in the form of a decision not to impose stringent regulatory limitations on GHG emissions when subsequent information shows that the harms from warming are serious and that more stringent controls should have been adopted, can be corrected by imposing more stringent controls later. There may, however, be a risk of critical non-linear damage thresholds, such that a small addition to the stock will trigger the threshold and cause serious environmental harm. Rapid global warming resulting from increased GHG emissions poses some small uncertain risks of such harms. Also, it may be argued that additions to the stock should be immediately stopped or minimized so as to avoid imposing an additional control burden on future generations. See Kolstad, C. D. (1996). Learning and Stock Effects in Environmental Regulation: The Case of Greenhouse Gas Emissions. *Journal of Environmental Economics & Management, 31*, 1.

97. See discussion and sources cited at pp. 110–111 *infra*.

98. The major U.S. federal environmental statutes, however, contain provisions granting general authority to the EPA or the Justice Department to take immediate enforcement action against activities that pose imminent and substantial risks of harm.

99. See generally Wildavsky, note 63.

100. If the type of regulation (for example, BAT regulation) allows an activity to go forward, e.g. pollute although to a reduced degree, some learning-by-doing regarding harm may still be possible.

101. See Wildavsky, *supra* note 63. The expected extent of learning-by-doing with regards to risks of harm under a BAT PP will depend on the circumstances.

102. See Stewart, R. B. (1980). Regulation, Innovation and Administrative Law: A Conceptual Framework. *California Law Review, 69*, 1256.

103. See Wildavsky, *supra* note 63.

104. Arrow, K. J., & Fisher, A. C. (1974). Environmental Preservation, Uncertainty, and Irreversibility. *Quarterly Journal of Economics, 88*, 312.

105. See, e.g. Peck, S. G., & Teisberg, T. J. (1993). Global Warming Uncertainties

and the Value of Information: An Analysis Using CETA. R*esource & Energy Economics 18*, 71 (stressing importance of resolving uncertainties about regulatory costs as well as harms); Montgomery & Smith, *supra* note 57 (rejecting appropriateness of strong PP presumptions and prescriptions for climate policy).

106. See Kelley, D.L, & Kolstad, C. D. (1998). Bayesian Learning, Growth, and Pollution. *Journal of Economic Dynamics & Control, 23*, 491; Miller, J. R., & Lad, F. (1984). Flexibility, Learning and Irrreversibility in Environmental Decisions: A Bayesian Approach. *Journal of Environmental Economics & Management, 11*, 161.

107. See Grossman, S. J., Kihlstrom, R. E., & Mirman, L. J. (1977). A Bayesian Approach to the Production of Information and Learning by Doing. *Review of Econonomic Studies, 4*, 533.

108. See Kelley & Kolstad, *supra* note 106.

109. See Epstein, L. G. (1980). Decision Making and the Temporal Resolution of Uncertainty. *International Economic Review, 21*, 269; Miller & Lad, *supra* note 106.

110. See Haneman, W. M. (1989). Information and the Concept of Option Value. *Journal of Environmental Economics & Management, 16*, 23.

111. See Kolstad, C. D. (1996). Learning Effects in Environmental Regulation: The Case of Greenhouse Gas Emissions. *Journal of Environmental Economics and Management, 31*, 1.

112. By the same token, as discussed below, learning-by-doing with respect to uncertain regulatory costs may in some circumstances justify initial imposition of regulatory controls.

113. See Wildavsky, *supra* note 63.

114. See Revesz, R. (1999). Environmental Regulation, Cost-Benefit Analysis, and the Discounting of Human Lives. *Columbia Law Review, 99*, 941.

115. The value of such information is discussed in: Howarth, R. B. (1995). Sustainability Under Uncertainty: A Deontological Approach. *Land Economics, 71*, 417.

116. Communication on the Precautionary Principle, *supra* note 10.

117. For example, the UNCED Text on Protection of Oceans, *supra* note 23, provides for "inter alia, the adoption of precautionary measures, environmental impact assessments, clean production techniques, recycling, waste audits and minimization, construction and/or improvement of sewage treatment facilities, quality management criteria for the proper handling of hazardous substances, and a comprehensive approach to damaging impacts from air, land and water." While there is nothing among these measures that specifically calls for reevaluation of regulatory decisions in light of new information, they at least demonstrate a recognition that continued information gathering is an important part of any "comprehensive approach" to regulation.

118. See Heyvaert, *supra* note 32. See also Harding & Fisher, *supra* note 61 at 11.

119. Communication on the Precautionary Principle, *supra* note 10.

120. See Geistfeld, *supra* note 21. See also Farber, D. A., 1998. *Eco-Pragmatism*. University of Chicago Press, Chicago.

121. See, e.g. Territo, M. (2000). Note, The Precautionary Principle in Marine Fisheries Conservation and the U.S. Sustainable Fisheries Act of 1996. *Vermont Law Review, 24*, 1352–1353.

122. See Part II, *supra*; Geistfeld, *supra* note 20.

123. See, e.g. Taylor, *supra* note 29; Wagner, W. E. (2000). Innovations in Environmental Policy: The Triumph of Technology-Based Standards. *University of Illinois Law Review 2000, 83*.

ACKNOWLEDGMENTS

Thanks to Steve Charest, Peter Gratzinger, and Rob Rees for research assistance, and to participants at the New York University School of Law Colloquium in Law, Economics and Politics for comments and suggestions. The support of the Filomen D'Agastino and Max E. Greenberg Research Fund at New York University's School of Law is gratefully acknowledged.

COMMENTS ON PAPER BY RICHARD B. STEWART

Richard O. Zerbe, Jr.

Professor Stewart has done his usual admirable job in laying out the dimensions of an important problem, the use of precautionary principle in environmental decision making under uncertainty, and in suggesting how current use of the principle is inappropriate both in domestic and international environmental policy. He focuses on versions of the strong precautionary principle. The weaker of the two versions states that activities that propose an uncertain potential for significant harm should be subject to best available treatment, BAT, to minimize the risk of harm, unless the activities proponents show they present no appreciable risk of harm. The stronger version states that such activities should be prohibited altogether unless its proponents can show no appreciable risk of harm.

Stewart maintains that in contrast to the strong precautionary principle, uncertain risks should be regulated under the same general decisional framework as risks that are well-characterized. In the face of uncertainty, regulatory agencies should resolve the uncertainty as best they can and treat the uncertain risk as the equivalent of a risk whose probability is known. In resolving uncertainty, decision makers should make their best estimate of the probability distribution of an uncertain risk, relying on the available evidence, scientific theory, expert judgment, and guidance from analogous regulatory problems and experience with them. They should not proceed, as with the precautionary principle, on the basis of "worst case" presumptions. Professor Stewart correctly labels these vague and draconian standards as an unsatisfactory response to uncertainty.

An Introduction to the Law and Economics of Environmental Policy: Issues in Institutional Design, Volume 20, pages 127–131.
© 2002 Published by Elsevier Science Ltd.
ISBN: 0-7623-0888-5

Addressing uncertainty has typically been a core task of risk assessment. Risk management is often viewed as a theory driven by value judgments, but in contrast benefit-cost analysis serves both assessment and management functions. Economists such as Pearce (1994) correctly worry there are no proper rules for incorporating economic trade-offs or scientifically manageable uncertainties into precautionary principles. So I ask what can benefit-cost analysis contribute to developing a sensible assessment and management approach to decision making under uncertainty?

Using the real option approach renders the possibility to incorporate a meaningful, measurable precautionary principle into benefit cost analysis when information is exogenous. This exemplifies the case in which information arrives as a random process. The greater the uncertainty the greater the value of waiting for additional information to arrive on the doorstep. Dixit and Pindyck (1994) have set out a canonical real options model to determine the threshold for action in the presence of uncertainty. The risk management problem is framed as one of expected value maximization. A solution is found for the maximum value of continuing the current activity – the value of the real option value – that is a function of the benefit of the new activity. The value of the option to delay the new activity becomes an added cost which it must bear. Explicitly, there is a value to delay that can be included as a cost of a new activity. The greater the uncertainty the greater will be the precaution.

The usual analysis assumes a sort of legal rights point of view that defines the status quo. The real option is a risk premium cost of changing the status quo. For example, suppose a polluter has a right to pollute and the issue is the value of stopping his polluting. The right comes into play in determining the direction of precaution. Precaution from the polluter's point of view requires that compliance costs be multiplied by the precautionary multiplier. Farrow (2001) has applied this approach.

For example consider the following derivation of benefits and costs for the Clean Air Act. These estimates are derived from a Monte Carlo simulation (Farrow, Ponce, Wong, Faustman & Zerbe et al., 2001). In this example only damages are stochastic.

Table 1.

Year	Mean Benefit (Damages Avoided) (billions)	Variance of Benefit	Mean Mitigation Costs	Variance of Costs
2010	110	$1.43 * 10^{21}$	27 billion	$4.12 * 10^{18}$

Given these figures the question is, what is the additional value gained by delay that would suggest precaution in applying the Act? The precautionary multiplier calculated assuming the polluters have the right to keep polluting is 3.37 (Farrow, 2001). This figure is multiplied by the costs and when this is done the total costs are now $91 million. This still gives a figure less than the benefits so precaution is insufficient to allow pollution to continue even considered from the polluters' right perspective.

The precautionary principle could also be calculated from an environmentalist's position, which assumes the status quo includes the Clean Air Act and the question is whether or not the Act should be repealed. Here the precautionary multiplier value equals 1.9 and this is multiplied by the benefits, which are already greater than the costs so again the suggestion is that the clean air act was worth its costs.

What about when we seek to determine the best course of action when there is no right? Or what if the starting point determines the outcome? Here again a precautionary principle can be calculated according to the framework below and as laid out in Zerbe (2001).

THE BENEFIT-COST FRAMEWORK

The standard benefit cost model says that gains are measurable by the WTP and losses from the WTA. Both are to be measured from a status quo legal starting point. Suppose that both parties A and B have a sense of economic ownership with respect to some property and that their rights are in dispute. What is the economically efficient decision?

Suppose their sense of economic ownership is between zero and 100%. This sense of ownership may be less than 100%. Let P_a and P_b represent the subjective sense of economic ownership by A and B, respectively. The economic efficiency principle states that the entitlement should go to the party to whom it is worth the most, which is correctly determined by considering both the WTP and WTA. The gains to A and B are measured by the WTP, and their losses are measured by the WTA. I have elsewhere shown that the right should go to A when the following condition is satisfied:

$$WTP_a + P_a(WTA_a - WTP_a) > WTP_b + P_b(WTA_b - WTP_b) \qquad (1)$$

where P_a and P_b represent the subjective probability of ownership by A and by B. The gain to A is A's WTP, weighted by the extent to which A does not have economic ownership. Similarly for B.[1] This is an interesting result, because it says that the *divergence* between the WTA B is B's WTA, weighted by the extent to which she does have economic ownership. The right goes to A when the gain to A from having the right is greater than the loss to B from being deprived and WTP is relevant to the decision concerning whom should receive the entitlement. When there is no economic or legal ownership this equation reduces to just:

$$WTP_a > WTP_b, \qquad (2)$$

which implies that the right should go to the highest bidder.

When the sense of economic ownership of A and B are equal, and both equal 100%, then the above condition reduces to suggest that the WTA determines ownership or:[2]

$$WTA_a > WTA_b, \qquad (3)$$

which says that the right should go to the party that values it the most, in a WTA sense. In general, as P_a and P_b increase, more weight will be given to the WTA.

These guidelines give us a rule when no party has the right to the option, which is: the right should go to the party who would pay the most to acquire it. The rule when both parties have economic ownership is: the option right should go to the party which would require the higher amount to give up the right were the party to own it.

RISK AVERSION

The two assumptions underlying the expected value model are risk neutrality and independence of risks. Professor Stewart would only apply risk aversion to risks that pose an appreciable risk of large harm from the viewpoint of society as a whole. He notes that neither science nor economics can answer the question concerning when regulators should be risk averse. I don't think this is correct. From the economic perspective, the sentiments of those potentially affected by a policy determine the relevant values to be used in decision making as far as they are discernable. If risks are spread so that no individual bears much risk, the assumption of risk neutrality is justified. This will not always be the case, however. If affected individuals are risk averse then such risk aversion should be incorporated into the valuation of benefits and costs. Once the level of risk aversion is known, certainty equivalents can be determined and incorporated into a benefit-cost, benefit-risk analysis. This sort of incorporation is consistent with Professor Stewart's emphasis on economic approaches as opposed to the current precautionary principle. Similarly if risks covary with other risks the individuals bear, then adjustments to expected value must be made. In the finance literature, adjustments for risks that covary with the market, can be assessed through the use of a risk adjusted interest rate. There is no reason this cannot be done for public investment. One can calculate betas and use the appropriate risk adjusted interest rate.

NOTES

1. In a famous anecdote, King Solomon, called on to decree which of two women should have ownership of the baby both of them claimed, proposed that the baby be cut in half. His proposal may be regarded as a clever device for determining their respective WTAs. The false claimant readily assented to the plan, but the real mother of the baby agreed to give it up so that it might live. The baby, then, went to the person who loved it the most (with love being defined as the person who attached the highest existence value to the baby).

2. Consider a contest between two parties over an entitlement, in which the first is willing to pay more than the second, but the second is willing to fight to the death for it. Equation 2 suggests that the one who is willing to fight harder should get it.

REFERENCES

Dixit, A., & Pindyck, R. (1994). *Investment Under Uncertainty*. Princeton: Princeton University Press.

Farrow, S. (2001). Using Risk Assessment, Benefit-Cost Analysis, and Real Options to Implement a Precautionary Principle. Working paper.

Farrow, S., Ponce, R., Wong, E., Faustman, E., & Zerbe, R. O. (2001). Facilitating Regulatory Design and Stakeholder Paricipation: The FERET Template with an Application to the Clean Air Act. In: P. Fischbeck & S. Farrow (Eds), *Improving Regulation: Cases in Environment, Health and Safety*. Washington, D.C.: Resources for the Future Press.

Pearce, D. (1994). The Precautionary Principle in Economic Analysis. In: T. O'Riordan & J. Cameron (Eds), *Interpreting the Precautionary Principle*. London: Cameron May.

Zerbe, R. O. Jr. (2001). *Efficiency in Law and Economics*. Aldershot, England: Edward Elgar.

PART 3:
OPTIMAL USE OF ECONOMICS
IN POLICY MAKING

CAN LAW AND ECONOMICS STAND THE PURCHASE OF MORAL SATISFACTION?

Richard O. Zerbe, Jr.

ABSTRACT

In recent years there has been a debate over whether or not moral sentiments should be included in normative economic analysis. This paper compares the standard normative criteria for benefit cost analysis, Kaldor-Hicks, that does not include moral sentiments with a modification that does called KHZ. The choice between these criteria should rest on which is the most acceptable and useful. The conclusion is that KHZ dominates KH even by the standards of KH itself and that its use illuminates certain problems in environmental law and economics such as comparing projects with compensation and those without and whether discount rates should be used in evaluating the far future.

1. INTRODUCTION[1]

For over sixty years the practical criteria for economic efficiency, the Kaldor-Hicks criteria, has excluded, or at least not properly included, moral sentiments in normative economic analysis. In part this is due to historical reasons, in part to normal inertia associated with any academic discipline and in part because it

An Introduction to the Law and Economics of Environmental Policy: Issues in Institutional Design, Volume 20, pages 135–172.
© 2002 Published by Elsevier Science Ltd.
ISBN: 0-7623-0888-5

was until recently not greatly important to include them. This is no longer the case. Existence values often reflect moral sentiments and such values can arise with respect to traditional environmental amenities as well as historic buildings that might, for example, be damaged in an earthquake. This paper asks if the value of moral satisfaction and moral harm should be included in benefit-cost analysis and, if so, what welfare criterion should be used to measure it. The issue is important in environmental economics because questions of existence value and of compensation for harm are prominent and can involve moral sentiments.

I conclude that moral sentiments are legitimately included in normative economic analysis as their inclusion improves the quality of analysis, clarifies issues in environmental economics and potentially improves decision making. This conclusion rests on using a suitable metric, such as the one I suggest, that captures normative sentiments for normative analysis. The implications of this conclusion are that: (1) the potential compensation criteria should be dropped in favor of a variant of Kaldor-Hicks which will better capture values actually relevant to decision making; (2) standing issues in both law and economics should rely (aside from questions of separation of powers and federalism in law) on how well the relevant sentiments can be measured; (3) in principle, existence values as a type of moral sentiment should be included in benefit-cost analysis; (4) a project with compensation is different from the same project without compensation and should be so recognized in efficiency analysis; and (5) criticisms of the use of discount rates because they do not sufficiently recognize future values arise because moral harm is not considered and are unconvincing once moral sentiments are included.

2. HOW MEASUREMENT PROBLEMS AFFECT LEGAL RIGHTS

I wish to suggest that common law doctrine disallowing relief or collection of damages, as for example in awards of damages for emotional distress, can be best understood as a problem in measurement. Courts will disallow damages or standing to sue for damages or other relief where damages or the value of relief can not be reasonably measured except in certain outrageous cases. I suggest that in economics as in law the rule should be to include all sentiments reasonably measurable.

2.1. The Legal Concept of Standing

Legal standing is especially relevant to economic analysis because it can provide a reference point for reasonable expectations and psychological ownership. Legal standing gives plaintiffs a right to sue. Although the particular details of the law of standing can be arcane and complex, standing can be understood, aside from consideration of separation of powers, as arising from measurement distinctions.[2] The doctrine of legal standing is derived from the "case or controversy" clause of Article III of the U.S. Constitution. In *Lujan v. Defenders of Wildlife* (*Lujan II*), the Supreme Court articulated three constitutionally required elements that plaintiffs must demonstrate before they are entitled to legal standing. The Court held that plaintiffs must show that: "(1) [they] ha[ve] suffered an 'injury in fact' that is: (a) concrete and particularized, and (b) actual or imminent, not conjectural or hypothetical; (2) the injury is fairly traceable to the challenged action of the defendant; and (3) it is likely, as opposed to merely speculative that the injury will be redressed by a favorable decision" (*Friends of the Earth v. Laidlaw*, 528 U.S. 167, 180 (2000)), citing *Lujan*, 504 U.S. 555, 560–561 (1992)).

The standing doctrine, however, also embraces several judicially self-imposed prudential limits on the exercise of federal jurisdiction. Courts often deny third parties legal standing because they are unable to demonstrate they have suffered a concrete "injury in fact" or that their injuries are traceable to the defendant.[3] Courts also require that a plaintiff's complaint fall within the "zone of interests" protected by the law invoked.[4] Less often, courts deny parties standing when a decision by the courts would interfere with constitutional separation of powers[5] or raise grave federalism concerns.[6]

By limiting the parties who can bring an action before a court of law, the doctrine of standing helps courts guarantee that "the party whose standing is challenged will adequately represent the interests he asserts" (Blackmun, J., *dissenting* in *Sierra Club v. Morton*, 92 S. Ct. 1361 (1972)). As the Court explained in *Baker v. Carr*, this "concrete adverseness . . . sharpens the presentation of issues upon which the court so largely depends for illumination of difficult constitutional questions." (*Baker*, 369 U.S. 186, 204 (1962)).

All the reasons for denying standing, except for separation of powers and federalism concerns may be regarded as resulting from measurement issues that arise in law as well as economics.[7] Clearly, a right that is hypothetical or conjectural will be difficult for a court with limited institutional capacity to measure. Moreover, where the causation of injury or the likelihood of its redressibility is unknown, courts are unable to measure the benefits and costs to parties involved. The disfavor in which the courts hold third party suits – suits in which one claims to represent someone else's interest – may be interpreted as a problem in measuring the primary person's interest solely based on the third party's claim.

Courts have not yet fully articulated measurement concerns as the reasons for standing issues. They should. Despite the long history of the standing doctrine, the definition of "injury in fact" remains remarkably murky. This ambiguity may stem in part from a use, non-use distinction applied by courts that is also important in environmental economics.[8] The Supreme Court has consistently refused to recognize non-use moral sentiments as sufficient grounds for standing. For example in *Sierra Club v. Morton*[9] (1972) the Court, in denying standing to the Sierra Club to raise claims relating to the construction of a road and ski resort in Mineral King and Sequoia National Park, distinguished between "mere interest in a problem" and injury. While the court acknowledged that "Aesthetic and environmental well-being, like economic well-being, are important ingredients of the quality of life in our society . . . the 'injury in fact' test requires more than an injury to a cognizable interest." (*Id.* at 734–735.) In *Morton*, the Court opined that the doctrine of standing's "attempt to put the decision as to whether [litigation is pursued] . . . in the hands of those who have a direct stake in the outcome" would be "undermined were we to construe the APA to authorize judicial review at the behest of organizations or individuals *who seek to do no more than vindicate their own value preferences through the judicial process*." (*Morton*, 405 U.S. at 740 (emphasis added).)[10] Thus, because in *Morton* the Sierra Club did not allege that it or its members would be affected in a direct personal way by park development, the Court denied standing under the APA. In *Valley Forge Christian College v. Americans United for Separation of Church and State, Inc.* (454 U.S. 464, 486 (1982)), Justice Rehnquist also dismissed the idea that standing could be based solely on injury to plaintiff's interests or moral sentiments. Rehnquist stated, "standing is not measured by the intensity of the litigant's interest or the fervor of his advocacy." (*Id.*) In many cases, when moral sentiments are fully recognized the distinction between mere interest and injury in fact disappears. The distinction is problematic because a court's decision to grant or deny standing based on a plaintiff's category of injury (mere interest or injury in fact) rather than on the measurability of a plaintiff's claim often leaves society without judicial remedies even when, in theory, those remedies could be efficiently distributed.

Judicial resolve to deny standing based on "mere interest," however, may be faltering, opening the way for a measurement-based theory of standing. In *Friends of the Earth v. Laidlaw Environmental Services* (528 U.S. 167 (2000)), the Supreme Court was willing to grant standing to the plaintiff on the basis of claims of injury that some justices found vague (Scalia, J., & Thomas, J., dissenting). In *Laidlaw*, the record of actual loss of use to FOE members was spotty. Moreover, FOE's claims were made in the face of a finding that no environmental harm occurred, on the basis of a subjective evaluation of environmental quality and individual fear of possible harm rather than actual harm.[11] The treatment of interest in the *Sierra Club v. Morton* and *Laidlaw* appears inconsistent. The court's approach in *Laidlaw* may indicate a greater willingness to give plaintiffs standing based in part on their non-use moral sentiments.

A distinction based on measurement differences would be less likely to suffer from illogical discontinuity. From a legal perspective, the question I should raise is whether or not a distinction based on measurement is better than one based on the type of injury – mere interest versus what the courts call economic harm, but which might more accurately be called market harm. If the distinction based on the type of injury fails to give standing to important sources of harm that are reasonably measurable, then failure to offer standing to those whose harm is based on moral sentiments will be inefficient.[12] In *Lujan II*, Defenders of Wildlife presented a series of nexus theories of injury. They argued that the "ecosystem nexus" "proposes that any person who uses *any part* of a 'contiguous ecosystem' adversely affected by a funded activity has standing even if the activity is located a great distance away," that the "animal nexus" theory would allow "anyone who has an interest in studying or seeing the endangered animals anywhere on the globe" standing; and that the vocational nexus approach would allow suit by anyone with a professional interest in such animals. (*Lujan II*, 504 U.S. at 565.) Finally, DOW presented the "vocational nexus" approach, "under which anyone with a professional interest in such animals can sue." (*Id.* at 565.) Ultimately, the *Lujan II* majority agreed that the Defenders of Wildlife did not present sufficient evidence of an injury in fact, since they only claimed an intent to visit affected sites at some indefinite future time, at which time they might be unable to observe endangered animals. Yet, Justices Kennedy and Souter, in their concurring opinion, were "not willing to foreclose the possibility that . . . a nexus theory similar to those proffered here might support a claim to standing" on a different set of facts (504 U.S. at 578) a possibility Justice Blackmun and O'Connor echoed in their dissent (504 U.S. at 592).

The ability of the courts to solve these problems in legal standing doctrine is a modern day test for the evolution towards common law efficiency. I suggest that a focus on measurement can help. I suggest that a party should have standing to have his or her gains or losses considered in a legal decision when the weight of public opinion, in the form of the regard for others, supports granting their gains or losses standing and those gains or losses are reasonably measurable (Zerbe, 2001b).

When determining whether a party can bring a claim for damages before the court, the threshold question, from a remedies perspective, is whether a party is entitled to damages at all. This is essentially a question of whether a court should grant the party standing. Courts hold that there cannot be recovery in the form of damages unless there has been an injury that the law recognizes. If there is an injury the law recognizes, the court will award compensatory damages in order to restore the party to the position he would have been in, had he not suffered the unlawful injury. The rule of compensatory damages most closely resembles WTA, rather than WTP. To determine the position a person would have been in but for an unlawful injury, one must recognize the substitution effect, income effects, and loss aversion.[13] In reality, courts frequently use measures of damages that are closer to WTP than WTA.

The explanation for the use of WTP rather than WTA is that the WTP is easier to measure. Courts usually employ market value as a measure of damages when that measure is a reasonable estimate of their value to the owner, as in the case of goods that are held for resale or exchanged in the course of business. Courts are less likely to use market value when the owner bought those goods for his own use or convenience.

While courts do not use market value for non-commercial goods, they fail to employ a 'pure' WTA measure. With few exceptions, courts have refused to provide damages for tort or contract[14] claims based purely on emotional damage or distress. Thus, they have generally refused to recognize sentimental value. This appears to be due to difficulty in measuring sentimental value. One exception to the rule denying a claim of sentimental value is found in *Campins v. Capels*. In the case, the plaintiff's home had been burglarized resulting in the theft of a wedding ring and three 'national racing championship rings.' The defendant had received the rings from thieves and then melted them down so that he could recast them and sell the new jewelry. The sentimental value claim was only for the championship rings. After noting that the plaintiff's 'sentimental' attachment was neither 'mawkishly emotional' nor an 'unreasonable emotional' attachment, the court acknowledged that there was no 'readily available' market for the buying and selling of championship rings. Thus, virtually all of the rings' value stemmed from their emotional significance. The court appeared to feel that the jury could put reasonable sentimental value on the championship rings; such value was not dependent on claims of sentiment towards other people which might be difficult to evaluate but on a value which most could recognize. The court held that the plaintiff could not demonstrate sentimental value by arguing that he would not have sold the rings at some price.[15]

2.2. Measuring Damages to Moral Sentiments: Intentional Infliction of Emotional Distress

Emotional damages or damages to moral sentiments are notoriously difficult for courts with limited institutional capability to measure. Since measurement presents such a dilemma, courts have employed judicially created standing mechanisms that funnel only the most deserving cases into the courtroom. The most common claim of damage to moral sentiments, the tort of intentional infliction of emotional distress, illustrates the courts attempts to cope with measurement issues. The tort of intentionally inflicting emotional distress is defined as outrageous conduct by the defendant intentionally causing or recklessly disregarding the probability of causing emotional distress.[16] Because conduct must be outrageous, only cases that shock the public's collective conscience are given standing. This limits cases for emotional damages to those where the regard for others would stand firmly behind allowing recovery. Moreover, in many jurisdictions, plaintiffs must show physical injury before recovery for emotional distress is allowed by the court. When plaintiffs claim both physical and emotional damage, the court's measurement task is less onerous because physical damage provides a quantitative base by which to gauge emotional damage. Furthermore, claims involving some element of physicality are usually more egregious than claims involving only emotional damage, and thus physicality may act as a screen to rid the court of frivolous or the least important claims.

Despite these judicial attempts to limit cases involving emotional damages to only the most deserving and easily measured, the system is still fraught with measurement problems. In cases involving emotional damages, the task of quantifying emotional damages falls to the jury. However, as Cass Sunstein (1992) relates, in intentional infliction of emotional distress cases,

> Monetization is extremely difficult. Significant arbitrariness is entirely to be expected. Similar cases may well give rise to dramatically different awards. How does a jury know what amount would provide an employee, or a student, with adequate compensation for quid pro quo . . .?[17]

The jury's determination must be made in a virtual vacuum, since no comparative valuation is allowed. Moreover, the jury's valuation problem is exacerbated by Judges' unwillingness to allow experts to address measurement of emotional damages before the jury. This unwillingness to allow expert testimony on the subject of emotional damages also spans to punitive damages. As court in *Voilas v. General Motors Corp.* holds,

> Extending the rationale of the opinions in these other analogous areas leads to the logical conclusion that expert testimony on punitive damages is neither desirable nor necessary, and indeed, would invade the sacrosanct role of the jury. The Court is satisfied that the trier of fact in the present matter is the proper, and only, quantifier of punitive damages; in essence, there is no science or methodology to awarding such damages where the jury can evaluate . . . and determine how to punish accordingly.[18]

Courts reluctance to allow juries valuation tools is ironic, given their careful attention to giving only the most deserving and measurable cases standing.

3. THE CONCEPT OF STANDING IN ECONOMICS

Whittington and Maccrae (1986) explain the concept of economic standing as
a method of describing who is able to have their values count in economic
analysis. This seems to be a contradiction to the prevailing if unarticulated
notion under the standard Kaldor Hicks (KH) criteria that every person has
standing to have their values counted in an economic analysis.[19] Elsewhere
Zerbe (1991) has shown that legal rights can be used to determine economic
standing as well and can be taken as pragmatic indicators of general sentiment.
Under any paradigm, economic standing is determined in the first instance by
ownership. Ownership is usually not difficult to determine. Ownership defines
the reference point and in economic analysis entitles negative changes from
that point to be treated as losses and measured by the willingness to accept
payment for a change (WTA), and positive changes to be treated as gains and
measured by the willingness to pay to acquire the goods or rights (WTP).

Under KH an action is said to be efficient if the total WTP to acquire the
goods or rights resulting from that action exceeds the total of the WTA payment
in exchange for goods or rights given up and if the potential winners could
hypothetically compensate the potential losers. Anyone affected by a proposed
action will have either a WTP or WTA so that implicitly all affected persons
are arguably to be included in the economic efficiency analysis. The issue of
excluding certain people or sentiments did not traditionally come up.

Except that KH is not so inclusive in practice. A notable example is the treatment of foreigners in the context of benefit-cost analysis. Typically benefits
and costs imposed on foreigners are ignored (Harberger, 1978). So in practice
foreigners do not usually have economic standing in benefit-cost analyses. The
standard best practice is to conduct a benefit-cost analysis from a national
perspective. But in practice when one jurisdiction, a city or a state, is conducting
its own benefit-cost analysis, it ignores effects on other jurisdictions unless they
are strong enough to involve consequences for the original jurisdiction.[20]

In spite of common practice, the general rule that follows from the logic of
the KH criteria is that all goods for which there is a WTP are economic goods.
Questions of economic standing are really questions of how to count values,
i.e. problems of measurement. For example, for many projects the value of
information gained from determining distributional effects will not be worth the
cost. While economics can justify ignoring distributional effects in many
instances, I suggest it should do so on measurement grounds instead of principle.

Under KH, distributional effects are not usually counted. The KH assumption
is that compensation to those harmed by a project and changes to the income
distribution are to be ignored as part of efficiency analysis.[21] If equity is

considered at all, and usually it is not, it is considered separately from efficiency. This has led to a failure to recognize equity as just another economic good and a separation between efficiency and equity that is mistakenly based on principle rather than convenience of practice.

As economic analysis has extended to a broader variety of subjects and a broader array of values, issues of standing have grown apace. In a policy affecting the costs of immigration, should the benefits the children of illegal aliens receive from public education in the U.S. be included in an analysis of educational policies? Should the benefits criminals receive from crime be counted in considering policies affecting criminal behavior? Should existence value of environmental amenities be counted?

Legal standing can often be used to inform economic standing and resolve measurement issues. For instance, where legal and illegal behavior arises, the law can be used as a proxy for a more extensive measurement. For example, Zerbe (1998, 2001a) suggests standing should be denied to thieves to have the value of stolen goods to them count, except where the law of theft is itself being considered (Zerbe, 1991, 2001a). It is not the illegality per se that matters in economics. Where illegality is clear, and it can be taken as an expression of sentiment of a broad part of the population, then the denial of standing to thieves is just an efficient way of summarizing the results of a broader poll. That is, if one is considering whether or not thieves should be required to return stolen goods, even though they are worth as much to them as to the owners, one could estimate that were the sentiments of the population polled broadly, the result would favor denying thieves standing and returning the goods. Similarly, in considering whether or not it is worth spending additional money to prevent a murder, one may be safe in ignoring the value of murder to the murderer. This is unlikely to be greater than the loss of others opposed to murder. The failure to adopt this approach leads Posner into untenable positions (see Zerbe, 2001a).

Standing concerns which sentiments should count in economic analysis. This issue is the subject of this essay. The questions to be considered are: (1) How are moral sentiments to be treated? (2) Are projects with compensation different from those without? (3) Can harm arise when it is unknown to those affected? and (4) most importantly, should an aggregate welfare measure such as KHZ be used instead of KH?

4. HOW SHOULD MORAL SENTIMENTS BE VALUED?

Moral sentiments are those that concern the welfare of other creatures. To ask if moral sentiments should be included in normative analysis seems peculiar;

on the face of it moral sentiments would seem to be a crucial ingredient. The rule suggested here is to include all sentiments reasonably measurable. Yet important parts of such sentiments, the income distribution and compensation for harm, have historically been omitted from efficiency analysis. Similarly a variety of theoretical as well as practical arguments have been advanced for ignoring existence values (Hausman, 1993).

The argument for including moral sentiments in benefit-cost analysis is developed here by considering the arguments that have been made against such inclusion and the advantages for inclusion. The basic test used is to ask which among choices of normative measures of value appears to be the most useful and acceptable. The conclusion is that a version of an aggregate WTP–WTA measure holds the most promise. This measure I have called KHZ (Zerbe, 2001a).

The KHZ criteria is met when the sum of the WTP's for a change exceed the sum of the absolute value of the WTA's. KHZ assumes simply that: (1) all values count, or more precisely all goods and sentiments for which there is either a WTP or a WTA are economic goods; (2) gains and losses are to be measured by the WTP and WTA respectively and from a psychological reference point that is determined largely by legal rights; and (3) transactions costs of operating within a set of rules are included in determining efficiency. The rationale for these assumptions may be found in Zerbe (2001a). Here I focus on the rationale for counting all values.

5. ATTACKS ON THE INCLUSION OF MORAL SENTIMENTS

Six arguments have been advanced against inclusion of moral sentiments in benefit-cost analysis. It is said that the inclusion of moral sentiments results in: (1) the acceptance of projects that fail to pass a potential compensation test (PCT) (Winter, 1969; Milgrom, 1993); (2) different weights being given to different individuals (Quiggin, 1997); (3) the inclusion of purely redistributive policies (Quiggin, 1997); (4) the inclusion of undesirable sectarian sentiments (Quiggin, 1997); (5) double counting (McConnell, 1997); and (6) measurement problems. I consider these in turn.[22]

5.1. Failure of the Potential Compensation Test

The strongest argument against inclusion of moral sentiments in calculations, at least to economists, will be concern that such inclusion leads one to accept projects that do not pass the PCT (Milgrom, 1993). The PCT requires that winners from a project are hypothetically able to compensate the losers while retaining some of their gains. No actual compensation is required.[23]

Assume with Milgrom that one party, B, cares about transfers to person A but that A does not care about transfers to B. The recognition of B's moral sentiments can suggest acceptance of a project with a positive net surplus that does not, however, pass a PCT. Suppose that A gains from the project and B loses. A hypothetical transfer from A to B would result in B hypothetically gaining less than A loses. B's moral sentiments cause him to lose from A's loss, so that B can not in fact be hypothetically compensated. For example, consider a project which costs $160 and from which A has a gross gain of $100 and bears none of the costs. Her net gain is $100 and B has a gross gain of $100, but pays the entire $160 for a loss of $60. The net surplus from the project will be $40 ($100 – $60). Suppose, following Milgrom, that B's gain from the project is given by $50 + 0.5 times A's net surplus. If A hypothetically were to transfer $60 to B as compensation, this would hypothetically fail to do the job as the $60 hypothetical gain to B is partly offset by A's loss. B's hypothetical loss is now $30, as B loses $30 from A's loss of $60. If B then transfers an additional $30, B's loss would fall to $15, A would have a surplus of $10 and the project no longer has a positive net surplus. The actual rather than hypothetical transfer of money would result in both a failure to pass the PCT and to achieve positive net benefits.

The lesson Milgrom draws from this is that one should exclude moral sentiments from benefit-cost analysis. There are several difficulties with this argument, the strongest of which is that it is incorrect as a practical matter.[24] Another is that the KH measure does not assure passage of the PCT.

5.2. Under What Conditions Will The Potential Compensation Test be Failed

Without loss of generality consider a pure distributive project.[25] In a move that transfers Y to users from altruists users gain Y and altruists lose $Y - \alpha Y$, where α represents the percentage value altruists place on users welfare. Zerbe (2001b) shows that no failure of PCT will occur where $\alpha > 1$.

He also shows that α is greater than 1. Redistribution is a public good (or bad). The term α is not the value that an altruist gives to the user's income which I assume will always be less than 1. Rather it is the summation of the weight over all affected altruists. If 160 altruists are charged $1.00 each to transfer $160 to the users they will value each dollar transferred by some percentage less than 1. If this is for example 10%, α will be 16 and αY will be $2,560. The weight given over all individuals is the sum of each affected individual weight. This will be greater than 1 as long as the number of people times this weight is greater than one.

Milgrom's example is essentially 2 person and he assumes an α less than one. Suppose instead that α is > 1.5. It is then impossible to construct his example. For example, if the cost of the project in Milgroms' example is $160 and α is say greater than 1.7, then both users and altruists gain so that the PCT does not arise.

5.3. KH Does Not Pass the Potential Compensation Test

Moreover, using KH itself does not assure passing the PCT. The standard benefit-cost test, the Kaldor test, sums the compensating variations (CVs). For almost twenty years it has been known that a positive sum of CVs is a necessary but not a sufficient condition for the passage of a compensation test (Boadway & Bruce, 1984). So projects evaluated according to the traditional criterion may not pass the compensation test. Symmetrically the Hicks' test, the sum of equivalent variations (EVs), is a sufficient but not a necessary condition for passage of such a test. To use the Hicks' test will result in rejection of projects that do pass the compensation test. KH is then an imperfect guide to passage of compensation tests.

More importantly, Baker (1980, p. 939) has pointed out a more interesting and legally relevant failure of the KH to pass a PCT. He notes that when rights are in dispute, the usual case in matters at law, the sum of the expectations of the parties will normally exceed the value of the right so that no potential compensation is possible. For example, suppose a piece of property worth $120 to Ronald and $100 to Richard is in dispute between Richard and Ronald but each believe with 80% probability that they own the property. The total value of expectations is $176 and the winner could not in principle compensate the loser. If the property is awarded to Ronald, he has a gain of $24 and Richard a loss of $80. There is no gain from which to potentially compensate Richard. As long as the sum of expected values is greater than the actual value the project can not pass the PCT. The inability to determine the efficient allocation is an indictment of the PCT but not of a variant of KH that abandons the test.[26]

5.4. Weights

Quiggin (1997) notes that giving people credit for altruistic sentiments means giving them additional weight in a social welfare function. The benefit-cost analysis would involve "the application of unequal weights to different individuals depending on the extent to which they are objects of altruistic concern. ... Yet benefit-cost analysis is based on the idea that benefits should count equally, no matter to whom they accrue" (1997, p. 149).

I would say rather that the assumption in benefit-cost analysis is that the marginal utility of income is the same for all. The adoption of the assumption of equal marginal utilities of income as part of the Kaldor-Hicks criteria arose to avoid interpersonal comparisons. Unequal weights that arise from altruism do not violate this assumption. When the demand for oranges increases, oranges receive more weight in a benefit-cost analysis that concerns oranges but this is no different from the increased weight received by the object of moral concern when the demand for moral satisfaction increases. Since the weights arise from WTPs, that is endogenously, they are as other goods.

5.5. Purely Redistributive Policies

Quiggin correctly points out that the adoption of an aggregate WTP–WTA criterion results in the inclusion of purely redistributive projects. He then notes the objection that benefit-cost analysis is rarely applied to purely distributive projects. This is true but as Quiggin notes, and as I noted earlier (1991), such inclusion can improve benefit-cost and policy analysis and is required by any criteria based on aggregate WTP–WTA such as KHZ. If our standard is usefulness and acceptability, historical practice is a weak reason to fail to adapt.

The KH criteria arose out of discussions among prominent British economists during the late 1930s.[27] Before this time it was generally assumed that each individual had an "equal capacity for enjoyment" and that gains and losses among different individuals could be directly compared (Mishan, 1981, pp. 120–121; Hammond, 1985, p. 406). Robbins (1932, 1938), disturbed this view by arguing that interpersonal comparisons of utility were unscientific. Economists accepted this and attempted to develop a welfare measure that would avoid interpersonal comparisons and which was more broadly applicable than Pareto efficiency. Kaldor (1939, pp. 549–550) acknowledged Robbins' (1938, p. 640) point about the inability to make interpersonal utility comparisons on any scientific basis, but suggested it could be made irrelevant. He suggested that where a policy led to an increase in aggregate real income;

> ... the economist's case for the policy is quite unaffected by the question of the comparability of individual satisfaction, since in all such cases it is possible to make everybody better off than before, or at any rate to make some people better off without making anybody worse off.

Kaldor goes on to note (1939, p. 550) that whether such compensation should take place "is a political question on which the economist, qua economist, could hardly pronounce an opinion."

Thus, it came to be thought that including considerations of the income distribution or of compensation would involve interpersonal comparisons, that

such comparisons could be avoided by excluding considerations of compensation or of the income distribution, and that the measure of efficiency could be
made more scientific. Yet, in their desire to avoid interpersonal comparisons,
economists failed to make their analysis more scientific and instead created
additional problems. First, any measure of efficiency that is normative is based
on moral assumptions. It can only be scientific in the sense of being consistently
applied. Second, the test simply assumes that all individuals' gains and losses
are to be treated equally and receive a weight of one and that any change in these
weights should be made by politicians or non-economists. So there exists within
the criteria interpersonal comparisons. Yet let us suppose that giving a weight of
one and leaving further value judgments to decision makers does have an appeal
of impartiality. To include compensation or income distribution considerations
does not diminish this impartiality nor require any other weight than equal weight
for individuals values. Ask yourself if there are cases in which one would be
willing to pay to affect some degree of compensation for the losses of others.
Since the answer for some will be yes, then the enactment of compensation is
itself an economic good. There will be a WTP or WTA for the purchase of this
good. These WTP and WTA measures will receive a weight of one across different individuals just as is done for other goods. Thus, to include compensation
or changes in income distribution as economic goods requires no interpersonal
comparisons except the requirement already a part of KH to treat all equally.

5.6. Sectarian Preferences or Immoral Sentiments

Moral sentiments can be negative as well as positive. Quiggin notes, "the unattractive policy implications of this conclusion may be illustrated by the case
when some individuals have sectarian preferences, characterized by altruism
toward members of the their own racial or religious group and zero or negative
altruistic WTP for others" (Quiggin, 1997, p. 151). It is at least ironic that
the most extended and harshest criticisms of economic efficiency outside the
profession are its failure to consider moral sentiments. Elsewhere (1998) I have
argued that benefit-cost analysis should take preferences as they are, and have
considered the implications at length (Zerbe, 1998, 2001a). This sort of problem
is considered extensively in the context of criticism of utilitarianism or of benefit-
cost analysis at length in the philosophy and legal literature (Dworkin, 1980,
1986; Posner, 1981; Smart & Williams, 1973). But this sort of problem does not
arise necessarily from moral sentiments but just from "bad utility", that is from
values at variance with the norm. These issues are thus raised in considering
matters unconnected with moral sentiments such as how to value goods in the
hands of criminals or the WTP by consumers of illegal drugs and so forth, as

well as by many problems of moral sentiments. The protection against the weight of immoral sentiments is the weight of moral sentiments. Quiggin finds this "by no means a straightforward solution" (1997, p. 151). I disagree.

First, legal rights will determine the starting point for the determination of whether a WTP or WTA measure will be used. To the extent to which existing rights grants a WTA measure as the cost of an immoral arrangement such as slavery, the costs will be very large in economic measurement.

In addition, it is reasonable to take the law determining rights, where it is clear, as furnishing information that serves as a shortcut for a weighting of values. Thus the law against theft may be taken to imply that the value of goods in the hands of the thief is zero. It is not that the thief's valuation of the stolen good is zero but that she has no right to it so that in a full analysis the WTP of those who would deny value to the thief will be greater than the value of the goods to the thief. Society has a WTP (or rather a WTA) to find theft illegal. Now when, however, one is considering whether or not theft should be illegal then the values of criminals in this matter should be explicitly counted as the law is in question.[28]

Benefit-cost analysis is not an efficient instrument for determining the "right thing to do" nor even for determining the decision but is rather a tool for furnishing information for the decision process. To not count bad values, without, as it were, legal authorization, is to substitute values of the analyst and to dilute the reliability of the analysis.

5.7. Double Counting

McConnell (1997) suggests that whether or not existence value should be included depends on the motivation for the sentiments. Diamond and Hausman (1993) similarly show that, for the type of altruism McConnell calls non-paternalistic, to include existence value is to double count. This appears to be a mistaken interpretation. What these authors have shown instead is that the size of existence value will depend in particular on the availability of substitutes and this will vary with the type of moral sentiment.[29] Existence values should always be included but may be close to zero in cases where substitutes are abundant.

Following McConnell (1997) suppose there are two groups, users and altruists and that users are not altruists nor altruists users. A project will be desirable when

$$W = NG_u + NG_a > 0 \qquad (1)$$

where NG_u and NG_a are the net gains to the users and altruists respectively and W represents the overall welfare gain. McConnell then shows that where one cares about the general welfare of others, (non-paternalistic altruism) the moral sentiments of the altruists, that is NG_u, should not be included so that the relevant condition is just

$$W = NG_u - Ca > 0. \tag{2}$$

Where Ca are those costs borne by the altruists.

This conclusion implicitly but mistakenly assumes that moral satisfaction may be purchased at no cost other than the money transferred. In calculating WTP from compensated demand functions, the relevant expenditure functions are a function of a vector of prices of substitute and complementary goods. Where the altruists care about the general well being of other persons defined by those persons own preferences a relevant substitution is direct compensation. Assuming direct compensation may be made at zero costs, the demand function is perfectly elastic and the existence value is zero so that McConnell's conclusion holds that existence value should be ignored.

But perfect substitutes are rarely available. In general the existence value for the altruists will be the cost of achieving the same gain to the users through the best available alternative project (say direct compensation). The consumer surplus to the altruists will be limited by the costs of carrying out compensation or by the extent of their altruism. Suppose the relevant alternative is the direct transfer of cash. In transferring cash to these consumers there will be a loss due to the costs of determining the injured, the extent of injury, and the costs of carrying out the actual transfer. We will call this cost ϕ. Existence value will then equal:

$$NG_a = NG_u/(1 - \phi) - NG_u - C_a = NG_u \phi/(1 - \phi) - C_a \tag{3}$$

Where C_a is the portion of the project's costs borne by the altruists.

The total value of the project will be the net gain to the users plus the existence value, or

$$W = NG_u/(1 - \phi) - C_a \tag{4}$$

When the cost of providing a substitute good, ϕ, is zero, Eq. (4) reduces to McConnell's condition shown above in Eq (2).

Where ϕ is greater than 1, there is no possibility of gaining value from a cash transfer. In this case substitute projects are not available so the existence value is given by the altruists willingness to pay, usually called the warm glow parameter, times the net gain. If α is the aggregate warm glow parameter, the net gain to the altruists will be αNG_u. In general the existence value can never be greater than the smaller of αNG and $NG_u[\phi/(1 - \phi)]$. The total value of the project will be the net gains to the users plus the existence value. This will be the smaller of

$$NG_u/(1 - \phi) - C_a$$
$$NG_u + \alpha NG_u - C_a = NG(1 + \alpha) - C_a \tag{5}$$

The fact is that existence value should always be considered and should be included when it can be measured but that in those rare cases where perfect substitutes are available it will be zero. It is easy to see also that when the cost parameter is high, for example greater than 1, and the number of altruists large, existence value can be huge.

5.8. Measurement

One may legitimately point to problems in measuring moral sentiments. The courts have denied standing to sue in certain environmental cases in which moral sentiments were at issue and have generally disallowed claims based on sentimental values or emotional distress. I suggest these exemptions are due to costs of measurement.

The distinction among individuals and among sentiments that needs to be drawn, both in law and in economics, is between values that are reasonably measurable and those that are not. By reasonably measurable I mean measured sufficiently for the purposes at hand. Consider the value of property that the government proposes to take. The value of the property to the owner is measured as its market value both in law and in economics (Posner, 1986). In principle, however, the correct economic measure, as with any loss, is the owner's WTA. As a practical matter this will be difficult to determine.[30] To offer WTA measures will encourage holdup and unverifiable assertions of value. Thus, the panel of experts (Arrow et al., 1993, pp. 4601–4614) suggests that the "willingness to pay format should be used instead of the compensation required because the former is the conservative choice" (p. 4608), and this is the rule that is followed in practice.[31] It is arguably correct that the WTP measure should be used as a matter of practicality, not because it is the conservative choice, but because measurement problems are liable to be more severe for the WTA.

Measurement considerations, however, do not justify excluding moral sentiments as a class in principle. Such sentiments are not necessarily unmeasurable. For practical considerations measurement is needed only to the extent relevant for policy decision and this need not be full or very exact measurement. For environmental goods, contingent valuation studies have been used to determine values that often involve moral sentiments, as in the case of existence values. These can be applied also to other values such as the value of compensating those who suffer a project's costs. Harberger (1978) has suggested a defensible method of determining the value of moral satisfaction that is consistent with traditional benefit-cost valuations in situations with income distribution changes where there are available perfect substitutes. The maximum value of a change is limited to the costs of achieving the same change by the most efficient alternative mechanism.

6. ADVANTAGES OF INCLUSION OF MORAL SENTIMENTS

6.1. The Analysis Of Legal Entitlement

To analyze the example of Ronald and Richard in the previous example under KHZ, first consider legal rights. Suppose first that legal rights are clear and that Ronald in fact legally owns the property so that Richard's claim is mistaken or fraudulent. An analysis built on legal rights will then find that the award of the property to Ronald is legally correct and is economically efficient. There will be no gain nor loss except for transactions costs. The decision to reaffirm Ronald's legal right is worthwhile as long as the transactions costs are less than the $120 value to Ronald. That is, if no legal determination of rights occurs, the loss is $120. The decision to affirm Richard's right is justified under the KHZ measure but is not, arguably, under KH.

Suppose instead that the ownership right to property is unclear, and that both parties have a convincing prima facie claim. They each have a degree of economic or what may be better called, psychological ownership which represents their expectations.

Let P_R represent the ownership expectations of Ronald and P_r represent the ownership expectations of Richard. The gains to R and r are measured by the WTP and their losses by the WTA. The gain to R is R's WTA weighted by the extent to which he does have economic (psychological) ownership. Similarly for r, the right goes to R when:

$$WTP_R(1 - P_R) + WTA_R(P_R) > WTP_r(1 - P_r) + WTA_r(P_r) \qquad (6)$$

Where P_R and P_r represent expectations expected as probabilities. Eq. (6) can be expressed as:

$$WTP_R + (WTA_R - WTP_R)\,(P_R) > WTP_r + (P_r)(WTA_r - WTP_r) \qquad (7)$$

This is an interesting result as it says that the *divergence* between the WTA and WTP is relevant to the decision concerning whom should receive the entitlement. In a contest between two parties over an unclear entitlement in which the first party has the higher WTP but the second is willing to fight to the death for it, Eq. (7) suggests that the one most willing to fight should get it.

Suppose Ronald's WTA is $140 and Richard's is $200, while the WTPs are $120 and $100 as before and ownership remains at 80% probability for each person. The value to Ronald of receiving good title is his WTP of $120 plus 80% of the divergence between is WTA and WTP of $120, for a total of $136. The value to Richard is $180 so, even though Ronald has the higher WTP, the entitlement should go to Richard. Where there is no prior psychological ownership, Eq. (7) reduces to

$$WTP_R > WTP_r \tag{8}$$

That is, the right should be auctioned off.

When the economic ownership of both R and r are 100%, then (7) reduces to:

$$WTA_R > WTA_r \tag{9}$$

When R has the right and r does not, the condition is just:

$$WTA_R > WTA_r \tag{10}$$

Elsewhere I (2001a) show the value of these legal rules. They follow naturally from an aggregate WTP-WTA formulation based on rights but are difficult to justify based on KH.

6.2. An Aggregate Welfare Measure Dominates KH

The KHZ test dominates KH with respect to the PCT test. I ask would the move from KH to KHZ itself, pass the PCT test. The answer is yes. First, consider the problem suggested by Baker in which KH fails PCT when uncertain rights are considered. Suppose that no moral sentiments are involved. KHZ counts efficiency gains ignored by KH so there is a net positive WTP to move from KH to KHZ, and the PCT is satisfied.

Similarly, when one considers moral sentiments there will be a net positive WTP to move from KH to KHZ as long as the *sum* of the value that all others attach to a wealth transfer from altruists to users is greater than 1 and the waste from the transfer is less than 100%. The sum is found by multiplying the number of altruists by the weight each gives to the value of a transfer to a user. If each altruist gives a weight of say 1/10 of one cent, than the sum of the altruistic value over all altruists is the number of altruists, N, times the 0.1 cent. When N is greater than 1000, the value of a dollar of wealth transferred would be greater than 1. When this issue is considered for society at large, N will be large, and a decision by society to move from a KH to a KHZ test would pass the PCT. Milgrom's example is essentially 2 person and he assumes an α less than one. Suppose instead that α is > 1.5. It is then impossible to construct the Milgrom example so that there are losers as long as the transfer is from altruists to users and the costs are less than αY. If, for example, the costs are \$160 and α is 1.7 the project which transferred \$160 to users at a cost to altruists of \$165 would produce a net benefit of \$272 to the altruists and of \$160 to the users.

Finally, it is easy to show that when moral sentiments are ignored, one may choose a project which passes PCT, not counting moral sentiments, but whose NPV is negative, counting moral sentiments, over one which is has a positive NPV even if it does not involve moral sentiments. This does not seem a good choice.

6.3. Are Projects with Compensation Different?

No criticism of the KH criteria is more widespread than that they neglect distributional effects. The views of the former Solicitor General of the United States, Charles Fried (1978, p. 93f) are representative. He sees the economic analysis of rights as using a concept of efficiency that is removed from distributional questions. He holds that economic analysis does not consider whether the distribution is fair or just. He then concludes from this that the fact that a given outcome is efficient does not give it "any privileged claim to our approbation" (1978, p. 94). The view that efficiency is unconcerned with distributional issues, or with fairness (e.g. 1984), is widespread in both law and economics. Economists generally pay little attention to criticisms from outside the profession. If acceptability of our criteria is important, however, this is a mistake.

Consider the following example:

> I give you the choice between 2 projects. One involves adding an additional bridge to abate traffic congestion around Seattle. No moral sentiments are involved. The other improves access to the Warm Springs Indian Reservation also around Seattle. The improved access will increase jobs on the reservation where unemployment is now running at 35%. The residents of Seattle care about the Native Americans on the reservation: in particular they gain half the net gain in moral satisfaction of the Native Americans. The residents also gain from their own improved access to the reservation. Their WTP is \$50 plus one-half of the gains to the Native Americans. The Native American WTP is just \$100. Table 1 below compares the two projects **when moral sentiments are ignored**.
>
> Under KH and PC the Native American project is superior since moral sentiments can not be counted. Suppose, however that the actual breakdown of benefits and costs for the NA project is as shown below in Table 2.

The Native American project is actually a very bad idea for the Native American population since they bear all of the costs and have a loss from it. Thus, in the example at hand, to exclude moral sentiment leads to acceptance of an inferior project that offends moral sentiment. Moreover to include moral sentiments makes it clear that with a different distribution of costs the Native American project is superior to either of the firsts choices as shown below.

To ignore moral sentiments is to ignore information that affects welfare. As I have shown, to ignore moral sentiments can lead to the acceptance of projects that reduce welfare although they pass the PCT and to acceptance of inferior over superior projects.

Table 1.

BRIDGE PROJECT		NATIVE AMERICAN PROJECT	
Benefits	$149	Benefits	$150
Costs	$140	Costs	$140
NPV	+ 9	NPV	+ 10

Table 2.

NATIVE AMERICAN PROJECT

	Native Americans	Altruists (Seattle Residents) [Not Counting Moral Sentiments]	Altruists (Seattle Residents) [Counting Moral Sentiments]	Total Not Counting Moral Sentiments	Total Counting Moral Sentiments
Benefits	$100	$50	$30	$150	$130
Costs	−140	0	0	−140	−140
Total	−40	50	30	+10	−10

Table 3.

NATIVE AMERICAN PROJECT WITH DIFFERENT DISTRIBUTION OF COSTS

	Native Americans	Altruists (Seattle Residents) [Counting Moral Sentiments	Total Counting Moral Sentiments
Benefits	$100	$100	$200
Costs	0	140	140
Total	+100	−40	+60

Any efficiency criteria will find its justification in its usefulness and its acceptability.[32] KH has been reasonably acceptable because it has the potential to increase wealth for all parties. Some parties will gain from a project and lose from other projects. As long as gains exceed losses all parties can gain over many projects. Yet this moral justification for KH applies a fortiori to a measure that includes moral sentiments. That is society will be better off and would pay a positive amount to have a regime that included moral sentiments in benefit-cost analysis decision making over one that did not.

6.3.1. The Municipal Incinerator

Here, I am concerned with the analysis of two projects, identical except that one provides compensation in the form of mitigation and the other does not. Consider a benefit-cost analysis of the efficient location of a municipal incinerator. The decision to locate the incinerator in the poorest neighborhood on the grounds of economic efficiency will typically raise issues of environmental justice. It may then be said that KH leads to an unjust result. A typical simplified analysis may look as follows.

Table 4. A Standard Benefit-cost Analysis.

	Project with No Mitigation	Project with Mitigation
Cost of Land	$500,000	$500,000
Present Value of Operating Costs	$100,000	$100,000
Costs of Incinerator	$1,000,000	$1,000,000
Costs from Environmental Health Damage to Poorer Residents	$200,000	0
Costs of Preventing Environmental Health Damage	0	$200,001
Benefits (Savings From Not Using Other More Expensive Locations)	$2,000,000	$2,000,000
Total Benefits	$2,000,000	$2,000,000
Total Costs	$1,800,000	$1,800,001
Benefits – Costs	**$200,000**	**$199,999**

The standard KH benefit-cost analysis would choose the project without mitigation. But this result typically arises because the analysis has not taken into account the moral sentiments of those who care about the equity effects but are not otherwise affected. One might give standing to these sentiments of justice in Table 5 below by considering the moral objection that others have to the imposition of costs onto the poor. Such an inclusion is required by KHZ.

Table 5. Benefits and Costs When Moral Sentiments are Included.

	Traditional Analysis	KHZ analysis
	KHZ Analysis (no mitigation)	KHZ (with mitigation)
Cost of Land	$500,000	$500,000
Present Value of Operating Costs	$100,000	$100,000
Costs of Incinerator	$1,000,000	$1,000,000
Environmental Health Damage	$200,000	$200,000
Moral Harm to Current Generation	X	0
Benefits (Net Costs Avoided From Not Using Best Alternative Site)	$2,000,000	$2,000,000
Total Benefits	$2,000,000	$2,000,000
Total Costs	$1,800,000+X	$1,800,001
Benefits – Costs	**$200,000-X**	**$199,999**

This is shown in Table 5 where the WTP is by members of the community at large who do not live in the poorer neighborhood. The benefits are the net costs avoided by not building in the best alternative site.

The project with mitigation will be superior as long as the moral harm is greater than $1, that is $X > 1$. As long as others (those not directly affected) care sufficiently about the welfare of those harmed, that is as long as they care sufficiently about environmental justice, then the project with mitigation is superior. Since these sentiments about equity are part of KHZ, they should in principle be taken into account. Hence under KHZ, a project to locate the incinerator in a poor neighborhood without mitigation of damages to those residents who will suffer is different from the same physical project when the damages are mitigated. The same conclusion applies to products with and without compensation under KHZ. That is, under KHZ but not under KH, projects that involve no mitigation or compensation, are different from ones that provide mitigation only, compensation only or some combination of mitigation and compensation. These differences are obscured by the traditional analysis.[33]

7. STANDING AND THE DISCOUNT RATE PROBLEM

In benefit-cost analysis, future benefits and costs are discounted using an interest rate referred to by economists as the discount rate. There are a number of issues in considering the problem of discounting, particularly when discounting beyond the lives of the decision makers (Aherne, 2000). Here I am concerned with only one, the widespread criticism of the use of discounting in benefit-cost analysis on the grounds that it is unethical to discount the benefits to be gained and the costs to be borne by future generations (e.g. Parfit, 1992, 1994; Schultze et al., 1981). It is said that the utility of future generations should count equally with the utility of the present generation (Schultze et al., 1981; Pearce, 1989). For example, Parfit (1992, p. 86) argues that "the moral importance of future events does not decline at n% per year. . . ." This sort of criticism has been noted with favor by economists (e.g. Schultze et al., 1981; Pearce et al., 1989), lawyers (Plater et al., 1998, pp. 107–109), and philosophers (Parfit, 1992, 1994). Similarly Brown (1991) notes that ". . . discounting imperils the future by undervaluing it."[34]

Consider the following example of the sort of problem with which these critics are concerned:

> A nuclear project is being considered that produces benefits of about $65 billion at a cost of about $30 billion but, in addition, produces a toxic time-bomb that will cause enormous environmental costs sometime in the far future.[35] (I remove questions of uncertainty from this example). Suppose that current waste-disposal technology will contain this waste for 500 years after which it escapes its sarcophagus but will remain toxic for 10,000 years. The estimated cost of the future environmental damage in constant, year 2000 dollars will be about $16 trillion, about the size of the current U.S. GDP. The present value of these damages discounted at a 3% real social rate of time preference (SRTP), assuming the waste escapes at the first opportunity 500 years from now, is about $12 million, not insignificant, but far far less than the damage that will occur in 500 years and far too small to affect the results of the benefit-cost analysis. Discounting these damages then results in the project going forward as the benefits are said to exceed the costs by almost $35 billion.

It is said that this project would be unfair to future generations and on this basis it is argued that the use of discount rates is immoral.

A commonly proposed solution to the problem of unethical harm to future generations is to use low, or even negative, discount rates (e.g. Schultze et al., 1981) or not to use discount rates at all (Parfit, 1994). This sort of argument is, I believe, a moral plea about what our sentiments should be towards future generations, but not an effective statement about what or whether discount rates should be used. The proposed solution of using no or low discount rates is ad hoc and, if generally applied, will lead to other ethical problems – for example, the adoption of projects that give less benefit to both present and future generations.[36]

This argument for unacceptability is not based on the preferences of future generations, which one cannot know exactly, but on our own preferences, based on our empathy with future generations. Under KHZ one can give standing to moral sentiments about future generations as long as these can be reasonably measured. This allows a solution to the ethical dilemma of the discount rate problem that acknowledges ethical concerns as valid and seeks an ethical solution, while acknowledging the values that commend use of a discount rate. To use a discount rate below the rate at which people will trade off present for future consumption, i.e. a rate of time preference, will lead to economic inefficiency by justifying investment with insufficient returns. To use a rate that is too low attempts to cope with inequity by adjusting prices. The result is that an inequity appears to be an inefficiency.

The economic efficiency of the project will then depend on the sentiments of the present generation. For example, the present generation may feel that future generations should be free of problems caused by the current generation. Evidence from Kunreuther and Easterling (1992, p. 255) and from Svenson and Karlsson (1989) suggests that, at least as regards nuclear waste disposal, individuals tend to place a high weight on future consequences. On the other hand, the present generation may find that compensation for environmental harm is unwarranted, given a likelihood that future generations will be wealthier than the present one.

If the law grants those with moral objection to future damage a right (legal standing) to veto the project, the economic reference point incorporates this right so that the measure of the value of failing to provide compensation would be through the WTA. If there is no such right then the provision of compensation is a gain for those others who desire it and the correct measure is the WTP.

7.1. Discount Rate Example

Table 6 shows the KHZ solution to the discount rate problem. Where the current generation cares about compensation, the relevant choice set includes a project A, with compensation and project B, without compensation. For project B there will be a WTP (or WTA) associated with loss to the current generation from providing no compensation. This amount is the X in Table 6 and it is entered as a loss.[37] Where people care nothing about providing compensation, the same two projects now labeled C and D should be considered. In this case harm to future generations should not be counted at all – to have one's values counted is the issue of economic and legal standing.[38]

The KH value of the project is \$34.988 billion whether or not future generations are compensated.[39] KHZ departs from KH in this respect in that it requires us to consider four projects, one in which future generations are compensated and one in which they are not, and one in which the future generations have standing to have their values counted and one in which they do not. An interesting result of Table 6 is that when the current generation cares about compensating the future generation, the project with compensation, project A, is always superior to the project without compensation, project B. This is true even if the amount the current generation is willing to provide for compensation is less than the amount required, ignoring transactions costs. For compensation itself will just be a transfer, and its failure will generate some harm for the current generation as long as they care about compensation.

Table 6. A Discount Rate Problem Resolved.

When People Care About Harming Future Generations		When People Don't Care About Harming Future Generations		
	A The Project **With** Compensation (billions)	B The Project **Without** Compensation (billions)	C The Project **With** Compensation (billions)	D The Project **Without** Compensation (billions)
Present Value of Benefits	$65	$65	$65	$65
Present Value of Costs	–$30	–$30	–$30	–$30
Present Value of Harm to Future Generations	–$0	–$0.012	–$0	$0.00 (no standing)
Present Value of Compensation for Future Generations	–$0.012	–$0	–$0.012	–$0
Present Value of Ethical Harm to Present Generation	0	–X	0	0
Net Present Value (billions)	**$34.988**	**$34.988-X**	**$34.988**	**≥ $35**

To keep Table 6 manageable assume (reasonably) that the future generation fails to have standing only when the current generation does not care about the future generation. In project D the damage estimates to the future generation are excluded so discounting does not arise. In project C the future damage is counted but there is no loss of moral satisfaction associated with the failure to compensate the future generation. Table 6 is simplified in that it does not take into account the administrative costs of providing the compensation. The project with compensation will entail additional administrative costs. The project with compensation will be superior as long as X, the value to the current generation of providing compensation exceeds the administrative costs A. That is the project with compensation is superior as long as $X > A$.

8. THE ISSUE OF COMPENSATION

Compensation may be made by mitigation, replacement, restoration, or by reparations.

Suppose first that compensation of the future generation is possible through mitigation. Let $X(\overline{D} - C)$ be the present value of the moral damage the current generation suffers from knowing that future generations will suffer a loss with present value, D, damages for which they are provided compensation C. The effect of compensation C is to provide moral satisfaction. Moral damage or satisfaction may be general or particular. As noted earlier the more particular the higher will be existence value, ceteris paribus. I make the following assumptions:

$$X(D - 0) = X(D) = 0$$

$$X'(D - C) < 0, \ C < D$$

$$X''(D - C) > 0, \ C < D$$

$$A(0) > 0$$

$$A'(C) > 0$$

$$A''(C) < 0$$

The assumptions state first that satisfaction achieved by providing zero compensation is zero and that the value of the satisfaction from compensating the future generation rises as the level of compensation rises, but at a decreasing rate. Administrative costs involve some fixed costs even with zero compensation and marginal administrative costs increase with the level of compensation but at a decreasing rate. I assume that the moral damage suffered by the current generation may also be considered as potential gain from compensation.[40] If the present value of the damage done by a project is $X(\overline{D} - C)$, then the moral benefit of compensation may be thought of as a function of the level of compensation, C, given $\overline{D}$.[41]

For the project to be better than the status quo, it must also be the case that NPV is positive. Let NPV_0 be the net present value without considering harm to future generations. When the future generation has standing so that the estimate of their damage is counted and the current generation has standing for their moral harm, then, the compensated net present value, NPV_c, will be:

$$NPV_c = NPV_0 - D - A(C) - \overline{D} \tag{11}$$

That is, the net present value including consideration of the harm to future generations is equal to the NPV without such consideration, less the damage to future generations, the administrative cost of compensation and the residual damage to the moral sentiments of the current generation, $X(\overline{D} - C)$.[42] This will be positive as long as $NPV_0 > (D + A(C) + X(\overline{D} - C))$. The optimum level of C will be found where NPV is at a maximum. This level of compensation, C*, will occur where $A'(C) = X'(\overline{D} - C)$ subject to the usual second order conditions.

The project with compensation will be preferred to the project without compensation as long as

$$X(D) > A(C^*) + X(\overline{D} - C^*) \tag{12}$$

for some positive value of C. That is the moral harm from no compensation is greater than the administrative costs of compensation plus the residual moral harm after compensation.

NPV_c will be maximized at a level of compensation C^* which may be zero, representing no compensation, $\overline{D}$, representing complete compensation, or some intermediate level, representing partial compensation. Consider three cases. The first is where zero compensation is optimal. This will result if $X(C) < A(C)$ for all values of C. The costs of compensating a future generation outweigh the benefits, $C = 0$ and, assuming no infrastructure for compensation has been created (so that $A(0) = 0$),

$$NPV_c = NPV_0 - \overline{D} - X(\overline{D} - 0) \tag{13}$$

If infrastructure for compensation has already been established, zero compensation will be optimal if $X(C^*) < A(C^*) - A(0)$. It is no surprise that if the infrastructure for compensation is in place, it is more likely that compensation will take place. This is a non-trivial matter when considering the potential difficulties in compensating people living far in the future

The second case to consider is where there is full compensation for the future damage, so that $C^* = \overline{D}$. This will result if the marginal cost of increasing compensation from C to $\overline{D}$ is less than the marginal reduction in the damage to moral sentiments, subject to second order conditions which require that the NPV be positive.

In this case,

$$NPV_c = NPV_0 - A(\overline{D}) - X(\overline{D} - C) \text{ or}$$
$$NPV_c = NPV_0 - A(\overline{D}) \tag{14}$$

The final term, $X(\overline{D} - C)$ disappears here as $C = \overline{D}$ and there is no residual damage to the moral sentiments of the current generation. It might be that there is some discontinuous benefit from achieving full compensation relative to that achieved by, say, 99% compensation.[43] This possibility would make it more likely that the optimal compensation was 100%. The net value of full compensation may be greater than partial compensation.

Some insight is gained into the conditions necessary for complete compensation by considering the derivative of NPV_c at some level C_D near $\overline{D}$.

$$\left. \frac{dNPV_c}{dC} \right|_{C = C_D} = -1 - AC'(C_D) + X'(C_D) \tag{15}$$

If increasing compensation from C_D to $\overline{D}$ will increase a project's NPV this derivative must be positive. This suggests, if that $AC' > 0$, that $X'(C)$ must be positive as C approaches D, or that full compensation offers some important moral advantage over, say, 99% compensation so that a discontinuous increase in $X(C)$ occurs in moving from $(100 - \varepsilon)\%$ compensation to 100% compensation.

The third case is where some intermediate level of compensation, C*, is optimal. This situation may be seen as an internal solution to the problem

$$\max_{c} \text{NPV}_C = \text{NPV}_0 - C - A(C) - (X(D) - X(C))$$

$$\frac{d}{dC} = -1 - A'(C) + X'(C) = 0$$

$$X'(C) = 1 + A'(C) \tag{16}$$

In addition to this first order condition, the second order condition for maximization requires that $|X''(C)| > |A''(C)|$ if both second derivatives are negative, or that the marginal change in moral satisfaction from an additional dollar of compensation fall faster than the marginal cost of administering compensation. If it is assumed that $X(0) = 0$ and $A(0) > 0$, $X(C)$ must be sufficiently large for small values of C to overcome the gap before turning sharply toward horizontal. This is shown in Fig. 1 below.

If an internal solution is optimal, the NPV of the project will be

$$\text{NPV}_c = \text{NPV}_0 - C^* - A(C^*) - X(\overline{D} - C^*) \tag{17}$$

Comparison of these three cases leads to the observations that:

(1) If $X(\overline{D}) < A(\overline{D})$, no compensation is preferred to full compensation. That is, if the cost of full compensation outweighs the benefits to the moral sentiments of the current generation, then there should be no compensation.
(2) If $A(C) > X(C)$ for all C, no compensation is preferred to any compensation.
(3) If $X'(C) > A'(C)$ for all values of C, then full compensation is preferred, subject to the second order requirement that the NPV_c is positive.

Returning to Table 6, Project A will be superior to B as long as people care sufficiently about the future generation. Where people do not care, the project without compensation will always be superior.[44] Note that one is not required to determine the magnitude of the WTP of the current generation to compensate future generations. One is instead only required to determine that there exists some WTP for it that is greater than either the administrative costs of providing the compensation or the value of the damage itself.[45]

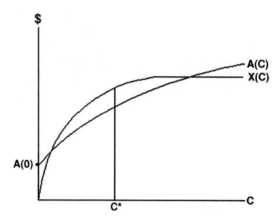

Fig. 1. The Purchase of Moral Satisfaction.

The moral issue is not the discount rate or the use of benefit-cost analysis, but what people care about. Ethical concerns are better incorporated directly. This keeps the accounting clear.

8.1. When Compensation is Not Possible

There are three reasons compensation may not be possible: (1) the amount of compensation can not be reasonably determined; (2) the injury is not compensable; and (3) there is no method to provide compensation even if the amount is known. When compensation can not be provided, the moral sentiments of others remain important. Thus in the nuclear waste example compensation may not be possible because no institution can be devised with reasonable guarantees of being able to provide it in the far future. In this case the moral harm from the project which might be considerable should be included as part of the benefit-cost analysis. Colleagues tell us, for example, that in the benefit-cost analyses of clean up of nuclear wastes at Hanford, WA and Rocky Flats, CO they did not include such moral harm and recommended short term fixes rather than more permanent clean up.[46]

In this case it is not the amount of compensation actually required for those injured that is directly relevant here. Rather, it is the amount of compensation the current generation thinks is correct. This is information that is likely to be obtainable at least in principle. (In the absence of good information about actual damage, however, people may have no opinion.) Even when compensation is not possible, it is generally possible through contingent valuation survey to determine at least in principle the WTP or WTA of "others" who have moral sentiments about the project.

10. CONCLUSION

Ronald Coase (1988) pointed out that law and economics can involve two different sort of tasks: economists taking into account institutional structures in economic analysis or economists using economic theory to consider what efficient law would be. He thought the former was the more neglected area. This paper combines economic reasoning with institutional considerations. Economics and law inform each other. Normative economic analysis requires a grounding in rights; yet at the margin such analysis can also suggest additional rights (Heyne, 1988). This is nowhere better illustrated than in environmental law and economics.

The conclusion is that moral sentiments should be included in normative economic analysis as with all sentiments, limited as a practical matter by the ability to appropriately measure them. That is moral sentiments have standing subject to appropriate measurement.

This will only happen when the profession abandons the KH criteria in favor of one that drops the potential compensation criteria. Milgrom's argument for ignoring moral sentiments exposes the failures of KH analysis. I suggest a modification of KH efficiency that, inter alia, treats all goods for which there is a WTP or WTA as economic goods. Goods not considered are simply those whose value can not be reasonably measured. I suggest that the KH approach lends itself to the introduction of the concept of economic standing that combines valuation with a specification of legal rights. I suggest that this approach distinguishes between a project which provides compensation and one that does not. I suggest further that it illuminates the issue of discounting benefits and costs occurring in the far future.

Among the implications of our approach is that existence value, as a type of moral sentiment, should be included in economic valuation. McConnell's argument that existence value should not be included in the case of non-paternal altruism arises because he fails to note that moral sentiment can be purchased in various markets. Thus its value will be limited by the administrative costs of effecting transfers by the most efficient means. The distinctions that McConnell raises in general are best thought of as a matter of the extent of substitutes for the goods in question.

As a practical matter it is not efficient for benefit-cost analysis to consider all relevant goods and affected individuals so any analysis will fail to meet the requirements of theoretical perfection whether for KH or KHZ. Yet, in doing practical analysis it is always desirable to have the better theoretical template in mind so that decisions about practice can be well considered and not ad hoc. Our purpose has been to contribute to this template.

NOTES

1. This paper relies on the recent book, (Zerbe, 2001a).
2. I limit the discussion of legal standing to cases where a right to sue exists under the Constitution, a statute, a regulation or the common law and could be exercised but for the plaintiff's inability to demonstrate a concrete, actual, redressable injury in fact. While environmental statutes such as the Federal Water Pollution Control Act, §505, 33 U.S.C. § 1365 and the Clean Air Act, §304, 42 U.S.C. § 7604 contain provisions granting standing to all citizens or all persons, courts in practice limit standing to plaintiffs who meet the injury-in-fact, causation, and redressabiltiy requirements articulated in *Lujan II*.
3. See *Warth v. Seldin*, 422 U.S., 490 for an in-depth explanation of the general rule against third-party standing. In *Warth* organizations and individuals in Rochester, New York brought suit against the town of Penfield to enjoin application of its exclusionary zoning ordinance but the Court held that "Petitioners must allege and show that they personally have been injured, not that injury has been suffered by other, unidentified members of the class to which they belong and which they purport to represent." *Id.* at 499. Traditionally, the Court has permitted third-party standing in limited instances including First Amendment overbreadth cases (see *Broadrick v. Oklahoma*, 413 U.S. 601 (1973)) in order to ensure that an overbroad statute does not chill First Amendment rights of those not before the court; in cases where immigrants seek to assert the citizenship rights of a parent (see *Miller v. Albright*, 118 S.Ct. 1428 (1998)); and in cases where criminal defendants seek to challenge the exclusion of jurors based on race (see *Powers v. Ohio*, 499 U.S. 400 (1991)).
4. The purpose of the "zone of interests" test is "to exclude those plaintiffs whose suits are more likely to frustrate than to further statutory objectives." Clarke v. Securities Indus. Ass'n, 479 U.S. 388, 397 (1987).
5. See *Allen v. Wright*, 468 U.S. 737 (1984).
6. See *Los Angeles v. Lyons*, 461 U.S. 95 (1983).
7. One who does not adequately represent interests makes it more difficult to measure these interests.
8. This distinction divides uses such as recreation and natural resource consumption from non-uses such as aesthetic and existence values.
9. See *Sierra Club v. Morton*, 405 U.S. 727, 732 (1972).
10. The distinction between "mere interest" or moral sentiments and concrete harm was also a factor in denying standing to environmental groups in *Lujan v. Sierra Club* (*Lujan I*), 504 U.S. 902 (1992). These cases, particularly *Lujan II* raises the possibility that Congress can create standing for "mere interest" by transforming it into concrete injury. Congress may create standing by offering a bounty to winners who would otherwise lack standing, thus making their interest measurable. However, this may not be possible if the Supreme Court determines that "mere interest" does not fulfil Article III's requirement for an injury in fact.

11. Prior to *Morton*, the Supreme Court indicated in dicta that it would be willing to consider non-economic harm to plaintiffs in cases where these non-economic interests fell within the zone of interest of the contested statute. In *Ass'n of Data Processing Service Orgs v. Camp*, 397 U.S. 150, 154 (1970), the Court stated that the zone of interest "at times, may reflect 'aesthetic, conservational, and recreational as well as economic values." (citations omitted). Because a statute's zone of interest may include aesthetic, conservational or recreational values, the Court stated in dictum that "standing may stem from [non-economic injury] as well as from the economic injury in which petitioners rely here." *Id.* at 154.

12. In 1997, Shintech, a large Japanese chemical corporation, proposed to build a plastics plant in Convent, Louisiana. According to Robyn Blummer writing for the St. Petersburgh Times as reported by the Denver Rocky Mountain News (Oct. 10, 1998, p. 62A) they passed state and federal emissions standards and agreed to hire substantially from the local community. They were opposed by Greenpeace on Title VI of the Civil Rights Act. Under this Act, EPA may withhold a permit if there is a "disparate impact" on minorities. As a result the Company has withdrawn its plans to build in the Convent location. A web search using Google and the name Shintech turns up conflicting reports about the sentiments of local residents. Some reports say that residents wanted the plant because they wanted the jobs. Reportedly, manufacturers are responding to Title VI rules by avoiding minority communities. Whatever the truth of the matter, it raises an issue of to what degree the moral sentiments of indirectly affected persons should count where the directly affected public disagrees. Certainly such sentiments by others can not be taken in law or economics as representing those directly affected. See Zerbe (1991, 1998) for material relevant to this sort of issue.

13. If these effects were ignored in the calculation of damages, a party would not be restored to the position he would have been before the injury.

14. Courts have been more willing to award damages for emotional loss or disappointment when the subject matter of the contract was emotional well-being or happiness.

15. Even in jurisdictions which allow sentimental value, courts have been unwilling to allow recovery for sentimental value unless a good has virtually no market value. In *Carye v. Boca Raton Hotel & Club*, the plaintiff's jewelry was stolen while entrusted to the defendant. While the jewelry had been accumulated over the course of the plaintiff's 48-year marriage, and the court did not dispute the jewelry had sentimental value, it denied recovery for sentimental value because the market value of the rings was considerable.

16. See *M. H. By and Through Callahan v. State*, 385 N.W.2d 533 (Iowa, 1986).

17. Sunstein, Cass, "Assessing Punitive Damages (With Notes on Cognition and Valuation in Law)," 107 Yale Law Review 2071, 2134 (1998).

18. 73 F. Supp. 2d (D. N. J. 1999).

19. KH is the standard measure of normative efficiency in economics.

20. This restriction to less than a national jurisdiction would not be acceptable to some economists.

21. As Kaldor explained where a policy results in an increase in aggregate income, the policy is quite unaffected by the question cf the comparability of individual satisfaction, since in all such cases it is possible to make everybody better off than before, or at any rate to make some people better off without making anybody worse off. Kaldor goes on to note (1939, p. 550) that whether such compensation should take place "is a political question on which the economist, qua economist, could hardly pronounce an opinion."

22. In addition, Tim Swanson in a verbal communication raises the issue whether or not the inclusion of moral sentiments allows equilibrium analysis. Elsewhere (Zerbe, (2001b) I show that an internal equilibrium is reached as long as altruism is not extreme.

23. Since one may care about transfers, e.g. compensation for a government taking, as a matter of principle or for self regarding reasons such as fear that a bad precedent may be applied also to oneself, the restriction of altruism alone is too narrow.

24. See Zerbe (2001), "A Normative Metric for the Purchase of Moral Satisfaction"

25. One objection to including moral sentiments may arise from these examples in which net benefits are always highest when the altruists pay the entire costs. Is this fair? Fairness is also a moral sentiment and there is no reason to assume that its incorporation will result in results that are less fair than current practice.

26. A move from a legal regime that does not use (KH or KHZ) efficiency as a rule for legal decision to one that does would pass the potential compensation test, For example a rule that inefficiently awarded the property to Richard would result in a loss of $120 instead of just $100 so there would be a net social WTP of $20 to move to a regime that used an efficiency criteria.

27. These are: Robbins, Hicks, Kaldor and Harrod, all writing in the *Economic Journal*.

28. For more on this matter see Zerbe 1998 and 2001.

29. Moral sentiments about others or other things may be expressed in four ways: One may care about: (1) the general positive well-being (utility) of others (love) or non-paternalistic (altruism); (2) the well being of others but believe it is promoted only by their consumption of particular goods (paternalistic love); (3) the consumption of particular goods unrelated to the well being of others; (4) the existence of particular goods regardless of use by any person. These cases can be discussed simply on the basis of the extent to which substitutes exist for the good in question.

30. The market value would be a lower bound on the WTA, because if the WTA was below market value, the current owner would have sold.

31. See also Arrow, 1996.

32. One objection to including moral sentiments may arise from these examples in which net benefits are always highest when the altruists pay the entire costs. Is this fair? Fairness is also a moral sentiment and there is no reason to assume that its incorporation will result in results that are less fair than current practice.

33. Elsewhere Zerbe (1998) has shown that moral sentiments concerning the justice of who pays will also affect the economic analysis of a project's viability.

34. Shrader-Frechette has argued that both the decision and the process by which it is made require informed consent. This is not possible when decisions affect future generations. See Ahearne (2000).

35. Cases in which this sort of issue has risen include Baltimore Gas & Electric v. Natural Resources Defense Council, Inc. 462 U.S. 87, (1983); Pacific Gas and Electric Co et al. v. State Energy Resources Conservation and Development Commission, 461 U.S. 190, (1991). see also 123 U. S. 45 (1999).

36. For example, consider two projects with initial costs of $100. Project A has benefits of $150 in the first period. Project B has benefits of $150 in 100 years. With negative or sufficiently low discount rates project B is preferred. Project A however, may result in greater wealth in 100 years so that it is superior for both the current generation and the 100th year generation. One may object that these future benefits to the 100th year generation associated with project A that arise from reinvestment of proceeds need to be counted. This is not required, however, where the discount rate is

equal to the growth rate and only serves to show the peculiar adjustments that would need to be used to get the best decisions with too low or negative discount rates.

37. The X might be determined by a contingent valuation survey.

38. But see Whittington and Macrae (1986).

39. For a discussion of the practicality of compensation see Lind (1999).

40. A potentially important difference here is that one view of the issue may represent WTP while the other represents WTA, but this is left for future investigation.

41. Looking at the issue from this point of view yields the following diagram:

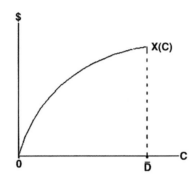

42. The compensation provided by the current generation is a transfer to the future generation and thus would enter Eq. (2) as both a positive and a negative were it to be included. So that it would not affect the result expressed in Eq. (2).

43. Indeed, there need be nothing special about 100%. Such a discontinuity could occur at any level providing some sort of moralistic focal point.

44. It would be superior as long as future individuals have no standing. See Whittington and Macrae (1986).

45. One might ask in arriving at the analysis of Table 6 what has been discounted? When compensation is possible, the discount rate used to determine the size of the compensation investment today to compensate in the future will depend on the estimate of what rate of return can be obtained not the time preference rate. The damage to future generations, or more accurately our assessment of this damage, has been discounted. The sentiments of the current generation if they prevail for a lifetime are discounted over the lifetime of the current generation. The administrative costs of providing a mechanism for compensation have been discounted as has the cost of compensation.

46. Conversation with Howard McCurdy of American University, Steve Tarlton of the Colorado Department of Public and Environmental Health, Max Powers of the Department of Ecology of the State of Washington, and Elaine Faustman of the University of Washington.

ACKNOWLEDGMENTS

I wish to thank Douglas Allen, Allen Bellas, Amber Dufseth, Mark Funke, Michael Hanneman, Daniel Huppert, Stewart Jay, and David Layton for useful comments. Stewart Jay helped me understand legal standing. Allen Bellas suggested several places where greater clarity was needed and did most of the work on the mathematics model associated with the future generation problem. This work was supported in part by the Center for the Study and Improvement of Regulation at Carnegie Mellon University and the University of Washington and in part by the Earthquake Engineering Research Centers Program of the National Science Foundation under Award Number EEC-9701568. This article builds on Zerbe (2001a).

REFERENCES

Ahearne, J. F. (2000). Intergenerational Issues Regarding Nuclear Power, Nuclear Waste, and Nuclear Weapons. *Risk Analysis, 20*(6).

Arrow, K. J., Cropper, M., Eads, G., Hahn, R., Lave, L., Noll, R., Portney, P., Russell, M., Schmalensee, R., Smith, V., & Stavins, V. (1996). *Benefit-Cost Analysis in Environmental Health and Safety Regulations: A Statement of Principles.* Washington, D.C.: American Enterprise Institute.

Arrow, K. J., Solow, R., Portney, P., Leamer, E. E., Radner, R., & Schuman, H. (1993). Report on the National Panel on Contingent Valuation. *Federal Regulations.* 58(10, January 15), 4603, 4601–4614.

Baker, E. C. (1980). Starting Points in the Economic Analysis of Law. *Hofstra Law Review, 8*, 939.

Boadway, R. W., & Bruce, N. (1984). *Welfare Economics.* New York: Basil Blackwell.

Brown, P. (1991). Greenhouse Economics: Think Before You Count: a Report from the Institute for Philosophy & Public Policy. Volume 11.

Coase, R. H. (1988). *The Firm, the Market, and the Law.* Chicago: University of Chicago Press.

Diamond, P. A., & Hausman, J. (1993). On Contingent Valuation Measurement of Non-use Values In: J. A. Hausman (Ed.), *Contingent Valuation: A Critical Assessment. Contributions to Economic Analysis* (Vol. 220, pp. 417–435). Amsterdam; London and Tokyo: North-Holland; distributed in the U.S. and Canada by Elsevier Science, New York.

Dworkin, R. (1980). Is Wealth a Value? *Journal of Legal Studies, 9*, 191.

Dworkin, R. (1986). *Law's Empire.* Cambridge: Belknap Press.

Dunford, R. W., Johnson, F. R., Sandefur, R. A., & West, E. S. (1997). Whose Losses Count In Natural Resources Damages? *Contemporary Economic Policy, 15*(4), 77(11).

Fried, C. (1978). *Right and Wrong.* Cambridge MA: Harvard University Press.

Hammond, P. (1985). Welfare Economics. In: G. Fiewel (Ed.), *Issues in Contemporary Microeconomics and Welfare.* New York: Macmillan.

Harberger, A. (1978). On the Use Of Distributional Weights in Social Cost-Benefit Analysis. *Journal of Political Economy, 86*, 635.

Hausman, J. A. (Ed.) (1993). Contingent Valuation: A Critical Assessment. *Contributions to Economic Analysis, 220.*

Heyne, P. (1988). The Foundations of Law and Economics. In: R. O. Zerbe, Jr. (Ed.), *Research in Law and Economics* (Vol. 11, pp. 53–71). Greenwich: JAI Press.

Kaldor, N. (1939). Welfare Propositions In Economics and Inter-Personal Comparisons Of Utility. *Economic Journal, 49*, 549.

Kunreuther, H., & Easterling, D. (1992). Are Risk-Benefit Tradeoffs Possible In Siting Hazardous Facilities? *American Economic Review, 80*(2), 252–256.

Lesser, J., & Zerbe, R. O. Jr. (1998). A Practitioner's Guide To Benefit-cost Analysis. In: F. Thompson & M. Green (Eds), *The Handbook of Public Finance*. New York: Marcel Drekker.

Lind, R. (1999). Analysis for Intergenerational Decisionmaking. In: P. Portney & J. Weyant (Eds), *Discounting and Intergenerataional Equity* (pp. 173–180). Washington, D.C.: Resources for the Future.

Loomis, J. (1995). Measuring the Benefits of Removing Dams and Restoring the Elwha River: Results of a Contingent Valuation Survey. Working paper, Department of Agriculture Resources, Colorado State University, Fort Collins, CO.

McConnell, K. E. (1997). Does Altruism Undermine Existence Value? *J. of Envir. Economics and Management, 32*, 22–37.

Milgrom, P. (1993). Is Sympathy an Economic Value? Philosophy, Economics, and the Contingent Valuation Method. In: J. A. Hausman (Ed.), *Contingent Valuation: A Critical Assessment. Contributions to Economic Analysis* (Vol. 220, pp. 417–435). Amsterdam; London and Tokyo: North-Holland; distributed in the U.S. and Canada by Elsevier Science, New York.

Mishan, E. J. (1981). *Introduction to Normative Economics*. New York: Oxford University Press.

Parfit, D. (1992). An Attack On The Social Discount Rate. In: C. Mills (Ed.), *Values and Public Policy*. Fort Worth: Harcourt Brace Javanovich.

Parfit, D. (1994). The Social Discount Rate. In: R. E. Goodwin (Ed.), *Politics Of The Environment*. Aldershot: Edward Elgar.

Pearce, D., Markandya, A., & Barbier, E. (1989). *Blueprint for a Green Economy*. London: Earthscan Publications.

Plater, Z., Robert, J. B., Abrams, H., Goldfarb, W., & Graham, R. L. (1998). *Environmental Law and Policy: Nature, Law, and Society*. St. Paul: West Publishing Co.

Posner, R. (1981a). *The Economics of Justice*. Cambridge, MA: Harvard University Press.

Posner, R. (1981b). A Reply to some Recent Criticisms of the Efficiency Theory of Common Law. *Hofstra Law Review, 9*, 775.

Posner, R. (1984). Wealth Maximization and the Judicial Decision Making. *International Review of Law and Economics, 4*(2), 131–135.

Posner, R. (1986). *Economic Analysis of Law* (3rd ed.). Boston: Little-Brown.

Quiggin, J. (1997). Altruism and Benefit-Cost Analysis. *Australian Economics Papers, 36*, 144–155.

Robbins, L. (1932). *An Essay on the Nature and Significance of Economic Science*. London: Macmillan.

Robbins, L. (1938). Interpersonal Comparisons of Utility: a Comment. *Economic Journal, 48*, 635af.

Schultze, W. D., Brookshire, D. S., & Sandler, T. (1981). The Social Rate of Discount For Nuclear Waste Storage: Economics of Ethics. *Natural Resources Journal, 21*, 811–832.

Smart, J. C., & Williams, B. (1973). *Utilitarianism – For and Against*. Cambridge: Cambridge University Press.

Sunstein, C. (1992). What's Standing After Lujan? *Michigan Law Review, 91*, 163.

Svenson, O., & Karlsson, G. (1989). Decision Making, Time Horizons, and Risk in the Very Long Run Perspective. *Risk Analysis, 9*, 385–398.

Whittington, D., & Macrae, D. Jr. (1986). The Issue of Standing in Benefit-cost Analysis. *Journal of Policy Analysis and Management, 9*(2), 201–218.

Winter, S. G. (1969). A simple remark on the second optimality theorem of welfare economics. *Journal of Economic Theory, 1*, 99–103.

Zerbe, R. O., Jr. (1991). Comment: Does Benefit-cost Analysis Stand Alone? Rights and Standing. *Journal of Policy Analysis and Management, 10*(1), 96–105.

Zerbe, R. O. Jr. (1998a). An Integration of Equity and Efficiency. *Washington Law Review, 73*(April), 349–361.

Zerbe, R. O. Jr. (1998b). Is Cost-Benefit Analysis Legal? Three Rules. *Journal of Policy Analysis and Management, 17*(3), 419–456.

Zerbe, R. O. Jr. (2001a). *Efficiency in Law and Economics.* Aldershot, England: Edward Elgar.

Zerbe, R. O. Jr. (2001b). A Normative Metric for the Purchase of Moral Satisfaction? Working paper, Evans School of Public Affairs, University of Washington, Seattle.

COMMENTS ON PAPER BY
RICHARD O. ZERBE, JR.

Andreas Kontoleon

Zerbe offers a very insightful contribution to the issues surrounding the inclusion of moral sentiments in environmental decision making. The paper introduces a variant of the Kaldors-Hicks criterion to better accommodate such values in economic analysis. It concludes that moral sentiments (such as various forms of altruism) should be included in normative economic analysis since their inclusion is beneficial both to the quality and accuracy of the analysis but also to our understanding of issues in environmental economics. For example, Zerbe shows how confusions and criticisms regarding the use of discounting in economic analysis could be mitigated and addressed by sufficient recognition of future values, interpreted as a form of intergenerational altruism.

Zerbe's offers a comprehensive discussion of the various attacks on the inclusion of moral sentiments that have been raised by economists and other theorists. His analysis of the "double-counting" argument (that economic analysis should not include non-paternalistic or 'pure' altruistic preferences) provides an alternative view to that held by welfare theorists such as McConnell (1997). Yet, the discussion is limited and could be generalised to address views held by other economists which utilise micro-welfare foundations to show under what conditions various types of altruistic preference are compatible with economic analysis (for example views expressed in Johansson, 1992; Jones-Lee, 1992; Johansson-Stenman, 1998).

In addition, the arguments raised by an increasing number of authors (for example Blamey, 1998; Spash, 2000) on the 'legitimacy' of using moral sentiments in economic analysis could be addressed. The common core of

An Introduction to the Law and Economics of Environmental Policy: Issues in Institutional Design, Volume 20, pages 173–175.
© 2002 Published by Elsevier Science Ltd.
ISBN: 0-7623-0888-5

these arguments is that non-use values are not 'proper' economic values in that they are not the outcome of individual choices involving trade-offs. The implication of these arguments is that such non-use values are not specific to the environmental issue at hand but are general expressions of one's moral sentiments.

Finally, Zerbe addresses the arguments raised by Dunford et al. (1997) that demand for knowledge about the environmental resource and/or its injury are required for one's non-use value to have legal standing. Zerbe's response implies that individuals have preference over classes of environmental goods (not particular types of environmental resource) and thus they would suffer a legitimate loss in non-use value from a damage to a particular environmental asset even if they had no prior knowledge of the asset and/or the injury (a similar view is expressed by Randall, 1997). There are several objections to this reasoning. First, the fact that some individuals have resorted to developing such heuristics to be able to cope with the complexity of the real world tells us something about their preferences. Second, accepting general rather than specific knowledge of environmental resources allows for the aggregation population to be overwhelmingly large. This may be reasonable for some unique natural resources but is not convincing for resources with many substitutes. Third, accepting that individuals care about classes of environmental resource poses problems on interpreting how people make choices over specific resources when asked to do so. That is, if, for example, people care about 'all species', on what basis can their intensity of preferences (i.e. their values) differ for one particular species to another? Would this mean that individuals would have the same value for any member of the class of resources? If not, then on what basis would these values differ other than individuals have different orderings for such specific preferences? Zerbe's argumentation also implies that prior knowledge of the resource is not required since society "owns" the resources managed by trustees. Yet this view ignores a crucial difference between non-use values for public resources and private property. Non-use values do not exist independent of individual perception. Hence, losses in non-use values require some prior knowledge whereas losses in use values do not. Also, justifying legal standing on property rights is troublesome since property rights are often not clear (e.g. to whom do these rights extend to?). To clarify his point Zerbe provides an interesting example of a rich individual who owns many firms which are run by managers. Though the wealthy individual does not have knowledge of his specific firms he/she would receive a legitimate welfare loss if were to find out that one of his/her enterprises went bankrupt. There are two problems with this example: first the individual has the *private* property right over all his/her firms. The individual receives use value from his/her wealth. Value from privately held resources is not a matter of perception (it arises from personal benefits enjoyed by the individual) while non-use values over commonly owned resource arise from the knowledge that certain environmental flows accrue to others. Hence, it seems that prior knowledge is a requirement for non-use values to exists independently of a contingent valuation study. Further, we can interpret individuals with no demand for knowledge of the resource as having 'waived' their right to the resource and thus as not having standing.

REFERENCES

Blamey, R. K. (1998). Decisiveness, Attitude Expression and Symbolic Responses in Contingent Valuation Surveys. *Journal of Economic Behavior and Organization, 34*(4, March), 577–601.

Jones-Lee, M. W. (1992). Paternalistic Altruism and the Value of Statistical Life. *Economic Journal, 102*, 80–90.

Johansson, P. O. (1992). Altruism in Cost-Benefit Analysis. *Environmental and Resource Economics, 2*, 605–613.

Johansson-Stenman, O. (1998). The importance of ethics in Environmental economics with a focus on existence values. *Environmental and Resource Economics, 11*(34), 429–442.

Spash, C. L. (2000). Ethical Motives and Charitable Contributions in Contingent Valuation: Empirical Evidence from Social Psychology and Economics. *Environmental Values, 9*(4), 453–480.

Randall, A. (1997). Whose Losses Count? Examining Some Claims about Aggregation Rules for Natural Resources. *Contemporary Economic Policy, 15*(4, October), 88–97.

INDIVIDUAL PREFERENCE-BASED
VALUES AND ENVIRONMENTAL
DECISION MAKING:
SHOULD VALUATION HAVE
ITS DAY IN COURT?

Andreas Kontoleon, Richard Macrory
and Timothy Swanson

ABSTRACT

The paper focuses on the question of the extent to which individual preference-based values are suitable in guiding environmental policy and damage assessment decisions. Three criteria for "suitableness" are reviewed: conceptual, moral and legal. Their discussion suggests that: (i) the concept of economic value as applied to environmental resources is a meaningful concept based on the notion of trade-off; (ii) the limitations of the moral foundations of cost-benefit analysis do not invalidate its use as a procedure for guiding environmental decision making; (iii) the input of individual preferences into damage assessment is compatible with the basic foundations of tort law; (iv) using individual preference-based methods provides incentives for efficient levels of due care; (v) determining standing is still very contentious for various categories of users as well as for

An Introduction to the Law and Economics of Environmental Policy: Issues in Institutional Design, Volume 20, pages 177–214.

aggregating non-use values. Overall, the discussion suggests that the use of preference-based approaches in both the policy and legal arenas is warranted provided that they are accurately applied, their limitations are openly acknowledged and they assume an information-providing rather than a determinative role.

1. INTRODUCTION

The role of individual preferences and cost-benefit analysis (CBA) in environmental decision making has been extensively debated by economists (e.g. Kopp, 1991, 1992; Freeman, 1993), lawyers (e.g. Daum, 1993; Shavell, 1993; Boudreaux et al., 1999; Posner, 1980, 1983; Kennedy, 1981) and philosophers (e.g. Hubin, 1994; Sagoff, 1994; Dworkin, 1980; Kelman, 1981). Yet, despite the voluminous literature, the discussion remains disordered and confused. Further, though CBA has been widely used in practise (primarily in the U.S. but now increasingly in the EU as well) its actual influence on policy has been relatively limited.[1] Despite this limited acceptance and checkered history, the European Union is now considering a new Directive on Civil Liability that might imply bringing valuation into European courtrooms. Does this make sense? Should valuation have its day in court?

One of the reasons for the ongoing controversy over valuation and CBA within academic circles, and for the hesitancy in using CBA in the policy and legal arenas, can be traced to an entanglement of distinct issues.[2] For example, commentators typically confuse issues of measurement (e.g. 'Are estimates of individual values valid?') with conceptual issues – (e.g. 'Is the economic concept of value coherent?') or with moral issues (e.g. 'Are decision makers and the courts morally obligated to consider individual preferences?' or legal issues (e.g. 'Do economic values adhere to the current legal framework of damages?'). It is far beyond the limits and scope of this paper to comprehensively present and review all the various aspects of the debate, and all the forms of confusion that exit. Instead the aim of this paper is to provide an eclectic survey of the issues concerning this debate. More specifically we focus on the question of the extent to which individual preference-based values are suitable in guiding policy and damage assessment decisions related to environmental resources. Three criteria for "suitableness" are reviewed here: (1) the conceptual; (2) the moral; and (3) the legal.[3]

The conceptual issues that are most relevant to this discussion have to do with the notion of 'value' as understood in economics. Within an economic context individual *preferences* over environmental goods and services are manifested through individual *choices* which in turn are used by the economist to infer individual *economic values*. Further, most economists agree that individuals make choices from which we can infer both so-called 'use' and 'non-use values' for environmental resources.[4] Here we will focus on the conceptual debate relating to non-use values (see for example Quiggin, 1998, 1993). That is, is the economic concept of non-use value sufficiently coherent to be used in environmental policy and damage assessment or is it fundamentally flawed and unsuitable?

Beyond conceptual considerations the role of individual preferences in environmental decision making will be discussed on moral and legal grounds. The main question under discussion concerns whether the concept of economic value is compatible with the moral basis of environmental decision making or with the legal framework for damage assessment decisions. That is, if the concept is coherent and consistent, is it also socially acceptable and/or legally viable?

One last note on the organisation of the issues. We have intentionally left out of the discussion the issues on measurement. These refer to the general question on whether individual economic values (as expressions of the intensity of individual preference) can be validly and adequately measured. No doubt these issues are very important.[5] Yet, they are not the most fundamental ones. For the sake of argument the discussion that follows accepts that economic values are readily and validly measurable. Instead, we proceed with the more fundamental issues of the debate which concern the conceptual, moral and legal validity of using preferences in environmental decision making.

The organisation of the paper is as follows: The following section briefly classifies and reviews types of decision making processes in accordance with the manner and degree to which they rely on individual preferences or expert opinion. Section 3 deals with various objections that have been raised with respect to the adequacy of the concept of economic value as applied to environmental policy decisions. Section 4 turns to the fundamental moral question of whether policy makers and courts are morally obligated to utilise information from individual preference-based values. Finally Section 5 reviews some key legal issues surrounding the debate on the use of individual preference-based techniques for environmental policy and liability decisions.

2. CLASSIFICATION OF APPROACHES TO ENVIRONMENTAL DECISION MAKING

One way of classifying different approaches to environmental decision making is in accordance with their degree of reliance upon individual preference-based values. All forms of decision making rely on individual preference to some extent. The major difference lies in the manner in which society (or the policy maker or the courts) decides whose preferences matter. Some approaches rely on stakeholder groups, others on selected juries/focus groups or on large random surveys drawn from selected populations. In each case the preferences of the select group matter, but are handled in different ways. In some approaches to decision making, the preferences of the groups are supposed to determine the result of the policy or legal process, in other approaches, preferences are only part of that process. If a society moves away from other approaches to decision making and towards preference-bases approaches, then it is extending and deepening its reliance upon individual preferences in environmental policy making. In this section we review possible approaches and indicate the direction of increasing reliance. Table 1 depicts a classification of various decision making methods along a 'preference reliance' spectrum. As we move from left to right along the spectrum the reliance on individual preferences and economic values in the decision making process diminishes. A brief account of these methods is presented in the following sections.

2.1. preference-based *Valuation Methods*

Preference-based valuation methods can be split into formal valuation methods and environmental pricing techniques.[6] The former are used to assess standard (neo-classical) *welfare measures* while the latter focus on market prices that are assumed to reflect economic scarcity and thus are in essence *efficiency* or *market prices*.

Valuation techniques are classified into revealed and stated preference techniques. *Revealed preference* valuation techniques (including travel costs, hedonic pricing and wage differential approaches) rely on information from individual consumption/ purchasing behaviour occurring in markets related to the environmental resource in question (surrogate markets). The price differential of the good (purchased in the surrogate market), once all other variables that affect choice apart from environmental quality have been controlled for, will reflect the purchaser's valuation of that particular level of environmental quality. These methods have the appeal of relying on actual/observed behaviour but their main fundamental drawbacks are the inability to estimate non-use values[7] and the dependence of the estimated values on the assumptions made on the relationship between the environmental good and the surrogate market good.[8] *Stated Preference* techniques (including contingent valuation, choice experiments, and contingent ranking) are used in situations where both use and non-values want to be estimated and/or when no surrogate market exists from which environmental (use) value can be deduced. These techniques use questionnaires to develop a hypothetical market through which they elicit values (both use and non-use) for the environmental good under investigation. Stated preference techniques do not suffer from the same technical limitations as revealed preference-based approaches and can also be applied to non-use values. Yet, the hypothetical nature of the market constructed has raised numerous questions regarding the validity of the estimates (Navrud, 2000).

Table 1 then lists three categories of *environmental pricing techniques*. The first method relies on the use of market prices of directly related goods and services as surrogate values for environmental amenities. The quality of the environmental good is treated as an input into the production function of various goods and services (outputs). Changes in these environmental inputs may lead to changes in productivity or production costs which, in turn may lead to changes in prices and output levels which can be observed and quantified (Dixon et al., 1988). These approaches have been referred to as 'dose-response' techniques.[9] The second set of pricing techniques relies on data from *actual* costs of maintaining or preventing environmental degradation as a proxy for environmental value.[10] The third set of pricing methods is similar to above but relies on *potential* (as opposed to actual) costs as proxies for environmental value. These include methods as such *'shadow-project appraisal'*.

Pricing techniques have been widely used since they rely on real price data and can provide useful information for appraisal purposes. Yet they suffer from serious limitations. The dose response approaches do not account for either behavioural adaptations or price responses (Navrud, 2000) which can lead to over or underestimation of environmental damage. Potential cost approaches produce ad-hoc values that may bear little relationship to true social values. Actual and potential cost techniques disregard the benefits of change in the quality of environmental resource and only provide cost information. This is inadequate for a complete cost benefit analysis (Lovett et al., 2001).

In sum, valuation and pricing techniques both rely in individual preferences (through hypothetical or surrogate markets or through price information). Yet, the latter do not capture total social net value since they rely on price data to provide information on only the costs of environmental change. This places valuation techniques higher up the 'preference reliance' scale.[11] Also, stated preference approaches are the only methods available to capture both use and not-use values. In this respect they top the 'preference reliance' spectrum.

2.2. Participatory and or Deliberative Approaches

Participatory approaches have been suggested as an alternative decision making process that could possibly avoid some of the limitations of valuation techniques while allowing a platform for individual preferences to feed into environmental decisions.[12] The citizens jury approach is one of the most explicit applications of participatory decision making processes that has been used on several occasions in the U.S. and Europe.[13] The approach has been modelled after the criminal law system where a " group of randomly selected citizens, when exposed to good information presented by witnesses from differing points of view, is able to make good judgements on public policy matters even though in terms of training and experience there are many people more competent than they" Crosby (1995). The citizen jury (also referred to as value juries – e.g. Brown et al. (1995) method was developed by the Jefferson Centre (in Minnesota, USA), a non-profit, non-partisan facilitation organisation. A randomly selected group of about a dozen jurors, designed to represent a micro-cosm of their society, is impanelled to study a specific local or regional public policy issue. The facilitating organisation develops a narrow 'charge', which is presented to jurors at the beginning of the process. The charge generally contains a clear statement of the problem to be addressed, often asking jurors to chose between three or four pre-selected options, and subsequent follow-up questions to consider. The jurors, who are paid for their time, participate in hearings over 4–5 days, facilitated by a neutral moderator. They hear from "witnesses" presenting a wide range of views on the issue. Jury members may question witnesses. The jurors then deliberate and issue findings and recommendations to policy makers. The process is designed, like a criminal jury, to examine a narrowly defined charge. Jurors receive limited background information and training, and the process does not promote critical inquiry into issues outside the limited mandate (Tickner & Ketelsen, 2001; Renn et al., 1995). As the decisions are made by majority vote, minority positions may not be adequately considered in the jury discourse. And, of course, currently these jury decisions have no legal weight but may or may not have a direct, formal input into the policy-making system. Indeed, the use of the term 'jury' is to some extent unfortunate in that it may imply a body with the power to decide a particular issue. It is both preferable and more legitimate to view such mechanisms as a method of providing information input to the policy process.

Consensus conferences and planning cells are two mechanisms that are very similar to citizen juries. They differ from the latter in that they engage citizens in examining broadly-defined questions of regional or national importance (see Dienel & Renn, 1995; Joss & Durant, 1994).[14]

Further scenario workshops, focus groups sessions, and other such models of deliberative decision making have been used as vehicles for goal-setting and alternative assessment. In Europe, several governments have undertaken "scenario workshops" to develop future visions for a country or region. They involve different groups (residents, government, academics, business, etc.) and address broad questions, such as "how to develop a sustainable community" or "how to address toxic contamination." Often goals are set and strategies are developed to achieve those goals. In the U.S., sustainable community planning exercises have been undertaken in various locations (Tickner & Ketelsen, 2001). Citizen Advisory Committees (CACs) have been used in the U.S. and Canada (since the early 1980s) to provide advice to federal, state and local government on implementing environmental law. As in citizen juries, a charge is given to the CAC (usually by the governmental agency responsible to resolving the problem at hand), yet its members are usually appointed. Members include interest groups or representatives of the constituency affected by the environmental issue. The main function of the CAC is to achieve some form of reconciliation between the participants rather than being instrumental in solving a particular problem (see Vari, 1995; Lynn & Kartez, 1995). It suffers from the 'small numbers' problem akin to all similar participatory methods but has the advantage of allowing public participation in a procedural stage where no preliminary decisions have been made. Thus, its scope is not restricted to the final decision but can include the definition of goals and constraints.

2.3. Expert-Based Approaches

Expert-based approaches vary enormously and defy real classification. However, it is clear that they rely least on the individual preferences of those involved. Instead experts are usually thought to rely more on experience and scientific evidence to reach decisions. For these reasons, expert-based policy making falls at the end of the preference-based spectrum in Table 1 (Appendix).

Multi-Criteria Analysis (MCA) is an example of a structured decision-making approach sometimes used in expert-based decision making.[15] MCA requires policy makers, experts and/or stakeholders to identify a set of decision-making criteria and a scoring scale for each criterion. The various decision criteria are then weighted (alternative means of doing this are possible). The scoring of alternative environmental decision policies against the weighted criteria are then considered and the choice of the most appropriate alternative is made. MCA techniques have been more popular in European countries compared to the U.S., mainly because MCA purports to account for many policy objectives including distributional concerns which are much more stressed in European countries.

The *Delphi* technique informs policy experts by means of surveying groups of experts.[16] The experts are preferably selected from various fields and are typically interviewed more than once. The size of the panel varies considerably from under 10 to a few hundred. At each interview round they are presented with the evaluations of the other experts and are asked to re-assess their opinion based on this new information. The method is used to either obtain a consensus or a characterisation of the distribution of experts' valuations. (Pearce & Mourato, 1998). The results of such an exercise provides information that can assist in ranking environmental resources (on ecological criteria) or in undertaking some form of cost-effectiveness analysis.[17]

In sum, expert-based decision making is grounded on many criteria, and the manner in which these criteria contribute to decision making varies in many ways as well. Hence, this approach to decision making is the least reliant upon the structured identification and application of individual preferences.

3. THE VALIDITY OF THE CONCEPT OF ECONOMIC VALUE

Having described the various ways in which preferences can inform policy making, we now turn to the substantive issue of the conceptual nature of economic value. At the conceptual level the debate over the use of individual preferences in environmental decision making derives from the debate over the meaning and the validity of the concept of economic value in general, and as applied to environmental issues in particular. Comprehensive coverage of these topics can be found in Foster (1996), Crowards (1995), and Kopp (1992, 1991).

One source of confusion in the literature can be traced to the differential usage of similar terms. For economists the term 'value' has a very specific and limited meaning. For a moral philosopher, however, both individual and societal values are treated and articulated in a quite distinct way from preferences, and certainly cannot be equated in any way with preferences.

For economists, individual preferences are important in so far as they allow people to make choices over goods or more generally 'over states of the world'. The economist's definition of value is an inherently instrumentalist and anthropocentric concept that is based on the idea that people make choices under various constraints (e.g. income, time, information, etc.). Hence, economic value implies the notion of a 'trade-off': value is the 'amount' that has to be given up in order to get something else.[18]

In the most extreme case, critics have argued that the economic concept of value is inherently flawed when applied to environmental resources and thus should have no place in environmental decision making. The main line of attack revolves around the idea that people simply 'don't have values' for such resources in the way perceived by the economist, and that values for environmental resources cannot be defined in economic terms.

But this argument confuses different meanings of the term 'value'. Economic values are simply attributed characteristics based on (actual or stated) choices. As

Kopp (1992) points out, many critics (e.g. Gregory et al., 1991) erroneously assume that economists believe that people have values for "things." Yet economists merely assume that people *make choices* over bundles of things and value is merely the realisation of choice, i.e. what you give up to get something else.

There is nothing in well established economic theory that limits the object of choice to physical private goods. People do make choices in everyday life that do involve trade-offs between levels of environmental quality. Using these observed choices the economists can estimate the economic value for *using* environmental resource. These estimates lead to measures of use value: values that are related to the observed uses of the services provided by natural assets.

A particular form of value that has been at the centre of much debate is the so called *non-use values* (NUVs). The general/intuitive idea of NUVs as the value associated with no direct use of an environmental resource is usually attributed to Krutilla (1967). In economic terms NUVs are best conceptualised as a form of a pure public good. (e.g. McConnell, 1983). The conception of NUV acknowledges that one's welfare can be enhanced from a particular natural resource without engaging in any observable behaviour. Note that the economic conception of economic value does not invalidate other types or conceptions of value (see Turner, 2000 for a review of various conceptions of value). Yet, these are the discourse of other sciences. "Value pluralism" may be important but is beyond the domain of economics. It is the role of policy makers – not the economist – to rank the importance of other forms of values.

Objections to the concept of economic value as applied to environmental resources

The economic definition of environmental use values as well as the conception of NUVs as forms of pure public goods has raised various objections, the most important of which are reviewed below.

'Slippery Slope' Argument

Some (e.g. Rosenthal & Nelson, 1992) have argued that perceiving NUVs as pure public goods may lead to a dangerous 'slippery slope': almost any 'good' may have a public good component and by including NUVs in CBA or damage assessment the task would become daunting. We agree that in principle anything could have a pure public good component and should thus be included in any environmental decision making process. Yet this would cause problems (e.g. over estimation of damages) only if the estimates from NUVs would be *equally* large for all environmental resources (i.e. the value would not vary with the nature of the good or damage). Yet, there is no evidence that suggests that values are in fact non-good specific and do not vary with the nature or size of the good (see Carson, 2001)

'Complexity of the Good' Argument

Others have used some form of the 'complexity of the good' argument (e.g. Vatn, 2000; Vatn & Bromely, 1994; Clarke et al., 2000; Green, 1997; Jacobs, 1997) which acknowledges that economic value is a valid concept but one that is *not* valid for environmental goods since these are too complex to be 'commodified'. Vatn and Bromely (1994) offer a very convincing defence of this position based on cognition, incongruity and composition problems. Yet, these arguments seem misplaced in that the economic conception of value does not 'commodify' natural resources but simply treats them as objects of choice.

Incommensurability, incomparability and lexicographic preference arguments

The concept of economic value has been attacked by an array of arguments claiming *incommensurability, incomparability and lexicographic preferences* (Beckerman & Pasek, 1996; Lockwood, 1999; Rekola et al., 2000; Spash, 2000, 1997). These arguments support the view that environmental resources are not proper objects of choice and cannot be used to undertake trade-offs. Related to these arguments is the claim that people's preferences over these resources may change according to whether the individual is consulted as an individual (e.g. in a CV study) or as a citizen (e.g. in a citizen jury) (Sagoff, 1994; Blamely et al., 1995; Common et al., 1997; Spash, 2000; Martinez-Alier et al., 1998; Edwards, 1992).

One of the implications of the line of reasoning found in the above arguments is that other social goods such as health or education would also not be compatible with an economic framework of choice. Yet clearly, people *do* make choices over matters of health, education and the environment however complex the nature of the choices may be. The majority of these arguments concern choices made in stated preference studies. They argue that empirical evidence from these studies suggests that people do not make trade-offs over environmental resources and thus the concept of economic value is inappropriate and unsuitable for assisting environmental policy decision. What is important to note here is that these arguments fail to demonstrate that people in *any setting* (either actual or hypothetical) do not make trade-offs over environmental resources. There is abundant *revealed* preference data where individuals make choices over environmental resources (or public goods in general) as well as data on actual consumer choices motivated by commitment and a sense of moral responsibility (e.g. donations to charities or environmental organisations).[19]

Thus, the concept of economic value is a meaningful concept based on the notion of trade-off and opportunity cost. Whether such a value should be used in policy and damage assessment decisions is discussed in the following sections.

4. MORAL ISSUES IN USING INDIVIDUAL PREFERENCE IN POLICY DECISIONS.

Having touched upon the conceptual validity of the economic notion of environmental valuation we can now turn to even more fundamental levels of the debate. This concerns the debate over the moral and the legal validity of using preference-based values in policy and damage assessment respectively.[20] In this section we discuss whether rational and moral decision makers would or should consult an account of economic benefits and costs in the course of policy making (Randall, 2002; Copp, 1985).[21] Economists justify the use of cost-benefit analysis in environmental decision making on the basis of Welfarism: CBA is seen as an empirical test of whether proposed public actions would increase preference satisfaction.[22] The economists' argument rests on attempting to argue that welfarism is the most adequate moral theory for public decision making and that CBA is justified as the direct implementation of the 'correct' moral theory.

Most critics against individual preference-based decision making (e.g. Sagoff, Spash) focus on criticising the welfarism and consequentialism on which CBA rests. Although this is done quite successfully[23] they fail to realise that undermining the moral theory of a procedure does not undermine the validity or the moral relevance of the procedure itself.

Hubin (1994) argues that the *procedure* of CBA (and the use of preference-based techniques) would be justified even if there exists flaws in the underlying moral theory.

He best summarises his argument by reference to an analogy to the validity of democratic procedures. Democratic moral theory – the theory that the right action *is* just that action approved by the majority – is the moral foundation of democratic electoral procedures. Yet, philosophers since Plato have (quite easily) shown that democratic moral theory is fundamentally problematic. "But this is not concern for the democrat; she has never felt that her conviction to democratic institutions committed her to democratic moral theory. Rather, the democrat sets about justifying democracy by appeal to other more plausible moral theories. The proponent of CBA should do likewise" (Hubin, 1994, p. 177).

Hubin (1994) further demonstrates that commonly accepted moral theories (consequentialism, contractualism, deontology) would accept that information derived from preference-based approaches is morally relevant and useful. The fact that the information incorporated in a CBA is deemed morally significant and useful by most currently held moral theories does not mean that such information *is* morally relevant. Yet, Hubin argues that the currently accepted moral theories are representative of the range of plausible moral theories. That means that it is reasonable to expect that whatever moral theory turns out to be correct, it is likely to assign some positive moral value to the justification of intrinsic preferences. Therefore, it is likely to be valid to take information about the degree to which such preferences are satisfied to be morally relevant information. Hence preference-based information should be considered to be valuable inputs into public decision making processes (see also Randall, 2002).

If the information from preference-based approaches is morally relevant, then what is the appropriate role for CBA in policy making? Most economists take the more modest stance that CBA merely provides information to decision makers which is to be treated as an advisory form of input to any decision making process (e.g. Arrow et al., 1996; Kopp, 1992). The role of CBA should, thus, be seen as providing information to the decision making process and not to be determinative in its self.[24] Hence, individual preferences can provide input towards finding more general rules of action (heuristics) rather than determining the details of a particular decision. Alternatively, information from individual preference-based approaches can be used as a decision making method subject to constraints. For example, CBA tools can be used in so far as this use does not infringe upon a set of basic well-defined human rights. We can view the role or use of constitutions in liberal societies as embodiments of such constraints.

In sum, the inadequacy of welfarism as a moral theory does not invalidate the use of preference-based approaches (such as cost-benefit analysis) as a procedure for guiding environmental policy decisions. This limitation does imply, however, that such approaches should be confined to an 'advisory' or 'information-providing' role in environmental decision making.

5. SHOULD VALUATION HAVE ITS DAY IN COURT?

As set forth in the introduction, the policy issue of greatest concern here is whether it makes sense to extend the use of preference-based approaches to natural resource damage (NRD) assessment determination in Europe. That is, should valuation have its day in court in the EU, or is it a flawed approach to public decision making in this context? Having concluded that preference-based approaches can be justified in policy making but only on limited or constrained basis, we turn now to assess what this implies regarding their use in the context of NRD assessment. As is the case in the debate on the validity of CBA tools in environmental policy decisions generally, a sizeable part of this debate concerns measurement issues (e.g. Shavell, 1993). The objections raised for the use of individual preference-based values mainly concern estimates of so called NUVs. The concerns raised are mostly the same as those found in the general debate on the validity of using stated preference techniques in CBA. Yet, there are two particular arguments raised in relation to using stated preferences estimates for damage assessment. The first concerns accuracy. Some have argued (e.g. Desvousges et al., 1993; Johnson et al., 2001) that damage assessment requires a much higher degree of accuracy that that required for CBA. Errors in welfare estimates for CBA may or may not influence realised outcomes, and realised benefit and costs are usually distributed broadly across many gainers and losers in the population. In contrast, the damages estimated for a NRD assessment may be born by a single or a few responsible parties.

The second point of concern has to do with the costs required to undertake a 'state-of-the-art' CBA. Some have argued (e.g. Shavell, 1993) that in many cases the cost of undertaking the study may exceed the damage itself and thus CBA may not pass a CBA itself!

Shavell (1993) believes that inclusion of preference-based estimates of loss would be costly and increase the bias and risks of the legal procedures, whereas exclusion would not greatly harm incentives when "the true elements of loss are not very large" (p. 379). Yet this line of reasoning breaks down if we accept that NUVs are a large component of natural resources value. Also, it tells us little about whether inclusion of preference-based values is legally justified if measurements of such values could be undertaken cheaply and accurately.

That is, the more fundamental issues with respect to using NUVs and individual preference in damage assessment are not issues of measurement but concern the problem of whether individual preference-based values are compatible with the legal framework of damage assessment. We turn now to assess these issues.

5.1. Is Valuation Consistent with Compensation?

Daum (1993) examines the extent to which damages calculated using preference-based techniques correspond to ordinary legal definitions of compensable damage and loss. Daum argues that though the *ex ante* use of preference-based values for the determination of benefits may be valuable for policy decisions, it does not follow that it is equally useful or desirable to use these methods *ex-post* for the measurement of damages. According to Daum the model of damage calculation embedded in tort law for determining compensation is not compatible with the type of damages that are derived from (stated) preference-based techniques for two reasons: first, stated preference studies are always carried out after the damage has occurred and does not reflect pre-existing values independent of the accident and of the valuation process, and second, stated preference studies simply do not estimate real economic value but something else (e.g. a sense of moral duty). The latter statement is a measurement issue and thus will not be dealt with further here. The first charge, however, is much more substantial. Economists do recognise that WTP to avoid damage is a different welfare concept than the value of damages to an environmental resource after the occurrence of harm. This simply means that stated preference techniques should be designed so as to capture the change in the value of the asset as a result of harm as opposed to estimating WTP to avoid damage. Thus, Daum's point that NUVs and stated preference techniques do not capture the appropriate concept of "compensation for loss" can be rectified by developing stated preference studies with the requirements of the legal system in mind.

Finally Daum argues that preference-based non-use values are not only incompatible with the standard legal notion of compensation but are also *unnecessary* in determining restoration levels (if that should be the prescribed remedy). Remediation of the resource can be applicable to damage to environmental resource: under such a rule the defendant would be liable for the cost of restoring the resource to its condition prior to the accident and is also liable for the interim loss in use values. And Daum concludes that the calculation of costs of restoration requires science (i.e. experts) and not individual preferences (i.e. for Daum the amount people are WTP to prevent harm to a resource has nothing to do with the actual costs of restoring that resource after it has been damaged). Yet, Daum's reasoning does not account for loss in non-use values or for situations where restoration is not feasible (irreversibilities). It also tells us very little about the determination of the type and level of restoration. A resolution of these issues is likely to require some reference to individual preferences. Therefore both compensation and remediation are likely to be consistent with the use of valuation techniques, if properly constructed and applied.

5.2. Is Valuation Consistent with Incentives?

For the *lawyer*, environmental damage cases fall into the domain of tort taw in which the role of damages has a dual role: (a) to compensate the victims for the loss suffered; and (b) to serve as an incentive for the tortfeasor to take cost-justified care to avoid damages (Brookshire & McKee, 1994). The deterrence role is usually described by reference to the so called 'Hand-rule', which provides for incentives to avoid damage to environmental assets to the point where the cost of care is equal to the expected cost of the damages. When the full amount of damages are not calculated (as in the case when NUVs are omitted), then this elementary incentive mechanism breaks down (Posner, 1970; Stephen, 1988; Hirsch, 1979).

For the *economist,* environmental damages are based on the diminished value of the services (both consumptive and non-consumptive) provided by the natural resource as a result of the harm caused. The values measured for these reductions in services represent the monetised change in individual's utility as a result of the injury to the resource. If the value of the diminished NUVs is not included in the damage award, then the award does not reflect the complete loss in monetised well-being to those members of society who benefit from the resource. The prospective efficiency of damage awards in inducing the optimal quantity of due care on the part of those undertaking risky activities rests on the damage award accurately reflecting society's loss once the accident has occurred (Kopp, 1991; Shavell, 1984, 1987). Hence setting the correct 'price' signal is crucial. Not using preference-based values would most likely under-estimate this signal (since non-use values would most likely be excluded) thus leading to inefficient levels of due care. It thus seems that the economist's rationalisation for using individual preferences is compatible with the requiremtn of economic efficiency.

This view has been contested by several authors mainly from the legal profession, e.g. Cummings and Harrison (1994), Daum (1993), Boudreaux et al. (1999) who question the record of success of using individual preference-based techniques to promote efficient levels of environmental protection. Yet these criticisms have almost exclusively been based on arguments that have to do with measurement issues (i.e. the totality of economic values for the environmental cannot be adequately measured) and thus do not challenge the use of valuation techniques at the legal or conceptual levels.

Even if we accept that measurements of economic values form stated or revealed preference techniques are not accurate, the use of this information is still justified if the losses avoided by such use exceed those from not using the information. Seen in this way, those who oppose the use of individual preferences in the determination of damage assessments should demonstrate that the information has no merit. Thus, utilising even imprecise information in damage assessment cases can improve the decision-making process. Skewed indicators of individual preferences can be still useful indicators, provided the ways in which they are skewed are understood. Even prior to its repair, the Hubble telescope was apparently returning valuable information despite the distortions produced by the improper design of the telescope (Hubin, 1994, p. 185).

5.3. The Debate Over Standing

Finally, the most critical issue concerning the use of preference-based techniques in court concerns the question: "Whose preferences matter?" Though there is an extremely voluminous literature on various issues associated with estimating the (unit) value of environmental damages, there has been disproportionately less discussion on the issues of standing, that is the issues involved in determining whose preferences are to be included and whose to be excluded in a natural resource damage assessment.[25,26,27]

Generally, we can say that we should count whoever has suffered a *real* loss. Determining this population is relevant for both the purposes of sampling and aggregating. Sampling will produce an estimate of unit average damage. Aggregation will produce the total amount of damages. The choice of the relevant affected population will affect the estimated shape of the demand function but, more importantly, the choice of population will have an even greater effect on the estimated level of damages. Hence, if we were merely interested in unit mean values, then the problems of defining the relevant population are not so severe. Yet, in environmental damage assessment aggregate values are what matter and hence determining who should be included in the aggregation population can have profound consequences for the outcome of the litigation process.[28]

The economic conception of standing is much broader than the legal definition. It implies that everyone who experiences a real welfare loss should be included in the aggregation population (Whittington & Macrae, 1986). Legal standing is a much less inclusive concept and includes those individuals that can pursue a lawsuit or other cause of action against another party. When property rights are certain, determining legal standing is straight forward. Yet, in cases involving natural resource damages property rights over the resources involved are often uncertain and hence standing is far from an unambiguous matter. What must be resolved is not which individuals have experienced a welfare loss but which individuals have experienced a *compensable* welfare loss. Commentators have tried to discern the legal and economic constraints that delineate the appropriate "welfare space" for assessment of natural resource damages.[29] All those individuals in the appropriate welfare space that experience a loss consistent with these economic and legal requirements/constraints are to be included in aggregation. Yet, there is considerable debate on the nature and extent of these constraints. A categorisation and clarification of some of these issues follow below. Only a selection of the issues involved is presented.[30] The issue of standing is still very much open both in the courts and in the academic journals.[31] The exposition highlights some of the misunderstandings and disagreements between economists and lawyers rather than purporting to offer a definitive resolution.

5.3.1. Standing and Use Values

For the case of use values, determining the population for sampling and aggregation is less contentious. All individuals who can reasonably claim an expectation of possible or potential future use may be included in the population. There are some disagreements between the economic and legal conception of standing over certain categories of individuals such as children, 'rubberneckers' (those who go and observe damaged natural sites and clean-up/restoration operations), and tourists and foreigners (e.g. illegal aliens). Disagreement also exists over how to count individuals that claim damages when the facts (or experts) attest that there is no physical injury to the environmental site (e.g. individuals who continue not to use a damaged recreational site because they are unaware or not convinced that the site has been adequately resorted). Rulings in court cases have varied on whether and in what way preferences of such individuals are to be counted in the aggregation process. For a theoretical discussion of these issues and references to contrasting case rulings in the U.S. see Dunford et al. (1997) and Randall (1997), Whittington and Macrae (1986), Trumbull (1990), Zerbe (1991, 1998, 2001). For an exposition of how opposing parties and courts deal with such issues of standing for use values in practice see Chapman and Hanemann. (2000) who report on The American Trader Case, one of the few cases that used individual preference-based values that was not settled out of court.[32]

5.3.2. Standing and NUVs

The issues surrounding the issue of standing for estimating unit and aggregate NUVs are much more contentious. Apart from measurement issues, the main problems of standing for non-use damages are two. The first raises concerns over the 'legitimacy' of various motivations (namely altruism and moral commitment) leading to individual NUVs while the second concerns the extent of the welfare space that defines compensable losses in non-use values.[33] Whilst, the criticism on the 'illegitimacy' of motives for NUV can be dismissed on grounds of misconception of modern utility theory, the issues raised as to whose non-use preferences are to be counted are much more serious. In practice the courts in the U.S. have been inconsistent in defining the relevant population of non-users[34] while one of the few legal disputes in the U.K. that considered non-use values also produced conflicting results.[35]

The difficulties with determining the population of non-users are not confined to cases where property rights are poorly defined (for example who should have standing and who should compensate whom for loss in NUV from the extinction of a species in a habitat with open access?). Even if property rights are well defined, as is the case when the trustees act on behalf of society, the problems of defining who is included in this 'society' remain.

NUV was defined as the value one obtains from a natural resource when no present or future direct personal use is realised or intended. It is best to think of NUVs as not held over natural objects themselves but over the flows (or uses) these resources generate. Since non-use value by definition excludes personal enjoyment of these uses, it can be inferred that NUVs are derived from the knowledge that certain flows from a natural resource benefit certain other constituents (other people in the current or future generation leading to altruistic and bequest values or nature itself leading to existence and intrinsic values).[36] Hence, human *perception* or some *knowledge* about the resource is an important part of the definition of NUVs and has been the basis for the debate over standing.

Dunford et al. (1997) and Johnson et al. (2001) have argued that demand for knowledge about the resource and/or its injury are required for one's NUV to have legal standing. The authors acknowledge that since NUV leave a very poor behavioural trail, the courts are uncertain as to who has in fact experienced a loss in NUV (and thus who has standing) as the result of an injured natural asset. They argue, however, that observing the demand for information about the resource and/or its injury can provide a good indication as to who in fact has experienced such a loss and thus who should be compensated. They suggest using marketing questionnaire techniques (similar to those used in stated preference methods) to ascertain the percentage of people in a society (which could extend to the national level) that have some prior knowledge of the resource and some current or potential demand for information about the injury. They argue that it is only these individuals that should be granted legal standing. The rationale of the argument is that people with no prior demand for information about the resource and/or its injury in fact do not have true non-use values. That is, the lack of such demand for information tells the court something about the true preferences of these individuals. NUVs were defined as being a matter of conception and conception, their argument goes, involves some prior knowledge. Information acquisition activities involve opportunity costs are thus are indicators of one's interest in (or intensity of and preferences for) a particular natural resource. Respondents in CV studies that have not (endogenously) acquired such information nevertheless receive (exogenous) information from the study itself. The authors in essence are claiming that expressed non-use values from individuals with no prior or no intended demand to acquire information are somehow "induced", "constructed", "hypothetical" or even "fictional" preferences and that the subsequent estimated losses would not have occurred if the respondent had not been sampled. The usefulness of the estimated values from such individuals for damage assessment is questionable. This raises the familiar issues of the role of information in stated preference studies.[37] Though the literature provides ambivalent guidance in resolving these informational issues, the crux of their arguments point to an important distinction between economic and legal standing for NUVs. The emphasis on supplying information to respondents makes sense in "traditional" non-use value studies designed to help policy makers evaluate the potential benefits of policy alternatives. These are *ex ante* studies of proposed changes and thus neither the entire number of constituents of a society nor the sample used in a stated preference study can have knowledge of the proposed changes. Further, measures of awareness and knowledge may be poor indicators of voting behaviour, regulatory mandates, or budget-allocation decisions and may have little to contribute to determining economic standing. It does not necessarily follow, however, that supplying information to respondents is also appropriate when assessing *ex post* compensation for actual welfare losses from a sample of respondents representing the general population (Dunford et al., 1997). Hence, attempts by natural resource trustees to measure aggregate losses in NUVs over informationally unrepresentative sub-samples of larger populations may be inconsistent with the *revealed* knowledge and concerns of that population (Johnson et al., 2001, p. 61).

Economists are divided over the necessity of positive (actual or potential) information demand as a precondition for real compensable losses in NUVs (e.g. see Zerbe (2001, 1998) and Randall (1997) arguing against while Moran (2000) arguing in favour of it). In the former camp there are two counter-arguments worth mentioning.

First, it has been argued that individuals have preference over *classes of environmental goods* (not particular types of environmental resource) and thus they would suffer a legitimate loss in NUV from a damage to a particular environmental asset even if they had no prior knowledge of the asset and/or the injury (Randall, 1997; Zerbe, 2001). Randall describes the existence of such preference emerging as a form of heuristic to deal with the realities of an overwhelming complex world and incomplete knowledge: people care about a class of things implies caring about particulars in that class. People that have such a class in their utility function once informed about the injury to a particular member of this class *may* suffer a utility loss. There are several objections to this reasoning. First, the fact that some individuals have resorted to developing such heuristics tells us something about their preferences. Second, accepting general rather than specific knowledge of environmental resource allows for the aggregation population to be overwhelmingly large. This may be reasonable for some unique natural resources but is not convincing for resources with many substitutes. Third, accepting that individuals care about *classes* of environmental resource poses problems on interpreting how people make choices over *specific* resources when asked to do so. That is, if, for example, people care about 'all species', on what basis can their intensity of preferences (i.e. their values) differ for one particular species to another? Would this mean that individuals would have the *same* value for *any* member of the class of resources? If not, then on what basis would these values differ other than individuals have different orderings for such specific preferences? Fourth, economists fail to appreciate/distinguish that such types of meta-preferences can allow for one to have economic but not legal standing. The purpose of damage assessment is to obtain compensation for injuries to *specific* natural resources. Thus general knowledge of 'the environment' is not sufficient for *legal* standing. While it may be good public policy to protect the environment (economic standing), there is no basis for crediting unaware citizens with *compensable* welfare losses (Johnson et al., 2001).[38]

Secondly, one may argue that prior knowledge of the resource is not required since society "owns" the resources managed by trustees (Zerbe, 1998). Yet this view ignores a crucial difference between NUVs for public resources and private property. NUVs do not exist independent of individual perception. Hence, losses in NUVs require some prior knowledge whereas losses in use values do not. Also, justifying legal standing on property rights is troublesome since property rights are often not clear (e.g. to whom do these rights extend to?).

Zerbe (2001, 1998) provides an argument similar to that found in Randall but base it in the context of rights. He argues that individuals care about environmental wealth in general and that once they are informed about the damage to a particular environmental resource they may suffer a real and legitimate loss in non-use value. He provides an interesting example of a rich individual who owns many firms which are run by managers. Though the wealthy individual does not have knowledge of his specific firms he/she would receive a legitimate welfare loss if were to find out that one of his/her enterprises went bankrupt.[39] There are two problems with this example: first the individual has the *private* property right over all his/her firms. The individual receives use value from his/her wealth. Value from privately-held resources is not a matter of perception (it arises from personal benefits enjoyed by the individual) while non-use values over commonly-owned resource arise from the knowledge that certain environmental flows accrue to others. Hence, prior knowledge is a requirement for non-use values to exists *independently* of a CV study. Further, we can interpret individuals with no demand for knowledge of the resource as having 'waived' their right to the resource and thus as not having standing.

5.4. Conclusion – Should Valuation have its Day in Court?

The previous sections provided a brief exposition of some legal and economic theory arguments for the inclusion of NUVs in damage assessment. It was shown that the need to include such values can be debated on both efficiency (economic) and tort law (legal) grounds. On balance there appears to be no logical or moral grounds for excluding this information, if it is used for the limited purpose of aiding improved decision making. Yet, considering the conceptual and measurement issues that have concerned many lawyers and economists, one would be enticed to ask whether these values, and NUVs in particular, are sufficiently large to necessitate their inclusion in damage assessment. If the court could somehow know a priori that the NUVs for a particular case would be small, it could avoid the complications and costs of their estimation.

For the purposes of answering this question, the literature on NUVs has emphasised the uniqueness or 'specialness' of the resource in question and the irreversibility of the loss or injury as criteria for generating large NUVs. In addition the literature suggests that NUVs may be small in cases where recovery from an injury is quick and complete, either through natural processes or via restoration acts. Yet there are problems in giving operational meaning to the idea of uniqueness. In economic terms, uniqueness would be reflected in the absence of substitutes and a low price elasticity of demand. Yet Freeman (1993) points out that there is no threshold on price elasticity that distinguishes between the presence or absence of close substitutes. Similarly, long-term injury with slow recovery (e.g. restoring a whale population) could give rise to NUVs that are of the same order of magnitude as those with irreversible injury (Freeman, 1993). These issues are not yet resolved which signifies the need for ongoing comparative research that tries to identify factors which could give a priori indications when NUVs are bound to be small.[40]

In sum, we believe that the input of individual preferences in damage assessment is compatible with the basic foundations of tort law since it promotes both the compensatory and deterrent role of damages. Though the assignment of property rights that would give rise to non-use values is problematic when the environmental resources in question are privately owned, the assignment of such rights for publicly owned resources is quite sound.

6. CONCLUDING REMARKS

The role of individual preferences and cost-benefit analysis in environmental decision making has been extensively debated by economists, lawyers and philosophers Yet, despite the voluminous literature, the discussion remains disordered and confused. Further, though CBA has been widely used, in practise its actual influence on policy has been relatively limited. Despite this limited acceptance and tainted history, the European Union is now considering a new Directive on Civil Liability that might imply bringing valuation into European courtrooms. The paper examined whether valuation should have a role in environmental policy and legal processes. The paper provided only an eclectic survey of the issues concerning this debate by focusing on the question of the extent to which individual preference-based values are suitable in guiding environmental policy and damage assessment decisions. Three criteria for "suitableness" were reviewed: conceptual, moral and legal.

The preceding discussion of these issues suggests that: (i) the concept of economic value as applied to environmental resources is a meaningful concept based on the notion of trade-off and opportunity cost. It was argued that a substantial portion of the criticism on the validity of the economic concept of value is ill-targeted since it is based on a misconceived understanding of the nature and scope of the concept of economic value; (ii) the limitations of the moral foundations of CBA do not invalidate its use as a procedure for guiding environmental decision making. It was further argued that individual preference-based approaches would be relevant to policy makers operating under a broad range of accepted moral theories (consequentialism, contractualism, deontology); (iii) the input of individual preferences into damage assessment is compatible with the basic foundations of tort law for determining compensation since it promotes both the compensatory and deterrent role of damages; (iv) using individual preference-based methods in damage was shown to provide incentives for efficient levels of due care; (v) the most critical issue concerning the use of preference-based techniques in court concerns the questions of standing. It was shown that determining the relevant population that has experienced a compensable welfare loss is still very contentious for various categories of users as well as for aggregating non-use values.

Overall, the discussion suggests that the use of preference-based approaches in the both policy and legal arenas is warranted provided that they are accurately applied, their limitations are openly acknowledged and they assume an information-providing rather than a determinative role.

NOTES

1. Throughout the paper we will be referring to preference-based approaches as cost-benefit analysis (CBA). For a concise history of CBA (Pearce, 2000).

2. There are other reasons apart from disagreements or lack of proper understanding over measurement, conceptual, moral and legal issues related to the use of individual preference-based values. Scientists often feel threatened by economic estimates in that their own input or importance in the decision-making process is diminished. Also, regulators have shown antipathy towards CBA stemming from the need to avoid controversy and from the anxiety that their regulatory discretion and flexibility will be somehow diminished (Pearce, 1999). Finally, CBA has been criticised for neglecting distributional issues. Though this is partly true (see for example Zerbe, 1998, for how CBA can include distributional concerns) the fact remains that CBA fails when it comes to address issues of equity. Yet, CBA is not the only decision-making tool-kit that fails to account for multiple policy objectives (such efficiency, equity, etc.). See Section 4 on suggestions for the appropriate (albeit limited) role of CBA.

3. We acknowledge that there is considerable overlap between the three levels of the discussion (conceptual, moral and legal). Yet, the issues involved in each level are sufficiently different that a separate discussion is warranted.

4. Very roughly, the former refer to values associated with the direct in-situ use of the services provided by environmental resources (e.g. recreation) while the latter refer to individual values that are not associated with any current, potential or future personal use of any such services.

5. A striking illustration of the range of results produced by CBA techniques is given by Stirling, 1997. The author analysed over thirty published CBA studies of the external environmental costs of coal-fired power stations whose individual results were often expressed with a high degree of precision. But taken as whole, the results were so varied that they had to be expressed on a log scale table, with the highest values some 50,000 times the lowest. One message for policy makers is at least to be aware of the uncertainties involved, and to be clear about underlying assumptions.

6. For an introductory discussion of these techniques (Bateman, 1999; Freeman, 1993; Dixon et al., 1988).

7. See Larson (1992) for an alternative view.

8. See Freeman (1993) for a thorough discussion.

9. Three such techniques have been widely used: *'changes-in-productivity'* approaches where impacts on environmental quality are reflected in the changes in the productivity of the systems involved and these, in turn, are used to assign values. The physical changes in productivity (e.g. crop yield) are valued using market prices for inputs and outputs. *'Loss of earnings'* approaches measure the impacts on environmental quality from changes in human productivity. The value of lost earnings and of medical costs created from the degradation in the quality of some environmental resource (e.g. water poisoning) is used under such approaches as a proxy for environmental value.

'Opportunity cost' approaches are based, as the term suggests, on the concept of opportunity costs: the value of using an environmental resource for a particular purpose is approximated with the value in forgone income from alternative uses of that resource. (see Dixon et al., 1988; Freeman, 1979, for a detailed exposition of such approaches).

10. This set includes *'cost-effectiveness'* analysis where a predetermined goal or objective regarding the quality of an environmental asset is set and then the most cost effective means of achieving it are chosen and *'preventive or mitigation expenditure'* approaches where the value of an environmental recourse is approximated by the cost of the preventive measures that people are willing to pay to avoid any damage to it or from the cost savings obtained from a reduction in maintenance cycles due to reduced damage rates.

11. A substantial part of the deliberations in many environmental liability cases in the U.S. have centred around the differences between pricing/costing and valuation techniques. For example, in the American Trader oil spill case the defence brought the very concept of 'consumer surplus' into dispute and argued that reliance on existing market price and cost data would suffice for a decision on damages to be reached. (see Chapman & Hanemann, 2000, for a detailed account of these arguments between the legal defence and the economists which were acting as expert witnesses in this case). It is, thus, useful to elaborate further on why economists argue that pricing techniques do not provide adequate measures of the benefits (loss) experienced by society from reduced (increased) damage in environmental resources. Cummings (1991) has shown that the market prices used by policy makers and the courts since the 1950s in the U.S. do not reflect economic values. He argues that violations of the assumptions of perfectly competitive markets and mobility of agents are the root of the problem. Hanemann and Keeler (1995) have further shown that even without such violations, market prices fail as a measure of value for *non-marginal* changes in environmental resources. This has been understood by economist since Hotelling's (1938) exposition of how the correct measure of value for non-marginal changes in the allocation of market goods is the change in consumer surplus. This is given by the area under two relevant demand curves or equivalently by people's willingness to pay for reduced damages (or the willingness to accept to tolerate these damages). WTP to prevent damage may be larger, smaller or equal to estimates from pricing or maintenance cost techniques. For marginal changes or for goods that are perfectly divisible, market prices work adequately as measures of welfare. When one uses market prices to measure the marginal value for a divisible market good, heterogeneity in preferences becomes irrelevant, and aggregation is trivial. At the margin, all consumers who face the same price have the same marginal value, regardless of their preferences, income or other commodity or individual attribute. All that the policy maker or the courts needs to know about peoples marginal value of the good is provided in the market price. There is no need for further knowledge about the actual demand curve. In addition, since all individuals have the same value at the margin, aggregation of marginal value across consumers is relatively simple. This is not so for non-divisible goods with non-marginal changes. In this case, knowledge of the demand curve is required in determining individual welfare changes and preference heterogeneity becomes important in obtaining aggregate welfare estimates. This recognition has led to an important paradigm shift which moved the central focus of valuation in economics away from market prices and towards demand curves as the core repository of value. (Hanemann & Keeler, 1995, pp. 5–6). Such curves

or functions are behavioural relations, and the key implication of the behavioural shift is that economics re-affirmed itself as not merely the study of markets but more broadly the study of human preferences and behaviour.

12. Economists have recognised some of the appealing features of these methods and are attempting to develop 'hybrid' methods that combine economic and participatory approaches. Notable examples are the Market Stall approach (Macmillan et al., 2000) and the Valuation Workshop approach (Kenyon & Hanley, 2000).

13. Examples of the use of such techniques in the United Kingdom include a 1997 citizen's jury organised by the Welsh Institute for Health and Social Care on the subject of genetic testing for common disorders in the National Health Service. Moreover, the first attempt to apply the Danish model of consensus conferences involving a cross-section of the lay public was the 1994 three day Conference on plant biotechnology organised by the Science Museum in London and funded by the Biotechnology and Biological Sciences Research Council. See Royal Commission on Environmental Pollution (1998) 21st Report Setting Environmental Standards Cm 4053 HMSO, London.

14. The lay panel in planning cells is the main actor in the process, determining the expert panel that provides the information and the questions to be asked. The process consists of three steps: education and reception of information on the topic so that the panel members can formulate specific questions to be explored; processing of information through panel discussions, hearings, and questioning of experts; and group deliberations and findings. (Dienel & Renn, 1995; Sclove & Scammel, 1999; Fixdal, 1997). The planning cell procedure draws from Multi-attribute Utility Theory to elicit values, criteria, and attributes and the assignment of relative weights to the different value dimensions. Participants are asked to rate each decision option on each criterion that they deem important. Each criterion is weighted against each other criterion resulting in a matrix of relative weights and utility measures for each option and each criterion. Both tasks (the transformation in utilities and the assignment of trade-offs) are performed individually and in small groups (Dienel & Renn, 1995). The process is facilitated by a neutral third party. Results are generally widely distributed in the media and are the basis for further local hearings. Consensus conferences generally address broader issues than normally addressed by experts, and they issue broader recommendations. A Norwegian lay panel on genetically modified foods, for example, found that such foods were not needed because the selection and quality of food was already sufficient and there was too much uncertainty about the potential impacts of these foods on health and the environment. (Tickner & Ketelsen, 2001). For a review of applications of consensus conferences and planning cells in Europe and the U.S. (Dienel & Renn, 1995).

15. See for example Nijkamp and Voodge (1984) for an introduction to MCA.

16. The technique was developed by the Rand Corporation during the 1950s and 1960s (Pearce & Mourato, 1998).

17. Kuo and Yu (1999) use the Delphi technique to assist selection of which areas to e designated as national parks in Taiwan while Macmillan et al. (1998) use this method for cost-effective analysis of woodland ecosystem restoration.

18. Money is merely used to simplify matters by providing a single metric against which all states of the world can be traded-off.

19. See Zerbe (1998, p. 425) for similar arguments against the "citizen vs. consumer" argument.

20. The separation of the discussion between the moral relevance of individual preferences for policy decisions and legal compatibility for damage assessment reflects

the general debate in the existing literature that acknowledges that individual preferences and expert opinion may have differential roles or varying degrees of validity in these two fields. This debate primarily has focused on the use of individual preferences that lead to so called non-use values for environmental resources and is summarised in Table 2. That is, though most economists would agree that inclusion of use values is equally valid for both policy and damage assessment decisions, there is no such consensus regarding NUVs.

21. Of course there are other moral issues associated with the use of CBA. For example, some ethical philosophers argue that it is morally objectionable to debase the environment and render it as 'saleable' good. The argument is associated with the other familiar arguments that environmental goods are subject to lexicographic preferences. The justification here of the 'lexical' argument is made on moral rather than conceptual grounds. This view neglects and ignores the opportunity costs involved in conserving the environment. Philosophical discussion has concerned arguments on whether and on what basis environmental resource have some superior or higher order moral status over other (public) goods that would allow for a moral justification for ignoring costs (see Randall, 2002; and Pearce, 2000 for a discussion).

22. The Potential Pareto Improvement criterion: CBA as an empirical test for PPIs. In essence PPI implements welfarism (Randall, 2002).

23. The most effective points against the moral foundations of CBA include: (a) CBA moral theory assigns a morally unjustified status to the current state of affairs; (b) it fails to accord the appropriate role to considerations of distributive justice; (c) it fails to accord the proper status to future generations and to those individual/agents (human and non-human) lacking the cognitive abilities to express WTP/WTA; and (d) CBA moral theory endorses a naive form of subjectivism (Hubin, 1994).

24. Reasons for rejecting an unrestricted/decisive role for CBA include: (a) CBA *itself* does not allow any role for side constraints on government action (e.g. CBA itself would not allow for a constitution); (b) CBA only captures economic values. Non-economic (e.g. intrinsic, non-anthropocentric values) are not captured (see Turner, 2000); (c) the reliance on WTP/WTA skews the analysis in favour of those with greater initial endowments; (d) CBA is indifferent to matters of distribution (this is a consequence of the fact the CBA is rooted in consequentialist moral theories) (Hubin, 1994). These are 'result oriented' objections to CBA (i.e. there are objections directed against the kind of choices made as a result of strict application of CBA). Yet, there may be even more fundamental 'process-oriented' objections. For example, most would object to dictatorial procedures even if they did reach the same results as democratic ones.

25. The term standing to refer to the issue of who is to be counted in CBA has been coined by Whittington, and Macrae (1986).

26. Considering that determining the relevant population determines both the estimated demand function (required to estimate unit damage values) and the subsequent estimated aggregated values such lack of comparative attention is in fact irregular. One explanation for this could be that such issues are of political or normative nature and should not be the subject matter of economics. Of course, economics and CBA is unavoidably laden with value judgements and hence such an assertion bears little weight.

27. Though the emphasis in U.S. regulation has shifted from monetary to in-kind compensation, the present discussion of standing is still relevant.

28. The Eagle Mine case is typical of the relative importance of the standing issue over the issue of estimating average unit damages. In this case the state of Colorado

sought damages for the release of hazardous substances into groundwater. What is interesting is that although both the trustees and the defendants' estimates of unit average damages coincided, their estimates of aggregate damages differed by several orders of magnitude (see Kopp & Smith, 1989 for more details).

29. The term "welfare space" is attributed to Trumbull (1990).

30. For a more comprehensive view of the debate see Dunford et al. (1997), Randall (1997), Johnson et al. (2001), Zerbe (1991, 1998, 2001) Trumbull (1990), Whittington and Macrae (1986) Kopp and Smith (1989). The discussion in these papers assume that non-use values are invariant across individuals. The issue they discuss is how to identify who has standing and then impute the same average value to the specified population. The discussion concerns "who counts" in aggregation. Yet, the issue of standing can also be viewed as a matter of degree. That is, individuals may have partial and full standing. Here the issue is "how much weight do we assign to each individual or group of individuals". Such forms of 'non-temporal discounting' can be performed using income weights, distance decay assumptions or other variables affecting WTP. For a discussion see Bateman et al. (2000), Moran, 2000, Pearce (2000), Trumbull (1990) Johnson et al. (2001), Pate and Loomis, (1997) and Sutherland and Walsh, (1985).

31. "Of all the issues of CBA few are misunderstood more", Trumbull (1990, p. 201).

32. Discussion of other natural resource cases in the U.S. can be found Brown et al. (1983), Kopp and Smith, (1993) and Ward and Duffield, (1992).

33. The issues of standing for use values (children, rubberneckers, tourists, foreigners etc.) mentioned above also apply to NUVs and become even more troublesome.

34. In the Nestucca oil spill case, for example, the populations of Washington and British Columbia were used for estimating damages, while in the case of the Exxon Valdez spill, the population of the entire United States was held to be the potentially affected population. In a more recent case, Montrose Chemical Corp. v. Superior Court, the Trustees defined the potentially affected population as the English-speaking households in California (Zerbe, 1998).

35. See Moran (2000) for a description of how the issue of standing over NUVs was handled in a case between Thames Water Utilities and the U.K. Environment Agency over ground water abstraction damages.

36. The concept of intrinsic value should not be interpreted as meaning the value something has in and of itself irrespective of any human 'valuer'. Such a metaphysical conception of value may have philosophical basis but is of no practical merit. That is, it is entirely irrelevant in a framework that involves making choices. Instead, intrinsic value can be interpreted in an anthropocentric manner, in that a human agent must acknowledge such a value. Hence, 'trees do have standing' if people have preference for granting such rights (on this issue see Stone, 1974).

37. For an overview of these issues see Munro and Hanley (2000), Chilton and Hutchinson (1999), Blomquist and Whitehead (1998), Blomquist and Whitehead (1998), Cameron and Englin (1997), Boyle, K. J. et al. (1995), Whitehead and Blomquist (1991), and Bergstrom, Stoll and Randall (1990).

38. Note that there is also ample empirical evidence that WTP from non-users declines and eventually is reduced to zero when demand for information is absent. Various studies have shown that NUV have declined with distance and familiarity with the resource. See Bateman et al. (2000), Moran (2000), Pate and Loomis (1997), Smith and Desvousges (1986), Peters et al. (1995) and Sutherland and Walsh (1985).

39. Presumably the individual in the hypothetical example has inherited his/her wealth since otherwise the individual would have engaged in information acquiring information in order to build his/her fortune.

40. Meta-analytic research of existing valuation studies could be potentially useful to address these issues (see e.g. Loomis & White, 1996).

ACKNOWLEDGMENTS

This work has been partly funded by the European Commission's Energy, Environmental and Sustainable Development Programme (under the EMEREG and BIOECON projects). The authors would like thank Prof. David W. Pearce, Prof. Richard B. Stewart, Prof. Richard O. Zerbe for helpful comments to this paper.

REFERENCES

Ajzen, I., Brown, T., & Rosenthal, L. H. (1996). Information Bias in contingent valuation: effects of personal relevance, quality of information, and motivational orientation. *Journal of Environmental Economics and Management, 30*, 43–57.

Agnolucci, P. (2001). Comparing Contingent Valuation and Expert Opinion studies: an application to the Management of European Remote Mountain Lakes. University College London, Department of Economics, Thesis for the MSc in Environmental and Resource Economics.

Arrow, K., Solow, R., Leamer, E., Portney, P., Radner, R., & Schuman, H. (1993). Report of the NOAA Panel on Contingent Valuation. *Federal Register, 58*, 4601–4614.

Arrow, K. J., Cropper, M. L., Eads, G. C., Hahn, R. W., Lave, L. B., Noll, R. G., Portney, P. R., Russell, M., Schmalensee, R., Smith, V. K., Stavins, R. N. (1996). Is There a Role for Benefit-Cost Analysis in Environmental, Health, and Safety Regulation? *Science*, (Vol. 272, No. 5259), Issue of 12 April 1996, 221–222.

Arrow, K. (1951). *Social Choice and Individual Values*. New York: Wiley.

Barendse et al. (Eds) (1998). NOW International Conference, Beyond Sustainability integrating behavioural, economic and environmental research. Conference Report, 19–20th November 1998, Amsterdam, Netherlands.

Bateman, I. J. (1999). Environmental impact assessment, cost-benefit analysis and the valuation of environmental impacts. In: J. Petts (Ed.), *Handbook of Environmental Impact Assessment*, (Vol. 1) *Environmental Impact Assessment Process, Methods and Potential*. Oxford: Blackwell Science.

Becker, G. (1993). Nobel Lecture: the economic way of looking at behaviour. *Journal of Political Economy, 101*, 385–409.

Beckerman, W., & Pasek, J. (1996). Plural Values and Environmental Valuation. CSERGE Working Paper GEC 96–11.

Beierle, T. C. (2000). The Quality of Stakeholder-Based Decisions: Lessons from the Case Study Record. RRF Discussion Paper 00–56.

Bergstrom, J. C., Dillman, B. L., & Stoll, J. R. (1985). Public environmental amenity benefits of private land: the case of prime agricultural land. *Southern J. Agric. Econ., 17*, 139–149.

Bergstrom, J. C., Stoll, J. R., & Randall, A. (1990). The impact of information on environmental commodity valuation decisions. *Am. J. Agric. Econ.*, *72*, 614–621.

Bishop, R. C., & Welsh M. P. (1992). Existence values in Benefit Cost Analysis and Damage Assessment. *Land Economics*, *68*, 405–417.

Bishop, R. C. (1978). Endangered Species and Uncertainty: The Economics of Safe Minimum Standard. *American Journal of Agricultural Economics*, *60*(1), 10–18.

Blamey, R. K. et al. (1995). Respondents to Contingent Valuation Surveys: Consumers or Citizens? *Australian Journal of Agricultural Economics*, *39*, 263–288.

Blamey, R. K. (1998). Decisiveness, Attitude Expression and Symbolic Responses in Contingent Valuation Surveys. *Journal of Economic Behaviour and Organisation*, *34*(4), March 1998, 577–601.

Blomquist, G. C., & Whitehead, J. C. (1998). Resource quality information and validity of willingness to pay in contingent valuation. *Resource and Energy Economics*, *20*, 179–196.

Boadway, R. W., & Bruce, N. (1984). *Welfare Economics.* Oxford, England: Basil Blackwell.

Bonnieux, F., & Rainelli, P. (2000). Contingent valuation methodology and the EU Institutional Framework. In: I. Bateman & K. Willis (Eds), *Valuing the Environment Preferences.* Oxford University Press.

Boudreaux, D. J., Meiners, R. E., & Zywicki, T. J. (1999). Talk is cheap: the existence value fallacy. *Environmental Law*, *29*(4), 765.

Boyle, K. J. et al. (1995). Validating contingent valuation with surveys of experts. *Agricultural and Resource Economics Review*, *2*, 247–253.

Brans, E. H. P., & Uilhoorn, M. (1997). Liability for Ecological Damage and Assessment of Ecological Damage. Background Paper for EU White Paper on Environmental Liability.

Breedlove, J. (1999). Natural Resources: Assessing Non-market Values Through Contingent Valuation. Congressional Research Service, Report for Congress.

Brookshire, D., & McKee, M. (1994). Is the Glass half empty, is the Glass half Full? Compensable Damages and the Contingent valuation. *Natural Resource Journal*, *34*, 51–72.

Brookshire, D. S. et al. (1986). Existence values and Normative Economics. *Water Resource Research*, *22*, 1509–1518.

Brown, T. C., Peterson, G. L., & Tonn, B. E. (1995). The values jury to aid natural resource decisions. (Speculations.) *Land Economics*, *71*(2), 250–260.

Brown, G. M., Jr., Congar, R., & Wilman, E. A. (1983). Recreation: Tourists and Residents, Chapter 4 in U.S. National Ocean Service. Assessing the Social Costs of Oil Spills: The Amoco Cadiz Case Study. National Oceanic and Atmospheric Administration Report NTIS PB84–100536, Washington, D.C.

Cameron, T., & Englin, J. (1997). Respondent Experience and Contingent Valuation of Environmental Goods. *Journal of Environmental Economics and Management*, *33*(3), 296–313.

Carlsen, A. J. et al. (1994). Implicit Environmental Costs in Hydroelectric Development: An analysis of the Norwegian master Plan for Water resources. *Journal of Environmental Economics and Management*, *25*(3), 201–211.

Carson, R. T., Flores, N. E., & Meade, N. F. (2001). Contingent valuation: controversies and evidence. *Journal of Environmental and Resource Economics*, *19*, 173–210.

Carson, R. T. et al. (1994). Contingent valuation and Lost Passive Use: Damages from the Exxon Valdez. RFF Discussion Paper 94–18.

Chapman D. J., & Hanemann, W. M. (2000). Environmental Damages in Court: The American Trader Case. U.C. Berkeley – Dept. of Agri. and Resource Econ. Working Paper No. 913.

Chilton, S. M. & Hutchinson, W. G. (1999). Exploring Divergence between Respondent and Researcher Definitions of the Good in Contingent Valuation Studies. *Journal of Agricultural Economics, 50*(1), 1–16.

Ciriacy-Wantrip, S. von (1968). Resource Conservation: Economics and Politics (3rd ed.). Berkely, CA: *University of California Division of Agricultural Economics, 60*, 10–18.

Christopher, L. L. (1994). The role of property rights in economic research on U.S. wetlands policy. *Ecological Economics, 1*(11), 27–33.

Clark, J., Burgess, J., & Harrison, C. (2000). I struggled with this money business – Respondents' Perspectives on Contingent Valuation. *Ecological Economics, 33*(1), 45–62.

Common, M, Reid, I., & Blamey, R. (1997). Do Existence Values for Cost Benefit Analysis Exist? *Environmental and Resource Economics, 9*(2), 225–238.

Copp, D. (1985). Morality, Reason and Management Science: the rationale of cost-benefit analysis. In: E. Paul, J. Paul & F. Miller (Eds), *Ethics and Economics* (pp. 128–151). Oxford: Blackwell.

Crosby, N. (1995). Citizens Juries: One Solution for Difficult Environmental Questions. In: O. Renn, T. Webler & P. Wiedemann (Eds), *Fairness and Competence in Citizen Participation: Evaluating Models of Environmental Discourse* (pp. 157–174). Technology, Risk, and Society series. Dordrecht; Boston and London: Kluwer Academic.

Crowards, T. (1995). Non-use values and Economic Valuation of the Environment: A Review. CSERGE Working Paper GEC 95–26.

Cummings, R. G., & Harrison, G. W. (1994). Was the Ohio court well informed in their assessment of the accuracy of the contingent valuation method? *Natural Resource Journal, 34*, 1–36.

Cummings, R, G. (1991). Legal and Administrative Uses of Economic Paradigms: A Critique. *Natural Resource Journal, 31.*

Damage Assessment and Restoration Programme (1999). Discounting and the Treatment of Uncertainty in Natural Resource Damage Assessment. Technical Paper 99–1, NOAA: Maryland.

Daum, J. F. (1993). Some Legal and Regulatory Aspects of Contingent Valuation. In: J. A. Hausman (Ed.), *Contingent Valuation: A Critical Assessment.* North-Holland, Amsterdam, London, New York, Tokyo.

Desvousges, W. H. et al. (1993). Measuring Natural Resource Damages with contingent Valuation: Tests of Validity and Reliability. In: J. A. Hausman (Ed.), *Contingent Valuation: A Critical Assessment.* North-Holland, Amsterdam, London, New York, Tokyo.

Diamond, P. A., & Hausman, J. A. (1994). Contingent valuation: is some number better than no number? *Journal of Economic Perspectives, 8,*(4), 45–64.

Diamond, P. (1996). Testing the internal Consistency of Contingent valuation surveys. *Journal of Environmental Economics and Management, 30,* 337–347.

Dienel, P. C., & Renn, O. (1995). Planning cells: A gate to fractal mediation. In: O. Renn et al. (Eds), *Fairness and Competence in Citizen Participation* (pp. 117–140). Boston: Klewer Academic.

Dixon, J. A. et al. (1988). *Economic Analysis of the Environmental Impacts of Development Projects.* London: Earthscan.

Dunford, R. W., Johnson, F. R., Sandefur, R. A., & West, E. S. (1997). Whose Losses Count In Natural Resources Damages? *Contemporary Economic Policy, 15*(4), 77(11).

Dworkin, R. M. (1980). Is Wealth a Value? *The Journal of Legal Studies, 9*(2), 191–226.

Edwards, S. F. (1992). Rethinking Existence Values. *Land Economics, 68,* 120–122.

English et al. (1993). Stakeholder Involvement: Open Processes For Reaching Decisions About The Future Uses Of Contaminated Sites. Working Paper, Waste Management Research and Education Institute, University of Tennessee, Knoxville.

Fixdal, J. (1997). *Consensus Conferences as Extended Peer Groups. Presented at Technology and Democracy: Comparative Perspectives.* Centre for Technology and Culture, University of Oslo, Norway, January 17–19.

Fleming, J. (1983). *The Law of Torts* (6th ed.). Perth: The Law Book Company.

Foster, V. (1996). Do Non-Use Values Exist? The State of the Debate. Mimeo, CSERGE.

Freeman, M. A. (1979). *The Benefits of Environmental Improvement.* Washington, D.C.: Resources for the Future.

Freeman, M. A. (1993). *The Measurement of Environmental and Resource Values: Theory and Methods.* Washington, D.C.: Resources for the Future.

Fuguitt, D., & Wilcox, S. J. (1999). *Cost-benefit analysis for public sector decision makers* (pp. xiv, 325). Westport, Conn. and London: Greenwood, Quorum Books.

Green, C. (1997). *Are Blue Whales Really Simply Very Large Cups of Coffee?* Middlesex University Flood Hazard Research Centre.

Gregory, R. et al. (1991). Valuing Environmental Resource: A Constructive Approach. *Decision Research*, Eugene, Ore.

Haddock, D. D. et al. (1990). An Ordinary Economic Rationale for Extraordinary Legal Sanctions, 78. *California Law Review, 1,* 17.

Hanemann, W. M. (1992). Natural Resource Damages for Oil Spills in California. In: J. Ward & J. Duffield (Eds), *Natural Resource Damages: Law and Economics* (pp. 555–580). New York: John Wiley.

Hanemann, W. M. (1996). Theory versus data in the contingent valuation debate. In: D. J. Bjornstad & J. R. Kahn (Eds), *Contingent Valuation of Environmental Resources* (pp. 38–60). Brookfield, VT: Edward Elgar.

Hanemann, W. M., & Keeler, A. G. (1995). Economic Analysis in Policy Evaluation, Damage, Assessment and Compensation: A Comparison of Approaches. University of Berkeley Working Paper No. 766.

Hirsch, W. Z. (1979). *Law and Economics: An Introductory Analysis.* London: Academic Press.

Hubin, D. C. (1994). The Moral Justification of Benefit/Cost Analysis. *Economics and Philosophy, 10,* 169–194.

Jacobs, M. (1997). Environmental Valuation, Deliberative Democracy and Public Decision-Making Institutions. In: J. Foster (Ed.), *Valuing Nature? Economics, Ethics and Environment* (pp. 211–231). London: Routledge.

Johansson, P. O. (1992). Altruism in Cost-Benefit Analysis. *Environmental and Resource Economics,* (2), 605–613.

Johnson, F. R. (2001). Role of Knowledge in Assessing Nonuse Values for Natural Resource Damages. *Growth and Change, 32*(1), 43–68.

Joss, S., & Durant, J. (1994). *Consensus Conferences.* London: National Museum of Science and Industry.

Kahneman, D., & Knetsch, J. (1992). Valuing Public Goods: The Purchase of Moral Satisfaction. *Journal of Environmental Economics and Management, 22,* 55–70.

Kelman, S. (1981). Cost-Benefit Analysis: An Ethical Critique. *Regulation, 5*(1), 33–40.

Kenyon, W., & Hanley, N. (2000). Economic and participatory approaches to environmental evaluation. Discussion paper 00–15, Economics Department, University of Glasgow.

Kenyon, W., & Edwards Jones, G. (1998). What Level of Information Enables the Public to Act Like Experts When Evaluating Ecological Goods? *Journal of Environmental Planning and Management, 41*(4), 463–475.

Kennedy, D. (1981). Cost-Benefit Analysis of Entitlement Problems: A Critique. *Stanford Law Review, 33,* 387–445.

Kohn, R. E. (1993). Measuring the Existence Value of Wildlife: Comment. *Land Economics, 69*(3), 304–308.

Kontogianni, A. et al. (2001). Integrated stakeholder analysis in non-market valuation of environmental assets. *Ecological Economic, 37,* 123–138.

Kopp, R., & Pease, K. (1996). Contingent valuation: economics, law and politics. RFF Working Paper.

Kopp, R., & Smith, V. K. (1989). Benefit Estimation Goes to Court: The case of natural damage assessments. *Journal of Policy Analysis and Management, 8,* 593–612.

Kopp, R. J., & Smith, V. K. (Eds) (1993). *Valuing Natural Assets.* Washington, D.C.: Resources for the Future.

Kopp, R. J. (1991). The Proper Role Of Existence value in Public Decision Making. Resourses for the Future, Discussion Paper QE 91–17.

Kopp, R. J. (1992). Ethical Motivations and Non use Values. Resources for the Future. Discussion Paper QE92–10. Washington, D.C.: Resources for the Future.

Krutilla, J. (1967). Conservation Reconsidered. *American Economic Review, 57*(4), 777–786.

Kuo, N. W., & Yu, Y. H. (1999). An Evaluation System for National Park Selection in Taiwan. *Journal of Environmental Planning and Management, 42*(5), 735–745.

Kuitunen, M., & Törmälä, T. (1994). Willingness of Students to favour the Protection of Endangered Species in a Trade-off Conflict in Finland. *Journal of Environmental Management, 42,* 111–118.

Larson, D. M., & Loomis, J. B. (1994). Separating Marginal Values of Public Goods From Warm Glows in Contingent Valuation Studies. Department of Agricultural Economics, University of California, Davis.

Larson, D. M. (1992). Can Non Use Value be Measured from Observable Behavior? *American Journal of Agricultural Economics (proceeding), 74,* 1114–1120.

Levaggi, R. (1998). Resources Allocation in the Internal Market: Whose Preferences Count? *Studi-Economici, 53*(65), 105–122.

Lewis, O. A. (1994). Contingent Valuation of Natural Resource Damages. At: http://www.lawinfo.com/forum/conval.html

Linstone, H. A., & Turoff, M. (1976). *The Delphi Method, Techniques and Applications.* London: Addison-Wesley Publishing Company.

Lockwood, M. (1999). Preference Structures, Property Rights, and Paired Comparisons. *Environmental and Resource Economics, 13,* 107–122.

Lockwood, M. (1997). Integrated value theory for natural areas. *Ecological Economics, 1*(20), 83–93.

Loomis, J. et al. (1994). Do reminders of substitutes and the Budgets Constraints Influence CV estimates? *Land Economics, 70*(4), 499–506.

Loomis, J. B., & White, D. S. (1996). Economic Benefits of Rare and Endangered Species: Summary and Meta-analysis. *Ecological Economics, 18*(3), 197–206.

Loomis, J. B. (2000). Contingent valuation Methodology and the U.S. Institutional Framework. In: I. Bateman & K. Willis (Eds), *Valuing the Environment Preferences.* Oxford University Press.

Lynn, F. M., & Kartez, J. D. (1995). The redemption of citizens advisory committees: A perspective from critical theory. In: O. Renn, T. Webler & P. Wiedemann (Eds), *Fairness and Competence in Citizen Participation: Evaluating Models for Environmental Discourse.* Boston: Kluwer Academic Publishers.

Macalister, E. et al. (2001). Study On The Valuation And Restoration Of Biodiversity Damage For The Purpose Of Environmental Liability. Final report of project B4–3040/2000/ 265781/MAR/B3. EU Commission.

Macmillan, D. C., Harley, D., & Morrison, R. (1998). Cost-effectiveness analysis of woodland ecosystem restoration. *Ecological Economics, 27*(3), 313–324.

Macmillan, D. C., Hanley, N., Philip, L., & Alvarez-Farizo, B. (2000). Valuing the non-market benefits of wild goose conservation: A comparison of individual interview and group-based approaches. University of Aberdeen, University of Glasgow.

Madariaga, B., & McConnell, K. E. (1987). Exploring Existence Value. *Water Resources Research, 23*(5), 936–942.

Margolis, H. (1982). *Selfishness, Altruism, and Rationality: A Theory of Second Choice.* Cambridge, Cambridge University Press.

Martinez-Alier J., Munda, G., & O'Neill, J. (1998). Weak Comparability of Values as a Foundation for Ecology and Economics. *Ecological Economics, 26*(3), 277–286.

Mas-Colell, A., Whinston, M. D., & Green, J. R. (1995). *Microeconomic Theory.* Oxford University Press.

McConnell, K. E. (1983). Existence and Bequest Value. In: R. D. Rowe & L. G. Chestnut (Eds), *Managing Air Quality and Scenic Resources at National Parks and Wilderness Areas.* Boulder. Colorado: Westview Press.

McConnell, K. E. (1997). Does Altruism Undermine Existence Value? *Journal of Environmental Economics and Management, 32*(1), 22–37.

Moran, D. (2000). Accounting for Non-Use value in options appraisal: environmental benefits transfer and low flow alleviation. In: *Economic Valuation of Water Resources: Policy and Practice.* CIWEM.

Munro, A., & Hanley, N. D. (2000). Information, Uncertainty, and contingent valuation. In: I. Bateman & K. Willis (Eds), *Valuing the Environment Preferences.* Oxford University Press.

Murphey, J. (2000). Noxious Emissions and Common Law Liability: Tort in the shadow of regulation. In: J. Lowry & R Edmunds (Eds,) *Environmental Protection and the Common Law* (pp. 51–76). Hart Publishing, Oxford.

Navrud, S. (2000). Strengths, weaknesses and policy utility of valuation techniques and benefit transfer, Invited Paper for the OECD-U.S.DA workshop The Value of Rural Amenities: Dealing with Public Goods, Non-Market Goods and Externalities. Washington, D.C., June 5–6.

Navrud, S., & Pruckner, G. J. (1997). Environmental Valuation-To Use or Not to Use? A Comparative Study of the United States and Europe. *Environmental and Resource Economics, 10*(1), 1–26.

Nijkamp, P., & Voogd, J. H. (1984). An Informal introduction to multicriteria evaluation. In: G. Fandel & J. Spronk (Ed.), *Multiple Criteria Decision Methods and Applications.* Berlin, Springer-Verlag.

Norton, B. (1992). Sustainability, Human Welfare and Ecosystem Health. *Environmental Values, 1,* 97–111.

Nozick, R. (1974). *Anarchy, State and Utopia.* New York: Basic Books, Inc.

Nyborg, K. (2000). Homo Economicus and Homo Politicus: interpretation and aggregation of environmental values, *Journal of Economic Behaviour and Organisation* (Vol. 42, No. 3, 305–322.

Opaluch, J. J., & Grigalunas, T. A. (1992). Ethical values and personal preferences as determinants of non use values: Implications for natural resource damage assessment. Staff paper, Department of Resource Economics, University of Rhode Island.

Pearce, D. W. (1999). *Economics and the Environment.* Edward Elgar, Cheltenham, U.K.

Pearce, D. W. (2000). Cost-Benefit Analysis and Environmental Policy. In: D. Helm (Ed.), *Environmental Policy: Objectives, Instruments, and Implemenation.* Oxford, OUP.

Pearce, D. W., & Mourato, S. (1998). The Economics of Cultural Heritage. World Bank Support to Cultural Heritage Preservation in the MNA Region, CSERGE Report 1998.

Penn, T. A (2000). Summary of the Natural Resource Damage Assessment regulations under the United States Oil Protection Act. NOAA Report.

Peters, T. et al. (1995). Influence of choice set considerations in modeling the benefits from imrpoved water quality. *Water Resource Research, 31*(7), 1781–1787.

Posner, R. A. (1980). The Ethical and Political Basis of the Efficiency Norm in Common Law Adjudication. *Hofstra Law Review, 8*(3), 487–508.

Posner, R. A. (1983). *The Economics of Justice*. Cambridge, MA: Harvard University Press.

Posner, R. A. (1972). *Economic Analysis of Law*. Boston: Little Brown.

Quiggin, J. (1993). Existence Value and Benefit-Cost Analysis: A Third View. *Journal of Policy Analysis and Management, 12*(1), 195–199.

Quiggin, J. (1998). Existence Value and the Contingent Valuation Method. *Australian Economic Papers, 37*(3), 312–329.

Quiggin, J. (1997). Altruism and Benefit-Cost Analysis. *Australian Economic Papers, 36*(68), 144–155.

Randall, A. (1991). The Economic Value of Biodiversity. *Ambio – A Journal of the Human Environment, 20*(2), 64–68.

Randall, A., & Stoll, J. R. (1983). Existence value in a Total Valuation Framework. In: R. D. Rowe & L. G. Chestnut (Eds), *Managing Air Quality and Scenic Resources at National Parks and Wilderness Areas*. Boulder. Colorado: Westview Press.

Randall, A., & Farmer, M. C. (1995). Benefits, Costs, and the safe minimum standard of conservation. In: D. W. Bromely (Ed.), *Handbook of Environmental Economics* (pp. 26–44). Cambridge, MA: Basil Blackwell, Ltd.

Randall, A. (1997). Whose Losses Count? Examining Some Claims about Aggregation Rules for Natural Resources. *Contemporary Economic Policy, 15*(4), 88–97.

Randall, A. (2002). Benefit-Cost Considerations Should Be Decisive When There is Nothing More Important at Stake. In: D. Bromley & J. Paavola (Eds), *Economics, Ethics, and Environmental Policy: Contested Choices*. Blackwell Publishers (2002 forthcoming).

Ready, R., & Bishop (1991). Endangered Species and the Safe Minimum standard. *American Journal of Agricultural Economics, 72*(2), 309–312.

Rekola, M. et al. (2000). Incommensurable preferences in contingent valuation: commitments to nature and property rights, presented in EAERE conference, Crete.

Renn, O., Webler, T., & Wiedemann, P. (1995). The pursuit of fair and competent citizen participation. In: O. Renn et al. (Eds), *Fairness and Competence in Citizen Participation*. (pp. 339–368). Boston: Klewer Academic.

Rorty, A. O. (1992). The Advantages of Moral Diversity. In: E. Frankel, F. D. Miller & J. Paul (Eds), *The Good Life and the Human Good*. New York: Cambridge University Press.

Rosenthal, D. H., & Nelson, R. H. (1992). Why Existence Value Should Not Be Used in Cost-Benefit Analysis. *Journal of Policy Analysis and Management, 11*(1), 116–122.

Rowe, R., Shaw, D., & Schulze, W. (1992). Nestucca Oil Spill. In: J. Ward & J. Duffield (Eds), *Natural Resource Damages: Law and Economics*. New York: John Wiley.

Royal Commission on Environmental Pollution (1998). 21st Report Setting Environmental Standards. Cm 4053 HMSO, London.

Sagoff, M. (1994). Should Preferences Count? *Land Economics, 70*(2), 127–144.

Samples, K. C., Dixon, J. A., & Gowen, M. M. (1986). Information disclosure and endangered species valuation. *Land Economics, 62*, 306–312.

Samuelson, P. (1948). *Economics: An introductory analysis*, New York: McGraw-Hill.

Sclove, R., & Scammell, M. (1999). Practising the principle. In: C. Raffensperger & J. Tickner (Eds),

Protecting Public Health and the Environment: Implementing the Precautionary Principle (pp. 252–265). Washington, D.C.: Island Press.

Schade, D. A., & Payne, J. W. (1994). How people respond to contingent valuation questions: A verbal Protocol Analysis of Willingness to Pay for Environmental Regulation. *Journal of Environmental Economics and Management, 26*, 88–109.

Sen, A. (1970). *Collective Choice and Social Welfare.* San Francisco: Hodlen Day.

Shavell, S. (1984). Liability for Harm Versus Regulation for Safety. *Journal of Legal Studies,* (June), 357–374.

Shavell, S. (1987). *Economic Analysis of Accident Law.* Cambridge, Mass.: Harvard University Press.

Shavell, S. (1993). Contingent Valuation of the Non-use value of Natural Resources: Implications for public policy and the Liability System. In: J. A. Hausman (Ed.), *Contingent Valuation: A Critical Assessment.* North-Holland, Amsterdam, London, New York, Tokyo.

Schade, D. A., & Payne, J. W. (1994). How people respond to contingent valuation questions: A verbal Protocol Analysis of Willingness to Pay for Environmental Regulation. *Journal of Environmental Economics and Management, 26*, 88–109.

Smith, K. V. (1987). Non-use values in benefit cost analysis. *Southern Econ. J., 54*, 19–26.

Smith, K. V. (1992). Environmental risk perception and valuation: conventional versus prospective reference theory. In: D. W. Bromley, & K. Segerson (Eds), *The Social Response to Environmental Risk.* Boston, MA: Kluwer Academic Publishers.

Smith, V. K., & Desvousges, W. H. (1986). *Measuring Water Quality Benefits.* Boston: Kluwer Nijhoff Publishers.

Spash, C. (1997). Ethics and Environmental Attitudes With Implications for Economic Valuation, *Journal of Environmental Management, 50*, 403–416.

Spash, C. L. (2000). Ecosystems, contingent valuation and ethics: the case of wetland re-creation, *Ecological Economics, 34*(2), 195–215.

Stephen, F. H. (1988). *The Economics of the Law*, Harvester Wheatsheaf.

Stirling, A. (1997). Limits to the Value of External Costs, *Energy Policy, 25*, 517–540.

Stone, C. D. (1974). *Do Trees have standing? Towards Legal Rights for Natural Objects.* Los Altos, California: William Kaufman.

Sunstein, C. R. (1993). Endogenous Preferences, Environmental Law, 22. *Journal of Environmental Legal Studies TUD., 217*, 252–253.

Sutherland R. J., & Walsh, R. G. (1985). Effect of Distance on the Preservation Value of Water Quality. *Land Economics, 61*(3), 281–291.

Swallow, S. K. et al. (1994). Heterogeneous Preferences and Aggregation in Environmental Policy Analysis: A Landfill Siting Case. *American Journal of Agricultural Economics, 76*(3), 431–443.

Swanson, T. M., & Kontoleon, A. (2000a). Nuisance. In: B. Bouckaert, G. De Geest & E. Elgar (Eds), *Encyclopaedia of Law and Economics* (Vol 2), Cheltenham, U.K.: Civil Law and Economics.

Swanson, T. M., & Kontoleon, A. (2000b). Why Did the protected Areas Fail the Giant Panda? The Economics of conserving endangered species in developing countries. *World Economics, 1*(4), 135–148.

Tickner, J., & Ketelsen, L. (2001). Democracy and The Precautionary Principle. *The Networker, 6*.

Turner, K. R. (2000). The Place Economic Values in Environmental Valuation. In: I. Bateman & K. Willis (Eds), *Valuing the Environment Preferences.* Oxford University Press.

Trumbull, W. N. (1990). Who Has Standing in Cost Benefit Analysis? *Journal of Policy Analysis and Management, 9*(2), 201–218.

Unsworth, R. E., & Bishop, R. C. (1994). Assessing Natural Resource Damages Using Environmental Annuities. *Ecological Economics, 11*, 35–41.

Vari, A. (1995). Citizens' advisory committee as a model for public participation: A multiple-criteria evaluation. In: O. Renn, T. Webler & P. Wiedemann (Eds), *Fairness and Competence in Citizen Participation: Evaluating Models for Environmental Discourse*. Boston: Kluwer Academic Publishers.

Vatn, A. (2000). The Environment as a Commodity. *Environmental Values, 9*(4), 493–510.

Vatn, A., & Bromley, D. W. (1994). Choices without Prices without Apologies. *Journal of Environmental Economics and Management, 26*(2), 129–148.

Viscusi, K. W. (1989). Prospective reference theory: toward an explanation of the paradoxes. *J. Risk and Uncertainty, 2*, 235–264.

Ward, K. M., & Duffield, J. W. (1992). *Natural Resource Damages: Law and Economics*. New York: John Wiley & Sons, Inc.

Walsh et al. (1989). Issues in nonmarket valuation and policy application: a retrospective glance. *Western Journal of agricultural Economics, 14*, 178–188.

Weber, R. P. (1985). Basic Content Analysis. Sage University Papers Series. Quantitative Applications in the Social Sciences, No. 07–049. Sage: Beverly Hills.

Wenstop, F. et al. (1994). Valuation of Environmental Goods with Expert Panels. In: J. Climaco (Ed.), *Multicriteria Analysis: Proceedings of the XIth International Conference on MCDM, 1–6 August 1994* (pp. 539–548), Coimbra, Portugal. Heidelberg and New York: Springer.

Whitehead, J. C., & Blomquist, G. C. (1991). Measuring contingent values for wetlands: effects of information about related environmental goods. *Water Resour. Res., 27*, 2523–2531.

Williams, B. (1985). *Ethics and the Limits of Philosophy*. Cambridge, MA: Harvard University Press.

Whittington, D., & Macrae, D. Jr. (1986). The Issue of Standing in Benefit Cost Analysis. *Journal of Policy Analysis and Management, 9*(2) 201–218.

Yankelovich, D. (1992). A Widening Expert/Public Opinion Gap. *Challenge, 35*(3), 20–27.

Zerbe, R. O., Jr. (1991). Comment: Does Benefit-cost Analysis Stand Alone? Rights and Standing. *Journal of Policy Analysis and Management, 10*(1), 96–105.

Zerbe, R. O., Jr. (1998). Is Cost-Benefit Analysis Legal? Three Rules. *Journal of Policy Analysis and Management, 17*(3), 419–456.

Zerbe, R. O., Jr. (2001). Can Law and Economics Stand the Purchase of Moral Satisfaction? Paper Presented at Symposium on Law and Economics, University College London Sep 5th, 2001

APPENDIX

See Table 1 on facing page and Table 2 on page 216.

Table 1. Spectrum of Environmental Decision Making Approaches (adopted from Dixon et al., 1988; Bateman, 1999, Navrud 2000; Renn et al., 1995; Beierle 2000; English et al., 1993; and Tickner & Ketelsen 2001).

Valuation Methods		*Pricing Techniques*			*Participatory/ Deliberative Approaches*		*'Expert'-based Methods*
Stated Preference	Revealed Preference	Market prices/ dose-response techniques	Value of *Actual* Expenditures Approaches	Value of *Potential* Expenditures Approaches	Mixed economic and participatory approaches	'Pure' participatory approaches	
• Contingent Valuation • Choice Experiments • Contingent Ranking	• Travel Cost Method • Hedonic Pricing methods • Wage Differential approaches	• Changes in productivity approaches • Loss of earnings approaches • Opportunity Cost Approaches	• Cost effectiveness analysis • Preventive or mitigation expenditure approaches	• Replacement Cost approaches • Relocation Cost approaches • Shadow-Project approaches	• Valuation Workshops • Market Stall	• Citizen Juries • Consensus Conferences • Focus Groups • Planning Cells • Citizens Advisory committees • Scenario workshops • Town meetings	• Multi-criteria Analysis • Delphi Method • Task Forces • Expert Panels

Strong reliance on individual preference based values ←————————→ Weak reliance on individual preference based values

Table 2.   Views on the Use of NUV's CBA in Policy and Damage Assessment Decisions.

In favour of using non-use values for CBA (policy recommendations) *and* legal judgements (internalising externalities). Argue in favour of the concept of non-use values	Non-use values should be used for CBA *but not* for legal settlements	Non-use values should be decomposed and only some of these sub-components can be used for CBA and/or legal settlements	Non-use values should be used in legal settlements *but not* for CBA.	Against the use of non-use values for both policy and legal recommendations. Criticise the concept of non-use values per se.
NOOA Panel (Arrow et al., 1993) Carson et al., 1994; Haneman, 1992; Kopp, 1991	Desvouges et al., 1993; Shavell, 1993	Milgrom, 1993; Madariaga and McConnell, 1987; Brookshire et al., 1986	Bishop and Welsh, 1992	Diamond and Hausman (1994)

COMMENT ON PAPER BY ANDREAS KONTOLEON, RICHARD MACRORY AND TIMOTHY SWANSON

Richard B. Stewart

Kantoleon, Macrory and Swanson discuss on the role of preference-based valuations in environmental decison making generally and natural resource damages (NRD) determinations in particular; their discussion contains three parts.

First, the authors briefly summarize a typology of different approaches to environmental decision making organized by their degree of reliance on individual preference-based valuations. These include revealed preference techniques, environmental pricing techniques, participatory and deliberative approaches, and expert-based approaches. While this classification scheme is not extensively developed or linked to the other parts of the paper, it is quite suggestive and illuminating. One hopes that the authors will develop their typology further and explore its applications and implications in future work. For example, an important question, noted briefly by the authors, is whether or not and why restrictions should be imposed on the valuation methods used for determining damages liability that are not applicable to regulatory determinations.

Second, the authors briefly defend the proposition that the "concept of economic value is a meaningful concept based on the notion of trade-off and opportunity cost," addressing various objections to the use of economic valuation of environmental resources. They conclude that "information derived

An Introduction to the Law and Economics of Environmental Policy: Issues in Institutional Design, Volume 20, pages 215–218.
© 2002 Published by Elsevier Science Ltd.
ISBN: 0-7623-0888-5

from preference-based approaches is morally relevant and useful" in environmental decision making. Few, even ardent environmentalists, would disagree with such a carefully qualified proposition.

Third, the authors devote most of the article to the question whether preference-based valuations, particularly with respect to non-use value, should have a role in NRD assessments for purposes of imposing liability for resource injury. The question that they pose is: "Should Valuation Have its Day in Court?" Their answer is affirmative. The authors posit at the outset that the economic value that individuals place on preservation or other non-use values can be validly measured in economic terms, thus bypassing a controversial set of issues. Although they do not discuss which of the several valuation techniques in their typology may validly and most appropriately be used for determining non-use values, they seem to assume that survey techniques employing contingent valuation methodology (CVM) will be used. The authors discuss a number of important problems in using the non-use valuations elicited, including questions of standing (whose preferences should count), whether a person must have prior knowledge of the injured resource in order to have his preferences included, and whether there are reliable means for disaggregating preferences for individual environmental resources from preferences for more general classes of environmental goods. The authors conclude that preference-based non-use valuations should be employed in NRD determinations in order to promote both incentive and compensation objectives. I have two principal difficulties with their analysis and conclusion.

First, I am not persuaded that issue of measurement validity can be so neatly separated from the issues of law and policy involved in using CVM survey results to determine NRD. In major part, the values that underlie preservation of environmentally significant resources are collective in character rather than purely individual. Environmental goods are inherently collective and are often recognized by individuals as such. Individuals value the preservation of many, probably most, highly valuable environmental resources – national parks, rare wilderness areas or the blue whale – in substantial part because they are public in character and are also valued as public goods by others. In these circumstances, the premise of CVM – that values for such goods can be determined solely in terms of each individual's discrete, independent preferences – is unconvincing, especially where the survey respondents often have little or no familiarity with the resource in question or with public discussion or dialogue regarding its significance relative to other environmental and non-environmental public goods. Accordingly, the value of the most cherished environmental goods (where non-use components of value are most important) is not appropriately determined by summing isolated individual preferences elicited through survey techniques, but must include a process of dialogue involving the relevant publics or their representatives. Some of the valuation techniques other than CVM which the authors discuss that involve participatory and deliberative elements – such as citizen juries – could be an appropriate means for conducting such a dialogue and might thus serve as an appropriate basis for determining the collective non-use value of significant environmental goods. An analogue might be damages awards of pain and suffering by juries in personal injury cases in the United States.

Second, the authors recommend that NRD liabilities include non-use valuations without sufficiently critical examination of the appropriate objectives of a legal regime for dealing with injury to public natural resources and without considering alternative means of securing those objectives. It is common U.S. practice, and apparently an emerging principle of international law, for NRD to include: (a) the costs of appropriate restoration of the injured resource or of acquisition and protection of alternative resources providing equivalent ecological services; and (b) interim lost use values pending completion of restoration, which are normally determined through revealed preference methodologies. In such a regime, the argument that compensation objectives require imposition, in addition to the above elements of liability, damages for lost non-use values is weak at best. The government is the plaintiff. Governments have superior risk-pooling capacities. Further, corrective (as opposed to distributive) justice principles only properly apply, in my view, to injuries caused by and suffered by private parties, but not those incurred by the government. Further, restoration of the injured resource and the services that it provides will in almost all case protect long-run preservation values. The concept of interim lost-non use values is too elusive to merit much concern or effort. Further, as the authors note, the transaction costs associated with use of CVM studies to measure non-use values are quite large. These costs include not only the costs of conducting rival studies, but of litigating difficult and hotly-contested issues regarding the validity and reliability of the results. Even on the view (with which I disagree) that individuals' non-use preferences for a wide variety of discrete environmental goods can be validly and reliably measured through CVM surveys, the compensation arguments for including non-use values in NRD are too weak to justify incurring these large transaction costs.

Incentive objectives, on the other hand, are undeniably important in the design of a legal regime for dealing with injuries to public natural and environmental resources. As noted above, however, there remains a serious question whether a liability regime based on restoration costs and lost non-use values would be so inadequate from an incentive perspective to justify the high transactions costs involved in determining, case-by-case, damages based on lost non-use values.

Even if there would be a serious incentives gap if non-use valuations were not included, there are alternative to case-by-case judicial determinations of NRD liabilities that could be devised to meet that gap. These include civil penalties for causing resource injury, or a system, enforced through administrative or judicial proceedings, of scheduled damages similar to workers compensation schemes. In both instances, the amount of penalty or compensation could be determined or bounded ex ante through scaling based on the general type and value of the resource injured and the magnitude of the spill or release in question. While these alternatives, which I favor, would involve up-front administrative costs, they would avoid the large transaction costs involved in attempting to fix NRD liabilities based on lost non-use values determined ex post case by case for each particular injured resource. They would also achieve greater consistency in NRD determinations. Further, these methods could be used to determine the entire amount of NRD, not just the non-use component.[1] Some of the alternative approaches to resource valuation surveyed by the authors at the beginning of their article, including citizen juries and expert panels, might well be used in the design of such schedules.

NOTE

1. See Richard B. Stewart, Liability for Natural Resource Damages: Beyond Tort. In: Richard L. Revesz & Richard B. Stewart (Eds), *Analyzing Superfund: Economics, Science and Law* (*Resources for the Future*. Washington D.C. 1995).

SECTION B:
THE LAW AND ECONOMICS
OF INSTRUMENT CHOICE
AND DESIGN

PART 1:
OPTIMAL INSTRUMENT CHOICE

TOWARD A TOTAL-COST APPROACH TO ENVIRONMENTAL INSTRUMENT CHOICE

Daniel H. Cole and Peter Z. Grossman

ABSTRACT

Most theories of environmental instrument choice focus exclusively on differential compliance costs. But compliance costs comprise only part of the total costs of environmental protection. Administrative costs – particularly the costs of measuring emissions and monitoring compliance – can differ significantly between environmental instruments. Those administrative cost differentials may offset the compliance cost advantages commonly associated with economic instruments, such as tradeable permits and effluent taxes. Moreover, measurement and monitoring constraints may increase ex ante uncertainty over the differential costs and benefits of alternative regulatory policies. That uncertainty may militate against selecting regulatory instruments that appear superior from the perspective of models focusing exclusively on compliance-cost differentials.

1. INTRODUCTION

Much of the extensive theoretical literature on the efficiency of instruments for environmental regulation is predicated on the presumption of *ex ante* uncertainty about the *ex post* costs and benefits of policy choice. Beginning

An Introduction to the Law and Economics of Environmental Policy: Issues in Institutional Design, Volume 20, pages 223–241.
ISBN: 0-7623-0888-5

with Weitzman (1974), the literature has centered on the factors that might lead regulators to favor a price-based over a quantity-based instrument, or vice versa.[1] Although Weitzman did not prescribe exact types of price or quantity instruments, many scholars see the issue as a binary choice problem pitting a price-based effluent tax regime against a quantity-based regime of tradeable emissions permits. The comparison of only these two alternatives reflects a normative presumption that only such "economic" instruments have any possibility of producing an efficient outcome. Other potential alternatives, such as non-tradeable emissions quotas or more general taxation arrangements (such as input or production taxes) are ruled out as inherently inefficient (Tietenberg, 1985; Stewart, 1996) and even anti-democratic (Ackerman & Stewart, 1985; Stewart, 1992; Sunstein, 1997).

Moreover, most of the literature relies on an important but unwarranted assumption: that cost and benefit functions, although they may be subject to uncertainty, are identical regardless of the regime that is chosen; that is, price and quota systems are assumed to face the same cost and benefit curves with the same expected values. Most crucially, the models assume that no regime will be subject to greater or lesser *uncertainty* than another. In other words, the variance is assumed to be *in*variant with the choice of regulatory regime. Under these assumptions, virtually all existing economic theories of environmental policy suggest that instrument choice should be determined simply by the relative elasticities of the curves.

Environmental instrument choice in the real world is, however, more complicated than most existing theories suppose because the assumptions on which those theories are based do not always obtain. In the real world, the costs of administering – that is, monitoring and enforcing – one regime might be quite different from the costs of administering another regime. Technological and institutional factors may make one regime not just *more* costly but infeasible, in which case the most efficient instrument must be some alternative that appears inferior from the perspective of theories that focus exclusively on differential compliance costs. These considerations apply not only to comparisons of effluent tax regimes and tradeable permit programs but also to comparisons between "economic" instruments generally (including both taxes and tradeable permits) and command-and-control regulations and general Pigovian taxes. Though much of the theoretical literature predicts greater efficiency from the former, the theory is seriously incomplete.

A number of studies suggest that economic instruments may not always be the most efficient choice. Burrows (1980) notes that "the differential in information, administration, and enforcement costs tend to favour the regulation of uniform standards rather than charges." Hahn and Axtell (1995) show that even in theory, if monitoring is assumed to be imperfect, it is not clear that market-based approaches to environmental protection will entail lower total costs than approaches that utilize technological standards or Pigovian taxation. Ogus (1982, p. 40) has noted that "specification standards" can entail lower monitoring costs than "emissions standards" because compliance is determined not by measuring emissions but simply by verifying the correct installation of emissions-reduction equipment or processes. More recently, the same author (Ogus, 1994, p. 253) has pointed out that the high enforcement costs some-times associated with command-and-control regimes must "be balanced against the administrative burden of collecting taxes in a larger number of cases and dealing with the proceeds." Indeed, Opschoor and Vos (1989, p. 51) observe that more than half of all revenues generated by Germany's system of effluent taxes are consumed by the costs of administering the tax. Nevertheless, many economists still employ simplifying assumptions about administrative costs, creating a strong bias in favor of economic instruments.

The incorporation of administrative-cost differentials between alternative environmental instruments results in a more complete theory that explains more cases. Conventional models of environmental instrument choice that focus exclusively on differential compliance costs have no trouble explaining cases in which economic instruments are more efficient than traditional regulatory regimes. But those models cannot explain other cases, for example, those in which economic instruments fail to achieve exogenously-determined environmental protection goals because efficiently-functioning market institutions are absent (see Cole & Grossman, 1999). A total-cost model would, however, be able to explain both cases, as well as others.

The key factor in determining the comparative efficiency of alternative approaches to environmental protection are transaction costs – specifically, for purposes of this paper, measuring, monitoring, and enforcement costs. Much of the theoretical literature comparing environmental instruments assumes away these costs as insignificant. As Russell et al. (1986, p. 3) put it, economists tend to assume "perfect (and incidentally, costless) monitoring." Or they assume that measuring and monitoring costs are constant across instruments, so that those costs do not affect the analysis. These assumptions conflict with numerous studies that find sizeable differentials in measuring or monitoring costs from one environmental protection instrument to another, depending on technological and institutional circumstances (Office of Technology Assessment, 1995; Driesen, 1998; Anderson et al., 1977; Solow, 1971).

There are, indeed, circumstances in which command-and-control regulations or simple taxation schemes are less costly to monitor and measure than either effluent taxes or permit-based pollution quotas. This monitoring-cost differential may be so great in some cases as to more than offset the compliance-cost differential that typically favors such "economic" instruments. In other words, while effluent taxes or tradeable pollution permits may reduce the *compliance* costs of pollution control, they may, in some cases, entail higher *total* costs because of higher monitoring costs.

This paper offers a theoretical framework to explain why cost-curves will differ depending on regime choice, and applies that framework to a current regulatory effort. The model is elaborated in the following section. Section 3, then, discusses its implications for environmental policy generally, and in particular for the Kyoto Protocol to the United Nations Framework Convention on Climate Change, which would institutionalize a tradeable permitting approach to greenhouse gas emissions reduction. The framework described in this paper provides reason to question whether tradeable permitting is the most efficient approach to reducing greenhouse gas emissions internationally, considering the technological constraints and institutional defects that plague many (if not most) of the parties to the convention.

2. THE COSTS AND BENEFITS OF REGULATORY INSTRUMENTS

Theoretical Considerations

All considerations of the economic efficiency of regulatory instruments begin, and actually end, with the basic condition that to maximize social welfare, pollution control should be expanded until the marginal social benefit (MSB) of control equals the marginal social cost (MSC). In a world of complete information this would be easy to formulate, and the lowest cost means of achieving it would be straightforward. In fact, in a world of complete information the form of regulation should not matter; the lowest cost would always be achieved (see Coase, 1960). However, in the real world of incomplete information the efficiency condition is difficult to achieve. As many have noted, there is always going to be some degree of *ex ante* uncertainty about the price and quantity of pollution control that will satisfy the basic efficiency condition, MSB = MSC. Weitzman (1974, p. 480) observed that in regulatory cases uncertainty arises because of an "information gap." Regulators and engineers can only estimate around random variables that are more or less difficult to quantify. Consequently, the ideal level of pollution control that will equate marginal social benefit with cost (and so maximize social welfare) simply cannot be known at the time the regulatory regime is designed and implemented. This would be true even if the universe of polluters were small; complete information about costs and benefits could only be roughly estimated *ex ante*.[2]

In general, both costs and benefits are subject to this information gap (Stavins, 1996). The *source* of this gap is, however, an important factor that is often neglected. With respect to environmental pollution, the main impediments to complete information are the difficulty and costliness of: (1) reliably measuring pollution discharges (and the short- and long-term effects of those discharges); and (2) monitoring polluter behavior. There may also be considerable uncertainty, and so uncertainty as to costs, with respect to enforcement. Economic models typically assume reliable, low-cost enforcement as it currently exists in polities like the United States. In some political systems, however, enforcement is irregular at best, subject to influence or corruption, and dependent on other factors that lower the reliability of enforcement and raise costs to society. Put more generally, the principal reasons for uncertainty are the transaction costs associated with the regulatory process.

High monitoring and enforcement costs may also be related to technological constraints. It can be very difficult and costly to measure both the amount of effluent that a polluter emits and the environmental damage that it causes. This may be especially problematic when regulators require continuous, real-time emissions measurement. In a quota-based system of tradeable pollution permits, if data cannot be reliably gathered on an ongoing basis, it may be impossible for regulators to know whether firms are adhering to their quotas.[3] Moreover, in the absence of monitoring it is unlikely than any market for permits would exist because there would be no incentive for owning them. Quotas, whether marketable or not, can be effective only if they are enforceable; and they are enforceable only if the regulator can monitor compliance. Who would bother to comply with a quota that the authorities could not enforce because they had no way of monitoring compliance? In this circumstance, the economic value of quotas would quickly fall to zero. Moreover, they would be completely ineffective as a pollution-control device.

Institutional factors, too, may raise the costs or reduce the benefits of pollution-control efforts. If laws are enforced only randomly, profit maximizing firms will weigh the expected cost of compliance against the expected cost of non-compliance. If the probabilities of enforcement are low enough, or the penalties for non-compliance are small, there will be substantial non-compliance, raising the social costs from additional pollution damage.[4] This is the case regardless of the regulatory instrument chosen. However, other institutional factors, such as weak or nonexistent market institutions, may raise the costs of controlling pollution with some instruments more than others. If market prices are, for example, inaccurate indicators of value, it may be very difficult to determine the efficient level for effluent taxes. Or, if polluters are subject to soft budget constraints, so that their continued existence does not depend on profitability in the market (see Kornai, 1986), *no* form of price regulation is likely to affect pollution levels.[5] Similarly, if trading and contracting are generally costly and uncertain, the utility of tradeable permit programs for reducing compliance costs is questionable, even if monitoring technology is adequate and cost-effective.

For this analysis, we assume a social-cost function comprised of three components: (1) compliance costs, which include primarily abatement costs for regulated polluters; (2) administrative costs, which include costs of measuring, monitoring, and enforcing regulations for government; and (3) the cost of damage stemming from the absence of pollution control. The latter category includes the costs of coping with the spillover affects of pollution that will be associated with the failure to comply, measure, or monitor accurately, as well as the cost of residual pollution that is accepted by society as an efficient by-product of production.

We also assume with Watson and Ridker (1984) that the marginal cost and benefit functions are non-linear. This assumption is warranted because: (1) total control costs – the sum of compliance costs and administrative costs – tend to rise at a faster rate as society attempts to gain higher levels of pollution control; and (2) social benefits tend to fall off more quickly at high levels than they do at moderate levels of abatement. The assumption carries an additional implication that forecast errors are multiplicative, not additive as most theoretical analyses have assumed. Such multiplicative forecast errors seem particularly likely to occur when there are measurement or enforcement problems. As measurement fails, for example, enforcement becomes more costly and random, the advantage to firms of noncompliance grows, and damage costs begin to expand rapidly.

An important theoretical issue arises, however, once curves are assumed to be non-linear. In Weitzman's (1974) basic model, the choice between a price instrument and a quantity instrument depended on the relative elasticities of the cost and benefit curves. By assuming non-linearity, the instrument likely to maximize social welfare, by producing the efficient level of pollution abatement, could change depending on where regulators believe the marginal benefit and cost curves intersect. Even from the perspective of standard analysis, then, non-linearity can alter the policy equation.

Costs and benefits are both subject to uncertainty, and so can only be estimated within a range. Probabilities are described by a distribution function, which assigns a positive probability to all outcomes within the range. The variance of this range is assumed to increase as a greater degree of control is desired. Thus, at low levels of control the degree of uncertainty is relatively less than it is at high levels of control.

The variance would appear especially sensitive to the quality and quantity of monitoring and measurement, and the more an instrument relies on monitoring and measurement for its success, the greater the variance is likely to be. It is axiomatic that the greater the difficulty in measurement, the greater the potential range of error.

In fact, measurement problems may increase either or both compliance and damage costs as too much control is demanded or as too much damage ensues. Consider a situation, for example, in which real-time measurement of effluent discharges is costly because cost-effective monitoring technology is not available.[6] As a consequence, firms and/or regulators must increase labor hours or other inputs to achieve more reliable measurement and consistent monitoring of discharges (assuming that labor and other inputs can serve as substitutes for unavailable technology), adding to marginal compliance and/or administrative costs. But the reality of measurement error also means that there is a positive probability of larger (or smaller) marginal damage costs. As the need for measurement increases, the probability of larger and larger error, and compounding of errors, also increases. Consequently, the range of cost estimates could expand quickly and dramatically as larger and larger increments of control are demanded.

For simplicity's sake, it will be assumed in all figures that the marginal benefit curve for all instruments is certain but the marginal cost curve is uncertain, and that regulators always seek to maximize net social benefits. In Figs 1 and 2, the choice will be limited to two quota-based regulatory instruments, X and Y, which correspond to a tradeable permit regime and a non-tradeable quota or command-and-control regime. For both instruments, there are *ex ante* uncertainties about costs because of measurement, monitoring, and enforcement problems.

We assume that the regulatory agency must choose both an instrument (either X or Y) and specify a quota limit, q*, expected to achieve the regulatory goal. Given uncertainty we assume that regulators initially set the quota at the point where the marginal social benefit curve intersects the mean of the range of the marginal social cost curves ($\bar{x}$), specified at point A on Fig. 1.[7] However, since all points within the range including the *extrema* are possible, the *ex ante* estimate of social losses would include the sum of the expected losses in each state. If actual costs prove to be lower than the mean, so that it is shown *ex post* that the marginal cost curve MSC2 is correct and the truly efficient point is where MSB = MSC (at B), then the social loss is described by the area, ABC, representing unrealized efficient gains. *Ex ante* regulators would calculate the

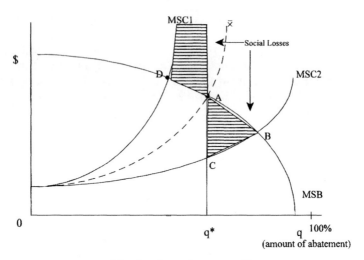

Fig. 1.   Quota Instrument X.

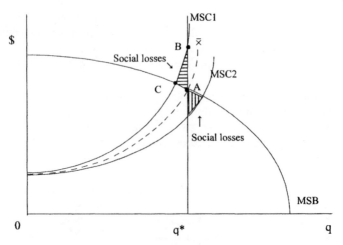

Fig. 2.   Quota Instrument Y.

expected value of the loss at point B: equal to the area ABC multiplied by the probability that MSC = MSB at B.

If, however, the mean is well below the *ex post* cost curve, and MSC1 is the correct representation of marginal costs (and the efficient point is at D), then any attempt to meet q* will be very costly, if it is attainable at any finite price. Net social costs would be prohibitively high. And regulators when estimating *ex ante* the efficiency of alternative instruments, must include the expected value of the shaded area on the graph in computing the expected social cost of using Instrument X.[8]

Figure 2 shows curves for the alternative non-tradeable quota-based instrument, Instrument Y. There is greater *ex ante* certainty about the cost of this instrument because it is cheaper and easier to monitor and measure. This reflects the fact that monitoring compliance with technology-based command-and-control instruments is relatively straightforward; regulators need only to ensure that the technology is installed and operating. As Maloney and Yandle (1984, p. 247) have noted, the installation and operation of the technology itself becomes the standard of compliance; actual emission rates need not be known, and therefore need not be measured. Consequently, monitoring for technology-based command-and-control regulations may well be less expensive than monitoring for tradeable permitting programs, where regulators must be able to monitor actual emissions to enforce compliance at any given time. In the early 1970s, for example, regulators managed to inspect, at some finite cost, every major stationary source of air pollution in Southern California once each month to ensure that required pollution-control technologies were properly installed and operating (Willick & Windle, 1973). Because of then-existing technological constraints, however, they could not have continuously monitored individual emissions at any finite cost, to ensure compliance with tradeable permit quotas (Cole & Grossman, 1999, pp. 920–921).

Command-and-control regimes usually entail higher *compliance* costs than tradeable permit programs because polluters with high costs of control are required to install the same equipment, reducing emissions by the same amount, as polluters with low costs of control. Thus, the mean of the range of marginal control costs is actually higher in Fig. 2 than it was in Fig. 1. The quota is set where MSB equals the mean of MSC, and therefore q* represents a lower level of pollution control than we saw in Fig. 1. The variance is, however, considerably smaller in Fig. 2 than in Fig. 1, and so too, are the expected losses. Moreover, even if the actual marginal cost curve is at its highest level, MSC1, q* will be attainable, with the loss represented by the shaded area, ABC.

As these graphs reveal, it may be well be the case that Instrument X (represented in Fig. 1) is *potentially* a more efficient instrument than Instrument

Y (represented in Fig. 2), but only if the range of costs in the first case can be narrowed. Otherwise, the uncertain and probably higher expected monitoring costs of Instrument X may more than offset its compliance-cost efficiency advantage. In that case, Instrument Y – the non-tradeable quota – would be more efficient, all things considered, than Instrument X – the tradeable quota. Instrument Y would maximize net social benefits.

The same reasoning may make a particular price instrument more efficient than another. Suppose, for example, society has a choice of two price options: an effluent tax on emissions levels (Instrument V) or an output tax on a polluting industry (Instrument W). Clearly, if firms have differential costs of pollution control, the latter tax will entail higher *compliance* costs. If, however, there is great *ex ante* uncertainty about emissions levels, or if enforcement of emissions limitations is more costly, the output tax may prove easier to monitor and measure. If so, it would minimize expected social losses, and so constitute the most efficient instrument in terms of *total* costs.[9]

This is illustrated in Fig. 3. Here regulators set a price target (p^*) which is assumed to equate MSB with MSC at an efficient level of control (q^*) shown at point A, again given in the figure as the mean of the range of estimated MSC.[10] The tax is set to achieve that. Where the real private cost is higher than expected, however, a tax will lead to a real incremental cost of abatement that will produce the level of control at $q1$. Given the full social cost curve, MSC1, the efficient level should actually be at B where MSB = MSC. The taxed price is too low, and social losses will be equal to the area of BCD. If MSC2

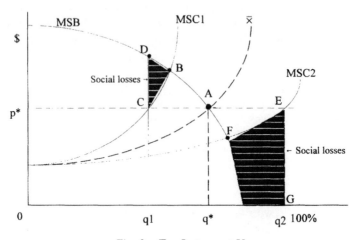

Fig. 3.   Tax Instrument V.

is closer to the actual level of cost, then the tax will be too high, and social losses from overinvestment in pollution control, EFG, will obtain. The full value of social costs must include the expected value of both of these areas, and it must be factored into the expected cost of the effluent program. Moreover, as before, expected social costs might well be lower with a different tax instrument, one with a higher mean but a smaller variance. This second tax instrument, though less efficient in world of low transactions costs, would be a more efficient alternative, as shown in Fig. 4.

An effluent tax may raise another problem. With Instrument V and marginal cost schedule MSC1, when price is set a p* the efficient level of pollution control should be reached at B; but since p = MSC at C, there will be no incentive to abate further than q1. But often a tax instrument is intended to produce, like a quota, a specific or, at least, an approximate level of abatement that regulators deem likely to improve public welfare. Assuming Instrument V and MSC1, even if the "efficient" level at B is reached, regulators will remain well short of the level of abatement they would have anticipated around the mean of the range. Indeed, if the other instrument, W, is chosen, the level of abatement will be closer to the expected abatement level, no matter where the actual cost falls within the range. Thus, the alternative instrument (W), which has no positive probability of abatement less than q1 (Fig. 4), may be preferred regardless of the fact that there will also be some positive probability that Instrument V would provide more abatement at lower cost.[11]

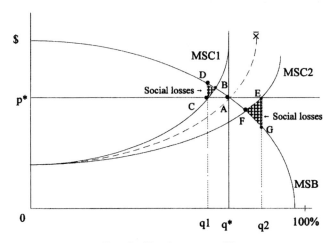

Fig. 4. Tax Instrument W.

Finally, the analysis can be extended to show instances, as in Fig. 5, in which a *quota* instrument of one type may be more efficient than a *price* instrument (or vice versa). Consider the case of effluent taxes, where a price is set equal to p* as in Fig. 3. Because of measurement and enforcement difficulties, the range of costs is considerable, and marginal costs are described by MSC1P and MSC2P. Again, the social costs must include the expected value of the social losses. Moreover, even if the expected mean of marginal costs of the second instrument, say a command-and-control quota, (described by MSC1Q and MSC2Q) is greater than the mean for the tax instrument, the smaller variance of the latter may well make it the more efficient choice. Again, lower *monitoring* costs may mean lower *total* costs, even if *compliance* costs are higher. Further, if MSC1P and MSC1Q were both to obtain, and the price instrument would be at p*, q and the quota at p, q*, there would be significant foregone social benefits by choosing the former (tax) over the latter (command-and-control) regime.

This analysis is, of course, static. There is no reason to expect that circumstances would remain the same over time. Technological improvements and changing institutional factors could well alter relative monitoring and compliance costs, and favor a change in regime. When the U.S. Clean Air Act was first enacted in 1970, for example, Congress could not have relied on effluent taxes, tradeable permits, or any other regime that depended on low-cost, precise, and continuous emissions monitoring because the necessary technology did not then exist. By the time the Clean Air Act was amended in 1990, however, technological improvements particularly the innovation of cost-effective continuous emissions monitoring systems (along with certain institutional changes, such as the Environmental Protection Agency's increasing economic

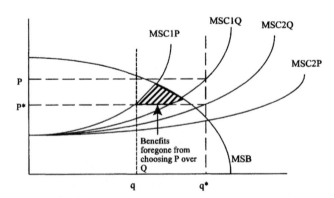

Fig. 5. Some Quota is More Efficient than Some Tax.

expertise) made emissions trading a lower cost alternative to command-and-control for *some* combinations of pollutants and sources.[12]

Even today the United States does not have the technological capability to cost-effectively monitor all pollutants from all sources. This continues to limit the utility of tradeable permit and effluent tax schemes as alternatives to technology-based command-and-control regulations. Consider, for example, the problem of controlling nitrogen oxide emissions from cars and trucks. It is beyond existing technological capability to continuously and cost-effectively monitor the emissions of each individual car and truck. Although emissions trading and effluent charges have succeeded in controlling emissions of some pollutants, including nitrogen oxides, from certain kinds of stationary sources such as coal-fired electric power plants, for mobile sources some kind of industry-wide technology-based standard would almost certainly entail lower monitoring costs, and lower total costs, than a tradeable quota system, according to which each car driver is assigned a certain level of allowable emissions.

3. IMPLICATIONS FOR POLICY MAKING

The consensus in the literature favoring so-called "economic instruments" effluent taxes and tradeable quotas for environmental protection is based on studies that compare only the compliance or abatement costs of alternative instruments. They do not compare the *total* costs, which are the sum of compliance/abatement costs, administrative (including monitoring) costs, and residual pollution costs. Consequently, they provide an insufficient basis for concluding that, in any specific case or across the general run of cases, effluent taxes or tradeable quotas are preferable to non-tradeable quotas or other Pigovian taxes. They may well be more efficient in many cases, but the studies are inconclusive because they fail to account for monitoring and other administrative costs, which may in some cases make traditional regulatory instruments preferable.

Despite this, the consensus favoring "economic instruments" has greatly influenced environmental policy making in recent years, both domestically and internationally. To the extent environmental protection policies are premised on the misperception that economic instruments necessarily entail lower total costs than traditional regulatory approaches, those policies are not well founded. They may, consequently, have negative environmental and economic consequences.

Consider, for example, the 1997 Kyoto Protocol on global greenhouse gas emissions. Under pressure from the American delegation, the parties to the Protocol decided, with very little deliberation, to rely on tradeable quotas as a primary mechanism for achieving the Protocol's emissions-reduction targets.[13] The ostensible goal was to minimize the costs of achieving those targets.

The parties to the Kyoto Protocol were certainly right to think that emissions trading would ensure lower compliance/abatement costs. But as we demonstrated in Section 2, lower compliance/abatement costs do not necessarily mean lower *total* costs. In some cases, abatement cost savings may be offset (or more than offset) by higher monitoring costs.[14] As noted earlier, the costs of monitoring individual, point-source emissions to ensure compliance with changeable, tradeable quotas are likely to exceed the costs of monitoring to ensure compliance with technology-based non-tradeable quotas. Simply put, it is cheaper and easier to check whether a scrubber is installed and operating than to continually measure actual emissions to ensure that each plant is in compliance with changeable, tradeable quotas at each and every point in time (see, for example, Maloney & Yandle, 1984, pp. 246–247). The question becomes whether those higher monitoring costs are at least offset by the abatement cost savings that tradeable quotas provide. In the absence of any comparison of the relative costs of compliance, administration, *and* residual pollution, the parties to the Kyoto Protocol had insufficient basis for concluding that emissions trading would achieve the Protocol's goals at lower *total* cost than alternatives, including technology-based non-tradeable quotas.[15]

The Kyoto Protocol provides only some general guidelines for emissions accounting, verification, and reporting. It directs signatories to use standard methods to measure and estimate their national greenhouse gas emissions, and it includes some general provisions regarding technology transfers, which could potentially be used to ameliorate monitoring deficiencies. According to critics such as Breidenich et al. (1998, pp. 324, 327), however, these guidelines are inadequate to ensure compliance. Moreover, as Tietenberg et al. (1999, p. 51) have cautioned, "[w]ithout the appropriate administrative structures and procedures, a tradeable allowance system could not only fail to achieve the objectives of the global warming convention, but could make the problem worse. If entitlements were transferred without ensuring that the appropriate compensating reductions were achieved, total emissions could rise, thereby violating one of the fundamental premises of the programme" (see also Gardner, 2000, p. 162).

This is not to say that tradeable quotas are inappropriate for reducing global greenhouse-gas emissions, only that the case for emissions trading has yet to be made. What is required is a comparative assessment of the *total* costs – the expected value of the sum of compliance/abatement costs, monitoring costs, enforcement costs, and residual pollution costs – of a tradeable quota scheme and alternative instruments. Until such a total-cost assessment is made, the case for tradeable quotas remains underdetermined.

4. CONCLUDING REMARKS AND SUGGESTIONS FOR FURTHER RESEARCH

It is clear that there is no universal, first-best approach to achieving environmental protection goals in this second-best world. The determination of the *best* approach is situational, dependent on institutional, technological, and other factors. As modeled in this paper, the best approach is that which achieves society's environmental protection goals at the lowest total cost, where "total cost" is defined as the expected value of the sum of compliance, administrative, and residual pollution costs. Focusing on compliance-cost minimization, without regard to administrative and residual pollution costs, is insufficient and indefensible in policy making.

Other factors deserving further attention in designing a total-cost model of environmental instrument choice include the differential incentives for technological innovation, both for pollution control and monitoring, under alternative regulatory instruments. Differential incentives to innovate could significantly affect *ex ante* total-cost calculations. Many theorists (including Magat, 1979; Ackerman & Stewart, 1980, pp. 1335–1336; and Stewart, 1993, p. 2063) claim that economic instruments create greater incentives to innovate than traditional regulatory regimes. There is, however, precious little empirical evidence, by which to judge. Jaffe and Palmer (1996, p. 4) note that "[a]lmost all of the existing literature on environmental regulation and R&D is theoretical in nature." Moreover, most existing studies focus only on the incentives of regulated industries to innovate. Those industries are not, however, the only potential sources of pollution-control innovations (see Repetto, 1983, pp. 276–277). Jaffe, Newell and Stavins (2000, p. 48) recognize that "if third parties can invent and patent better equipment, they can – in theory – have a ready market." Indeed, Cole and Grossman (1999, p. 911 n. 56) maintain that technology-based standards under the 1970 Clean Air Act "created positive incentives for independent environmental protection industries to innovate new pollution-control technologies." More generally, Driesen (1998, p. 304) contends that "demanding traditional regulation sometimes described as technology forcing often provides significant incentives to innovate." So, the issue of differential incentives to innovate under alternative regulatory regimes remains unsettled. Most importantly for present purposes, any differential effect on incentives to technological innovation, and on *actual* technological innovation, would likely be very difficult to calculate *ex ante*, further increasing the range of uncertainty over costs and benefits.

Finally, there is a dearth of empirical information on the costs of monitoring under various environmental-protection regimes. Such information is plainly needed to construct a robust and reliable total-cost model of environmental instrument choice.

NOTES

1. Some of the major contributions to this literature besides Weitzman's seminal piece include: Adar and Griffin (1976); Fishelson (1976); Portney (1990); Stavins (1996); Watson and Ridker (1984); Yohe (1978).

2. This paper assumes, at least with respect to theory, that ex post costs and benefits are known. This is not necessarily the case. For example, the EPA could only produce a very wide range of estimates in its attempts to quantify the net benefits that have been realized from the Clean Air Act of 1970. See Cole and Grossman (1999) for a discussion.

3. Cole and Grossman (1999) argue that it was the availability of cost-effective continuous emissions monitoring technologies that made tradeable permitting a feasible, and relatively efficient, policy choice for controlling sulfur dioxide emissions in the United States, under the 1990 Clean Air Act Amendments.

4. This paper looks at the broad costs and benefits of a regulatory regime. From the firm standpoint, the amount they are willing to spend on regulatory compliance will be determined by the expected penalty from non-compliance. If indeed the probability of enforcement is low, then the marginal penalty of non-compliance falls, and the willingness to incur compliance costs falls as well (see Harford, 1978; Hahn & Axtell, 1995).

5. This proposition is strongly supported by evidence from former communist countries, such as Poland, which instituted numerous environmental taxes, all of which failed to affect pollution emissions because of endemic soft budget constraints. Central planners nearly always compensated state-owned enterprises for any environmental fees and fines incurred. Even when they did not, the taxes did not affect polluting behavior because, ultimately, profits or losses had little bearing on enterprise survival (see Cole, 1998, pp. 146–153).

6. This situation obtained in the U.S. with respect to the administration of the Clean Air Act in the 1970s (Cole & Grossman, 1999).

7. This is intended as a simplifying assumption. In reality, policy makers are likely to set the quota limit not at the mean of the range of ex ante uncertainty but at the mean of the range of expected social losses stemming from that uncertainty. In many cases, the mean of the range of social losses might be offset, one way or the other, from the mean of the range of *ex ante* uncertainty. There are other cases, however, in which such a decision rule would make the choice of a emissions quota exceedingly problematic, as in the situation described by Fig. 1 and explained in note 9 below.

8. Consider a mathematical example: Assume there is a probability of 0.5 that the efficient equilibrium (MSB = MSC) will be D, and a probability of 0.5 that the equilibrium will be at point B. Note that the mean is at A even though the probability that MSC = MSB is at A = 0. However, the expected loss (L) will be $E(L) = 0.5$ (area ABC) + 0.5 (area between MSC1 and q^*). Since the last area is infinite, the loss function cannot even be calculated.

9. An output tax would raise output prices and so lower total output. Unlike an efficient effluent tax, however, an efficient output tax should not create any incentive for regulated firms to reduce pollution emissions; to the contrary, it would create a disincentive to invest in emissions reduction technology because emissions reduction

will not save regulated firms on taxes (unless the output tax is combined with tax -deductions for environmental investments). If, however, output taxes are adjusted to account for environmental damage costs, such instruments still would satisfy the basic Pigovian requirement that externalized costs be internalized.

10. As explained in note 7 above, we are using the simplifying assumption that policy makers will set the pollution-reduction target at the mean of the range of *ex ante* uncertainty. Figure 4 is an example where a more realistically set target would be at the mean of the range of expected social losses stemming from that uncertainty. This would shift the curve toward the MSC2 curve.

11. This might depend on social risk preferences.

12. For a study of the historical evolution of air pollution policy under the Clean Air Act, see Cole and Grossman (1999).

13. The Protocol permits member countries to select their own mechanisms for complying with emissions-reduction targets, but endorses a global emissions-trading regime.

14. This may be the case for economically advanced, as well as developing, countries. Fraschini and Cassone (1994, p. 102) conclude, for example, that the absence of economic instruments from Italy's water-pollution control regime are due to the "quite backward" state of emissions monitoring and enforcement technologies in that country.

15. This analysis is not to be taken as an endorsement of the need for any particular reduction in global greenhouse-gas emissions. We only assume the goal of reducing such emissions for purposes of the analysis.

ACKNOWLEDGMENTS

The authors are grateful to participants in panel discussions at the 1999 Southern Economics Association Conference in New Orleans and the 2001 Workshop on the Law and Economics of Environmental Protection at University College London.

REFERENCES

Ackerman, B., & Stewart, R. B. (1985). Reforming Environmental Law. *Stanford Law Review, 37,* 1333–1365.

Adar, Z., & Griffin, J. M. (1976). Uncertainty and the Choice of Pollution Control Instruments. *Journal of Environmental Economics & Management, 3,* 178–188.

Anderson, F. R. et al. (1977). *Environmental Improvements through Economic Incentives.* Baltimore: Johns Hopkins University Press/Resources for the Future.

Berg, A. (1994). Does Macroeconomic Reform Cause Structural Adjustment? Lessons from Poland. *Journal of Comparative Economics, 18,* 376–409.

Breidenich, C. et al. (1998). The Kyoto Protocol to the United Nations Framework Convention on Climate Change. *The American Journal of International Law, 92,* 315–330.

Burrows, P. (1980). *Economic Theory of Pollution Control.* Cambridge, Mass.: MIT Press.

Coase, R. (1960). The Problem of Social Cost. *Journal of Law and Economics, 3,* 1–44.

Cole, D. H., & Grossman, P. Z. (1999). When is Command-and-Control Efficient? Institutions, Technology, and the Comparative Efficiency of Alternative Regulatory Regimes for Environmental Protection. *Wisconsin Law Review,* 887–938.

Cole, D. H. (1998). *Instituting Environmental Protection: From Red to Green in Poland.* Basingstoke, U.K.: Macmillan Press Ltd.

Dales, J. H. (1968). *Pollution, Property, and Prices.* University of Toronto Press.

Driesen, D. (1998). Is Emissions Trading an Economic Incentive Program? Replacing the Command and Control/Economic Incentive Dichotomy. *Washington & Lee Law Review, 55,* 298–350.

Fishelson, G. (1976). Emissions Control Policies Under Uncertainty. *Journal of Environmental Economics & Management, 3,* 189–197.

Fraschini, A., & Cassone, A. (1994). Instrument Choice in Water Pollution Policy in Italy. In: H. Opschoor & K. Turner (Eds), *Economic Incentives and Environmental Policies* (pp. 89–112). Dordrecht: Kluwer.

Gardner, A. F. (2000). Environmental Monitoring's Undiscovered Country: Developing a Satellite Remote Monitoring System to Implement the Kyoto Protocol's Global Emissions-Trading Program. *New York University Environmental Law Journal, 9,* 152–216.

Hahn, R. W., & Axtell, R. L. (1995). Reevaluating the Relationship Between Transferable Property Rights and Command-and-Control Regulation. *Journal of Regulatory Economics, 8,* 125–148.

Harford, J. (1978). Firm Behavior Under Imperfectly Enforceable Pollution Standards and Taxes. *Journal of Environmental Economics and Management, 5,* 26–43.

Jaffe, A. B., Newell, R. G., & Stavins, R. N. (2000). Technological Change and the Environment. Working Paper 7970. Cambridge, Mass.: National Bureau of Economic Research.

Jaffe, A. B., & Palmer, K. (1996). Environmental Regulation and Innovation: A Panel Data Study. Working Paper 5545. Cambridge, Mass.: National Bureau of Economic Research.

Kornai, J. (1992). *The Socialist System: The Political Economy of Communism.* Princeton: Princeton University Press.

Kornai, J. (1986). The Soft Budget Constraint. *Kyklos, 39,* 3–30.

Magat, W. A. (1979). The Effects of Environmental Regulation on Innovation. *Law and Contemporary Problems, 43,* 4–25.

Maloney, M. T., & Yandle, B. (1984). Estimation of the Cost of Air Pollution Control Regulation. *Journal of Environmental Economics and Management, 11,* 244–263.

Office of Technology Assessment (1995). *Environmental Policy Tools: A Users Guide.* Washington: Government Printing Office.

Ogus, A. (1982). The Regulation of Pollution. In: G. Richardson, A. Ogus & P. Burrows (Eds), *Policing Pollution: A Study of Regulation and Enforcement* (pp. 30–68). Oxford: Clarendon Press.

Ogus, A. (1994). *Regulation: Legal Form and Economic Theory.* Oxford: Clarendon Press.

Opschoor, J. B., & Vos, H. B. (1989). *Economic Instruments for Environmental Protection.* Paris: OECD.

Portney, P. R. (1990). Economics and the Clean Air Act. *Journal of Economic Perspectives, 4,* 173–181.

Reitze, A. W., & Schell, S. D. (1999). Self-Monitoring and Self-Reporting of Air Pollution Releases. *Columbia Journal of Environmental Law, 24,* 63–135.

Repetto, R. (1983). Air Quality Under the Clean Air Act. In: T. C. Schelling (Ed.), *Incentives for Environmental Protection* (pp. 221–331). Cambridge, Mass.: MIT Press.

Russell, C. S., Harrington, W., & Vaughn, W. J. (1986). *Enforcing Pollution Control Laws.* Washington: Resources for the Future.

Solow, R. M. (1971). The Economist's Approach to Pollution and Its Control. *Science, 123,* 498–503.

Stavins, R. N. (1996). Correlated Uncertainty and Policy Instrument Choice. *Journal of Environmental Economics & Management, 30,* 218–232.

Stewart, R. B. (1996). The Future of Environmental Regulation: United States Environmental Regulation: A Failing Paradigm. *Journal of Law and Commerce, 15*, 585–596.

Stewart, R. B. (1993). Environmental Regulation and International Competitiveness. *Yale Law Journal, 102*, 2039–2106.

Stewart, R. B. (1992). Models for Environmental Regulation: Central Planning versus Market-Based Approaches. *Boston College Environmental Affairs Law Review, 19*, 547–559.

Sunstein, C. (1997). Free Markets and Social Justice. Oxford: Oxford University Press.

Tietenberg, T. H. (1985). *Emissions Trading: An Exercise in Reforming Pollution Policy.* Washington: Resources for the Future.

Tietenberg, T. et al. (1999). International Rules for Greenhouse Gas Emissions Trading: Defining the Principles, Modalities, Rules, and Guidelines for Verification, Reporting and Accountability, U.N. Doc. UNCTAD/GDS/GFSB/Misc. 6.

Watson, W. D., & Ridker, R. G. (1984). Losses from Effluent Taxes and Quotas under Uncertainty. *Journal of Environmental Economics & Management, 11*, 310–326.

Weitzman, M. (1974). Prices vs. Quantities. *Review of Economic Statistics, 41*, 477–491.

Willick, D., & Windle, T. (1973). Rule Enforcement by the Los Angeles County Air Pollution Control District. *Ecology Law Quarterly, 3*, 507–534.

Yohe, G. W. (1978). Towards a General Comparison of Price Controls and Quantity Controls Under Uncertainty. *Review of Economic Studies, 45*, 229–238.

COMMENTS ON PAPER BY DANIEL H. COLE AND PETER Z. GROSSMAN

Anthony Ogus

I begin with my conclusion. This is an excellent paper: it probes a very important dimension to instrument choice; it is well-structured and elegantly written; it has important policy implications which to a large extent coincide with my own intuitions (Ogus, 1998, pp. 780–783).

As regards the principal methodological issue, the extension of the models to cover monitoring and enforcement costs, the authors are preaching to the converted. It is obvious, to a lawyer, that these are very important variables, without reference to which any analysis is seriously incomplete. It also reinforces the significance of the "law and economics" approach, in contrast to "pure" economic analysis: the costs of institutional arrangements lie at the heart of the matter.

Two aspects of the paper nevertheless came to me as a surprise. The first is the assertion that these matters have been disregarded in the literature. It may be that the "purer" economic models have not included the full cost variables, but economists have certainly taken account of them. In pulling off my shelf a text written over twenty years ago on the economics of pollution control, I found that it deals with monitoring and enforcement costs in the context of comparing effluent fees with command-and-control (Burrows, 1979, pp. 133–135). And, albeit with less rigour, the consideration of these costs features in Richard Stewart's well known law journal articles from the same period (e.g. Stewart, 1981) and in my book on regulation (Ogus, 1994, pp. 155, 167, 253).

An Introduction to the Law and Economics of Environmental Policy: Issues in Institutional Design, Volume 20, pages 243–244.
© 2002 Published by Elsevier Science Ltd.
ISBN: 0-7623-0888-5

My second surprise relates to the comparison made in Fig. 2 between effluent charges and command-and-control. Why have the authors selected, as representing the latter, technological standards (or "specification standards" as I would prefer to call them)? For the last two decades or so they have been out of fashion, not the least because they inhibit technological innovation. Emission standards remain the most often-used form of environmental standard. And if the authors had compared the administrative costs of emission standards with effluent charges, they would have had to address an aspect which, to my knowledge, has been largely ignored. In theory, a charging system requires monitoring of all discharges, because the amount levied will depend on the amount discharged. With an emission standard, on the other hand, it is simply a question whether or not a firm has exceeded the permitted level and for this purpose random sampling will presumably suffice.

REFERENCES

Burrows, P. (1979). *The Economic Theory of Pollution Control.* Oxford: Martin Robertson.

Ogus, A. (1994). *Regulation: Legal Form and Economic Theory.* Oxford: Clarendon Press.

Ogus, A. (1998). Corrective Taxes and Financial Impositions as Regulatory Instruments. *Modern Law Review, 61,* 767–788.

Stewart, R. (1981). Regulation, Innovation and Administrative Law: A Conceptual Framework. *California Law Review, 69,* 1256–1377.

THE CHOICE OF INSTRUMENTS
FOR ENVIRONMENTAL POLICY:
LIABILITY OR REGULATION?

Marcel Boyer and Donatella Porrini

ABSTRACT

We address in this paper the problem of comparing and choosing among different policy instruments to implement the incentive objective of an efficient deterrence of environmental degradation and the remedy objective of an efficient clean-up of damages and a proper compensation of victims. Two main instruments are considered, namely the assignment of legal liability for environmental damage, such as in the American CERCLA and in the European White Paper, including extended liability provisions, and the design of an incentive regulation framework. Our results derive from a formal and structured analytical approach to modeling the economic interactions between different decision makers such as governments, firms, regulators and financiers.

1. INTRODUCTION

Different policies have been considered to implement a proper internalization of environmental externalities: taxes, quotas, subsidies, marketable emission permits, assignment of liabilities, etc. This addresses directly the problem of comparing and choosing among different policy instruments to implement a

An Introduction to the Law and Economics of Environmental Policy: Issues in Institutional Design, Volume 20, pages 245–267.
ISBN: 0-7623-0888-5

given set of environmental protection objectives.[1] Considering a law and economics approach, the chosen instrument must address an incentive objective (the efficient deterrence of environmental degradation) and a remedy objective (the efficient clean-up of damages and the proper compensation of victims).

We intend to compare in this paper two instruments, namely the assignment of legal liability for environmental damage and the design of an incentive regulation framework in the context of a political economy theory of environmental policy. A system of liability assignment can provide compensation to victims while internalizing the social costs of harm producing activities,[2] by identifying the cause of environmental harms, assessing the behavior of the actors responsible for such harms, and quantifying the harms for plaintiffs.

In a world of perfect or at least complete information, the law and economics approach suggests that this first instrument is an efficient method to solve the problem of internalizing the potential effects of environmental accidents. *Ex ante*, the firm, its owners and operators, face the proper incentive to take the efficient level of precaution and, *ex post*, the individuals harmed by pollution receive a proper and complete compensation, possibly through an insurance provider. But in practice, the allocation of individual responsibility seems to have caused delay in the clean up of damaged sites and contributed little to the objective of deterrence,[3] in particular when "judgement-proof" firms were involved.[4]

The following reasons have been suggested to explain this result. First, a specific polluter could in many cases be difficult to identify. A disease or a reduction in health could be attributed to a number of different factors besides the pollution. Even if a link between a pollutant and the disease could be established, it turned out to be difficult in many cases to determine which firm was responsible for the damage. Second, compulsory insurance contracts that the firms were induced or forced to buy turned out to be incomplete or insufficient because it was, in many cases, difficult to determine the probability of accident and the distribution of the loss caused by environmental accidents,[5] hence making the pricing of the contracts more difficult. Third, the polluter ended up in some cases to be insolvent and unable to pay for clean-up or compensation costs because of an increasing number of smaller firms operating in dangerous activities and because of the increasing costs and penalties of environmental accidents. Moreover, additional problems can arise in an incomplete information context, as analyzed by the economic literature: the asymmetric information about the firm's technology or accident preventing efforts implies that a rent must be given up to the stakeholders of the firm and the choice of a specific environmental policy affects this rent.

For all these reasons reducing the efficiency of a liability assignment system, it is appropriate to consider alternative instruments such as a regulatory framework. Of course this second instrument can present similar problems of implementation, such as informational problems (very often the level of effort to reduce the probability of environmental accidents is a private information of the firm) and capture problem (the regulator is often subject to "influence" by the firm itself or by political pressure). The problem is then to determine the circumstances or situations in which one instrument is better than the other.

2. THE PROBLEM OF THE CHOICE OF INSTRUMENTS

As a result of the large number of instruments that have been considered to implement a given set of environmental policy objectives, the relative efficiency of these policy instruments has become an important question in environmental economics, as shown by the recent surveys of Cropper and Oates (1992), Segerson (1996) and Lewis (1997). Although most of the discussions of the choice of instruments still use a benevolent social welfare maximizer paradigm, the necessity of looking at political economy factors underlying the choice of instruments has gained some ground at least since the early contribution of Buchanan and Tullock (1975). However, dissatisfaction remains. Lewis (1997, p. 844) wrote: I see the next progression in [environmental regulation] as being a positive analysis asking which kind of environmental policies will be implemented under information and distribution constraints when special interests try to intervene to affect policy.

Boyer and Laffont (1999) provided some preliminary steps in developing a formal political economy of environmental economics. They argued that economists' general preferences for sophisticated incentive regulation mechanisms must be reconsidered in a political economy approach explicitly considering the private information of economic agents, giving rise to policy sensitive and socially costly informational rents, and the incomplete contract nature of constitutions. When the different parties can contract without constraints, we know from the revelation principle that any policy instrument is equivalent to a revelation mechanism which is typically a command and control procedure. In such a mechanism, the different agents communicate truthfully their private information to an authority who then recommends proper actions. Once an optimal revelation mechanism has been obtained, it can be implemented through various policy instruments or institutions which by definition implement the same allocation. Hence, the question of instrument choice in such a context is empty.[6]

Nevertheless, the choice between instruments remains a meaningful problem insofar as one assumes the existence of constraints on instruments or of constraints on contracting possibilities. In the first case, various constrained instruments can be compared.[7] In the second case, different instruments, equivalent in a complete contracting framework, have different impacts when imperfections elsewhere in the economy are introduced.[8]

A systematic analysis of instrument choice in environmental policy should then be conducted in well-defined second best frameworks, all of which are shortcuts of an incomplete contract analysis. Political economy features can be viewed as a special case of this methodology. This is the object of our comparison between two major instruments: a legal instrument based on an extended liability framework for environmental damage and a regulatory instrument based on an incentive regulation framework subject to capture by the regulated firms. In both case, asymmetric information (moral hazard) is assumed making the first best allocation infeasible.

In the next two sections, we consider and discuss a real case application of the first instrument consisting in assigning a CERCLA (Comprehensive Environmental Response, Compensation and Liability Act, 1980, 1985, 1996) type liability, typically a strict, joint and several liability, on the owners and operators of the firm that is responsible of a catastrophic environmental disaster. In the following sections, we will tackle the analysis of regulatory instruments and the comparison of instruments.

3. THE LIABILITY SYSTEMS:
U.S. CERCLA AND EUROPEAN *WHITE PAPER*

In the eighties, the U.S. Congress enacted CERCLA and created a Superfund for the quick and effective clean-up of dangerous waste sites.[9] The U.S. liability system for environmental damages is a system that considers all owners and operators retroactively, strictly, jointly and severally liable for all damages through a system of extended liability. In spite of a secured interest exemption clause protecting financial institutions, holding indicia of ownership on the firm's assets, the U.S. courts have repeatedly considered secured lenders as owners or operators, insofar as their involvement in the operations of the firm exceeded the level warranted to secure their interest. This critical level was lowered over time and lenders' liability turned out to be more common than expected or intended.

A form of lenders' liability system was defined by the courts decisions, for example through the following cases: *United States v. Mirabile, United States v. Maryland Bank & Trust, United States v. Fleet Factors*, and *Bergsoe Metal v. East Asiatic*. Also important in the definition of the extended liability system was the 1992 EPA so-called Final Rule which attempted to make more precise the scope of traditional lender activities avoiding Superfund liabilities and leading to the 1996 CERCLA amendments.

Another experience of extended liability, to overcome the problem of *judgment-proof* firms, was the financial responsibility solution. By some rules, the potential polluters were required to demonstrate financial resources adequate to compensate for the environmental damage that they could cause. A financial assurance rule was for example authorized both by CERCLA and OPA (Oil Pollution Act) for waterborne vessels that carry oil or hazardous substances.[10] We can find such applications of the financial responsibility solution in many activities: in off-shore oil facilities; in underground petroleum storage tanks; solid waste landfills; in hazardous waste treatment, storage and disposal facilities; in wells to protect drinking water quality; in coal and hardrock mines; and in nuclear reactors and radioactive disposal facilities.[11]

The U.S. liability system, administered by the courts and governed principally by state law, played an extensive role in regulating air pollution, water pollution, hazardous and solid waste disposal, and pesticide use, among other environmental risks. It provides a mechanism for compensating victims, property, and health injuries by a strict liability system. Alongside the tort system, there exists a system of private and public insurance, both for the firms' liability and for the consequences on individual health.

But the U.S. CERCLA liability system also raised many problems. First of all many potentially responsible parties can be involved and, although it could be appropriate to divide among polluting parties the amount of needed compensation, this creates incentive problems insofar as the strict and joint liability system can induce firms to devote resources to legal strategies rather than to prevent accidents. In any case, it is difficult to coordinate numerous parties with conflicting interests and to find an agreement on a cost allocation plan. Moreover, since the government must recover response cost by suing all the potentially responsible parties or by targeting some "deep pocket" ones, significant transaction costs may result.

In addition to this transaction cost problem, CERCLA liability system was not supported by a significant development of the insurance market.[12] Insurance policies covering lending institutions in case of environmental accident turn out in many case to be unavailable or prohibitively expensive to obtain. Of course, the unavailability and the high cost of these kinds of policy are connected with the fact that the potential liability remains difficult to ascertain given the roles played by the EPA, the courts or the Congress. Federal court decisions have pointed out the effects of this problem. First, the insurance policies typically do not fit the CERCLA retroactive liability system because they are claims-made policies in the sense that they cover claims made while the policies are in effect and not the claims made before or after the period for which the insurance contract is in force. Second, both the premium and the deductible in the policies are extremely high, only a few insurance companies in the U.S. have issued such policies and many lending institutions have opted for self-insurance.[13]

The European Community has been trying for many years to define a common system of assignment of liability for environmental damages. In 1993, the European Commission published the *Green Paper on Remedying Environmental Damage*.[14] It presented the broad concepts on which a liability system could be built and led to discussions on the future EC liability regime. Its purpose was not to establish the elements of a specific unified system, but to stimulate a Community-wide debate and also collect the opinions of the interested parties. The *Green Paper* contained a description of the issues relevant to designing a Community-wide liability system. It focused on the liability criteria, the definition of environmental damage, the insurability of environmental damage, the limitations of liability, the problem of reinstatement of the environment, and the possibility of compensation funds financed by industries.

In the same year, the Commission explored the concept of the EC joining the 1993 Council of Europe Lugano Convention, but a definitive decision did not follow because of the intention to issue a specific White Paper and a proposal of Directive. In November 1997, the *Working Paper on Environmental Liability* outlined the key elements of a proposed environmental liability directive[15] and in October 1998, a commitment to adopt a White Paper on Environmental Liability was stated.[16] The Commission published a detailed environmental liability model for the EC in March 1999[17] and finally the *White Paper on Environmental Liability* in February 2000.[18]

The *White Paper* aims at determining who should pay for the clean-up and restoration costs of the environmental damage resulting from human acts. The question whether the costs should be paid by society at large, through the tax system, or by the polluter, when it can be identified, was answered by the imposition of liability on the party responsible for causing such damage.[19] The liability system is essentially a strict (no-fault) and non-retroactive liability system. Liability is only effective for future damage where polluters can be identified, damage is quantifiable and a causal connection can be shown. The Commission justifies the choice of such a system as follows. First, the "polluter pays principle" is more efficiently applied if the polluter must pay for the damages regardless of fault. Second, the operator of an hazardous activity should bear the risk inherent in it. Third, it can be difficult for the victims to prove the fault of the operator because of a lack of knowledge. Fourth, a non-retroactive system allows a quicker consensus by restricting attention on care for future accident prevention only.

Given the general rule that the polluter must always be the first actor a claim is addressed to, the *White Paper* does not explicitly deal with the problem of lender's liability. But it states that the person (or persons) who exercises control of an activity by which the damage is caused (namely the operator), should be the liable party, with the specification that lenders not exercising operational control should not be liable. In the final part of the *White Paper* that deals with the overall economic impact of environmental liability in the European Community, it is stated that the liability system generally protects economic operators in the financial sectors, unless they have operational responsibilities. The application of the financial responsibility in the common environmental liability system in Europe is not very well defined. For example, insurance markets are seen, in the *White Paper*, as one of the possible ways to obtain financial security, together with bank guarantees and internal reserves, but the European insurance system is still considered in a sense underdeveloped and unable to offer this kind of solution. So the Commission explicitly affirms "the EC regime should not impose an obligation to have financial security" (point 4.9).

The main differences between the provisions in the U.S. CERCLA system and the ones in the *White Paper* are summarized in Table 1.

Table 1.

FEATURES	U.S. CERCLA PROVISIONS	E.C. WHITE PAPER PROVISIONS
REGIME OF LIABILITY	Strict liability	Strict liability
APPLICATION	Retroactivity	No retroactivity
LIABLE PARTIES	Several and joint liability	Mitigated several and joint liability
DAMAGE	Every damage, even damages to natural resources	Traditional damages and the contamination of sites
FUND	Creation of a Superfund to finance cleaning-up	No special fund created
LENDER'S LIABILITY	Many applications by the courts	No application
FINANCIAL RESPONSIBILITY	Many applications for many activities	Voluntary and to be developed

We can see that the *EC White Paper* liability system is similar to the U.S. system because both of them are based on a strict liability regime in the sense that the liability is assigned only on the basis of the fact that the actor has caused the damage, without reference to the actor's behavior, diligence or negligence. But they are also different in many aspects: while the CERCLA system is applied retroactively, the EC *White Paper* provide a non-retroactive application; instead of covering every damage including the damage to natural resources, the European system covers only traditional damages, such as personal injury, damage to property, and the decontamination of sites; in the U.S. system, the Superfund was created to quickly clean-up the environmental damage, while no such fund is established by the *White Paper*. Differences exist also in the definition of lender liability and financial responsibility.

4. THE ECONOMIC ANALYSIS OF THE EXTENDED ENVIRONMENTAL LIABILITY SYSTEM

We complete the description of the liability system applied in the U.S. and in Europe by considering in this section the major analytical contributions to the study of extending liability to lenders, in terms of both its capacity to induce a proper internalization of environmental risks and of its capacity to ensure the proper financing of environmentally risky activities.

The economic analysis of extending liability to lenders as an environmental policy relies in good part on the incomplete (asymmetric) information principal agent paradigm, where the lender is the principal and the firm is the agent.[20] In those contexts, the firm is assumed to have private information about the cost of carrying out a task or project (adverse selection) and/or about how much self-protection or preventive effort it chooses to undertake (moral hazard) to reduce the probability of environmental disasters.[21] The analyses allow for the comparison of the different levels of care and of financing emerging in the different liability systems.[22] Those analyses lead to an evaluation of the predicted impacts of the different liability regimes in terms of social welfare. The benefits in terms of better accident prevention care and of better financing of risky activities must be compared with the cost of care, the system administrative expenses and the expected level of the damages, that is, both the expected number of accidents and their severity.

Pitchford (1995) raised the question of how appropriate extending liability to the lenders is, given that under asymmetric information it is likely to change the financial contracts offered to the firm. A limited liability regime reduces the firm's benefits of taking precautions to reduce the probability of accident. On the other hand, the lenders made liable for the cost of accident will require a form of insurance premium as part of the cost of financing to compensate them for their expected liability level. So extending liability to lenders increases the probability of proper compensation for external victims of an accident but may increase the probability of accident since the insurance premium reduces the wedge between the firm's relative value in the two possible states of the world, accident or no accident. Pitchford concludes by suggesting that, if the lender cannot observe the precautionary behavior of the firm under a limited liability regime, then increasing the liability of the lender can lead to an increase in the probability of accident. A better compensation system is thus obtained at the expense of a larger probability of accident.[23]

Boyer and Laffont (1997) consider a situation with both moral hazard and adverse selection in a model with two principals, a lender and an insurer. In a typical situation, the firm has better knowledge of its profit potential and of its accident prevention activities than the lenders or the insurers and extending liability modifies lending conditions and financial contracting between a firm and a lender. Under complete information between lender and firm but incomplete between insurer and firm, a regime of extended full liability to the lender when the firm goes bankrupt is optimal both for lending level and for the accident prevention or safety level. The relation with the CERCLA system and the related jurisprudence, allocating responsibility according to the involvement of the bank into the management of the firm, is clear: the assignment of full extended liability is appropriate as long as the risks are well defined and the agency costs are small.

If the firm's profit level is not observable by the lenders but the firm's accident preventing activities are observable, the financial contract cannot depend on profits and the best liability regime for lenders is a partial extended one. Extending full responsibility for environmental damages to the lender would ensure a perfect internalization but leads to insufficient lending. The result is obtained in three steps. First, the authors characterize the financing contract that a social welfare maximizing regulator would offer to the firm. Second, they characterize the financing contract that a private profit maximizing financier would offer as a function of the extended liability rule. Third, they compare the two and make the second solution as close as possible to the first one by varying the level of lender liability.

Under moral hazard, the lender can observe the profit of the firm but cannot observe its level of prevention activities and again a regime of full extended liability can cause the lender to lend too little. The optimal level of partial lenders liability is a function of the characteristics of the firm and/or the project to be realized. The practical implications of the results of this analysis is that the responsibility system must be well defined *ex ante* because it interferes with the banks' lending policy under asymmetric information. If there are no significant agency costs, full responsibility of the bank ensures the internalization of the environmental accident costs. If agency costs are significant, partial extended liability can balance the need to internalize the risks and the reluctance of the banks to finance risky but valuable activities.

Other complementary contributions can be found in the economic literature, analyzing other aspects of extending liability to the lenders. The modeling in this field presents an increasing level of complexity in its attempt to represent real situations that can involve more than one agent and more than one principal, with particular asymmetric information problems in dynamic settings with renegotiation issues present.[24] It is then possible to take into account the impact of different regimes on the structure of financial contracts, on the working of financial markets, on the availability of credit, on the cost of capital and on the level of investments and financing.[25] The specific structure of asymmetric information considered is crucial for such analyses.

The economic literature on the efficiency of the financial responsibility solution is much more limited. Feess and Hege,[26] in different recent contributions, tried to demonstrate that financial responsibility, as a variant of mandatory insurance, can be an efficient instrument to face the problem of bankruptcy of the polluter and of the consequent insufficient level of precaution incentive. They consider the following asymmetric information problem: investors have difficulties to correctly anticipate environmental risk (adverse selection) and cannot monitor the care level (moral hazard) without suffering a cost. They show that financial responsibility can be more efficient than lender liability and standard strict liability given that the contract between the firm and the lender or insurer who assume residual liability is chosen to reduce the agency cost at the minimum and that the firm is always held fully liable for the damage, regardless of the fact that the damage can be fully paid out by the firm.

5. THE INCENTIVE REGULATION SYSTEM

We want to review in this section the second instrument, namely an environmental regulation system. A regulation system is based on an authority or an agency that can use a number of tools to control the likelihood of an environmental accident. The instrument most often used is the setting of standards. Under a mandatory technology or abatement standard, the regulator can order the firms to reduce their emissions by a certain percentage, to emit no more than a specified amount of a pollutant, and/or to install a particular abatement technology. These are examples of command and control mechanisms. As an alternative, there are incentive market-based regulatory instruments: emission taxation, marketable permits, offset trading, and other incentive regulation mechanisms.

The command-and-control activities such as standards and emission limits are typically controlled through the conduct of inspections, actions in federal courts, and negotiated settlements with polluters. The regulator can alternatively use incentive regulation, such as a system of tradeable permits which typically works as follows: a plant or firm is allocated a number of permits, each of them allowing the emission of a given amount of a pollutant; if the facility is able to reduce its emissions, preferably through the use of different inputs or of less polluting technologies, it can sell its remaining emission permits to another facility that is unable to meet its quota.

Starting in the 1970s, the U.S. regulatory regime employed a variety of approaches to address the risk of pollution trying to regulate though standards the emission of toxic substances. But the task of regulating the myriad of sources of toxic emissions overwhelmed regulatory agencies and caused many problems.[27] In specific cases, some statutes provided general authority to regulate all the substances posing an environmental risk. But the problem still remained to establish clear thresholds. By subjecting standards to a feasibility constraint, Congress directed the EPA to set standards under the Clean Air act following "the best technological system of continuous emission reduction which (taking into consideration the cost of achieving such emission reduction, and any non-air quality health and environmental impact and energy requirements) has been adequately demonstrated."[28] Moreover "factors relating to the assessment of the best available technology shall take into account the age of equipment and facilities involved, the process employed, the engineering aspects of the application of various types of control techniques, process changes, the cost of achieving such effluent reduction, non-water quality environmental impact (including energy requirements), and such other factors as the Administrator deems appropriate."[29]

Given these guidelines from the Congress, a major problem still remained: how to regulate the use of chemicals for which conducting scientific tests to determine their effects on human health would take a long time and therefore would expose many people to potentially serious risk? Given the importance of regulating the risky activities and to fix appropriate standard to control *ex ante* the danger of environmental disaster we can conclude by quoting the court in the case *Boomer v. Atlantic Cement Co*: "It seems apparent that amelioration of air pollution will depend on technical research in great depth; on a carefully balanced consideration of the economic impact of close regulation; and of the actual effect on public health. It is likely to require massive public expenditure and to demand more than any local community can accomplish and to depend on regional and interstate controls."[30]

6. THE ECONOMIC APPROACH TO ENVIRONMENTAL REGULATION

One advantage of the regulation instrument, as argued by Boyer and Laffont (1999), is that politicians could use their detailed knowledge of the economy to choose a more flexible and adapted regulation policy. But in so doing, they could pursue their private agendas. In fact regulators can be subject to different kinds of influence that makes the government regulation not always congruent to the public interest. As noted by Faure (2000), rent-seeking problems can emerge in the case of environmental regulation in many different ways such as lobby for barriers to entry or lenient standards, and this influences also the instrument choice.

In the economic literature, the early contributions to the regulation of environmental risks have considered models in which the regulator maximizes a welfare function decreasing with the level of damage and the level of abatement costs. The regulatory policy is typically formulated in a single period and remains in effect afterwards.[31] More recently the literature presents models that take into account asymmetric information and delegation problems. In Laffont (1995), regulation, as an environmental policy instrument, is considered in relation with the potential trade-off between the regulatory efforts, which induce greater focus on cost minimization, and the agent's incentive to take too much risk. Laffont uses the basic model of a regulated monopoly with two types of effort variables, one that decreases production cost and one that decreases the probability of accident. The optimal regulation, under incomplete information, provides incentive for safety care and leaves a rent to more efficient firms. To mitigate the rent, safety care effort is reduced. In the absence of safety issues, the rent is the same but the level of efforts for cost minimization is lower, inducing higher costs than if such care considerations are not present. Laffont then introduces a limited liability constraint with the consequence that a rent must also be left to the least efficient firms as the only way to induce proper safety care. All the phenomena analyzed call for low powered incentive schemes: weaker incentives for cost minimization to induce safety effort at a lower social cost.

Boyer and Laffont (1999) consider the problem of choosing an environmental policy in an incomplete contract political economy context. Their model is that of a regulated monopolist who is privately informed of the cost of realizing a public project, a decreasing function of the level of pollution it is allowed to generate. Regulation may be delegated to political parties. Given the asymmetric information problem about the firm's technology, a rent must be given up to those who have stakes in the firm. The choice of an environmental policy affects this rent. Since the different political parties may be considered as having different stakes in the firm, more precisely in its informational rent, the environmental policy conducted through the regulatory framework generates policy fluctuations that could be welfare reducing. The authors recast the problem of instrument choice for environmental policy in the general mechanism design literature within an incomplete contract approach to political economy. They compare different sets of alternative instruments. In each case, a cruder less flexible regulation instrument is compared with a more sophisticated market-based incentive regulation instrument.

They show why "constitutional" constraints on the policy instruments may be desirable even though they appear inefficient from a standard point of view. Their justification lies in the limitations they impose on the capacity of politicians to distribute rents. For instance, given the delegation of environmental policy to political majorities, a comparison is made between restricting majorities to choose a single pollution level, a typical command and control regulation, and letting them select a policy consisting in choosing a menu of pollution-transfer pairs, a typical incentive regulation. Boyer and Laffont characterize the conditions under which the higher discretion associated with the second policy is compensated by its greater efficiency potential. Other instrument choices are also investigated. The results are that in general the larger the social cost of public funds and the greater the variability of economic variables are, then the more valuable flexibility is and the greater the delegation of authority to politicians should be. However, the thinner the majority or the larger the informational rents are, then the more the politicians objectives are biased away from maximizing social welfare, providing justification for cruder environmental policies that leave politicians or regulators less discretion.

A major problem arising in regulatory framework is the possibility of collusion and capture of the regulator by the regulatees. Given that this possibility is common rather than exceptional, there will be a cost to be incurred to prevent such collusion or capture when the assumption of a benevolent regulator is relaxed. One such prevention strategy is to split the regulatory tasks among different regulators. This strategy to counter the regulators' discretionary capability to develop wasteful activities will generate at the same time an administrative cost and a cost in terms of reduced coordination. Laffont and Martimort (1999) show that competition between regulators relaxes collusion-proofness constraints and makes the regulatory regime more efficient in terms of social welfare.

7. LIABILITY vs. REGULATION

The law and economics literature has focused predominantly upon the role of legal institutions and common law rules in achieving efficiency and distributive goals,[32] in particular in the area of environmental policy.[33] But relatively little attention has been given to a comparative institutional analysis between different systems.

Liability system and regulation can be compared considering the common objectives of deterring degradation and compensating victims of environmental harms: *ex ante* giving the incentive to precautions and controlling the environmental risk, *ex post* covering the costs and compensating for the damage. The authorities responsible for meeting these objectives are the courts that can assign liability, and the regulatory agencies that fix standards and check their compliance.

In a (strict) liability system the victim files an action claiming a causal link between the defendant's conduct and the plaintiff's injury or disease. Strict liability is typically applied to risks created by abnormally hazardous activities against defendants for all injuries caused by their conduct. This system has the advantage of internalizing environmental risks both from the incentive and the compensation points of view. On the other hand, it has many disadvantages. First, the system relies upon a case-by-case adjudication system. Second, there may be problems in determining the causal link. Third, it may lead to inconsistent verdicts, generate long delays in court proceedings and may be more profitable to lawyers and experts than to the victims.

The regulation system is characterized by a centralized structure. Its advantages are based on the fact that it is well-suited to set a control relying upon standards: centralized search facilities, continual oversight of problems and a broad array of regulatory tools can make the regulation system capable of systematically assessing environmental risks implementing a comprehensive set of policies. On the other hand, regulatory agencies may not be well adapted to the nature of the underlying regulatory problems. Moreover, centralized command structure with specialist decisions can be subject to political pressure and to capture by the regulatees and to collusion under different forms.

As defined in Rose-Ackerman (1991, p. 54), "Statutory regulation, unlike tort law, uses agency officials to decide individual cases instead of judges and juries; resolves some generic issues in rulemakings not linked to individual cases, uses non-judicialized procedures to evaluate technocratic information, affects behavior *ex ante* without waiting for harm to occur, and minimizes the inconsistent and unequal coverage arising from individual adjudication. In short, the differences involve who decides, at what time, with what information, under what procedures, and with what scope".

The main differences between the liability system and the regulation system are summarized in Table 2.

Shavell (1984) suggests four determinants for comparing different systems. The first determinant is the difference in know-how between private parties and the regulatory authority. It may relate to the benefits of activities, the cost of reducing risks, and the probability and the severity of accidents. It clearly could happen that the nature of the activities carried out by the firms is such that the private parties have better knowledge of the benefits, of the risks involved and of the cost of reducing risks. In such a case a liability system is better because it makes the private parties the residual claimants of the control of risks. The less informed regulator could overestimate the risks (probability and/or severity) and impose too stringent standards or could underestimate the value of the activities or the cost of reducing risk. But of course, it may also happen that the regulator has better knowledge because of the possibility of centralizing information and decisions, in particular when knowledge of risks requires special replicable and reusable expertise. It has the advantage of committing public resources to produce public knowledge. In such a case, direct regulation is likely to be better.

A second determinant is the limited capacity of private parties to pay the full costs of an accident, either because of limited liability or of insufficient assets. A traditional liability regime does not provide private parties with proper incentives for care. A regulatory system can impose directly or indirectly the proper decisions on the firms. So, the greater the probability or the severity of an accident are and the smaller the assets of the firm are relative to the potential damages, then the greater the appeal of regulation.

Table 2.

CHARACTERISTICS	LIABILITY SYSTEM	REGULATION SYSTEM
ACTORS	Private parties (not always)	Public Authorities
ACTIONS	Suits	Fixing standards and controls
EFFECTS	Indirect way to modify behaviors by deterrent effects	Direct ways to modify behaviors by requirements
STRUCTURE	Decentralized	Centralized
FOCUS	Parties in the suits	Whole population
DECISION MAKERS	Judges	Specialists
INFLUENCE	Independent	Political pressure

Clearly, a liability system can be linked with a compulsory insurance for the losses in excess of the assets of the firm. Under significant informational problems, moral hazard and/or adverse selection, the problem of insufficient incentives for care remains. Although the compulsory insurance provision would provide sufficient resources for cleaning-up and for compensating victims, the number of accidents would be inefficiently large unless the insurer has the ability to monitor and control the care activities of the firms. A similar alternative would be an extended liability regime that imposes strict, joint and several liability on all the deep-pocket stakeholders (suppliers, partners and financiers) of the firm.

The third determinant is the likelihood with which the responsible parties would face a legal suit for harm done. This problem is particularly present in environmental risks: in many cases the victims are widely dispersed with none of them motivated to initiate a legal action, harm may appear only after a long delay, and specifically responsible polluters may be difficult to identify. Compared with a regulatory system, the liability system is more uncertain and provides lower incentives for risk control.

The fourth determinant is the level of administrative expenses incurred by the private parties and the public. The cost of a liability system includes the cost of efforts, the legal expenses, the public expenses for maintaining legal institutions. The cost of the regulatory system includes the public expenses for maintaining the regulatory agencies and the private costs of compliance. The advantages of the liability system here is that legal costs are incurred only if a suit occurs and, if the system works well in the sense that there are incentives for choosing the efficient level of care, the suits are few and therefore the costs are low. On the other hand, under regulation, the administrative costs are incurred whether or not the harm occurs because the process of regulation is costly by itself and the regulator needs to collect information about the parties, their activities and the risks.

Considering the four determinants, Shavell (1984) concludes that administrative costs and differences in knowledge favor liability, while incapacity to pay (or limited liability) and escaping suit favor regulation. In general, a liability system is more efficient when private parties possess better information and when accident has a low probability to occur. Regulation is better when harm is usually large, is spread among many victims or takes a long time to show up, when accidents are not very rare events, and when standards or requirements are easy to find and control.

A fifth determinant can be added to the above traditional ones: the possibility of capture and collusion between the enforcers and the parties. The enforcers may be influenced by external pressure in both systems. But it seems that the courts are less likely to be captured than the regulation agencies. If the external pressures in the case considered are likely to be very strong, a system based on assignment of liability would be better than a regulatory system.

Considering the above determinants in the specific case of environmental risks, we can now try to describe a model in which the choice between an instrument or the other is represented in terms of some key-variables.

8. MODELING THE CHOICE OF INSTRUMENTS

In order to compare the two policy instruments defined above, we must consider their impacts in terms of social welfare. Such an analysis must balance the social benefits from the risky activities or industries with the costs of precautionary care, the expected level of damages (probability and severity), the administrative expenses associated with the different instruments, and the social cost and benefit of the firms' economic profits derived from informational rents.

To compare the two instruments in terms of social welfare the model should explicitly include the following crucial features: the administrative costs of the two instruments; the asymmetric information between the public regulatory authority and the firms and between the private banks and the firms regarding the level of accident preventing activities (moral hazard) effectively implemented by the firms; the efficient financial contract; and finally the possibility of capture.

Boyer and Porrini (2001) consider a formal political economy model to illustrate the different conditions under which a political economy instrument operating *ex post* is welfare superior to another instrument operating *ex ante* to regulate environmental accidents. Three contexts are characterized in a principal agent paradigm. The first context corresponds to a benevolent regulator as the principal maximizing the proper social welfare function (the reference case). The second context corresponds to the case of a profit maximizing private financier as principal, subject to extended liability if and when the firms it finances go bankrupt following a catastrophic environmental accident; the financier must then pay for the total costs, clean up and compensation, of the accident not covered by the firm's assets. The third context is characterized by a captured regulator maximizing a biased social welfare function, modeled in a reduced-form fashion through an overvaluation of the firms' profits as a source of social welfare. In all three settings, the principal party (benevolent regulator, captured regulator, private financier) determines the level of care to be implemented and the level of financing of risky activities while suffering from an informational disadvantage in its relation with the firms. The following factors are explicitly considered: the differential cost between low and high levels of environmental protection activities and the associated accident probabilities, the social cost of public funds, the informational rent of the firm, the net profitability of the risky activities, the level of damages if an accident occurs, the bias factor in case of capture of the regulatory agency.

The main results derived by the authors are the following. A relatively large cost differential between high and low levels of care, that is a high cost of accident preventing activities, favors the 'extended lender liability' regime. In this case, the 'regulator subject to capture' regime would end up inducing too much care, or too few environmental accidents, and/or allowing the financing of too many risky activities, that is an overdevelopment of environmentally risky industries, because the social value of the additional rents or profits so allowed are not large enough to compensate for the social cost of the extra care activities. It is better in this case to have more accidents than to allow higher rents or profits.

A relatively low cost of public funds, that is an efficient non-distortionary taxation system, favors the 'regulator subject to capture' regime. In that case, the 'extended lender liability' regime would end up inducing too little care, or too many environmental accidents, and/or allowing the financing of too few risky activities, that is an underdevelopment of environmentally risky industries.

In Boyer and Porrini (2002), additional results are derived and illustrated through a set of graphs. Again the two instruments are respectively a liability system, characterized by a strict regime of liability assignment (as in the U.S. CERCLA system and in the E.U. *White Paper*) and an incentive regulation system. Their results can be summarized as the follow:

- comparing the differential cost between high and low level of accident preventing activities with the cost of social funds, larger values of the cost of social funds favors the private financier solution, while lower values of the differential cost of care favor the regulation subject to capture regime;
- comparing the differential cost between high and low level of accident preventing activities with the magnitude of damage, larger damage favors the regulation subject to capture regime but the larger the value of the differential cost of care, the higher the critical value of damage is above which the regulation subject to capture regime is preferred;
- comparing the capture factor with the cost of social funds, larger values of the cost of social funds favor the private financier solution, while lower values of the capture factor favor the regulation subject to capture regime;

- finally, comparing the probability of high profit with the cost of social funds, larger values of the cost of social funds favor the private financier solution, while lower values of the probability of high profit favor the regulation subject to capture regime.

9. CONCLUSION

The main conclusion of our analysis is that choosing between a regulation framework and a legal framework to implement an environmental protection policy, a crucial factor in public policy evaluation, is a difficult task requiring a formal and structured analytical approach to the modeling of the social and economic interactions between different decision makers such as governments, firms, regulators and financiers. This requires balancing many factors in a social welfare accounting framework, namely the social value of the environment-risky activities, the costs of care, the cost of public funds, the possibility of regulatory capture, the asymmetric information position of the different actors, the net social value of the informational rents they generate, the probability and severity of accidents, the financial market efficiency. On the other hand, our results show the power of such a formal analytical approach.

NOTES

1. See Buchanan and Tullock (1975), Yohe (1976), Boyer (1979), Noll (1983), Hahn (1990), Cropper and Oates (1992), Laffont (1995), Segerson (1996) and Lewis (1996).
2. See Calabresi (1970); Landes, Posner, (1987); Shavell (1987).
3. See Menell (1991).
4. See Shavell (1986).
5. See Priest (1987).
6. Boyer and Laffont (1999) mentioned that "such a question often arose in the literature because authors were not careful enough in defining their instruments. For example, Yohe (1976) correctly shows that the alleged difference between quotas and price controls in Buchanan and Tullock (1975) disappears when instruments are appropriately defined."
7. As argued by Boyer and Laffont (1999), this is the case of Weitzman's (1974) comparison of prices and quantities. Asymmetric information then calls for non-linear prices as optimal instruments. Another example is the case of non-convexities where linear taxes are dominated by quotas because quotas are in fact non-linear taxes.
8. As argued by Boyer and Laffont (1999), "this is the case in Buchanan's (1969) example of a polluting monopolist when the subsidies required to correct the monopolistic behavior are not available." The linear tax is then clearly dominated by a quota implementing the second best tax.
9. The Superfund enabled the government to begin cleaning-up of priority sites placed on the National Priority List (NPL) with money generated principally by taxes on cruel oil, corporate income, petro-chemical feedstocks, and motor fuels.

10. 33, U.S.C § 2702; 42 U.S.C § 9607 (a)(1). Codified in: 33 CFR, part 138.

11. See Boyd (2001).

12. See Staton, (1993).

13. A. *Johnson & Co. v. Aetna Casuality & Sur. Co.*, 933 F.2d 66 (1st Cir., 1991); *United States Fidelity & Guar. Co. v. George W. Whitesides Co.*, 932 F..2d 1169 (6th Cir., 1991).

14. Commission of European Communities, Communication from the Commission of the Council and Parliament: *Green Paper on Remedying Environmental Damage*, COM (93) 47 final, Brussels, 14 may 1993, OJ, 1993 C 149/12.

15. Commission of European Communities, *Working Paper on Environmental Liability*, 17 November 1997.

16. Commission Decision 2176/98 (24/9/98) on the review of policy and action in relation to the environment and sustainable development, *"Towards Sustainability"*, OJ 1998 L 275/12.

17. The draft is an internal document of the Commission of European Communities not officially published, presented by DG XI in March 1999.

18. Commission of European Communities, *"White Paper on Environmental Liability"*, COM (2000), 66 final, Brussels, 9 February, 2000.

19. See Pozzo (2000).

20. See Pitchford (1995); Boyer, Laffont (1996, 1997).

21. See Porrini (2001).

22. The law and economics literature *per se* has focused predominantly on the role of institutions and common law rules in achieving efficiency and distributive goals (Calabresi, 1970; Polinsky, 1980; Landes & Posner, 1987; Shavell, 1987; Tietenberg 1989; Kornhauser & Revesz, 1994; Segerson & Tietenberg, 1992). In this approach, extended liability has been analyzed in terms both of its capacity to provide (*ex ante*) incentives to avoid environmental degradation and of its capacity to ensure (*ex post*) the proper compensation of victims. The courts are then ultimately responsible for meeting these objectives.

23. The contribution of Pitchford was recently criticized by Balkenborg (2001) who stresses the critical role of relative bargaining power in determining when lenders liability can increase the probability of accident, and by Lewis and Sappington (2001) who stress that the damages can take more that two values, a minor change which has a significant effect on the efficiency of the lenders liability solution. See also the reply of Pitchford (2001).

24. See among others Boyd and Ingberman (1997), Dionne and Spaeter (1998), Gobert and Poitevin (1998), Gobert (1999).

25. See Heyes (1996).

26. See Feess (1999); Feess, Hege, (2000); Feess, Hege (2001).

27. See Mennel (1998).

28. 42 U.S.C § 7411 (a)(1).

29. 33 U.S.C § 1314 (b) (2) (B).

30. *Boomer v. Atlantic Cement Co.*, 257 NE2d 870 (NY, 1970).

31. See Roberts, Spence (1976); Kwerel (1977); Dasgupta, Hammond and Maskin (1980); Baron (1985).

32. See Calabresi (1970), Landes and Posner (1987), Shavell (1987).

33. See Polinsky (1980), Landes and Posner (1984), Tietenberg (1989), Kornhauser and Revesz (1994).

REFERENCES

Balkenborg, D. (2001). How Liable Should a Lender Be? The Case of Judgment-Proof Firms and Environmental Risk: Comment. *American Economic Review, 91*(3), 731–738.

Baron, D. P. (1985). Regulation of Prices and Pollution under Incomplete Information. *Journal of Public Economics, 28*, 211–231.

Boyd, J. (2002). Financial Responsibility for Environmental Obligations: Are Bonding and Assurance Rules Fulfilling their Promise? In: T. Swanson (Ed.), *Law and Economics of Environmental Policy*. JAI Press.

Boyd, J., & Ingberman, D. (1997). The Search of Deep Pocket: Is Extended Liability Expensive Liability? *Journal of Law, Economics, and Organization, 13*, 233–258.

Boyer, M. (1979). Les effets de la réglementation. *Canadian Public Policy/Analyse de Politiques, 4*, 469–474.

Boyer, M., Lewis, T. R., & Liu, W. L. (2000). Setting Standards for Credible Compliance and Law Enforcement. *Canadian Journal of Economics, 33*, 319–340.

Boyer, M., & Laffont, J. J. (1996). Environmental Protection, Producer Insolvency and Lender Liability. In: A. Xepapadeas (Ed.), *Economic Policy for the Environment and Natural Resources* (Ch. 1, pp. 1–29). Edward Elgar Pub. Ltd.

Boyer, M., & Laffont, J. J. (1997). Environmental Risk and Bank Liability. *European Economic Review, 41*, 1427–1459.

Boyer M., Laffont J. J. (1999), Toward a Political Theory of the Emergence of Environmental Incentive Regulation. *RAND Journal of Economics, 41*, 137–157.

Boyer, M., & Porrini, D. (2001). Law versus Regulation: A Political Economy Model of Instrument Choice for Environmental Policy. In: A. Heyes (Ed.), *Law and Economics of Environment* (pp. 249–279), Edward Elgar Pub. Ltd.

Boyer, M., & Porrini, D. (2002). Modelling the choice between Regulations and Liability in Terms of Social Welfare. CIRANO Working Paper. Series Scientifique, 2002s-13.

Buchanan, J. M. (1969). External Diseconomies, Correctives Taxes, and Market Structure. *American Economic Review, 59*, 174–177.

Buchanan, J. M., & Tullock, G. (1975). Polluters' Profits and Political Response. *American Economic Review, 65*, 976–978.

Calabresi, G. (1970). *The Cost of Accident*. New Haven: Yale University Press.

Cropper, M. L., & Oates, W. E. (1992). Environmental Economics: A Survey. *Journal of Economic Literature, XXX*, 1675–1740.

Dasgupta, P. S., Hammond, P. J., & Maskin, E. S. (1980). On Imperfect Information and Optimal Pollution Control. *Review of Economic Studies, 47*, 857–860.

Dionne, G., & Spaeter, S. (1998). Environmental Risks and Extended Liability: The Case of Green Technologies. WP 98–12, École des HEC, Montréal.

Faure, M. (2000). Environmental Regulation. In: B. Bouckaert & S. De Geest (Eds), *Encyclopedia of Law and Economics* (pp. 443–520). Edward Elgar Pub. Ltd.

Feess, E. (1999). Lender Liability for Environmental Harm: An Argument against Negligence Based Rules. *European Journal of Law and Economics, 8*, 231–250.

Feess, E., & Hege, U. (2000). Environmental Harm, and Financial Responsibility. *Geneva Papers of Risk and Insurance: Issue and Practice, 25*, 220–234.

Feess, E., & Hege, U. (2001). Safety Monitoring, Capital Structure, and 'Financial Responsibility'. *International Review of Law and Economics*, 220–234.

Gobert, K., & Poitevin, M. (1998). Environmental Risks: Should Banks be Liable? WP 98s–39, CIRANO.

Gobert K. (1999), Responsabilité des créanciers en matière environnementale. CIRANO.
Hahn, R. W. (1990). The Political Economy of Environmental Regulation: Towards a Unifying Approach. *Public Choice, 65,* 21–47.
Heyes, A. (1996). Lender Penalty for Environmental Damage and the Equilibrium Cost of Capital. *Economica, 63,* 311–323.
Kornhauser, L. A., & Revesz, R. L. (1994). Multidefendants Settlements under Joint and Several Liability: The Problem of Insolvency. *Journal of Legal Studies, 23,* 517–542.
Kwerel, E. (1977). To Tell the Truth: Imperfect Information and Optimal Pollution Control. *Review of Economic Studies, 44,* 595–601.
Laffont, J. J. (1995). Regulation, Moral Hazard and Insurance for Environmental Risk. *Journal of Public Economics, 58,* 319–336.
Laffont, J. J., & Martimort, D. (1999). Separation of Regulators Against Collusive Behavior. *Rand Journal of Economics, 30,* 232–262.
Landes, W., & Posner, R. (1984). Tort Law as a Regulatory Regime for Catastrophic Personal Injuries. *Journal of Legal Studies, 13,* 417–434.
Landes, W., & Posner, R. (1987). *The Economic Structure of Tort Law.* Cambridge: Harvard University Press.
Lewis, T. R. (1996). Protecting the Environment when Costs and Benefits are Privately Known. *RAND Journal of Economics, 27,* 819–847.
Lewis, T., & Sappington, D. M. (2001). How Liable Should a Lender Be? The Case of Judgment-Proof Firms and Environmental Risk: Comment. *American Economic Review, 91,* 724–730.
Menell, P. S. (1991). The Limitations of Legal Institutions for Addressing Environmental Risks. *Journal of Economic Perspectives, 5,* 93–113.
Menell, P. S. (1998). Regulation of Toxic Substances. In: P. Newman (Ed.), *The New Palgrave Dictionary of Economics and the Law* (pp. 255–263). London: MacMillan.
Noll, R. (1983). The Political Foundations of Regulatory Policy, *Journal of Institutional and Theoretical Economics, 139,* 377–404.
Pitchford, R. (1995). How Liable Should a Lender Be? *American Economic Review, 85,* 1171–1186.
Pitchford, R. (2001). How Liable Should a Lender Be? The Case of Judgment-Proof Firms and Environmental Risk: Reply. *American Economic Review, 91,* 739–745.
Polinsky, A. M. (1980). Resolving Nuisance Disputes: The Simple Analytics of Injunctive and Damage Remedies. *Stanford Law Review, 7,* 1075–1112.
Porrini, D. (2001). Economic Analysis of Liability for Environmental Accidents. *RISEC, Rivista Internazionale di Scienze Economiche e Commerciali, 48,* 189–218.
Pozzo, B. (2000). Verso una responsabilità civile per danni all'ambiente in Europa: il nuovo Libro Bianco della Commissione delle Comunità Europee. *Rivista Giuridica dell'Ambiente,* 623–664.
Priest, G. (1987). The Current Insurance Crisis and Modern Tort Law. *Yale Law Journal, 96,* 1521–1590.
Roberts, M. J., & Spence, M. (1976). Effluent charges and licenses under uncertainty. *Journal of Public Economics, 5,* 193–208.
Rose-Ackerman, S. (1991). Regulation and the Law of Torts. *American Economic Review, Papers and Proceedings, 81,* 54–58.
Segerson, K. (1996). Issues in the Choice of Environmental Instruments. In: J. B. Braden, H. Folmer & T. S. Ulen (Eds), *Environmental Policy with Political and Economic Integration.* Brookfield: Edward Elgar Pub. Ltd.
Segerson, K., & Tietenberg, T. (1992). The Structure of Penalties in Environmental Enforcement: an Economic Analysis. *Journal of Environmental Economics and Management, 23,* 179–200.

Shavell, S. (1984). Liability for Harms versus Regulation of Safety. *Journal of Legal Studies, 13*, 357–374.

Shavell, S. (1986). The Judgment Proof Problem. *International Review of Law and Economics, 6*, 45–58.

Shavell, S. (1987). *Economic Analysis of Accident Law.* Cambridge: Harvard University Press.

Staton, D. S. (1993). EPA's Final Rule on Lender Liability: Lenders Beware. *The Business Lawyer, 49*, 163–185.

Tietenberg, T. (1989). Invisible Toxic Torts: The Economics of Joint and Several Liability. *Land Economics, 11*, 305–319.

Weitzman, M. L. (1975). Prices vs. Quantities. *Review of Economic Studies, 41*, 477–491.

Yohe, G. W. (1976). Polluter's Profits and Political Response: Direct Control Versus Taxes: Comment. *American Economic Review, 66*, 981–982.

COMMENTS ON PAPER BY MARCEL BOYER AND DONATELLA PORRINI

Michael Faure

The topic addressed by Boyer and Porrini is a highly important one for the economics of accident law. It deals with the central issue whether we should use tort law and/or regulation to control externalities and if the conclusion would be that the use of both instruments should be combined, the question arises how such a combination should work.

A. CRITERIA FOR SAFETY REGULATION

The arguments in favor of safety regulation to control environmental risks are well known. The choice between regulation and liability rules has been thoroughly examined by Steven Shavell in 1984, in a paper in which he advances several criteria that influenced the choice between safety regulation and liability rules.[1]

1. Information Asymmetry as a Criterion for Regulatory Intervention

Information deficiencies have often been advanced as a cause of market failure and as the justification for government intervention through regulation.[2] Also, for the proper operation of a liability system, information on, e.g. the existing legal rules, the accident risk, and efficient measures to prevent accidents, is a precondition for an efficient deterrence. According to Shavell, the parties in an accident setting generally have much better information on the accident risk

An Introduction to the Law and Economics of Environmental Policy: Issues in Institutional Design, Volume 20, pages 269–280.
© 2002 Published by Elsevier Science Ltd.
ISBN: 0-7623-0888-5

than that possessed by the regulatory body.[3] The parties themselves have, in principle, the best information on the costs and benefits of the activity that they undertake and of the optimal way to prevent accidents. This "assumption of information" will, however, be reversed if it becomes clear that some risks are not readily appreciated by the parties in an accident setting. Therefore, for every activity the question that will have to be asked is whether either the government or the parties involved can acquire the information at the least cost.

2. Insolvency Risk

If the potential damages can be so high that they will exceed the wealth of the individual injurer, liability rules will not provide optimal incentives. The reason is that the costs of care are directly related to the magnitude of the expected damages. If the expected damages are much greater than the individual wealth of the injurer, the injurer will only consider the accident as having a magnitude equal to his wealth. He will take, therefore, only the care necessary to avoid an accident equal to his wealth, which can be lower than the care required to avoid the total accident risk.[4] This is a simple application of the principle that the deterrent effect of tort liability only works if the injurer has assets to pay for the damages he causes. If an injurer is protected against such liability, a problem of underdeterrence arises.[5]

Safety regulation can overcome this problem of underdeterrence caused by insolvency.[6] In that case the efficient care will be determined *ex ante* by regulation and will be effected by enforcement instruments which induce the potential injurer to comply with the regulatory standard, irrespective of his wealth.

In that case a problem might arise if the regulation was also enforced by means of monetary sanctions. Again, if these were to exceed the injurer's wealth, the insolvency problem would remain. Hence, if a safety regulation is introduced because of a potential insolvency problem, the regulation itself should be enforced by non-monetary sanctions[7].

3. The Threat of a Liability Suit

Some activities can cause considerable damage, but even so a law suit to recover these damages may be never brought. If this were to be the case, there would of course be no deterrent effect of liability rules. Therefore, the absence of a liability suit would again be an argument to enforce the duty of efficient care by means of safety regulations rather than through liability rules.[8] There can be a number of reasons why a law suit is not brought, even though considerable damages have been caused.

Sometimes an injurer can escape liability because the harm is thinly spread among a number of victims. As a consequence, the damage incurred by every individual victim is so small that he has no incentive to bring a suit. In particular, this problem will arise if the damage is not caused to an individual but to a common property, such as, e.g. the surface waters in which each member of the population has a minor interest. In addition, a long time might have elapsed before the damage becomes apparent; in this case much of the necessary evidence may be either lost or not obtained. Another problem is that if the damage only manifests itself years after the activity, the injurer might have gone out of business.

A related problem is that it is often hard to prove that a causal link exists between an activity and a type of damage.[9] The burden of proof of a causal relationship becomes more difficult with the increasing passage of time since the damaging incident took place. Often a victim will not recognize that the harm had been caused by a tort, but might think that his particular ailment, e.g. cancer, had a "natural cause", associated with a general ill health. For all these reasons a liability suit might not be brought and hence safety regulation is necessary to ensure that the potential polluter takes efficient care.[10]

4. Administrative Costs

When examining the pro's and contra's of liability versus regulation, the administrative costs of both systems should also be compared. Liability rules are clearly costly in terms of time for both parties and in court fees. A part of these costs is borne by the whole community, such as, e.g. the cost of the legal system, fees for the judges, etc. Regulation produces costs for the community, including the costs of making the regulation, setting the standards, passing the statutes, etc. and of subsequent enforcement.[11]

In this respect the liability system seems to have an advantage: the administrative costs of the court system are only incurred if an accident has actually happened. The main advantage of the tort system is that a lot of accidents will be prevented by the deterrent effect of being held liable and having to pay damages to the victim. In case of safety regulation, the costs of passing the regulation and of enforcing it are always there, whether there are accidents or not.

B. THE NEED TO REGULATE ENVIRONMENTAL POLLUTION

After having discussed the public interest criteria for regulation let us now look how these criteria relate to environmental pollution. If one takes the criteria for safety regulation and applies them to the potential risk caused by environmental pollution, there is no doubt that liability rules alone are not sufficient.

If one looks at the first criterion, that of information costs, it must be stressed that an assessment of the risks of a certain activity often requires expert knowledge and judgement. Small organizations might lack the incentive or resources to invest in research to find out what the optimal care level would be. Also, there would be little incentive to carry out intensive research if the results were automatically available to competitors in the market: this is the well-known "free rider" problem. This problem can partially be countered by legal instruments granting an intellectual property to the results of the research. However, the problem remains that it may not be possible for small companies to undertake studies on the optimal technology for preventing environmental damage. Therefore, it is often more efficient to allow the government itself to do the research on the optimal technology (e.g. in a governmental environmental research institute). The results of this research can then be passed on to the parties in the market through the regulation.

Also, the insolvency argument points in the direction of regulation. Pollution can be caused by individuals or firms with assets which are generally lower than the damages they can cause by the pollution. In this respect it should not be forgotten that even a small firm can cause harm to a large number of individuals or to entire ecosystems. The amount of damages caused by this emission can of course largely exceed his individual assets. Moreover, most firms have been incorporated as a legal entity and therefore benefit from limited liability. Hence, the individual shareholders are not liable to the extent of their personal assets, but a creditor of the firm can only lay claim to part on all of the total assets purchased in the firm by the shareholders.

Also the chances of a liability suit being brought for damage caused by wrongful pollution is naturally very low. The damage is often spread over a large number of people, who will have difficulties to organize themselves to bring a law suit. In addition, the damage could only become apparent some years after the emission took place. This will bring proof of causation and latency problems, which will only make it difficult for a lawsuit to be brought against the polluter.

For these reasons it is clear that some form of government regulation of environmental pollution is necessary.

C. SAFETY REGULATION IN PRACTICE

When Shavell's criteria for safety regulation are applied to the environmental risk, one can easily note that there is a strong argument to make that the efficient care to be taken to avoid environmental damage should also be fixed ex ante by regulation.

In many cases this regulation consists of licenses or permits in which an administrative authority fixes an emission standard which must be followed by the potential polluter. These licenses play a crucial role in environmental policy in most countries. An improvement of environmental quality will mostly be reached by imposing more stringent emission standards in administrative licenses. Hence, the general requirement that emissions are controlled through licenses and that the quality and quantity of the emissions are regulated by the conditions in this license, is a cornerstone of environmental law. Since these licenses are administrative acts, in most legal systems environmental law is considered to be a part of administrative law. Criminal law usually only comes into the picture to sanction a violation of administrative regulations or emission standards in the licenses.

Although environmental pollution is in the first place controlled through these administrative licenses, in individual cases there can still be damage to the environment. Then again liability under tort law comes into the picture and the question is raised of the influence of regulation on the liability system and *vice versa*.[12] These complementarities between tort law and regulation shall be discussed below.

Finally we can point at literature that generally examined the effectiveness of safety regulation in controlling environmental harm. Dewees demonstrated that in North America the quality of the environment has improved substantially as a result of regulatory efforts, not so much in response to legal action in tort.[13]

This empirical evidence of the success of regulation, compared to tort law, has also been stressed in the recent book of Dewees/Duff/Trebilcock.[14] They hold that the large regulatory effort to improve the environment has met with considerable success when measured by the reduction of emissions, but that it is more difficult to argue that the environmental regulations of the 1970s in the U.S. equally had a considerable influence on the ambient environmental quality. Moreover, they also stress that while environmental regulation is a determining factor in pollutant emissions and ambient concentrations, other non-regulatory factors such as economic growth and even the weather also influence environmental quality.[15]

D. LIABILITY AND REGULATION: EXCLUSIVITY?

We just showed that on the basis of Shavell's criteria, there is a strong argument to control the environmental risk through *ex ante* regulation. However, in individual cases there can still be damage to the environment. Then again liability under tort comes into the picture and the question has been addressed in the literature how regulation influences the liability system and vice versa. These complementarities between tort law and regulation have more particularly been addressed by Rose-Ackerman,[16] Faure/Ruegg[17] and Kolstad, Ulen and Johnson.[18] Rose-Ackerman also compared U.S. and European experiences in using regulation versus tort law in environmental policy.[19] The first point which is often stressed, is that the fact that there are many arguments in favor of *ex ante* regulation of the environment, does not mean that the tort system should not be used any longer for its deterring and compensating functions. One reason to still rely on the tort system is that the effectiveness of (environmental) regulation is dependent upon enforcement, which may be weak. In addition the influence of lobby groups on regulation can to some extent be overcome by combining safety regulation and liability rules. Moreover, safety regulation, e.g. emission standards in licenses, can be outdated fast, which equally merits a combination with tort rules.

Hence, from the above it follows that although there is a strong case for safety regulation to control the environmental risk, tort rules will still play an important role as well. Hence, the question arises what the influence is of regulation on the liability system and vice versa.

E. VIOLATION OF REGULATION AND LIABILITY

The first question to be answered in that respect is whether a violation of a regulatory standard should automatically be considered a fault (or negligence) under tort law and thus lead to liability of the licensee.

Assuming that the license sets the regulatory standard at the efficient care level, a violation of the regulatory standard should indeed lead to liability to give the licensee an incentive to spend on care. However, Shavell argues that the costs of following the regulatory standard are not the same for all injurers. Following the standard might be inefficient for some injurers. The injurers for whom following the regulatory standard would only be possible at high costs should not be held to follow this standard since it would create inefficiencies.[20] The question is whether this means that these injurers should not be held liable if they violate the regulatory standard.

This problem can be compared with the *bonus pater familias* standard used in tort law. Although a detailed individualization of standards of efficient care would be the optimal solution in a first best world this is often impossible given the costs of an individualized standard setting. Therefore, the legal system sets the required level of care at an average level, the so-called *bonus pater familias* standard. The same can be said for regulation. If various groups can be identified at low costs a separate standard for a certain group is efficient as long as the gains from selecting a further group outweigh the further administrative costs. In most cases, however, the regulator will not have the possibility of identifying atypical parties that might be able to avoid a loss at lower costs, for instance because they pose lower risks than normal. Therefore, a single regulatory standard will be used.[21]

Although one could, therefore, argue that a failure to satisfy the regulatory requirement should not necessarily result in a finding of negligence, so as to avoid some parties who pose lower risks taking wasteful precautions,[22] most legal systems generally consider a breach of a regulatory duty a fault. This is the case in Dutch environmental liability law.[23] One of the reasons for introducing safety regulation to prevent environmental damage is that the regulator will usually possess better information to evaluate the efficient standard of care than the parties involved. Hence, the regulation passes on information to the parties on the efficient standard of care. The regulation also gives information to the judge who has to evaluate the behavior of the injurer *ex post* in a liability case. The judge might lack the information necessary to find out whether in a particular case an injurer should not be held to follow the regulatory standard, for example because he posed a lower risk than usual. Therefore, particularly in environmental cases a judge will accept a finding of negligence as soon as a regulatory standard has been breached.[24] Thus, the statutory standards can be applied to define negligence.[25]

F. COMPLIANCE WITH REGULATION AND LIABILITY

Above we have explained that to a large extent the prevention of environmental harm is also achieved as a result of regulation. A lot of attention is paid in the literature to the relationship between regulation and liability. In this respect for instance the question arises whether following a regulatory standard excludes from liability.

In environmental law this is particularly important, since the conditions under which an emission of pollutants is allowed are mostly laid down in a permit. Obviously these permits and their effects on liability can take various forms. In some cases permits and regulations are fairly general, but in other cases they specify emission limits. It is especially in the latter case, that is when specifically allowed releases in a permit have been respected, that the industry often argues that as long as they follow the conditions of that licence, no finding of negligence in tort law is possible. This is often referred to as the 'justificative effect of a licence' or the 'regulatory compliance defense'. Therefore this issue merits discussion at this point.

The regulatory compliance defense is, however, rejected in legal systems, like in Belgium and in the Netherlands.[26] One can find an economic rationale for this rule. If compliance with a regulatory standard or licence would automatically result in a release from liability, the potential injurer would have no incentive to invest more in care than the regulation asks from him, even if additional care could still reduce the expected accident costs beneficially.[27] A first reason to hold an injurer liable (if the other conditions for liability are met), although he has followed the regulatory standard, is that indeed this standard is often merely a minimum. The complete 'compliance defense' prevents any incentive to take precaution in excess of the regulated standard.[28] Exposure to liability will give the potential injurer incentives to take all efficient precautions, even if this requires more than just following the licence. This, by the way, holds both under negligence and strict liability. Since the regulatory standard can not always take into account all efficient precautionary measures an injurer can take, testing the measures taken by the injurer even though the regulatory standard was followed, will provide additional incentives. Allowing a regulatory compliance defense would also largely remove the beneficial incentive effects of strict liability. As we argued above, strict liability has the advantage that it provides the injurer with incentives to take all efficient measures to reduce the risk (prevention and activity level), even if regulatory conditions were followed. This outcome has been shown formally by Kolstad, Ulen and Johnson[29] and more recently by Burrows.[30] They argue that the complete compliance defense prevents any precaution in excess of the regulated standard. If there is serious under-enforcement of standards, the role of liability as an incentive to take precautions remains important.

A second reason is that exposure to liability might be a good remedy for the unavoidable capturing and public choice effects that play a role when permits are granted. If a permit would always release from liability, all a plant operator would have to do, is get a good permit with easy conditions from a friendly civil servant. That would then exclude any law suit for damages from a potential victim. Obviously the capturing and public choice effects should be addressed also via direct tools. In this respect one can think about the liability, even under criminal law, of the licensor.[31] Liability of the licensor (and appropriate sanctions within administrative law) can provide incentives to civil servants to act efficiently when granting licences.[32] This, however, still requires tort law to take into account the fact that regulatory standards are not always set efficiently. If the optimal level of care is higher than the regulatory standard liability will efficiently provide additional incentives.

Finally, tort law can also be seen as a 'stopgap' for situations not dealt with by the statute.[33] This makes clear that the exposure to liability notwithstanding the permit is an important guarantee that the plant operator will take efficient care.

Therefore, following the conditions of a license or – more generally – regulatory standards, should not have a justificative effect in tort. The opposite may only be true if it were clear that the administrative agency took into account all potential harm of all interested third parties when setting permit conditions. Indeed, theoretically regulators and licensors are supposed to set standards in regulations and permits in a way that reflects political choices about the level of risks that maximises welfare. Hence, ideally, when setting objective pollution standards, regulators are supposed to weigh costs and benefits of different norms and choose the standard that delivers the highest social net benefit. In such a case, a judge in a civil liability suit should not be "second guessing" efficient agency decisions. It is, however, rare that agencies will be able to take *ex ante* all these interests and possible damages into account when setting permit conditions. Hence, as a general rule, following licenses or regulatory standards should not free from liability; the opposite would be the exception. This is the case both under a negligence as well as under a strict liability rule. Indeed, holding an injurer liable, notwithstanding he followed regulatory standards, will play an important role under a strict liability rule, since this will lead the injurer to take efficient care and adopt an efficient activity level, i.e. to take all efficient measures to reduce the potential accident costs, although this might require to do more (as far as precaution is concerned) than the regulation requires. Under a negligence rule this case law is also significant if the efficient care standard (which is assumed to be equal to the due care standard required by the legal system) is higher than the regulatory standard. The basic reason remains that efficient preventive measures can be taken above what is often prescribed in the norm. Requiring a potential polluter to take these efficient preventive measures thus increases social welfare.

G. IN SUM

To conclude, I believe that the issues I have addressed in a non-formal way above, have been addressed excellently in the paper by Boyer and Porrini in a more formal manner. The relationship between tort law and regulation and the optimal combination of both instruments to control externalities is undoubtedly a field on which much more should be published in the near future.

NOTES

1. Shavell, S. (1984). Liability for Harm versus Regulation of Safety, *Journal of Legal Studies*, 357–374; Shavell, S. (1984). A Model of the Optimal Use of Liability and Safety Regulation, *Rand Journal of Economics*, 271–280; and Shavell, S. (1987). *Economic Analysis of Accident Law*, Cambridge, Harvard University Press, 277–290.

2. See the basic article by Stigler, G. (1961). The economics of information, *Journal of political economies*, 213; and Schwartz, A., and Wilde, L. (1979). Intervening in markets on the basis of imperfect information: a legal and economic analysis, *University of Pennsylvania Law Review*, 630–682; and Mackaay, E. (1982). *Economics of information and the law*, Boston, Kluwer.

3. Shavell, S. (1984). *Journal of legal studies*, 359.

4. Shavell, S. (1984). *Journal of legal studies*, 360.

5. Shavell, S. (1986). The judgement proof problem, *International review of law and economics*, 43–58.

6. Since we assume risk-neutrality insurance does not play a role here. If insurance would come into the picture it could overcome the problems of under deterrence, provided that the moral hazard problem, caused by insurance, can be cured.

7. Shavell, S. (1985). Criminal law and the optimal use of non-monetary sanctions as a deterrent, *Columbia law review*, 1232–1262.

8. Shavell, S. (1984). *Journal of legal studies*, 363.

9. See Landes, W., and Posner, R. (1984). Tort law as a regulatory regime for catastrophic personal injuries, *Journal of legal studies*, 417.

10. For alternatives to liability suits see: Bocken, H. Alternatives to Liability and liability insurance for the compensation of pollution damages, *TMA* 4–87, 83–87 and *TMA* 1–88, 3–10.

11. Shavell, S. (1984). *Journal of legal studies*, 363–364.

12. Complementarities between tort law and regulation have been addressed by Rose-Ackerman, S. (1992). Environmental Liability Law. In: T. H. Tietenberg (Ed.), *Innovation in Environmental Policy, Economic and Legal Aspects of Recent Developments in Environmental Enforcement and Liability* (pp. 223–243), Brookfield, Elgon; Rose-Ackerman, S. (1992). *Rethinking the Progressive Agenda. The Reform of the American Regulatory State*, 118–131, New York, Free Press; and Rose-Ackerman, S. (1991). Regulation and the Law of Torts, *American Economic Review, Papers and Proceedings*, (May), 54–58.

13. Dewees demonstrated that in North America the quality of the environment has improved substantially as a result of regulatory efforts, not so much in response to legal action in tort (Dewees, D. (1992). The Comparative Efficacy of Tort Law and Regulation for Environmental Protection, *The Geneva Papers on Risk and Insurance*, 446–467; and Dewees, D. (1992). Tort Law and the Deterrence of Environmental Pollution. In: T. H. Tietenberg, (Ed.), *Innovation in Environmental Policy, Economic and Legal Aspects of Recent Developments in Environmental Enforcement of Liability* (pp. 139–164) Brookfield, Elgon.

14. Dewees, D., Duff, D., and Trebilcock, M. (1996). *Exploring the Domain of Accident Law; Taking the Facts Seriously*, New York, Oxford, Oxford University Press.

15. Dewees, D., Duff, D., and Trebilcock, M. (1996). *Exploring the Domain of Accident Law; Taking the Facts Seriously*, 307–323. New York, Oxford, Oxford University Press.

16. Rose-Ackerman, S. (1992). *Re-thinking the Progressive Agenda, the Reform of the American Regulatory State*, New York, The Free Press; Rose-Ackerman, S. (1992). Environmental Liability Law. In: T. H. Tietenberg (Ed.), *Innovation in Environmental Policy, Economic and Legal Aspects of Recent Developments in Environmental Enforcement and Liability* (pp. 223–243), Brookfield, Edward Elgar; and Rose-Ackerman, S. (1996). Public Law versus Private Law in Environmental Regulation: European Union Proposals in the Light of United States and German Experiences. In: E. Eide and R. Van den Bergh (Eds), *Law and Economics of the Environment*, (pp. 13–39), Oslo, Juridisk Forlag.

17. Faure, M., and Ruegg, M. (1994). Standard Setting through General Principles of Environmental Law. In: M. Faure, J. Vervaele and A. Weale (Eds), *Environmental Standards in the European Union in an Interdisciplinary Framework*, (pp. 39–60). Antwerp, MAKLU.

18. Kolstad, Ch. D., Ulen, Th. S., and Johnson, G. V. (1990). *Ex Post* Liability for Harm vs. *Ex Ante* Safety Regulation: Substitutes or Compliments?, *American Economic Review*, (Vol. 80), 888–901.

19. Rose-Ackerman, S. (1995). Public Law versus Private Law in Environmental Regulation: European Union Proposals in the Light of United States Experience, *Review of European Community and International Environmental Law, RECIEL*, (Vol. 4), 31–32; and Rose-Ackerman, S. (1984). *Controlling Environmental Policy: the Limits of Public Law in Germany and the United States*, Yale University Press, New Haven and London.

20. See Shavell, S. (1984). *Journal of legal studies*, 365–366; and Faure, M., and Van den Bergh, R. (1987). Compulsory Insurance for Professional Liability, *The Geneva Papers on Risk and Insurance*, 109–110.

21. Compare for the bonus pater familias standard, Posner, R. (1986). *Economic Analysis of Law*, (3rd ed.), Boston, Little Brown, 151–152; and Shavell, S.(1987). *Economic Analysis of Accident Law*, Cambridge, Harvard University Press, 74.

22. Shavell, S. (1984). *o.c., Journal of Legal Studies*, 365–366.

23. See Van Dam, C. C. (1990). *Zorgvuldigheidsnorm en aansprakelijkheid*, Deventer, Kluwer, 74–81.

24. Faure and Van den Bergh have also argued that an advantage of this system is that it gives victims incentives to prove that the regulatory standard has been breached. This makes the victim an enforcer of safety regulation. He can claim compensation under the negligence rule as soon as a causal relationship between the violation of the regulatory standard and his damage is established (Faure, M. and Van den Bergh, R., (1987). *o.c., Geneva Papers on Risk and Insurance*, 110–111). According to Dutch tort law, however, as one of the requirements of the general tort provision the so-called relativity requirement will also have to be met (in Germany referred to as the "Schutznormtheory)".

25. Rose-Ackerman, S., *Rethinking the Progressive Agenda*, 127.

26. For a comparative analysis of the question whether following a permit excludes criminal liability see, M. Faure/J. C. Oudijk, Die strafgerichtliche Überprüfung von

Verwaltungsakten im Umweltrecht. Ein rechtsvergleichender Überblick der Systeme in Deutschland, den Niederlanden und Belgien, *JZ*, 1994, 86–91.

27. Shavell, S. (1984). Liability for Harm versus Regulation of Safety, *JLS*, 365; Faure, M., and Van den Bergh, R. (1987). Negligence, Strict Liability and Regulation of Safety under Belgian Law: An Introductory Economic Analysis, *GPRI*, 110.

28. Burrows, P. (1999). Combining regulation and liability for the control of external costs, 19 *IRLE*, 242. Recently Schwartz added to the debate by discussing whether compliance with federal safety statutes should have a justificative effect in state tort cases. See Schwartz, A. (2000). Statutory Interpretation, Capture, and Tort Law: The regulatory compliance defense, *American Law and Economics Review* (ALER), 1–57.

29. Kolstad, C. D., Ulen, T. S., and Johnson, G. V. (1990). Ex Post Liability for Harm vs. Ex ante Safety Regulation. Substitutes or Compliments? 80. *American Economic Review (AER)*, 888–901.

30. Burrows, P. (supra note 28) pp. 227–244.

31. Faure, M., Koopmans, I., and Oudijk, J. (1996).Imposing criminal liability on Government Officials under environmental law: a legal and economic analysis, *Loyola of Los Angeles International Comparative Law Journal*, 529–569.

32. Note, however, that industry argues against such a liability of the licensor, claiming that this may entail the risk that licensors would be too reluctant in allowing emissions, if this could give rise to their liability (G. J. Niezen Aansprakelijkheid voor milieuschade in de Europese Unie in: *Ongebonden Recht Bedrijven* 2000, p. 171).

33. Rose-Ackerman, S. (1992). *Rethinking the Progressive Agenda. The Reform of the American Regulatory State*, 123; and Arcuri, A. (2001). Controlling environmental risk in Europe: the complementary role of an EC environmental liability regime, *Tijdschrift voor Milieuaansprakelijkheid* (TMA), 43–44.

PART 2:
OPTIMAL INSTRUMENT DESIGN

ENVIRONMENTAL DAMAGE INSURANCE IN THEORY AND PRACTICE

Michael Faure

ABSTRACT

In this paper, some particularities of environmental insurance are addressed. First, the paper summarizes the general conditions of insurability. Then, it is explained why insuring environmental liability may be difficult. Specifically, the necessity of an adequate risk differentiation may be difficult to obtain in cases of environmental liability. Then, it is explained how, at the theoretical level, a move towards different insurance schemes (notably first party or direct insurance schemes) provides a solution to problems of insurance of environmental liability. Although the move towards first party or direct environmental insurance may seem attractive at a theoretical level, nevertheless a variety of practical questions may arise. The alternative utilized in practice is not first party insurance (whereby victims would take out insurance coverage), but a form of direct insurance, whereby environmental damage is insured directly, that is to say as soon as damage occurs and irrespective of liability. The paper then discusses a recent example of such a direct environmental insurance, as it was applied in the Netherlands. Although this system seems to have considerable benefits, the major disadvantage lies in the fact that apparently all insurers in the Netherlands moved to this new environmental

An Introduction to the Law and Economics of Environmental Policy: Issues in Institutional Design, Volume 20, pages 283–328.

283

*damage insurance. This raises important questions as to the competitive-
ness of the particular market.*

1. INTRODUCTION

Many have pointed at the fact that the scope of liability of industrial operators
is increasing in Europe as well.[1] This expanding liability also seems to hit the
area of environmental liability and, in combination with an increasing tendency
towards intensified (command and control) regulation, some have argued that
the polluter pays double.[2] A recent white paper on environmental liability,
launched on 9 February 2000 by the European Commission, seems – to some
extent – to confirm this tendency: the white paper proposes a strict (although
non-retroactive) environmental liability for damage caused to biodiversity,
provided that it is caused by dangerous and potentially dangerous activities,
regulated by EC environment related law.[3]

One result of the seemingly expanding scope of environmental liability is
that – not surprisingly – this has led to great concern for the European liability
insurers. An often-heard reaction, with many changes in liability law, is that as
a result of these expansions, the environmental liability would become
uninsurable. Whether this expanding environmental liability indeed endangers
the insurability of the liability risk is not the topic I am going to examine in
this paper.[4] An interesting effect of these changes in liability law is, however,
that both insurers and industrial operators seem to be looking for alternatives
to liability insurance to cover environmental risks. Several alternatives have
been developed, going from the use of capital markets to the installation of
compensation funds.[5]

There is one particular alternative, which merits a closer look. It concerns
more particularly the tendency in some countries to move away from third party
liability insurance towards first party insurance. It is precisely this tendency that
I would like to examine more closely in this paper. The reason is that this
change from third party towards first party insurance seems at first blush to
correspond nicely with George Priest's suggestion that this would be an
appropriate remedy for the American insurance crisis.[6] Some insurers seem to
have taken Priest's warnings for a liability crisis seriously and the same is
apparently true for the remedies he proposed. One can, especially in the
environmental sphere, notice an increasing tendency towards first party
insurance. This is, by the way, not only the case in the environmental area, but
also, e.g. in occupational health and medical malpractice and an increasing use
of first party insurance is even also discussed in the traffic accident area.
Therefore, this phenomenon definitely merits a closer look.

In this paper I will first of all look at the theoretical differences between first party and third party insurance. The question arises how, at least in theory, the two differ as far as their ability to prevent environmental harm is concerned as well as the possibility to provide adequate compensation to accident victims (Section 2). Then we will take a practical example provided by the Netherlands since the Dutch insurers recently decided to more or less abolish environmental liability insurance and change radically towards environmental damage insurance. We will have a critical look at this policy and examine whether this is indeed an acceptable example, which should be followed by other countries (Section 3). Obviously at the policy level, several questions arise, which are important both for the liability and for the first party insurance area. These policy issues have to do with the question whether insurance coverage should be made compulsory and also relate to the importance of competition policy to guarantee fully competitive insurance markets (Section 4). All of these issues will be addressed using the economic analysis of tort and insurance. The reason for using the law and economics method is obvious: it provides the adequate tools to examine critically whether various insurance devices can lead to optimal prevention and compensation.

A few concluding remarks concludes the paper (Section 5).

2. FIRST PARTY vs. THIRD PARTY INSURANCE: THEORETICAL DIFFERENCES

Introduction

In the economics of accident law and insurance the reasons why persons seek insurance coverage have been explained. The utilitarian approach with respect to insurance has demonstrated that risk creates a disutility for people with risk aversion. Their utility can be increased in case of loss spreading or if the small probability of a large loss is taken away from the injurer in exchange for the certainty of a small loss.[7] The latter is of course, exactly the phenomenon of insurance. The risk averse injurer has a demand for insurance; he prefers the certainty of a small loss (the payment of the insurance premium) whereby the probability of a larger loss is shifted to the insurance company, thereby increasing the utility of the injurer.[8] It is remarkable that in this utilitarian approach of insurance, liability insurance is in the first place regarded as a means to increase the utility of a risk averse injurer, not so much as a means to protect victims, as is sometimes argued by lawyers.

The reason an insurance company can take over the risk of the injurer is well known: because of the large number of participants the risk can be spread over a larger group of people. The insurer only has to pay attention that he builds relatively small risk groups in which the premium is as much as possible aligned to the risk of the members of that group.

In addition to this utility-based theory of insurance which sees insurance as an instrument to increase the expected utility of risk averse persons through a system of risk spreading, Skogh has powerfully argued that insurance may also be used as a device to reduce transactions costs.[9]

Thus, one can easily apply these general principles to explain why there will be a demand for environmental liability insurance: it can provide protection for risk averse injurers. However, it has been argued that several conditions have to be met to keep (environmental) liability an insurable risk. Let us look briefly at these conditions of insurability of liability. The question we are interested in within the scope of this paper is obviously whether some of these conditions of insurability might be easier to meet in case of first party than in case of third party insurance.

Insurability of Environmental Liability: A Brief Overview

Let us look at those conditions which are generally listed in the literature, either positively as conditions which have to be met to guarantee the insurability of (environmental) liability or, formulated negatively, as factors which may endanger the insurability of (environmental) liability:[10]

Predictability, Uncertainty and "Insurer Ambiguity"
Obviously for every insurance scheme, also for environmental liability insurance, it is crucial that the insurer possesses accurate information on the likelihood that the event will occur (the probability) and on the possible magnitude of the damage once the accident will occur. These expectations on probability and magnitude of the loss are essential for the insurer to be able to calculate his so-called actuarially fair premium. Increased with the so-called loading costs (for among others administrative expenses) and, depending on the market structure, a profit margin, this will constitute the premium to be paid by the insured.

In this respect environmental liability insurance is obviously not different than any other type of insurance; for these general principles we can hence refer to the insurance economics literature. However, it is important to stress that one crucial element in insurance, already to make an accurate premium calculation possible, as we just indicated, is precise information on the probability that a certain loss will occur and the possibility to make a more or less accurate estimate of the potential magnitude of the damage. This is not only necessary in order to make an accurate estimation of the premium to be charged, but also to set aside a reserve in case the accident for which insurance coverage was seeked occurs.

If the insurer ideally has *ex ante* perfect information on the *predictability* of the probability and the magnitude of the damage, we call the particular risk insurable. It is precisely on the basis of statistics that the insurer will require information on the likelihood that the risk will occur with a particular insured; statistics may also provide information on the possible magnitude of the damage. Both these requirements may, however, be a problem in the case of environmental (liability) insurance. Several elements may negatively influence the *ex ante* predictability of the risk.[11] The *ex ante* information on the predictability of the risk is often low, given the relatively new character of environmental risks. Reliable statistics may sometimes be missing both with respect to the probability of the event occurring and with respect to the damage.[12] Hence, there may not be a "law of large numbers" to be applied. This obviously is not only a problem for environmental insurance, but occurs in every case where insurers are confronted with new risks, where reliable data may be missing.

The question therefore arises whether the predictability of the liability risk can be increased, even in the absence of reliable statistics, or whether in that case the particular risk should be judged as uninsurable. The literature has indicated that uncertainty concerning the probability or the damage is of course an element with which the insurer can, in principle, take account *ex ante*. If there is uncertainty, because of a lack of reliable statistics, this should not necessarily lead to the conclusion that a particular risk is uninsurable. We are then dealing with the concept referred to as "insurer ambiguity" addressed by Kunreuther, Hogarth and Meszaros.[13] They argue that the insurer can react to this uncertainty concerning either the probability of the event or the magnitude of the damage by charging a so-called risk premium to account for this unpredictability. Hence, an insurer can, in principle, also deal with a "hard to predict" event, by charging an additional premium.

Capacity

We have already mentioned that a requirement for insurability is that the insurer should have *ex ante* information on the predictability of the risk and on the magnitude of the damage. So far we have dealt with the importance of the predictability of liability. However, also the magnitude of the harm may constitute a problem.

There may be uncertainty as far as the possible magnitude of the harm is concerned. The insurer may react to this uncertainty by providing for an adequate reserve to be able to provide coverage for the environmental damage once it occurs. However, in many cases the expected loss may exceed the possibilities of the individual insurer. In that case the insurer can use various traditional insurance techniques to cope with this capacity problem. One possibility is to insure a similar risk jointly with a few insurers (so-called co-insurance); another possibility is re-insurance. One other solution, which is often used in cases of environmental liability insurance, is pooling of capacity by insurers. In many countries insurers have shared risks in mutual pools on a non-competitive basis to be able to provide coverage, also for risks with a relatively high potential magnitude. This is typically the case for the nuclear risk.

Moral Hazard

If risk is fully removed from the injurer and shifted to the insurer, the injurer will indeed miss the incentive for care taking that was exactly given to him by the deterrent effect of having to pay compensation in case of an accident. Marc Pauly has, by the way, indicated that in fact this behaviour of the injurer is not immoral but completely rational since he simply reacts to varying costs for his behaviour.[14] For the insurer of course the problem arises how nevertheless incentives can be given to the insured to behave in exactly the same way as if no insurance were available. This is, of course, the goal of an optimal control of moral hazard.[15]

In the literature two ways of controlling the moral hazard problem are indicated.[16] The first is a control of the insured and an appropriate adaptation of the premium; the second is exposing the insured partially to risk. A first best solution is a detailed control of the insured.[17] In that case the premium conditions would be exactly adapted to the behaviour of the insured and the premium would reflect the care taken by the insured. In an optimal world this should give the insured incentives to behave exactly as if no insurance were available and the premium would reflect the true accident risk. Of course, this first best solution is only possible in the ideal world where control by the insurance company would be costless and information on the behaviour of the insured readily available. In practice this is of course not the case. There are, however, some means for a control of the insured and a differentiation of premium conditions according to certain groups of risk. This can either be an *ex ante* screening with a higher premium for certain high risk groups or an *ex post* premium increase or change of policy conditions based on previous loss experience. This is the so-called experience rating. Much of insurance legislation is also aimed at reducing moral hazard.[18] Think in this respect, e.g. of the prohibition, contained in many insurance laws, to insure accidents which are caused with intent.[19]

A second best solution is exposing the insured partially to risk. This is considered second best because insurance should ideally exactly aim at removing risk from the injurer. Exposing the insured to risk will mean that some degree of risk aversion will remain. This has, on the other hand, the advantage that the insured injurer will still have some incentives for care taking although he is insured. This exposure to risk can be either at the lower level of damage or at the higher level. One could indeed think of a system with a deductible whereby a lower threshold applies or one could introduce an upper limit on coverage whereby the insured bears his own loss in case the damage exceeds the insured amount.

Adverse Selection
As indicated above, that the insurance is based on a system of loss spreading. Therefore the insurer needs a minimum number of similar risks he insures. At the same time risk pools have to be constructed as narrow as possible, meaning that the average premium in the risk pool should correspond with the risk of most of the members in the particular pool. Would this not be the case, then the average premium would be relatively high for low risk members, which would then leave the group. In that case the well-known phenomenon of adverse selection could emerge, which has been described in the seminal paper of Akerlof on the market for lemons.[20]

Adverse selection will, in other words, arise if potentially responsible parties fail to disclose their true risk profile appropriately, which may endanger the narrowing of risk pools. Rogge holds that in Belgium the financial capacity to insure would be limited[21] precisely because only bad risks would have a demand for insurance. If this can not be 'compensated' with good risks an incurable adverse selection would remain.[22] Thus, "lack of demand has been matched by lack of supply."[23]

The literature has indicated that the appropriate remedy for both moral hazard and adverse selection is risk differentiation. We will come back to the importance of risk differentiation in environmental insurance below.

Avoid Shifting the Risk of Causal Uncertainty
There is another trend in liability law, which may indeed seriously endanger the predictability, and therefore the insurability, of certain risks. This concerns the tendency to shift the risk of causal uncertainty to enterprises (and therefore to their insurers).[24] There are various situations that arose in Dutch case law to illustrate this tendency. The danger of shifting the burden of causal uncertainty to the enterprise is that the insurer of the specific employer or producer will be required to compensate for damage which, on the whole, had probably not been caused by the insured party.[25]

Retrospective Liability

There is one final danger, having to do with the structure of liability law, which may affect the insurability of (environmental) liability. This concerns those cases where liability expands in such a way that it is no longer foreseeable and that policy conditions can hence not be adapted ex ante. The question arises whether such a retrospective liability would lead to uninsurability.

At first sight one could argue that this certainly is the case. If the insurer were not aware that the behaviour of this insured party might potentially have been considered wrongful, no premium would have been charged for this risk, no preventive measures would have been required in the policy conditions and no reserves against losses would have been set aside. Indeed, insurance assumes that the insurer covers future risks, which are, at least to some extent, foreseeable. Insurance requires some degree of predictability. However, the mere fact that insurers of, say, industrial waste disposal sites in the 1960s have – as a matter of fact – not foreseen that the activities of their insured parties could lead to a liability in the future, does not make this event totally unforeseeable. The potential of a change in the scope of liability is an uncertain element which the insurer can, in principle, take into account *ex ante*. We are dealing here with the concept of ìnsurer ambiguity' addressed by Kunreuther, Hogarth and Meszaros.[26] If insurers could foresee the likelihood of a possible change in the liability system, they could react to uncertainty by estimating the probability that this event would occur and charge an additional risk premium to account for this legal uncertainty. In sum: in an *ex ante* perspective one can argue that nothing is totally unforeseeable or unpredictable; insurers can in principle cope with `hard to predict' events such as the introduction of retrospective liability by charging an additional premium. However, in an *ex post* perspective this message is not very helpful for insurers who, at the time, did not take this risk into account and now have to provide cover to enterprises for risks which the insurers considered apparently unforeseeable. Hence, no additional premium was charged and no reservations were made, which explains why the retrospective liability, which is now laid down, say, for soil clean-up costs, leads to major problems for insurers.[27] There is, hence, no problem of the uninsurability of retrospective liability as such, but only the simple fact that insurers did not take these risks into account when the policy was drawn up.

Summary

This brief overview of the conditions of insurability of the environmental liability shows that, provided that a number of conditions are met both in liability law and in the structure of the insurance policy, the environmental liability risk is insurable. This obviously is no surprise since one can notice liability insurance policies covering the environmental risk for a long time. However, the extent to which the liability risks are insurable is another issue. One could imagine that, e.g. the amount of the expected damage may be so large that it would exceed the capacity, even after using the possibilities of re-insurance and pooling. Also the unpredictability may be that large that insurers would be forced to charge a relatively high risk premium for which there might not be a willingness to pay. Some of these issues which may endanger the insurability of liability can, of course, be dealt with either in policy conditions or by the law. We cannot deal with these remedies for expanding liability any further in this paper.[28] But one could point, e.g. at the fact that insurers could try to protect themselves (to some extent) against the risk of retroactive liability by changing the cover in time. It has been argued that a so called claims-made policy, holding that the claim for damages must have been received by the insurer during the period of insurance cover, would be an appropriate remedy to all these cases where there is a long time lapse between the moment that the tort occurs and the moment that the harm manifests itself.[29] Moreover, the legislator could also choose efficient remedies. For instance, as far as the issue of causal uncertainty is concerned, the legislator (or case law) could introduce a proportionate liability rule, meaning that in case of uncertainty concerning the causal relationship, the victim can claim a proportion of its damage, equal to the likelihood that his loss was caused by a certain event. This proportionate liability has been defended as more efficient than the rigorous all or nothing approach or the reversal of the burden of proof.[30] Thus a proportionate liability rule could enhance the insurability of the liability risk.

Within the scope of this paper we are, however, not interested in the possible devices to increase the insurability of environmental liability. The question I would like to turn to now is whether, taking into account the factors that affect the insurability of the environmental risk just discussed, one could argue that the environmental risk (but the same applies to other risks as well) would, under some circumstances, be better insurable under a first party than under a third party insurance system, such as liability insurance.

Advantages of a First Party Insurance Scheme

Risk Differentiation: Theory

The advantages of a first party insurance scheme are more particularly to be found at the level of risk differentiation. Let us first discuss the importance of risk differentiation more closely and then address the question why risk differentiation may be easier in first party insurance.

George Priest has claimed that the adverse selection problem has caused an insurance crisis in the United States and that it can only be cured by an appropriate differentiation of risk.[31] If the insurance policy requires preventive action from the insured party and provides for a corresponding reward in the premium this should give optimal incentives to the insured for accident reduction. Thus, risk pools should be constructed as narrowly as possible so that the premium reflects the risk of the average member of that particular pool.[32]

A further differentiation of the risk is obviously only efficient as long as the marginal benefits of this further differentiation outweigh the marginal costs of such a differentiation.[33] Risk differentiation certainly does not mean that insurers would have to use an individual tariff in each case.[34] The possibilities for individual differentiation will inevitably also depend upon the value of the particular insurance policy. For mass insurance products with a low premium, risk differentiation can only take place in general categories. In professional liability insurance of enterprises, however, the benefit of detailed differentiation, rewarding an enterprise for preventive action, may well outweigh the costs.

It is, thus, not difficult to make an economic argument in favour of effective risk differentiation as a remedy for a growing liability risk. If good risks are not rewarded for preventive action, either they will not have an incentive for prevention or they will leave the risk pool, and consequently the risk pools will be unravelled, as described by Priest.

First Party Insurance: Theory
Liability insurance is a third party insurance, whereby the insurer covers the risk that his insured (the potentially responsible party) will have to compensate a third party. A first party insurance is a system whereby the insurance coverage is provided and compensation is awarded directly by the insurer to the victim. Whether such a first party insurance can be considered as an efficient alternative for third party liability can not be answered in general terms. It depends to a large extent on the details of such a system and more particularly on the question whether the first party insurance is combined or not with the liability of the potentially responsible party.

The underlying principle in a first party insurance is that the insurance undertaking, in principle, pays as soon as damage occurs, provided that it can be proven that the particular damage has been caused by the insured risk. Payment by the insurance undertaking occurs irrespective of the fact whether there is liability. The arguments advanced in the literature in favour of first party insurance are that the transaction costs would be lower and that risk differentiation might be a lot easier.[35] The reason is simply that with first party insurance the insurer covers directly the risk of damage with a particular victim or a particular site. The idea is that it is therefore much easier for the insured to signal particular circumstances, which may influence the risk to the insurer. The problem with liability insurance is that the insurer is always insuring the risk that his insured (the potential injurer) will harm a victim (a third party) of which the properties are unknown *ex ante* to the insurer. Moreover, under liability insurance there are lots of uncertainties, e.g. how the judge will interpret this specific liability of the insured. In the ideal world of first party insurance the insurer directly covers the victim, e.g. the risk. He can therefore monitor directly the risk and, in principle, provide a much better risk differentiation.[36]

If one would take the importance of risk differentiation and the risks of adverse selection as seriously as Priest did, the shift towards first party insurance seems to be most promising. One can – at least at theoretical level – understand why first party insurance would be beneficial to insurers: it is obviously much easier to control and assess *ex ante* the risk that a particular victim would suffer damage instead of assessing the risk that his insured potential injurer would cause harm to a third party and would thus be found liable. There are indeed many more uncertainties in third party liability, which could make an adequate risk differentiation more complex. Obviously within liability insurance risk differentiation is possible as well, in the sense that, as we have explained above, the insurer can adequately monitor whether his insured is a good or a bad risk (e.g. in the sense of taking preventive measures), which could then be rewarded with a lower or higher premium. But even if an ideal risk differentiation were applied, there are still many uncertainties in liability insurance. For instance, it will depend on the state of case law and the interpretation by judges whether the insured potential injurer will be found liable for a specific behaviour and a specific damage; it is apparently especially this uncertainty caused by judges, which the insurers dislike. Moreover with third party liability insurance it is *ex ante* not possible to know whether the insured injurer will cause damage to a very high income or a very low-income victim. That type of uncertainty can of course be avoided under first party insurance where – at least in the ideal situation – the victim chooses *ex ante* the insurance coverage he wishes according to his own demand and expected losses.

Notwithstanding these theoretical advantages, several questions still remain. More particularly the question arises how this scheme could be used to enhance the insurability of the environmental risk, which is the central topic of this paper.

Direct Insurance for Environmental Risks: A Few Questions

Let us now see how such a first party insurance could be used as an alternative for liability insurance to cover environmental risks. Several practical questions will arise.

First Party vs. Direct Insurance

As we mentioned above, a first party insurance is based on the principle that the victim seeks coverage directly from the insurer. Such a system can, of course, hardly be applied to the full extent in the environmental sphere. Theoretically, in a pure first party insurance it is the victim who takes out insurance coverage and who therefore also pays the premium. Such a pure first party insurance scheme would probably not be very practical for environmental risks, unless one imagines industrial operators seeking coverage, e.g. for the pollution that may occur on their own plant. In that case, the operator would, in first party insurance, seek coverage for the damage he may suffer himself. Otherwise first party insurance would mean that private citizens who fear to suffer environmental harm, e.g. in their garden, seek insurance coverage themselves.

However, there is an alternative which looks like first party insurance, and which is called direct insurance. In a direct insurance scheme one would imagine the potential polluter, e.g. the one who possesses a particular site, who seeks insurance coverage providing protection also to third parties who could suffer damage resulting from that particular site.

Such a direct insurance scheme is applied in some countries with respect to occupational health. In that case, e.g. the employer would take insurance coverage on behalf of his employees, who could claim directly on the policy. Under direct insurance the policyholder is not the victim, but the potential injurer. It is, however, not a liability insurance since the trigger to compensate is no longer liability, but the mere existence of damage.

For the environmental risk one could thus examine both a pure first party insurance, whereby potential victims cover losses themselves, and a direct insurance, whereby a potential polluter seeks insurance coverage also to the benefit of a third party (the victim). As we will see below, in practice there may be a combination of first party and direct insurance, in case a possessor of a particular site from which, e.g. soil pollution may occur, would seek coverage both for his own damage and for the damage which may be caused to third parties.

Scope of Coverage and Causation

The main difference between traditional liability insurance and either pure first party or direct insurance is that the coverage under the latter policies is no longer for liability. Hence, a consequence (and insurers claim an advantage) is that liability is no longer required to be able to claim on the policy. Traditional insurance would say that one would need an "accident". However, it is well known that, given the gradual nature of many pollution cases, these can hardly be considered as the sudden events, which are accidents. Therefore, in environmental insurance, damage will probably suffice, but it will be important to describe clearly what constitutes an insured damage which may gave a rise to a claim on the policy.

Depending on how this is formulated in the policy, the insured will have to claim that his damage is caused by the insured activity. In the hypothesis of a direct insurance, which benefits third party victims, problems could arise in case of multiple causes. In that case the victim will obviously have to claim and prove that his damage was caused by the particular insured risk.

However, with environmental damage insurance, causation does not seem to be a major problem (other than that it may always be a problem in any environmental liability case as well). This is different in cases where first party insurance solutions are presented for health damage, such as in case of medical malpractice or for occupational diseases. In all of those cases the difficulty will be that the question will have to be settled whether the personal injury suffered by the victim is actually caused by medical malpractice or by an occupational disease. This type of causation issues, relating to the fact that it should be clear that the harm is caused by the insured risk, may also arise in environmental insurance, but seem less serious than in cases where personal injury is insured.

Financing
Although less interesting from a theoretical perspective, a practical question that will always arise at the policy level is who should pay for a first party or direct insurance scheme. A consequence of a pure first party system is that it theoretically is the victim who seeks insurance coverage himself and who therefore finances the first party insurance. To the extent that this victim is also the industrial operator who caused the risk, polluter and victim will be the same and it will effectively be the polluter who finances the insurance. If one would, however, adopt a first party insurance scheme for "innocent" victims, politicians would probably argue that now victims are forced to finance, e.g. the clean-up of the polluted soil in their own garden (by paying the premium for the first party insurance) whereas the polluter should pay for this. The answer would obviously be that if the first party insurer of the "innocent" victim were allowed a right of redress against the polluter, the polluter pays principle would still be satisfied and the polluter would still be given appropriate incentives for prevention.

However, it is most likely that at the policy level discussions would primarily go in the direction of a direct insurance since under such a system the polluters finance the insurance coverage, also for the damage caused to third party victims.

Prevention

A question, which will obviously be asked, is how polluters are given appropriate incentives for prevention of environmental harm if the risk is fully covered under insurance. The answer here is obviously that this poses no difference with traditional liability insurance. All insurance systems potentially lead to the moral hazard problem, discussed above. But we equally indicated that an adequate risk differentiation can be the appropriate remedy to moral hazard. First party and direct insurance schemes were precisely advanced because they would enable a better risk differentiation. If that were the case, they could even lead to better results as far as prevention is concerned.

In practice the risk differentiation under first party and direct insurance would mean that, e.g. if a particular site were insured, the insurer would use all the *ex ante* monitoring and *ex post* devices to check the "ecological reliability" of the particular operator, which should provide optimal incentives for prevention. Hence, if the theoretical possibilities of risk differentiation are used in an optimal manner, first party or direct insurance schemes should cause no problem as far as prevention is concerned.

Relationship With Liability Law

An interesting issue, since liability is no longer the trigger for coverage, is of course the relationship to liability law. In many instances where first party of direct insurance schemes are advanced (such as in the areas of medical malpractice and occupational health), those schemes are often supposed to totally replace the liability system. One can notice, e.g. in the area of occupational health in many European systems a direct insurance scheme (providing coverage to all insured employees of one employer) with an immunity for liability of the employer, except in case of intent or gross negligence. In many countries these insurance systems are embedded in social security law.

A traditional argument, e.g. in the area of medical malpractice, against first party (patient) insurance schemes and in favour of tort law is that the incentives of the health care provider would be diluted without a liability system.[37] At the theoretical level there would be no reason to grant immunity to the polluter even if, e.g. a mandatory first party or direct environmental insurance scheme were introduced. Even if a third party victim were compensated by the insurer, it would still be possible to use liability law (via the right of recourse) to provide appropriate incentives for prevention with the polluter.

Therefore, there seems to be no reason to let a first party insurance scheme replace liability law.

Notwithstanding this theoretical starting point, one can often notice in practice that first party or direct insurance schemes are only accepted (it is to say their mandatory introduction) if, e.g. in the area of occupational health the potential injurer (the employer) receives some kind of "compensation" in the form of a partial immunity of liability.

Guarantee of Coverage

Finally one could ask to what extent a first party or direct insurance scheme can provide a guarantee that funds will indeed be available at the crucial moment when they are needed, e.g. when a polluted soil needs to be cleaned up. This question has important policy implications and has several aspects. First of all, it relates to the question whether such a first party or direct insurance scheme should be introduced compulsorily. Indeed, the market could work out wonderful insurance devices, but if no operator would make use of those, there would still be no guarantee that, given the insolvency risk, funds would effectively be available when they are needed. However, that is obviously not a particular problem of first party insurance, but a more general problem relating to the question whether at the policy level a duty to seek financial coverage for environmental risk should be introduced. We will briefly address this question in Section 4. An issue, which is related to direct insurance, is whether one can provide for a direct right of action for a third party victim. This concerns those situations where the insurance policy of the industrial operator provides that his coverage also extends to damage caused to third parties. Normally these third parties do not, given the privitiy of contract, have a claim on the insurance policy, unless they are formally made third party beneficiary of the policy, which provides for the direct insurance. The question hence arises whether one can see those third party beneficiary clauses in practice.

After having identified the potential benefits and several points of attention relating to first party and direct environmental insurance, let us now look at how these new policies are introduced in practice.

3. ENVIRONMENTAL DAMAGE INSURANCE IN PRACTICE

A Trend Towards First Party Coverage

At the end of the last section I indicated that although first party insurance might seem very attractive at first sight at the theoretical level, since it might better enable a narrowing of risk pools, still a lot of questions arise, especially in its application to the environmental risk. It is probably best to examine these questions by looking at a practical example where a first party insurance of polluted sites has been implemented.

Before doing so it is probably important to stress that this exposé on first party insurance is not merely theoretical, but does indeed have a certain practical relevance. Indeed, the general liability committee of the Committée Européen des Assurances (CEA) has executed a study on first party legal obligations for clean-ups and corresponding insurance covers in European countries.[38] This study shows that although the insurance situation between the European countries still differs to a large extent, first party insurance coverage seems to be available in several member states.[39] In one country, more particularly in the Netherlands, the insurers have deliberately chosen to provide coverage of polluted sites on a first party basis. The idea is that the first party coverage should replace the traditional environmental liability insurance. Hence, it seems interesting to take a closer look at the insurance situation in the Netherlands.

Environmental Damage Insurance in the Netherlands

Dissatisfaction With Existing Coverages
Starting point for the concern of the Dutch insurers constituted the fact that all the theoretical problems I discussed above in Section 2 concerning the insurability of environmental liability, had also played a major role in the Dutch environmental practice. This had to do with the fact that the environmental risk is the example of a "long-tail risk", whereby the insurer today could be confronted with events which occurred in a distant past and would lead to liability of the insured now. Insurers held that this generally endangered the predictability of the risk.

Most of these problems were hence related to the fact that environmental harm does not constitute a sudden event, as is the case with most "traditional" accidents which are insured in liability insurance.

The Dutch insurance market therefore used to have the environmental risk covered through a variety of insurance policies, of which the most important were:

- the liability insurance policy (AVB)[40] for the sudden risks and for occupational health risks which were related to the environment;
- the environmental liability insurance (MAS) for risks of a more gradual nature, non-including personal injury;[41]

- fire insurance for clean-up costs after fires (although the precise coverage of that policy was debated).

This environmental liability insurance was provided by an environmental pool, referred to as MAS.[42] In this MAS 50 (re)insurers work together to cover the environmental liability risk. The MAS was hence constructed as an environmental pool, although the individual insurers, which were connected to the MAS, contracted individually the MAS policy under their own label.

There was, however, a lot of critique on that system, which can be summarised as follows.[43]

- First of all, the whole division of coverage between the AVB and the MAS was based on the idea that the AVB would cover the sudden risks and the MAS would cover the risks of a more gradual nature. As is well known, in practice it was not always possible to make a clear distinction between sudden and gradual risks, which led to uncertainties concerning the scope of the coverages of both policies. This was obviously the result of the fact that the Dutch insurers did not decide to exclude the environmental risk all together from the traditional liability insurance of companies (AVB).[44]
- The environmental liability policy (MAS) was considered to be rather complex and had a very complicated procedure towards accepting insured. In addition, the policy was rather expensive and hence difficult to sell.
- A further problem was that neither the general environmental liability policy (AVB) nor the environmental liability insurance policy (MAS) provided any coverage for damage caused to the own site of the insured. This caused a problem for the insured, since according to the case law of the Dutch Supreme Court, companies could be held liable as well for a pollution of their own site. In addition the fact that pollution caused to the site of the insured was not covered inevitably again caused uncertainty concerning the scope of coverage. One could imagine cases whereby, e.g. polluted groundwater went from the site of the insured to a neighbouring site. This caused problems since the pollution to the site of the neighbour was insured whereas the pollution to the own site was not.
- Furthermore there were more uncertainties concerning the question whether clean-up costs were covered under the fire insurance policy. The fire insurance covers under the clean-up costs after fire also environmental damage that would have occurred, but only on the condition that there was a prior fire. In the fire insurance policy, no account was taken of the fact that after a fire also serious soil and water pollution could occur. Also, it

was not always clear whether the soil clean-up costs, resulting from the fire, were also covered under the fire insurance.

- Finally there were obviously all the traditional problems related to environmental liability, such as the question whether a specific damage was indeed caused as a result of an insured risk. Also the case law concerning environmental liability extended in such a way in the Netherlands that this has not been foreseen by insurers. Hence, the liability risk was increasingly considered unpredictable and hence uninsurable as far as environmental harm was concerned.

Environmental Damage Insurance: Main Features

This led the Dutch insurers association to present in 1998 a new product, being the environmental damage insurance (MSV).[45] This policy is available now since 1 January 1998 and it takes a revolutionary different approach than the traditional environmental liability insurance.[46]

This new environmental damage policy provides for various new elements. First of all it provides an integrated coverage of all the environmental damage which occurs on or from the insured site. Prerequisite is that it concerns pollution of the soil or of the water. The integrated coverage means that the new environmental damage insurance will replace the traditional pollution insurance (for sudden pollution) in the AVB and the liability insurance of the MAS (for gradual pollution).

The whole idea is that this coverage constitutes a direct insurance. In other words, the insured site is insured, even when clean-up costs have to be made on the site of a third party. Coverage takes place, as soon as the insured site is polluted as the result of the insured risk, irrespective of the fact that the insured could be held liable for the damage or not. In some cases the third party (the victim), moreover, receives a direct action on compensation on the basis of the environmental damage insurance policy. The trigger for compensation under this policy is therefore no longer tort law, but the insurance policy as it has been concluded between the insured and the insurance company. This, therefore, typically is a first party insurance or, as it is called in the Netherlands, a direct insurance. It is direct insurance to the extent that it also benefits third parties. It is indeed not the third party victim who purchases insurance (although the insured may be the victim), but someone who has responsibility for a site on or from which water or soil pollution may occur. The policy benefits third parties as well, at least when this is provided in the policy.

Obviously the environmental damage insurance can not set aside tort law, but the main advantage according to Dutch insurers is that the coverage is not triggered on the basis of liability. The advantage from the victim's point of view is obviously that coverage can be provided more rapidly and probably at lower transaction costs than through the court cases which are necessary as a result of liability law.

The environmental damage insurance as provided by the Dutch Insurers Association constitutes of several categories with different policies. The insured can opt for various insurance policies. This obviously shows that indeed first party insurance better enables an optimal risk differentiation since every insured will be able to purchase insurance coverage according to his own preferences. The damage on the insured location itself is insured. This at least provides coverage for clean-up costs and this is rather broadly defined. Also costs for the repair of damage are included. Soil pollution, which is also envisaged, obviously falls within the environmental damage insurance. Note, however, that under this policy it is only the costs of "clean-up" which are covered.[47] Also the damage caused from the insured site and suffered by third parties is covered. This obviously provides a broader scope of protection.

The insured remains in principle fully liable, although the third party (beneficiary) that would be protected under the new MSV policy could claim directly on the policy and would hence, in principle, not have an interest in using liability law. However, it might be that the insured has taken too limited coverage and that in that particular case the third party would still (have to) use liability law. In that case the liability itself is not covered (and an insolvency risk remains), but the MSV policy provides for legal aid assistance in a couple of specific cases.[48] This is, e.g. the case if the sum which is insured under the MSV coverage would not be sufficient, e.g. to pay for the clean-up costs incurred by the government. Or in case a third party would choose the liability law instead of direct action under the MSV policy.[49]

The main feature of the new environmental damage insurance provided in the Netherlands is obviously that it is no longer a liability insurance, but only a first party (or direct) insurance. The advantage for the insurer (and for the insured) is that the difficult road of liability law is excluded. Whether liability law will still be used is uncertain. Third parties could still use liability law, although it is obviously easier for victims to use the direct action provided under the MSV policy. There is, however, one important weakness, which unavoidably remains, being the fact that the environmental damage insurance is not compulsory. Hence, there may remain situations where companies in the Netherlands have purchased no insurance coverage at all or situations where only a basic coverage was taken for damage on the particular site, but only to a limited extent for damage to third parties.[50] In those cases, third parties claiming against the responsible party might still be confronted with an insolvent polluter. Moreover, the new MSV regime is exclusive, meaning that the coverage for (sudden) soil and/or gound water pollution has now been removed from the liability insurance policy. This means that if an insured would, e.g. only have taken an MSV coverage for the insured site and a loss would occur to a third party on another site, this third party would probably use liability law against the polluter. In that particular case the polluter can not call on his general liability insurance (AVB) since the environmental risks have now been completely removed from that policy as a result of the entering into force of the MSV.[51]

But that can obviously hardly be considered a weakness of the system of first party insurance: the insured obviously does not get more than what he pays for. Since the MSV is a general policy with a lot of options for the insured, premiums and amount of coverage can vary. The type of costs which are insured are, however, identified in the general policy and accordingly to the CEA study on first party insurance the total amount of coverage available under this new environmental damage insurance in the Netherlands would be 25 million Dutch guilders.

Evaluation
Mixed first party and direct insurance. If we turn back to the specific questions relating to first party/direct insurance one can notice that the Dutch scheme is apparently a mix of first party and direct insurance. It is principally a first party insurance in the sense that the insured site of the potential polluter is covered. It is also a direct insurance since the policy provides that not only the damage on the insured site itself is covered, but also the damage suffered by a third party.

The advantage for the insured is clearly that this first party insurance now provides coverage for the damage on the site of the insured as well. That is a major change compared to the past. Very often public authorities forced the polluters to clean up their own polluted soils. Formally this was not constructed as liability and therefore no insurance cover was available. As far as the insurance policy is concerned, it also covers the damage suffered by third parties, it effectively functions as a liability cover.

The trigger for coverage is no longer liability, but damage. There is a clear causation requirement in the policy, since it requires that the pollution must be the direct and exclusive result of an emission caused by one of the insured risks. Moreover, only the costs of remediation of the insured site are covered, which excludes the uncertainties involved with assessing environmental damage.

Advantages. As far as prevention via an optimal risk differentiation is concerned, insurers obviously claim that this policy for them has the major advantage that they can adequately control ex ante the quality of a particular soil and the production methods applied by the insured. On that basis an adequate assessment of the risk can be made.

However, some company lawyers have been critical of this new insurance product. They argue that in practice insurers are so critical in providing coverage that effectively only the very good risks would be able to receive coverage.[52] Hence, they claim that one cannot blindly argue that the environmental damage insurance would be a remedy for all insurability problems concerning the environmental risks. Indeed, a flip side of an effective *ex ante* monitoring is obviously that insurers could chose either only to cover the good risks or to cover bad risks only for a high premium. But one can hardly reproach the insurers to apply insurance economic principles correctly.

Moreover, the environmental damage insurance introduced by the Dutch insurers is a completely voluntary system, introduced without any government interference. A consequence is then also that the liability law is not affected by this regime at all. Victims suffering harm as a result of soil pollution can still sue the polluter. If his insured has accepted the possibility for third parties to claim, the victim can even claim directly on the policy as third party beneficiary. If, on the other hand, the MSV policy did not contain a third party beneficiary clause, the victim will, just as is the case today in the Netherlands, have to sue the liable polluter. A problem in that case is, however, as we indicated above, that the MSV is supposed to replace the liability insurance, meaning that a liability insurance will no longer be available, at least for the environmental risks covered under the MSV.

According to information provided by the Dutch Insurance Association this new product would work remarkably well. They claim that the interests of enterprises in this new environmental damage insurance is much larger than in the traditional environmental liability insurance. Whether this new product is actually a success yet is more difficult to judge. There is, however, undoubtedly an increasing interest of industry for this new environmental damage insurance. The fact that, hence, a wider financial security for environmental damage is provided in the Netherlands as a result of this product should definitely be considered as positive. Moreover from the victim's (mostly the government) perspective, the fact that the environmental damage insurance provides for a direct action for victims should be considered as positive as well. However, one should note that this third party action right only applies if the insured has accepted this.

Moreover, the advantage for the insured is that under this environmental damage insurance, also damage caused to his proper site is covered, which was obviously not the case under liability insurance.[53] If, at a policy level, one would therefore conclude that an environmental liability system should be combined with some form of guarantee that financial security is available one should at least leave the option open to industry to provide this financial security via environmental damage insurance. The Dutch example shows that this first party type coverage seems able to meet that end.[54] Moreover, the example seems to be followed in other countries as well.[55]

Disadvantages. However, there are inherent limitations to this new environmental damage policy as well. First of all, one should note that this new insurance product only applies to soil pollution, ground water pollution and the subsequent remediation costs. It is therefore not a solution in case of other types of environmental harm like noise, surface water pollution or air pollution. But obviously it is also more difficult to apply liability law to those types of environmental damage as well.

Another point of criticism is that the amount of coverage that would be available seems to be relatively low. The CEA study reports that the amount of coverage available would be 25 million Dutch guilders. This would be sufficient to cover the average clean-up costs in case of soil pollution, but not for the larger scale cases. However, in practice, policies are not always awarded for this maximum amount of 25 million.

The fundamental question one can also ask at the policy level is whether the introduction of this new environmental damage insurance was really necessary to deal with the uncertainties of environmental liability, as the Dutch insurers claimed. One could argue that the tricky aspects of environmental liability, such as causal uncertainty and the long-tail risks could be dealt with either by the legislator or through adequate policy provisions. Note that, as a reaction to the long-tail risk caused under liability insurance, the claims made coverage was introduced. But this also applies in the MSV. Moreover, the risk differentiation may theoretically be easier in first party insurance, but if liability insurers would monitor potential polluters adequately (it is doubtful whether that happened in the past) the risk differentiation in environmental liability could be as adequate as in first party insurance. One has the impression that the real reason for the Dutch insurers to move to the environmental damage insurance was that they wanted to exclude the uncertainties having to do with the involvement of the judge (will there be wrongfulness, liability in this particular case? Will he accept causation or not?).

The insurers may now have excluded many of these uncertainties, but these will remain in existence for the polluters who are still fully exposed to environmental liability. This leads to a particular tricky aspect of the new regime, which really is a source of worry. Here I refer to the fact that this new MSV is supposed to replace the general liability insurance, at least for the environmental risks covered under the MSV. As a result of this, industrial operators in the Netherlands today no longer have the possibility to obtain liability coverage for their environmental risks. Third parties who suffer harm may be protected under the new MSV, but only under the assumption that the MSV coverage is sufficient to cover their losses. If this is not the case, the third party will simply have to use tort law and bring a lawsuit against the polluter. But then the problem will arise that this polluter is no longer covered under liability insurance. The victim may then find his defendant to be judgement proof.

Therefore, at the policy level we should examine whether it would not be necessary to introduce a compulsory insurance scheme, be it liability or first party based. In addition, the fact that the Dutch insurers are collectively moving from liability to environmental damage insurance raises questions concerning the competition on the insurance market. These aspects deserve separate attention in the next section.

4. POLICY ASPECTS

At the end of the previous section we concluded on the basis of the Dutch example with the environmental damage insurance that neither this new mechanism, nor the traditional liability insurance, do provide an adequate protection against insolvency. This, hence, merits some attention for the more general issue, whether a duty to provide financial security should be introduced for environmental harm (A). In addition, we came to the remarkable finding that apparently in the Netherlands in the future, only the new environmental damage insurance will be provided on the market and no longer liability insurance for environmental risk. This poses questions concerning the importance of an adequate competition policy on insurance markets (B).

Compulsory Insurance?

Introduction
We have to consider carefully the risk that a potential polluter would simply chose a "hit and run" strategy, whereby he would pollute, irrespective of the potential liability, since the liability rule could not affect him, given insolvency.

In legal doctrine, compulsory liability insurance is often advanced as a means to protect the innocent victim.[56] The duty of an injurer to purchase liability insurance would be a good way to protect victims against the insolvency of the injurer.[57] Economists on the other hand see, as we mentioned above, different benefits of liability insurance. It is an instrument to increase the utility of a risk averse injurer. Whereas economists would therefore see liability insurance mainly as an instrument to serve the interests of the injurer, lawyers tend to rely more on the victim protection argument. Of course these views do not necessarily contradict each other. Sometimes they can go in the same direction in favour of a duty to introduce liability insurance. Let us now address the question under what circumstances the purchase of insurance should be made compulsory.

Economic Arguments
Increasing the expected utility. A first way to look at this question could be to turn back to the basic utilitarian literature on the benefits of insurance.[58] If insurance is indeed, beneficial, since it removes risk from risk averse persons and thus increases their utility, are not these benefits that large that they warrant the introduction of compulsory liability insurance? A problem with this argument is that the degree of risk aversion varies. A Rockefeller will probably not be averse towards the risk of losing U.S. $1000, but a low-income family father probably will be. The low-income father will therefore probably have a demand for insurance against this risk of a loss of U.S. $1000, whereas a Rockefeller probably would not. This straightforward example makes clear that the introduction of a duty to insure might be inefficient in as far as it forces some people to purchase liability insurance that would normally not have a demand for insurance. Insurance does not increase their expected utility. A generalised duty to insure might therefore create a social loss. Whether this is the case will, of course, depend upon the number of people that is actually harmed by the introduction of a duty to insure. This might indeed be outweighed by the fact that others will certainly have a benefit from insurance. This, however, does in itself, not justify the need to introduce a duty to insure. These risk averse individuals might indeed have purchased insurance coverage anyway. This means that the simple fact that insurance increases utility can as such not justify the introduction of a duty to insure as long as we assume that all individuals are perfectly informed about the risk to which they are exposed and the availability of insurance.

This argument also rather paternalistically assumes that insurance is, under all circumstances, beneficial to potentially responsible parties. The argument neglects the fact that, as we have explained above, discussing the basic theory of liability insurance, the insured has to pay a price to have the risk removed from him. This price will unavoidably be a lot higher than the actuarially fair premium, consisting in the multiplier of the probability × damage ($p \times D$). Depending upon the efficiency of the administrative working of the insurance undertaking, a high or low amount of loading costs will be added and depending upon the degree of concentration on the market, a profit margin could also be added. In addition taxes may increase the premium as well. Moreover, we have indicated that if there is uncertainty concerning the risk, the insurer might compensate this insurer's ambiguity with a risk premium. In sum, the premium charged may well be a lot higher than the actuarially fair premium, this is the objective value of the risk. For some responsible parties this premium will still be attractive, but for others it may not. Moreover, some may choose a balanced approach of, e.g. self insuring a basic expected damage and only taking insurance for the excess loss. Compulsory insurance generally neglects the fact that the demand for insurance may vary according to the individual risk situation (and financial possibility) of every responsible party. Other things being equal, there is therefore no reason for a regulatory intervention solely based on the fact that insurance may increase expect utility. A regulatory intervention would have the disadvantage that it is counter to the more differentiated demand of the responsible parties. However, this assumes that the responsible parties have a knowledge about their exposure to risk, the availability of insurance and make a well-informed decision accordingly. If this assumption is not met, the question could again be raised whether insurance should be made compulsory.

Information problems. Information problems might arise in case the potential injurer cannot make an accurate assessment of the risk he is exposed to and the benefits of the purchase of insurance. An underestimation of the risk would in that case lead to the wrongful decision of the injurer not to purchase liability insurance. The legislator could remedy this information problem by introducing a general duty to insure. This information problem is probably a valid argument to introduce a generalised duty to insure for motor vehicle owners. Maybe the average driver of a car underestimates the benefits of insurance. If there would be no information problem and the legislator would nevertheless introduce a duty to insure because this would be "in the best interest" of the insured, this would of course be mere paternalism.

If empirical evidence would exist that most polluters greatly underestimate the costs of the environmental damage they may cause, and the probability that they will be held liable for this damage, this would then lead polluters to reserve too few resources to cover their potential liability. If these conditions are met, and one can indeed assume that polluters underestimate the cost of environmental damage, this information deficiency may be considered an argument in favour of compulsory insurance. But again, the policy argument based on information asymmetry relates merely to the fact that the polluter would underestimate the potential benefits of insurance. There may, however, be another argument why the (uninformed) decision of a polluter not to insure may lead to underdeterrence. This policy argument is precisely related to the insolvency risk mentioned in the introduction.

Insolvency. Another reason to introduce compulsory insurance is indeed the argument often used by lawyers, being the insolvency argument. The argument goes that the magnitude of the harm will often exceed the individual wealth of an injurer, whereby a problem of undercompensation of victims will arise. Lawyers would, hence, push forward compulsory insurance as an argument to guarantee an effective compensation to the victim. This – more distributional – argument obviously may play a role in the context of environmental liability insurance as well. If an injurer would be found judgement proof and hence, e.g. a polluted site would be "orphaned" the costs would be borne by society.

It is, however, also possible to make an economic argument that insolvency will lead to underdeterrence problems which might be remedied through insurance. Indeed, this so-called "judgement proof" problem has been extensively dealt with in the economic literature.[59]

Insolvency may, however, pose a problem of underdeterrence. If the expected damage largely exceeds the injurer's assets the injurer will only have incentives to purchase insurance up to the amount of his own assets. He is indeed only exposed to the risk of losing his own assets in a liability suit. The judgement proof problem may therefore lead to underinsurance and thus to underdeterrence. Jost has rightly pointed at the fact that in these circumstances of insolvency, compulsory insurance might provide an optimal outcome.[60] By introducing a duty to purchase insurance coverage for the amount of the expected loss, better results will be obtained than with insolvency, whereby the magnitude of the loss exceeds the injurer's assets.[61] In the latter case the injurer will indeed only consider the risk as one where he could at most lose his own assets and will set his standard of care accordingly. When he is, under a duty to insure, exposed to full liability the insurer will obviously have incentives to control the behaviour of the insured. Via the traditional instruments for the control of moral hazard, discussed above, the insurer can make sure that the injurer will take the necessary care to avoid an accident with the real magnitude of the loss. Thus Jost and Skogh argue that compulsory insurance can, provided that the moral hazard problem can be cured adequately, provide better results than under the judgement proof problem. This is probably one of the explanations why for instance for traffic liability compulsory insurance was introduced. Uninsured and insolvent drivers who have little money at stake which they may lose compared to the possible magnitude of accidents they may cause, may have little incentives to avoid an accident. Insurers might better be able to control this risk and could force the injurer to take care under the threat of shutting him out of the insurance. Thus, the insurer becomes under a duty to insure the licensor of the activity.

Indeed, this economic argument shows that insolvency may cause polluters to externalise harm: they may be engaged in activities, which may cause harm, which can largely exceed their assets. Without financial provisions these costs would be thrown on society and would hence be externalised instead of internalised. Such an internalisation can be reached if the insurer is able to control the behaviour of the insured. As we have shown above, when discussing how risk differentiation can be applied to environmental liability insurance, the insurer could set appropriate policy conditions and an adequate premium. This shows that if the moral hazard problem can be cured adequately, insurance even leads to a higher deterrence than a situation without liability insurance and insolvency.

Of course, this argument in favour of compulsory insurance relies on a few assumptions and conditions, which will be discussed in further detail below. One is obviously that the argument is only valid if moral hazard can be controlled adequately and insurers also have appropriate incentives to do so. Another condition is that the insurance markets should be competitive. But one can notice that indeed both from a legal and from an economic point of view the potential insolvency of the injurer is a problem since it can both lead to underdeterrence and to undercompensation. Compulsory insurance may remedy both problems since it may provide adequate victim compensation and – if certain conditions are met – remedy the risk of underdeterrence.

An issue, which merits attention, is that one should raise the question whether compulsory liability insurance is the best instrument to remedy the insolvency problem. Indeed, although one usually refers to liability insurance in this context, several alternatives may exist that might be able to cure the insolvency problem at lower costs. A first obvious alternative would be to turn to (compulsory?) first party insurance instead of third party liability insurance, as we have argued above.

In sum, one could argue that compulsory insurance might theoretically be an adequate means to remedy the insolvency problem. However, this should not necessarily be liability insurance, but may well be the first party or direct insurance, discussed in this paper.

Potential Dangers of Compulsory Insurance
Moral hazard. After having discussed these three basic criteria for compulsory insurance, two other points cannot remain to be undiscussed either. Firstly, one should remember that with insurance there will always be the moral hazard

problem, discussed above. This means that even if a legislator decides to introduce compulsory insurance, he should not restrain the possibilities of an insurer to control the moral hazard problem. Otherwise compulsory insurance will create more problems than it solves. Nevertheless, there seem to be problems since the legislator often limits the possibilities to expose the insured to risk. Indeed, with compulsory insurance the duty to insure is often equal to the total amount of liability and deductibles are not allowed. Hence, the total risk is shifted to the insurer which means that the only instrument available for the insurer to cure the moral hazard problem is a monitoring of the insured. If this would seem difficult or very costly the introduction of compulsory liability insurance might indeed create problems. Shavell even goes as far as to state that if the moral hazard problem cannot be controlled, the only regulatory intervention with respect to insurance should be a prohibition of liability insurance.[62] In any case an introduction of compulsory insurance does seem problematic if the moral hazard problem cannot be controlled adequately.

Concentration on insurance markets. A second, related, issue is that until now we assumed that insurance markets are perfectly competitive and that thus premiums and policy conditions will be nicely tailored to the individual needs and the behaviour of the insured in order to control moral hazard optimally. In practice, however, many restrictions on insurance markets exist. We will give examples below. If monopolistic premiums can be set, an insurer will have fewer incentives to align his premiums to the individual behaviour of the insured and thus there is less control of the moral hazard problem.

From a policy viewpoint it also seems highly problematic to make liability insurance compulsory on concentrated insurance markets. Indeed, in that case the inefficiencies on the insurance market would be reinforced by making the purchase of insurance compulsory. Also, here the interest group theory of government can of course explain why insurers might want to lobby in favour of compulsory liability insurance. If they can already determine the supply side of the market through monopolistic premium setting all such insurers should strive for is that every possible injurer should be forced to purchase insurance coverage. Through this regulatory intervention a certain demand is then guaranteed as well.

Nevertheless, in practice, insurers are never enthusiastic concerning compulsory insurance, at least for the environmental risk. Cousy claims that this is related to the fact that as a matter of law under compulsory insurance the insurer can often not invoke defences against the third party beneficiary of insurance. Moreover there would be problems related to the implementation and actual carrying out of the obligation to insure.[63]

Further Warnings

Dependence upon insurance market. The refinements and potential dangers just presented show that although theoretically a compulsory insurance mechanism may be desirable, also for environmental liability more particularly to cure information problems and the underdeterrence risk as a result of insolvency, in practice one should be cautious with the advice to introduce compulsory insurance for environmental damage. There are some more reasons to formulate such a cautious warning. One is that the legislator should be aware of the fact that as soon as it introduces compulsory insurance, it becomes dependant upon insurers to fulfil this duty to insure. The practical possibilities of an effective enforcement of a duty to insure will obviously to a large extent depend upon the willingness to insure on that particular market. It will ultimately be the insurance market who will decide whether they are willing to cover a certain risk. This may in the end lead to the undesirable situation that the legislator would introduce a duty to take out compulsory insurance, but that the market would refuse to provide such coverage. Introducing a duty to insure leads to a high reliance of the policy maker upon the insurance market. This seems to have led to problems with the German Environmental Liability Act of 1990 (Umwelthaftungsgesetz) which requires the owner of an installation that can cause significant damage to take out liability insurance or to have sufficient financial guarantees.[64]

One should indeed realise that if one makes the availability of insurance coverage a prerequisite for the operation of an enterprise, insurance undertakings in fact become the licensor of the industry, which may be questionable from a policy point of view.[65] In fact the insurer becomes the 'environmental policeman'.[66] If that means, however, that the insurer, as a "policeman" controls the ecological performance of the insured company, there is of course nothing wrong with that. It may only be problematic if insurance companies will effectively be able to decide which companies may exercise their activities. This problem obviously especially arises in a monopolistic market.

This may, moreover, cause practical problems. Imagine that an insurer has stipulated in the policy conditions that coverage will be excluded in case of non-compliance of the insured written mandatory government regulation. This may well be, as we have suggested above, an effective instrument to control moral hazard. If, however, an accident happens under compulsory insurance the insured will not be able to call on this exclusion ground vis à vis the third party beneficiary of the liability insurance policy. The fact that defences in the insurance contract are not opposable to third parties is a well-known problem under compulsory insurance. The insurer will thus have to compensate the victim and may have a (statutory or contractual) legal right of recourse against the insured, provided that the latter is solvent. This is, as we explained, one of the reasons why insurers are reluctant against compulsory insurances.

Necessity of cooperation with insurers. One could obviously argue that these problems can be remedied if a good cooperation takes place between the policy maker and the insurance world, whereby the insurance world would inform the policy maker on the insurability of environmental damage. However, practice has shown that information provided by insurers concerning the insurability of a certain risk or with respect to the available amounts of coverage may not always be reliable. Below we will discuss the example of the Dutch insurers who decided collectively that the flood risk would be uninsurable. In fact it was only this monopolistic decision that caused the uninsurability.

There seems to be a trade off in that respect: introducing a duty to insure without any cooperation with the insurance world (which may have been the case in Germany) may lead to the catastrophic result that the government forces industry to take out a certain insurance coverage, whereby the market would not be willing to respond with the provision of such a coverage. However, a close cooperation between the insurers (usually represented via one insurance association) and the government only increases the risk of high concentration on insurance markets.

Environmental liability insurance "grown up"? A further problem is that the policy maker should equally realise that today insurance coverage for environmental harm is still a relatively young and inexperienced branch. If a differentiated offer of insurance policies is limited, one could again question whether it makes sense to introduce a mandatory insurance if such coverage could only be found to a limited extent (or without sufficient competition) on private insurance markets. Of course, the limited availability of insurance cover for the environmental risk today is to a large extent caused by the adverse selection problem: since too little companies had a demand for insurance an optimal risk spreading (via the law of large number) is not possible.[67] Moreover, only the bad risks will have a demand for insurance, which precisely causes the adverse selection problem. Hence, one could naively argue that this can be cured by forcing all polluters, good and bad risks to take insurance coverage. However, it seems strange to cure the limited availability of insurance for environmental risks, which may largely be due to its difficult-to-insure-nature by forcing all polluters to purchase coverage.

In this respect we can refer to the fact that a risk differentiation in environmental insurance in Europe still stands at the beginning of its possibilities and that far more possibilities exist to relate policy and premium conditions in an appropriate way to the ecological reliability of firms. Hence, one can really question whether today insurance firms are yet able to differentiate environmental risks in such a way that one can argue that moral hazard can be controlled optimally on competitive insurance markets. The cure to these problems is obviously not to make a poorly functioning insurance system compulsory.

Duty to accept? Obviously one could naively react with the suggestion that if insurance markets refuse to provide appropriate coverage, the policy maker should not only introduce a duty for industry to take out mandatory coverage, but also a duty for insurance companies to accept. Such a duty to accept would, so it could be argued, at the same time remedy the risk that the insurance undertakings would de facto become the licensors of industry. However, introducing a duty to accept certain industries as insured seems like an extremely dangerous path to go, given the importance of an effective control of moral hazard. One important instrument of insurers to control moral hazard is precisely to have the possibility to monitor ex ante the risk, which a particular insured may pose. This could ultimately lead an insurance undertaking to the decision that it considers the environmental risk a particular industry poses as too high. A logical consequence of the wish to have an optimal control of moral hazard should be the right of insurance undertakings to freely decide which potentially responsible parties to insure and which not. A duty to accept certain risks seems therefore to collide with the basic principles, which have to be respected to guarantee an effective functioning of insurance markets.

Policy Recommendation

From the above it follows that there are, indeed, arguments in favour of introducing a compulsory insurance scheme, based on possible information deficiencies and on the risk of underdeterrence as a result of insolvency. However, already theoretically, one can point at dangers as well, more particularly the fact that insurers may become the licensors of the activities which may cause the environmental harm. This should not be remedied through a duty to accept, since this may cause incurable problems of moral hazard. All these considerations are therefore arguments for a policy maker to be extremely cautious with the introduction of a regulatory duty to purchase insurance coverage.

From this discussion on compulsory insurance follows probably an important conclusion. Although it may be important from a theoretical perspective to introduce a duty for the permit holder to secure appropriate means, it seems more appropriate to look for a flexible system whereby the licensing administrative authorities could judge in individual cases whether the obligation to provide financial security has been met. Such a system, whereby it is left to the administrative authorities to decide the form and amount of the financial obligation seems more flexible and entails less of the risks and dangers of a generalised system of compulsory liability insurance.

But to reiterate clearly: the principle that liability should be covered through some form of financial assurance (not necessarily insurance) should be laid down in legislation. The authorities would then only have to fix the amount, taking into account the expected damage (this will allow for an individualisa-tion and differentiation), and they would have to check whether the type of financial assurance offered by the potentially responsible party will be adequate to meet his financial obligations. Here maybe the example of the Flemish Interuniversity Commission can be useful.[68] The draft decree on environmental policy chooses, not to introduce a compulsory insurance, but provided that an obligation can be introduced for the licensee of a classified activity to provide a deposit in order to guarantee that specific obligations shall be complied with. As proposed by the Flemish Interuniversity Commission, the authority to fix the amount and to control the offer of financial assurance could be the one who grants the licence at the start of the operation of the activity.[69] In terms of the IPPC Directive this would be the competent authority granting the permit.[70] By the way, Sweden has a compulsory insurance scheme for environmental harm which would be working pretty well. Some argue that this is precisely the case because this Swedish scheme is not based upon liability insurance, but upon direct insurance.[71]

In sum, the liability regime should be combined with some kind of obliga-tion to provide financial security if one can assume that an insolvency risk may emerge. But:

- this financial security should not necessarily be liability insurance;
- the policy maker could indicate that a wide variety of mechanisms may be used to provide this financial security;
- the type of financial security provided should not be regulated in a general matter, but it adequacy may be assessed, e.g. by the administrative authori-ties who can require financial security as a condition in the license;
- at the same time the administrative authorities can equally determine the required amount of financial security on a case by case basis;
- there should certainly not be the introduction of a duty to accept risks on liability insurers, since this may have negative effects on the control of moral hazard;
- the administrative authorities imposing such a duty to provide financial security should make sure that sufficient varieties of financial securities exist

on financial and insurance markets in order to avoid that governments or administrative authorities would become dependant upon the financial or insurance industry, which would then effectively become the licensor of industrial activities;
- the proposed regime corresponds with the proposals made by the Interuniversity Commission in the Flemish Region. These proposals were promulgated as the result of information provided by insurers and the regime moreover applies in the Flemish soil pollution decree. Hence, there is some empirical evidence, which shows that such a balanced and mitigated obligation to provide financial security in limited cases may work effectively.

Competition Policy

Importance of Competition on Insurance Markets
We noted that the Dutch insurers association advised all their members to move towards environmental damage insurance and to let the policy replace the traditional environmental liability insurance. The effect could be that liability insurance for environmental risk is no longer available on the Dutch market. Moreover, in the past the Dutch insurers association took an opposite position with respect to flood risk: it argued for a long time that these risks were uninsurable and even prohibited their members to provide coverage for flood risks. Above we showed that these anti-competitive pressures may be particularly dangerous since they will obviously limit a wide and differentiated offer of a variety of financial and insurance products to cover the environmental risk. It is interesting to have a closer look at some of these practices, given the fact that the boundaries of the application of European competition law to the insurance industry have recently been outlined in a report of the European Commission with respect to the functioning of a regulation concerning the insurance industry.[72] Certain practices with respect to environmental liability insurance are explicitly addressed in this report and therefore merit some attention here.

I should first of all point at the fact that Commission regulation no. 3932/92 of 21 December 1992 exempts many cartel agreements in the insurance world from the prohibition under (old) article 85 (3) of the EEC treaty, provided that certain strict conditions were met.[73] Already this exemption has been heavily criticised by law and economics' scholars, who argued that competition policy should be fully applied to insurance markets as well. They argued that an exemption from competition policy may lead to a premium increase, lower product differentiation and potentially even to an increase of accident costs (because the incentives to control moral hazard would be reduced).[74] But these criticisms will probably not affect the insurance industry who can benefit from the exemption. However, the recent report to the European Parliament of 12 May 1999 made clear that some current practices may not fall under the conditions for the exemption from the cartel prohibition:

Binding Decisions
The Dutch insurers association had issued in the 1950s a, literally called, "binding decision" for all of its members, prohibiting them to insure the flood risk and the earthquake risk (the latter being relatively small in the Netherlands, with the exception of the area around southern Limburg). Their argument was that these risks were technically not insurable and that therefore all the members should refrain from covering those risks. The argument concerning the uninsurability seems highly doubtful,[75] but also clearly violated the conditions of regulation 3932/92. Consideration 8 preceding the exemption clearly states that standard policy conditions may, in particular, not contain any systematic exclusion of specific types of risk without providing for the express possibility of including that cover by agreement. This is repeated in article 7 (1) (a) of the exemption. In its report to the European Parliament of 12 May, the Commission explicitly discusses these binding decisions.[76] The report states that as a result of the questions asked by the Commission, the Dutch association of insurers decided to bring its binding decisions in line with article 7.1 sub a by simply converting it into a non-binding recommendation, leaving each insurer free to extend cover to flood risks. However, this example shows that non-competitive practices are apparently not an exception in Dutch insurance practice.

Standard Form Policies
A point which inevitably needs to be addressed in the context of these changes in the environmental insurance structure in the Netherlands is to what extent these standard policy models are compatible with European competition policy. We already noticed that, e.g. concerning the liability insurance for companies in the Netherlands (AVB) a new model policy came into being in 1996 which has been offered to the market.[77] This AVB model policy suggested the change from a loss occurrence coverage to a claims-made coverage.[78] Also, as has just been indicated, the Dutch insurers association decided to change the coverage of environmental risks from liability insurance to first party insurance.

The question which inevitably arises is, whether it is compatible with competition policy to formulate an advice via a standard policy which is de facto followed by the whole market. Again, this issue is covered by the exemption regulation 3932/92 of 21 December 1992. Article 6 of the exemption regulation provides that the exemption shall apply to standard policy conditions on the condition that it is made clear that these policy conditions are purely illustrative and that they expressly mention the possibility that different conditions may be agreed upon. In addition, the standard policy conditions must be accessible to any interested person and provided simply on request. The standard policy conditions used by the Dutch insurers in environmental insurance certainly can meet this test. Both the new insurance policy for liability of companies (AVB) and the environmental damage insurance (MSV) mention explicitly that they are merely illustrative and that it stands free to any individual insurer to deviate from the text or the contents of the policy model.[79]

Formally the standard policy conditions provided by the Dutch Insurers Association are hence compatible with the exemption regulation. Nevertheless, one cannot deny that the use of standard form policies will inevitably have an effect of restricting competition. Indeed, although these standard policy models all indicate that they are merely illustrative and that any insurer can deviate from them, it is today in the Netherlands almost impossible de facto to get any liability insurance for the environmental risks which are now covered under the MSV (more particularly for soil pollution). However, these concerns will probably not be a great worry to Dutch insurers, since the used standard policy forms can easily pass the test of the exemption regulation.

Indeed, the Dutch insurers association did not formally force all insurers not to provide any liability cover for environmental risk any longer and the standard policies stipulate clearly that every insurer is free not to follow the recommendations made in the standard policy. Formally the Dutch practice is therefore compatible with the requirements of European competition policy. However, in practice one can notice that almost all insurers cooperating within the Dutch insurers association follow the recommendations in the standard policy in practice. The effect therefore is that no Dutch company will be able to purchase liability cover for soil pollution any longer on the Dutch market. The same is the case, by the way, for foreign insurance companies operating on the Dutch market. The only alternative for Dutch operators today is to seek liability cover abroad.

Even though formally complying with European competition policy, some scholars have pointed at the fact that this tendency to follow these recommended standard form policies leads to serious restrictions of competition in practice and to a reduced product differentiation.[80]

In sum: although we have argued that it is understandable that the Dutch Insurers Association wished to offer an environmental damage insurance on first party/direct insurance basis, it is, from a competition perspective, regrettable that the introduction of this new product was accompanied with the (in practice) abolition of environmental liability insurance in the Netherlands, at least for the issues covered under the MSV. The effect therefore is a reduced variety of insurance products for environmental risk.

5. CONCLUDING REMARKS

In this paper I dealt on the one hand with a number of well-known problems which may endanger the insurability of the environmental risk; on the other hand special attention was given to a new environmental insurance product introduced recently in the Netherlands, but gaining popularity in other countries as well. The reason for giving that much attention to this new product, referred to as an environmental damage insurance, is that the insurers claim that many of the problems that arise when insuring the environmental risk are in fact problems related to environmental *liability*. In other words: they claim that the environmental risk would be better insurable if the risk could be insured on a first party basis. Many of the problems that appear under liability insurance would then disappear, or at least be reduced, so it is held.

From a theoretical perspective the Dutch insurers certainly have a point: already in 1987 George Priest claimed that the American liability and insurance crisis was caused by a shift from first party to third party insurance. The obvious remedy for him was then to move back from third party (liability) insurance towards first party insurance. However, in reality the first party model seems indeed an appropriate remedy for problems in liability cover, but in fact only for just one aspect of the insurability of the environmental risk, being the issue of risk differentiation. Indeed, first party insurances are held to enable a better risk differentiation than traditional liability insurance. Of course this is an important issue since an adequate risk differentiation has always been advanced as the appropriate remedy for the dangers of moral hazard and adverse selection. But on the other hand, other problems concerning the insurance of environmental risk will remain. This concerns for instance the issue of a limited capacity in case of catastrophic environmental accidents, the risk of judges shifting causal uncertainty to an industrial operator and the danger of a retrospective application of liability laws. Indeed, also under first party insurance, capacity will not be unlimited; also a first party insurance requires a condition of causation (it must be proven that the damage was caused by the insured risk) and the problem of retroactive liability resulting from the "long-tail" character of the environmental risk may be cured both under third party and under first party insurance (at least to some extent) by the introduction of a claims made cover.

The Dutch insurers introduced their new environmental damage insurance with great enthusiasm, claiming that they have now been able to increase the manageability and predictability of the risk. Their basic idea is that it is much easier to monitor and predict the potential of damage *ex ante* with one particular site, than to predict the chances that one operator may be held liable to pay damages to a third party. One important factor of uncertainty (the judge) has thus been excluded. Although one can understand the enthusiasm of the insurers and, to some extent, the corporate world (since under first party insurance they also receive coverage for their own losses), there are clearly disadvantages as well. The major disadvantage is not related at all to the fact that the first party environmental damage insurance would not be an adequate insurance product, but to the fact that the introduction of this new product was accompanied with the abolition of liability insurance for the environmental risk. As a result of this, some victims may not receive compensation any longer, at least in those cases where the damage insurance does not provide sufficient coverage and the polluter is insolvent. We took this finding to look at the more general policy question whether there should not be a duty to secure financial coverage for the environmental risk. Given the risk of insolvency we held that a serious problem of underdeterrence could arise which could be solved by introducing some obligation to provide financial security to cover the environmental risk. However, such a financial security should not necessarily be liability insurance. In that respect the environmental damage insurance may play a role again. But in any event, the administrative authorities should then control, e.g. by fixing a specific duty in licence conditions, that the type of security offered will indeed be adequate to cover the financial consequences of environmental damage that may be caused by that particular licensee.

Moreover, the finding that this new environmental damage insurance was introduced on the Dutch market as a standard policy caused some worries as well. Apparently the cooperation between insurers in the Netherlands is that close that, although formally the standard policy only is an "advice", in practice operators cannot obtain coverage for the liability risk (for instance on a result of soil pollution) any longer. This example shows that restrictions of competition on the insurance market may seriously limit product differentiation and thus endanger the insurability of the environmental risk. This shows that an effective competition policy is of utmost importance to guarantee that a large variety of efficient insurance policies is indeed available on the market for the environmental risk. Otherwise changes such a the one we discussed for the Netherlands could limit rather than enhance the insurability of the environmental risk.

NOTES

1. See e.g. Spier, J. (Ed.). *The limits of liability, keeping the floodgates shut,* The Hague; Kluwer, and Spier, J. (1996). *De uitdijende reikwijdte van het aansprake-lijkheidsrecht,* pre-advies Nederlandse Juristenvereniging.

2. Referring to the fact that the polluter on the one hand has to invest heavily to be able to comply with the conditions of a licence and on the other hand has to pay as well for the remaining environmental damage. See in this respect especially Bergkamp, L. (1999). *De vervuiler betaalt dubbel,* Antwerpen, Intersentia.

3. For a comment on this White Paper see Rice, P. (2000). From Lugano to Brussels via Aarhus: environmental liability White Paper published, *Environmental liability,* 39–45; and Rehbinder, E. (2000). Towards a community environmental liability regime: the commissions White Paper on environmental liability, *Environmental liability,* 85–96.

4. In another study, it has been held that, provided that the economic principles of insurance are respected (such as some degree of forseeability and predictability of the risk) the mere fact of expanding liability should not as such endanger insurability. See Faure, M., and Grimeaud, D. (2000). *Financial assurance issues of environ-mental liability,* final version 4.12.2000, which contains an analysis of the insurability of the regime proposed by the white paper on environmental liability. This study has been made available electronically via the website of the Commission: http:\\www.europa.eu.int\com\environment\liability\followup.htm

5. For a nice overview of alternatives, see Tyran, J. R., and Zweifel, P. (1993). Environmental risk internalization through capital markets (Ericam): the case of nuclear power, *International Review of Law and Economics,* 431–444; Radetzki, M., & Radetzki, M. (2000). Private arrangements to cover large-scale liability caused by nuclear and other industrial catastrophes, *The Geneva Papers on Risk and Insurance,* 180–195.

6. Priest, G. (1987). The current insurance crisis and modern tort law, *Yale Law Journal,* 1521–1590.

7. See Arrow, K. (1965). *Aspects of the Theory of Risk-Bearing,* Helsinki, Yrjö Jahnssonin Säätiö; and Borch, K. (1961). The Utility Concept Applied to the Theory of Insurance, *The Astin Bulletin,* 245–255.

8. See Shavell, S. (1987). *Economic analysis of accident law,* Cambridge, Harvard University Press, 190.

9. Skogh, G. (1989). The transactions cost theory of insurance: contracting imped-iments and costs, *Journal of Risk and Insurance,* 726–732.

10. Obviously these conditions cannot be dealt with in detail; the reader is referred for further information to the literature cited in the footnotes. For an overview see Kunreuther, H., and Freeman, P. (2001). Insurability, environmental risks and the law. In: A. Heyes, (Ed.), *The Law and Economics of the Environment* (pp. 305–306), Cheltenham, Edward Elgar.

11. Monti rightly points out that the fact that there may be both factual and legal uncertainty, Monti, A. (2001). Environmental risk: A comparative law and economics approach to liability and insurance, *European Review of Private Law,* 59–62.

12. Rogge, J. (1997). *Les assurances en matière d'environnement,* Loose-Leaf, Kluwer, 4.

13. Kunreuther, H., Hogarth, R., and Meszaros, J. (1993). Insurer Ambiguity and Market Failure, *Journal of Risk and Uncertainty,* 71–87.

14. Pauly, M. (1968). The Economics of Moral Hazard: Comment, *American Economic Review*, 531–545.

15. See Wagner, G. (1996). Versicherungsfragen der Umwelthaftung. In: M. Ahrens, & J. Simon (Eds), *Uwwelthaftung, Risikosteuerung und Versicherung* (pp. 104–105), Berlin, Erich Schmidt Verlag.

16. See Shavell, S. (1979). On Moral Hazard and Insurance, *Quarterly Journal of Economics*, 541–562.

17. Spence, M., and Zeckhauser, R. (1971). Insurance, Information, and Individual Action, *American Economic Review*, 380–391.

18. Rea, S. (1982). Non-pecuniary loss and breach of contract. In: *Journal of Legal Studies*, (pp. 35–53).

19. See van Eijk-Graveland, J. C. (1998). *Verzekerbaarheid van opzet in het schadeverzekeringsrecht*, Zwolle, Tjeenk Willink.

20. See Akerlof, G. (1970). The market for 'lemons': quality, uncertainty and the market mechanism, *Quarterly Journal of Economics*, 488–500.

21. Although he mentions that some larger companies are (in 1996) able to get coverage up to 1 billion Belgian Francs.

22. Rogge, J. (1997). *Les assurances en matière d'environnement*, Loose-Leaf, Kluwer, 5.

23. Cowell, J. Compulsory environmental liability insurance. In: H. Bocken & D. Ryckbost (Eds), *Insurance of Environmental Damage*, 327.

24. See Frenk, N. (1995). Toerekening naar kansbepaling, *Nederlands Juristenblad*, 482–491.

25. Also Abraham, K. (1988). Environmental Liability and the Limits of Insurance, *Columbia Law Review*, (Vol. 88), 959–960; and Katzman, M. (1988). Pollution liability insurance and catastrophic environmental risk, *Journal of Risk and Insurance*, 89–90.

26. Kunreuther, H., Hogarth, R., and Meszaros, J. (1993). Insurer Ambiguity and Market Failure, *Journal of Risk and Uncertainty*, 71–87.

27. See also Zeckhauser, R. (1996). 19th Annual Lecture of the Geneva Association and Catastrophes, *The Geneva Papers on Risk and Insurance*, 5, who equally argues that retrospective liability may affect the predictability of risks; as well as Abraham, K. (1988). Environmental Liability and the Limits of Insurance, *Columbia Law Review*, (Vol. 88), 957–959.

28. See further Faure, M., and Hartlief, T. (1998). Remedies for expanding liability, *Oxford Journal of Legal Studies*, *18*, 681–706.

29. See Faure, M., and Fenn, P. (1999). Retroactive liability and the insurability of long tail risks, *International Review of Law and Economics*, 487–500.

30. Rosenberg, D. (1984). The causal connection in Mass Exposure cases: a 'public law' vision of the tort system, *Harvard Law Review*, 851–929; Shavell, S. (1985). Uncertainty over causation and the determination of civil liability, *Journal of Law and Economics*, 587–609. Also the Dutch Attorney General Hartkamp defended a market share liability in the DES case (*Tijdschrift voor Consumentenrecht*, 1992, 241–258). In addition Spier pleaded in favour of a proportionate liability for latent diseases in his inauguration address (Spier, J., *Sluipende schade*, Deventer, Kluwer, 1990), as did Akkermans, A., *Proportionele aansprakelijkheid bij onzeker causaal verband*, Deventer, Tjeenk Willink, 1997, in his recent dissertation; Robinson, G., "Probabilistic causation and compensation for tortious risk", *Journal of Legal Studies*, 1985, 798.

31. Priest, G., *o.c.*, 1521–1590. Priest has been criticized by Viscusi, who claims that there were other reasons for the product liability crisis in the U.S. than adverse selection on its own (Viscusi, W. K. (1991). The Dimensions of the Product Liability Crisis, *Journal of Legal Studies*, 147–177).

32. Abraham, K., *o.c.*, 949–951.

33. Generally the question is whether the benefits of particularization outweigh their costs, which has to be addressed in tort law when a standard of care is defined (Posner, R., *Economic Analysis of Law*, 5th edition, New York, Aspen Law & Business, 1998), but also when legal rules are made (Ehrlich, I., and Posner, R. (1974). "An Economic Analysis of Legal Rule-Making", *Journal of Legal Studies*, 257) or standards are set (Ogus, A. I. (1981). Quantitative Rules and Judicial Decision Making. In: Burrows, and C. Veljanovski (Eds), *The Economic Approach to Law* (pp. 210–225). London, Butterworth; Ogus, A. I. (1994). Standard Setting for Environmental Protection: Principles and Processes. In: M. Faure, J. Vervaele, and A. Weale (Eds), *Environmental Standards in the European Union* (pp. 25–37)).

34. On the costs of risk differentiation, see Bohrenstein, S. (1989). The Economics of Costly Risk Sorting in Competitive Insurance Markets, *International Review of Law and Economics*, 25–39; and Wils, W. P. J. (1994). Insurance Risk Classifications in the EC: Regulatory Outlook, *Oxford Journal of Legal Studies*, 449–467.

35. This argument is – again – especially advanced by Priest, G. (1987). The current insurance crisis and modern tort law, *Yale Law Journal*, 1521–1590.

36. Many other law and economics scholars have pointed at the advantages of first party insurance. See, e.g. Bishop, W. (1983). The Contract-Tort Boundary and the Economics of Insurance, *Journal of Legal Studies*, (Vol. 12), 241–266; and Epstein, R. A. (1995). *Simple Rules for a Complex World*, Cambridge, Harvard University Press, 221; and, in the product liability area Epstein, R. A. (1985). Products Liability as an insurance market, *Journal of Legal Studies*, (Vol. 14), 645–669.

37. See, e.g. Danzon, P. (1999). Alternative liability regimes for medical injuries", *Geneva Papers on Risk and Insurance*, 3–22; and see Koziol (1997). Die Arzthaftung im geltenden und kunftigen Recht, *Haftungsrechtliche Perspektiven der Ärztlichen Behandlung*, Linz, Universitätsverlag Rudolf Trauner, 21–35.

38. CEA, *Study on first party legal obligations for clean-ups and corresponding insurance covers in European Countries*, Paris, CEA, 21 October 1998.

39. See the summary tables in the CEA study, 32.

40. Aansprakelijkheidsverzekering bedrijven.

41. See on that policy Wansink, J. H. (1985). De nieuwe milieuaansprakelijkheidsverzekering, *Milieu en Recht*, 98.

42. Milieu-aansprakelijkheidsverzekering Samenwerkingsverband.

43. I am grateful to Mr. P. A. J. Kamp of Nationale Nederlanden who provided us some further insights for the reasons behind the changes towards first party insurance in the Netherlands.

44. For further details on these difficulties see Wansink, J. H. (1997). Hoe plotseling en onzeker is de verzekeringsdekking voor millieuaansprakelijkheidsrisico's? In: *Miscellanea. JurisConsulto vero dedicata*, Essays offered to Prof.Mr. J. M. van Dunné, Deventer, Kluwer, 451–460.

45. Milieuschadeverzekering.

46. For a good description of the new policy see Wansink, J. H. (1999). Verzekering en milieuschade als gevolg van vervoer/opslag van gevaarlijke stoffen, *Tijdschrift voor*

Milieuaansprakelijkheid, 77–82 as well as Janssen, C. A. (1998). Aansprakelijkheid voor milieuschade en financiële zekerheid naar toekomstig recht: nieuwe oplossingen. *Nederlands Recht*. In: L. F. Wiggers-Rust & K. en Deketelaere, (Eds), *Aansprakelijkheid voor milieuschade en financiële zekerheid*, Die Keure – Vermande, 111–112 and Drion, P. J. M. (1998/2). Milieu onder één dak: milieuschadeverzekering (MSV), *Verzekeringsrechtelijke Berichten*, 19–21.

47. Article II.1.1. defines as the scope of coverage: "Insured are the costs of remediation of the insured site. This remediation must apply to a pollution which is the direct and exclusive result of an emission, caused by one of the insured risks . . .".

48. This legal aid coverage, however, is not applicable if the insured took a low amount of coverage.

49. See Wansink, J. H. (1999). Verzekering en Milieuschade als gevolg van vervoer/opslag van gevaarlijke stoffen, *Tijdschrift voor Milieuaansprakelijkheid*, 81.

50. In principle, damage caused from the insured site to third parties is always covered, but the amounts of coverage may be limited.

51. Wansink, J. H. (1999). Verzekering en Milieuschade als gevolg van vervoer/opslag van gevaarlijke stoffen, *Tijdschrift voor Milieuaansprakelijkheid*, 81–82.

52. Niezen, J. (1998). Nieuwe milieuschadeverzekering – geen panacee, *Milieu en Recht*, 114.

53. Cousy, H. Recent developments in environmental insurance. In: F. Abraham, K. Deketelaere & J. Stuyck (Eds), *Recent Economic and Legal Developments in European Environmental Policy* (p. 240).

54. Also Bergkamp argues strongly in favour of first party and against environmental liability insurance (Bergkamp, L. (2000). The Commissions White Paper on Environmental Liability: A weak case for an EC Strict Liability Regime, *European Environmental Law Review*, 112–114).

55. Ranson reports that also in Belgium a 'direct' environmental insurance would be offered which would also cover gradual pollution (but would exclude ecological damage). See Ranson, D. (2000). Verzekering van milieuaansprakelijkheid, *Milieu- en Energierecht*, 68. The new policy would cover, like the Dutch example, also remediation costs. It would be offered by AIG and would provide coverage up to 1 billion BEF (see Kerremans, H. Aansprakelijkheid voor Milieuzorg en verzekeringsmogelijkheden in Milieuzorg in de Onderneming, I., *Juridische, fiscale en organisatorische aspecten*, Antwerpen, Standaard, 537–583).

56. Compulsory environmental liability insurance is proposed by several scholars, including Cowell, J. Compulsary environmental liability insurance. In: H. Bocken & D. Ryckbost (Eds), *Insurance of Environmental Damage* (pp. 317–330); and Monti, A. (2001). Environmental risk: A comparative law and economics approach to liability and insurance, *European Review of Private Law*, 65.

57. Some even argue that mandatory financial responsibility could help implement the precautionary principle by ensuring availability of resources to meet the costs of any future environmental damage, so Richardson, B. J. (2000). Financial Institutions for Sustainability, *Environmental Liability*, 61.

58. Arrow, K. (1965). Aspects of the Theory of Risk-Bearing, Helsinki, Yrjö Jahnssonin Säätiö; Borch, K. (1961). The Utility Concept Applied to the Theory of Insurance, *The Astin Bulletin*, 245–255; Borch, K. (1963). Recent Developments in Economic Theory and their Application to Insurance, *The Astin Bulletin*, 322–341; and Pratt, J. (1964). Risk Aversion in the Small and in the Large, *Econometrica*, 122–136.

59. More particularly by Shavell, S. (1986). The judgement proof problem, *International Review of Law and Economics*, 43–58.

60. Jost, P. J. (1996). Limited liability and the requirement to purchase insurance, *International Review of Law and Economics*, 259–276. A similar argument has recently been formulated by Polborn, M. (1998). Mandatory Insurance and the Judgement-Proof Problem, *International Review of Law and Economics*, 141–146; and by Skogh, G. (2000). Mandatory Insurance: Transaction Costs Analysis of Insurance. In: B. Bouckaert, and G. De Geest (Eds), *Encyclopedia of Law and Economics* (pp. 521–537). Cheltenham, Edward Elgar. Skogh has also pointed out that compulsory insurance may save on transaction cost.

61. See also Kunreuther, H., and Freeman, P. In: A. Heyer, (Ed.), *The Law and Economics of the Environment*, 316.

62. Shavell, S. (1986). The judgement proof problem, *International Review of Law and Economics*, 43–58.

63. Cousy, H. (1997). Recent developments in environmental insurance. In: F. Abraham, K. Deketelaere and J. Stuyck (Eds), *Recent Economic and Legal Developments in European Environmental Policy*, p 241; and Rogge, J. (1997). *Les assurances en matière d'environnement*, Loose-Leaf, Kluwer, 39.

64. See Wagner, G. (1991). Umwelthaftung und Versicherung, *Versicherungsrecht*, 249–260; and Wagner, G. (1992). Die Zukunft der Umwelthaftpflichtversicherung, *Versicherungsrecht*, 261–272.

65. This point is also made in the Green Paper, 13. See also Rogge, J. (1997). *Les assurances en matière d'environnement*, Loose-Leaf, Kluwer, 40.

66. So Monti, A. (2001). Environmental risk: A comparative law and economics approach to liability and insurance, *European Review of Private Law*, 65.

67. See Rogge, J. (1997). *Les assurances en matière d'environnement*, Loose-Leaf, Kluwer, 38.

68. See Bocken, H., and Ryckebost, D. (Eds) (1996). *Codification of Environmental Law, Draft decree on Environmental Policy*, London, Kluwer Law International, 106.

69. Bocken, H., Ryckbost, D., and Deloddere, S. (1996). Liability and financial guarantees. In: H. Bocken and D. Ryckbost (Eds), *Codification of environmental law*, Draft Decree on Environmental Policy, London, Kluwer Law International, 214–223.

70. See Council Directive 96/61/EC of 24 September 1996 concerning Integrated Pollution Prevention and Control, *Official Journal*, 10 October 1996, L 257/26.

71. So Bocken, H. Rechtstreekse verzekeringen ten behoeve van derden en andere wisseloplossingen voor aansprakelijkheid en aansprakelijkheidsverzekering. Een typologie. In: *Liber Amicorum René van Gompel*, 35.

72. See the report of the Commission to the European Parliament and to the Council of 12 May 1999 concerning the operation of the exemption regulation 3932/92 (com (1999) 92 final).

73. *Official journal*, L 398/7 of 31 December 1992.

74. For this criticism see Faure, M. and Van den Bergh, R., "Restrictions of competition on insurance markets and the applicability of EC antitrust law", *Kyklos*, vol. 48, 1995, 65–85.

75. And was criticised among others by Faure, M., and Hartlief, T. (1998). Een schadefonds als alternatief voor aansprakelijkheid en verzekering, *RM Themis*, 220–222.

76. See report, no. 18, p. 9.

77. See Faure, M., and Hartlief, T. (1996). Ontwikkelingen in de werkgever-saansprakelijkheid voor beroepsziekten: aanleiding voor een nieuwe AVB-polis?, A&V, 140–151. For a critical comment see Haazen, O. A., and Spier, J. (1996). Amerikaanse toestanden en de nieuwe aansprakelijkheidsverzekering voor bedrijven en beroepen, NJB, 45–50.

78. See Spier, J., and Haazen, O. A. (1996). *Aansprakelijkheidsverzekeringen op claims-made grondslag*, Deventer, Kluwer; and Wansink, J. H. (1996). Het polismodel AVB '96 en de dekking voor long-tail risico's, *A&V*, 120–122.

79. See also Haazen, O. A., and Spier, J. (1996). Amerikaanse toestanden en de nieuwe aansprakelijkheidsverzekering voor bedrijven en beroepen, *Nederlands JuristenBlad*, 45.

80. See Faure, M., and Van den Bergh, R. (2000). Aansprakelijkheidsverzekering, concurrentie en ongevallenpreventie. In: T. Hartlief and M. Mendel (red.), *Verzekering en maatschappij*, Deventer, Kluwer, 315–342.

REFERENCES

Abraham, K. (1988). Environmental Liability and the Limits of Insurance. *Columbia Law Review*, 88.

Akerlof, G. (1970). The market for 'lemons': quality, uncertainty and the market mechanism. *Quarterly Journal of Economics*, 488–500.

Akkermans, A. (1997). *Proportionele aansprakelijkheid bij onzeker causaal verband*. Deventer, Tjeenk Willink.

Arrow, K. (1965). *Aspects of the Theory of Risk-Bearing*. Helsinki, Yrjö Jahnssonin Säätiö.

Bergkamp, L. (1999). *De vervuiler betaalt dubbel*. Antwerpen, Intersentia.

Bocken, H., Ryckbost, D., & Deloddere, S. (1996). Liability and financial guarantees. In: H. Bocken & D. Ryckbost (Eds), *Codification of Environmental Law. Draft Decree on Environmental Policy* (pp. 214–223). London: Kluwer Law International.

Bohrenstein, S. (1989). The Economics of Costly Risk Sorting in Competitive Insurance Markets. *International Review of Law and Economics*, 25–39.

Borch, K. (1961). The Utility Concept Applied to the Theory of Insurance. *The Astin Bulletin*, 245–255.

Borch, K. (1963). Recent Developments in Economic Theory and their Application to Insurance. *The Astin Bulletin*, 322–341.

CEA, (1998). *Study on first party legal obligations for clean-ups and corresponding insurance covers in European Countries*. Paris: CEA, 21 October.

Council Directive 96/61/EC of 24 September 1996 concerning Integrated Pollution Prevention and Control. *Official Journal*, 10 October 1996, L 257/26.

Cousy, H. Recent developments in environmental insurance. In: F. Abraham, K. Deketelaere & J. Stuyck (Eds), *Recent economic and legal developments in European environmental policy*, 240.

Cowell, J. Compulsory environmental liability insurance. In: H. Bocken & D. Ryckbost (Eds), *Insurance of Environmental Damage* (p. 327).

Danzon, P. (1999). Alternative liability regimes for medical injuries. *Geneva Papers on Risk and Insurance*, 3–22.

Drion, P. J. M. (1998). Milieu onder één dak: milieuschadeverzekering (MSV), *Verzekeringsrechtelijke Berichten*, 2, 19–21.

Ehrlich, I., & Posner, R. (1974). An Economic Analysis of Legal Rule-Making. *Journal of Legal Studies*, 257.

Faure, M., & Van den Bergh, R. (1995). Restrictions of competition on insurance markets and the applicability of EC antitrust law. *Kyklos, 48*, 65–85.

Faure, M., & Hartlief, T. (1996). Ontwikkelingen in de werkgeversaansprakelijkheid voor beroepsziekten: aanleiding voor een nieuwe AVB-polis? *A&V*, 140–151.

Faure, M., & Hartlief, T. (1998). Remedies for expanding liability. *Oxford Journal of Legal Studies, 18*, 681–706.

Faure, M., & Hartlief, T. (1998). Een schadefonds als alternatief voor aansprakelijkheid en verzekering. *RM Themis*, 220–222.

Faure, M., & Fenn, P. (1999). Retroactive liability and the insurability of long tail risks, *International Review of Law and Economics*, 487–500.

Faure, M., & Grimeaud, D. (2000). *Financial assurance issues of environmental liability*, final version 4.12.2000.

Faure, M., & Van den Bergh, R. (2000). Aansprakelijkheidsverzekering, concurrentie en ongevallenpreventie. In: T. Hartlief & M. Mendel (Eds), *Verzekering en maatschappij* (pp. 315–342). Deventer: Kluwer.

Frenk, N. (1995). Toerekening naar kansbepaling. *Nederlands Juristenblad*, 482–491.

Haazen, O. A., & Spier, J. (1996). Amerikaanse toestanden en de nieuwe aansprakelijkheidsverzekering voor bedrijven en beroepen. *Nederlands JuristenBlad*, 45.

Janssen, C. A. (1998). Aansprakelijkheid voor milieuschade en financiële zekerheid naar toekomstig recht: nieuwe oplossingen. Nederlands Recht. In: L. F. Wiggers-Rust & K. en Deketelaere, (Eds), *Aansprakelijkheid voor milieuschade en financiële zekerheid* (pp. 111–112). Die Keure – Vermande.

Jost, P. J. (1996). Limited liability and the requirement to purchase insurance. *International Review of Law and Economics*, 259–276.

Katzman, M. (1988). Pollution liability insurance and catastrophic environmental risk, *Journal of Risk and Insurance*, 89–90.

Koziol, (1997). Die Arzthaftung im geltenden und kunftigen Recht, *Haftungsrechtliche Perspektiven der Ärztlichen Behandlung*, Linz, Universitätsverlag Rudolf Trauner, 21–35.

Kunreuther, H., Hogarth, R., & Meszaros, J. (1993). Insurer Ambiguity and Market Failure. *Journal of Risk and Uncertainty*, 71–87.

Monti, A. (2001). Environmental risk: A comparative law and economics approach to liability and insurance. *European Review of Private Law*, 51–79.

Niezen, J. (1998). Nieuwe milieuschadeverzekering – geen panacee, *Milieu en Recht*, 114. *Official journal*, L 398/7 of 31 December 1992.

Ogus, A. I. (1981). Quantitative Rules and Judicial Decision-Making. In: Burrows & Veljanovski (Eds), *The Economic Approach to Law* (pp. 210–225). London: Butterworth.

Ogus, A. I. (1994). Standard Setting for Environmental Protection: Principles and Processes. In: M. Faure, J. Vervaele & A. Weale (Eds), *Environmental Standards in the European Union* (pp. 25–37).

Pauly, M. (1968). The Economics of Moral Hazard: Comment. *American Economic Review*, 531–545.

Polborn, M. (1998). Mandatory Insurance and the Judgement-Proof Problem. *International Review of Law and Economics*, 141–146.

Posner, R. (1998). *Economic Analysis of Law* (5th ed.), New York: Aspen Law & Business.

Pratt, J. (1964). Risk Aversion in the Small and in the Large. *Econometrica*, 122–136.

Priest, G. (1987). The current insurance crisis and modern tort law. *Yale Law Journal*, 1521–1590.

Radetzki, M. (2000). Private arrangements to cover large-scale liability caused by nuclear and other industrial catastrophes. *The Geneva Papers on Risk and Insurance*, 180–195.

Ranson, D. (2000). Verzekering van milieuaansprakelijkheid. *Milieu- en Energierecht*, 68.

Rehbinder, E. (2000). Towards a community environmental liability regime: the commissions white paper on environmental liability. *Environmental Liability*, 85–96.

Rice, P. (2000). From Lugano to Brussels via Aarhus: environmental liability white paper published. *Environmental Liability*, 39–45.

Richardson, B. J. (2000). Financial Institutions for Sustainability. *Environmental Liability*, 61.

Robinson, G. (1985). Probabilistic causation and compensation for tortious risk. *Journal of Legal Studies*, 798.

Rogge, J. (1997). *Les assurances en matière d'environnement.* Loose-Leaf, Kluwer.

Rosenberg, D. (1984). The causal connection in Mass Exposure cases: a 'public law' vision of the tort system. *Harvard Law Review*, 851–929.

Shavell, S. (1979). On Moral Hazard and Insurance. *Quarterly Journal of Economics*, 541–562.

Shavell, S. (1985). Uncertainty over causation and the determination of civil liability. *Journal of Law and Economics*, 587–609.

Shavell, S. (1986). The judgement proof problem. *International Review of Law and Economics*, 43–58.

Shavell, S. (1987). *Economic analysis of accident law.* Cambridge: Harvard University Press, 190.

Skogh, G. (1989). The transactions cost theory of insurance: contracting impediments and costs. *Journal of Risk and Insurance*, 726–732.

Skogh, G. (2000). Mandatory Insurance: Transaction Costs Analysis of Insurance. In: B. Bouckaert & G. De Geest (Eds), *Encyclopedia of Law and Economics.* Cheltenham: Edward Elgar, 521–537.

Spence, M., & Zeckhauser, R. (1971). Insurance, Information, and Individual Action, *American Economic Review*, 380–391.

Spier, J. (1990). *Sluipende schade,* Deventer: Kluwer.

Spier, J. (Ed.) (1996). *The limits of liability, keeping the floodgates shut.* The Hague: Kluwer.

Spier, J., & Haazen, O. A. (1996). *Aansprakelijkheidsverzekeringen op claims-made grondslag,* Deventer, Kluwer.

Spier, J. *De uitdijende reikwijdte van het aansprakelijkheidsrecht.* pre-advies Nederlandse Juristenvereniging.

The report of the Commission to the European Parliament and to the Council of 12 May 1999 concerning the operation of the exemption regulation 3932/92 (com (1999) 92 final).

Tyran, J. R., & Zweifel, P. (1993). Environmental risk internalization through capital markets (Ericam): the case of nuclear power. *International Review of Law and Economics*, 431–444.

van Eijk-Graveland, J. C. (1998). *Verzekerbaarheid van opzet in het schadeverzekeringsrecht.* Zwolle: Tjeenk Willink.

Viscusi, W. K. (1991). The Dimensions of the Product Liability Crisis. *Journal of Legal Studies*, 147–177.

Wagner, G. (1991). Umwelthaftung und Versicherung. *Versicherungsrecht*, 249–260.

Wagner, G. (1992). Die Zukunft der Umwelthaftpflichtversicherung. *Versicherungsrecht*, 261–272.

Wagner, G. (1996). Versicherungsfragen der Umwelthaftung. In: M. Ahrens & J. Simon (Eds), *Umwelthaftung, Risikosteuerung und Versicherung* (pp. 104–105). Berlin: Erich Schmidt Verlag.

Wansink, J. H. (1985). De nieuwe milieuaansprakelijkheidsverzekering. *Milieu en Recht*, 98.

Wansink, J. H. (1996). Het polismodel AVB '96 en de dekking voor long-tail risico's. *A&V*, 120–122.

Wansink, J. H. (1997). Hoe plotseling en onzeker is de verzekeringsdekking voor millieuaansprake-
    lijkheidsrisico's? In: *Miscellanea. Juris Consulto vero dedicata* (pp. 451–460), Essays
    offered to Prof. Mr. J. M. van Dunné. Deventer: Kluwer.
Wansink, J. H. (1999). Verzekering en milieuschade als gevolg van vervoer/opslag van gevaarlijke
    stoffen. *Tijdschrift voor Milieuaansprakelijkheid*, 77–82.
Wils, W. P. J. (1994). Insurance Risk Classifications in the EC: Regulatory Outlook. *Oxford Journal
    of Legal Studies*, 449–467.
Zeckhauser, R. (1996). 19th Annual Lecture of the Geneva Association and Catastrophes. *The
    Geneva Papers on Risk and Insurance*, 5.

COMMENTS ON PAPER BY MICHAEL FAURE

James Boyd

My comments are limited to a small set of areas that I think deserve some further clarification.

First, I think that it is imperative to make clear the important difference between mandatory and voluntary insurance. If cost internalization (compensation) and deterrence are the policy goal, it is most natural to think of insurance – normatively – as being mandatory. Voluntary insurance, if cost internalization is the goal, cannot be counted on to get the job done. (Voluntary insurance being geared toward utility improvements in the presence of risk aversion, when risks are borne by the insuring party themselves.) The firms most at risk of externalizing costs will also be the firms least likely to voluntarily self-insure, since the costs of insurance will be highest for them. (I think Michael is familiar with my views on this subject. And to be clear, I am not angling for a reference to my views. I do think, however, that a clearer contrast or acknowledgment of the differences in motivation for mandatory vs. voluntary insurance should be provided up front in the paper).

Second, I found the distinction between first-party insurance and "direct insurance" to be somewhat muddy – early in the paper. Also, the paper's real focus is on the "direct" mechanism so that should be the focus of the introduction.

Third, in the discussion of the Dutch approach, and the "direct approach" generally, it is unclear what the real benefit of this system is to the insured. From the U.S. perspective, it is hard to see how eliminating the *protection* of liability standards would be an improvement. An improvement arises for them

An Introduction to the Law and Economics of Environmental Policy: Issues in Institutional Design, Volume 20, pages 329–330.
© 2002 Published by Elsevier Science Ltd.
ISBN: 0-7623-0888-5

only if a clear boundary can be drawn around claims (and thus costs limited). I remain somewhat confused on how this boundary is maintained in the Dutch, or any other system. Again, from the U.S. perspective, there is no obvious way to limit claim to a particular subset of individuals, defined ex ante. What happens if you injure a set of victims not in this pre-defined group?

Fourth, I think that for the non-European readers, a better understanding of the market concentration issues in European insurance is of real interest. Weak competition clearly changes the normative recommendations for insurance, so I think background on the competitive situation is important.

THE DESIGN OF MARKETABLE PERMIT SCHEMES TO CONTROL LOCAL AND REGIONAL POLLUTANTS*

Jonathan Remy Nash and Richard L. Revesz

ABSTRACT

In recent years, there has been a steady rise in the use of marketable permits in environmental regulation. They have been employed as tools to control both air and water pollution, and have been implemented on local, regional, and national scales. These trading regimes – based upon a single market in emission permits – do not control the distribution of emissions throughout the trading region or prevent the formation of "hot spots" of pollution. In this chapter, we propose a marketable permit scheme that is consistent with the attainment of ambient standards and that does not significantly interfere with the benefits of trading.

INTRODUCTION

The idea of establishing markets in pollution rights is not new. J. H. Dales (1968) strongly advocated this regulatory technique in the late 1960s. Thereafter, tradeable permit regimes became popular among academic economists. In a 1985 article that had considerable influence on the legal literature, Bruce

* Based upon our article 'Markets and Geography: Designing Marketable Permit Schemes to Control Local and Regional Pollutants', *Ecology Law Quarterly*, 28(2001), 569.

An Introduction to the Law and Economics of Environmental Policy: Issues in Institutional Design, Volume 20, pages 331–377.
© 2002 Published by Elsevier Science Ltd.
ISBN: 0-7623-0888-5

Ackerman and Richard Stewart (1985) strongly advocated tradeable permits as an alternative to command-and-control regulation.

In recent years, there has been a steady rise in the use of tradeable permits in the regulatory process. They have been employed as tools to control both air and water pollution, and have been implemented on local, regional, and national scales. While trading regimes have, to date, been used mostly in the United States, they are attracting increasing attention abroad, and there is now considerable interest in emissions trading on a global scale to address the problem of anthropogenic greenhouse gases.

The design of the trading regime most commonly advocated in the academic literature – based upon a single market in emission permits – is relatively straightforward. A policy maker determines the total number of emission permits that will be allocated to a region. These permits are then distributed among polluters, generally either by means of an initial auction or through grandfathering. Subsequently permits trade in an open market. Assuming that a robust market for permits arises, a tradeable emission permit regime reduces aggregate emissions to the chosen aggregate level at least cost.

Such a trading regime, however, does not control the distribution of those emissions throughout the region. As a result, it does not ensure that an ambient standard – specifying a maximum permissible concentration of the pollutant – would be met throughout the region. For example, to the extent that trades result in a disproportionate concentration of the pollutant in some portion of the region, ambient standards could be violated. More generally, a tradeable emission permit regime does not prevent the formation of "hot spots" of pollution – that is, location at which the damage caused by pollutants is particularly severe. Ackerman and Stewart (1985, p. 1350) acknowledge this shortcoming: "[T]he market system we have described could allow the creation of relatively high concentrations of particular pollutants in small areas within the larger pollution control region." They note that "[t]he extensive literature on marketable permits . . . points to a variety of feasible means for dealing with the hot spot problem," and recommend that "a long-run strategy for institutional reform should strive to take advantage of these more sophisticated market solutions to the problem of intraregional variation" (Ackerman & Stewart, 1985, p. 1351).

Over the years, commentators have advocated several alternative design structures to deal with the problem of ambient standard violations and hot spots. Each of these alternatives, however, has significant drawbacks, either providing only an incomplete solution to the problem or introducing complexity that could stand in the way of the efficient functioning of the market.

In this chapter, we propose an alternative that would avoid the violation of ambient standards and formation of hot spots without greatly increasing administrative complexity or introducing excessively high transaction costs. Our idea is to construct a market in tradeable emission permits under which trading would be entirely unfettered, with the sole exception that a prospective buyer and seller would have to receive approval from a website before they could consummate their trade. The website, administered by the government, would contain emissions data for all sources in the region. When a proposed trade was submitted for approval, the website would temporarily update its saved data to reflect the change in the geographic distribution of emissions that would result from the proposed trade. The website would then use an atmospheric dispersion model to predict the impact of the emissions from all the sources in the region – as modified by the proposed trade – on ambient pollution levels at various receptor points. The website would reject any trade resulting in the violation of an applicable ambient standard and would approve all other trades.

Part I of this chapter describes the functioning of the typical regime of tradeable emission permits. It examines how such a regime can lead to the violation of ambient standards and the formation of hot spots.

Part II describes the structure of three recent and prominent U.S. regulatory programs involving tradeable emission permits: the national sulfur dioxide trading program, regional trading in ozone precursors in the northeast, and local trading in sulfur and nitrogen oxides emissions in the Los Angeles metropolitan area. It explains why these programs are poorly suited to ensure the attainment of ambient standards or the prevention of hot spots. It also discusses the unsuccessful efforts to correct these shortcomings.

Part III evaluates three proposals that commentators have advanced as alternatives to the typical emission trading regime. We first examine a system in which the emissions market is divided into zones and interzonal trades are prohibited. Next, we analyze a system of markets in units of environmental degradation, under which permits do not entitle a holder to emit a fixed amount of pollutant, but rather to cause a fixed amount of environmental damage at a certain receptor point. Last, we discuss pollution offset markets, under which trading in emission permits does not occur on a one-to-one ratio.

Part IV presents our proposal for a tradeable emission permit regime designed to solve the problem of ambient standard violations. First, we explain its major elements. Second, we compare its relevant features to those of the three alternatives examined in Part III. Third, we highlight our proposal's impacts on the structure of the permit markets.

Part V discusses how our proposal would deal with a number of important issues. In particular, we analyze purchases of permits by sources that locate in the area after the establishment of the market; variable – as opposed to uniform – ambient standards throughout the region; auctions of permits, either at the time of the initial distribution or subsequently; and purchases of permits with the intent to bank them for use at a future time, retire them in order to improve environmental quality, or hold them as an investment with the goal of reselling them in the future.

I. TRADEABLE EMISSION PERMITS AND THE VIOLATION OF AMBIENT STANDARDS AND FORMATION OF HOT SPOTS

We first examine the structure of the most common tradeable pollution permit regime, in which trades take place in a single market, in units of emissions, and on the basis of a one-to-one trading ratio. Next, we explain how this regime can lead to the violation of ambient standards and the formation of hot spots.

Elements of a Market System

The design and implementation of a tradeable emission permit regime proceeds in several steps. First, the policymaker identifies the pollutant to be regulated, and the region over which the regulation will extend. In the paradigmatic regime, the entire region constitutes a single trading zone.

Next, the policy maker determines what aggregate level of emissions in a given year (or other time period) will be deemed acceptable. It then subdivides this amount into a number of discrete emission permits, each of which authorize the holder to emit a fixed amount of the regulated pollutant. The policy maker also defines the bundle of rights that accompanies each permit, including their longevity, whether the government can eliminate permits before their expiration, and whether unused permits can be retained for future use.

The policy maker then adopts a mechanism for allocating these permits among prospective polluters. For example, it can distribute the permits to existing polluters according to their prior emission history. If it chooses this grandfathering option for the initial allocation of permits, the policy maker can distribute the permits at no charge, or it can charge some pre-determined amount. Alternatively, it can distribute all the permits, or some subset, through an auction.

The policy maker must also devise rules for the subsequent trading of the permits in an open market. For example, should non-polluters be free to buy and hold emission permits, either for investment purposes or in order to retire them as a means of improving environmental quality?

Impact of Trading on Environmental Quality

In discussing the impact of a particular pattern of emissions on environmental quality, it is important to distinguish among three different types of pollutants. For "global" pollutants, this impact is independent of the location of the various sources; only the aggregate amount of emissions matters. Greenhouse gases are a prominent example of a global pollutant. For example, the adverse impact of a ton of carbon dioxide is the same regardless of whether it is discharged in the United States or China.

Most pollutants, however, do not have this feature. Instead, their harm depends on the location at which they are emitted. For "local" pollutants, the location of the emissions and the place where the harm occurs are relatively coextensive. "Regional" pollutants travel further – hundreds of miles and across state and national boundaries in the case of sulfur dioxide – but the affected region is defined by reference to where the emissions come from. Thus, for both local and regional pollutants, the location of the emissions determines the location and magnitude of the adverse environmental consequences. In the vocabulary of environmental economics, the emission of one unit of pollution at different locations results in "spatially differentiated" externalities.

For local and regional pollutants, the environmental impacts of trading are a function of two important sets of variables. First, the costs of pollution control faced by the various participants in the market determine the distribution of emissions among these sources. A polluter will reduce its emissions up to the point at which a further unit of reduction is more costly than the market price of a permit. A polluter that can reduce its emissions for less than the market price of a permit will invest additional resources in pollution control technology and sell permits. In contrast, a polluter with comparatively high costs of pollution control will save money by increasing its emissions and buying additional permits. Trading continues until each source's marginal cost of pollution reduction – the cost of an additional unit of emission reduction – is equal to the market price of the permit.

Second, emissions at different locations throughout a region have different effects on environmental quality. For example, in the case of air pollution, the height of the stack from which pollutants are emitted has an important effect on ambient air quality levels: typically, higher stacks produce greater impacts further from the source and smaller impacts closer to the source. The effects are also dependent on topographic and meteorological conditions, particularly wind patterns. Thus, for local and regional pollution, trades among sources within a region can have an impact on the levels of environmental quality at particular locations even though the aggregate level of emissions in the region remains constant.

If the public policy objective were to maximize social welfare, the optimal distribution of pollution concentrations in the region would depend on the shape of the damage function – the function linking the pollutant's concentrations to its adverse effects. If the damage function is convex (i.e. an additional unit of pollution concentration causes greater harm at higher concentrations), social welfare is maximized by spreading the pollutant concentrations throughout the region. Indeed, additional levels of pollutant concentrations cause greater harm in dirtier areas. In contrast, if the damage function is concave (i.e. an additional unit of pollution concentration causes greater harm at lower concentrations), social welfare is maximized by concentrating the pollution rather than spreading it around. In this case, additional concentrations of a pollutant cause a smaller adverse effect in areas that are already highly polluted.

Whereas disproportionate concentrations of pollution at a particular location can be either desirable or undesirable from the perspective of maximizing social welfare, they tend to be undesirable from the standpoint of distributional concerns. For example, the environmental justice movement worries about any disproportionate concentration of pollution because of fears that its impacts will be felt primarily by the most vulnerable members of society. This concern is independent of the shape of the pollutant's damage function. Even if the aggregate social welfare would be increased by concentrating pollution in a particular area, environmental justice advocates would nonetheless favor dispersion.

The academic literature has focused almost exclusively on the possibility that emissions trading will give rise to excessive concentrations of pollution at particular locations, and has expressed concern over two separate manifestations of this phenomenon: the formation of hot spots and the violation of ambient air quality standards. In the case of local pollutants, such hot spots are likely to be in locations in which the levels of emissions are particularly high. But geographic factors matter as well. For example, for a given level of emissions, a hot spot is more likely to develop if the area is close to a topographical barrier, such as a mountain range, and where prevailing winds blow the pollution toward the range rather than away from it. In the case of regional pollutants, what matters is not only the location of the emissions but also the manner in which emissions are transported. As a result, there may be a hot spot at a downwind location even if the emissions are not concentrated at any upwind location.

Environmental justice advocates, in particular, worry that marketable permit regimes will exacerbate hot spots. The oldest, most highly polluting factories are often located in lower income neighborhoods. If the initial allocation of permits is performed through grandfathering, these older plants will get a disproportionate number of permits and will be able to continue polluting at their historic rates without facing any additional costs. In contrast, under a command-and-control regime, they might eventually face more stringent standards. Moreover, older factories are more likely to have higher marginal costs of pollution reduction. As a result, they will purchase additional permits following the initial allocation, and increase their pollution rather than spend money on more effective pollution control technology.

In fact, even if older facilities were not located in predominantly poor or minority communities when the government first implements a trading program, "market dynamics" in the housing market may change that situation over time. In particular, the placement of a locally undesirable land use in a community may "convert" the community into one whose residents are more likely to be poorer and more likely to be of color. Thus, any concentration of pollution in a particular area, even if it was not originally a predominantly poor or minority area, could have adverse environmental justice consequences.

In addition to producing hot spots, trading regimes can also lead to violations of ambient standards. Several existing environmental programs feature such standards. The Clean Air Act, for example, establishes national primary and secondary ambient air quality standards (NAAQS), which provide protection against adverse effects on the public health and public welfare, respectively. Similarly, under the Clean Water Act, the states must promulgate ambient water quality standards.

The leading proposals for tradeable permit regimes do not contain a mechanism for ensuring that ambient standards are met. So, for example, if the ambient air quality level in an area is equal to the ambient standard before the transition to a trading regime, subsequent purchases of permits by sources in the area could lead to a violation of the ambient standards. As a result, trading regimes lose the strong rhetorical claim that they are simply a way of achieving the same level of environmental protection more cheaply than under command-and-control regulation. But a regulatory regime that does not ensure that such standards will be met cannot be said to provide the same level of protection as one that does, even if the aggregate amount of emissions are the same under both programs.

II. SHORTCOMINGS OF THE EXISTING TRADEABLE POLLUTION PERMIT REGIMES

This Part analyzes three U.S. regulatory regimes employing tradeable emission permits: the national sulfur dioxide trading program, regional trading in ozone precursors in the northeast, and local trading in sulfur and nitrogen oxides emissions in the Los Angeles metropolitan area. In each of these three programs, permits are generally traded in a single market, with no geographic restrictions. We highlight how these programs are not well designed to deal with local and regional pollution and discuss the unsuccessful efforts to ameliorate this problem.

The National Sulfur Dioxide Trading Program and the Control of Acid Rain

Design of the Trading Program
One of the principal causes of acid deposition, commonly known as acid rain, is the transformation in the atmosphere of sulfur dioxide emissions into sulfates. These sulfates, which are acidic compounds, can travel hundreds of miles as a result of prevailing winds before being dissolved into rain or snow.

Although EPA promulgated NAAQS for sulfur oxide in 1971, these standards are designed to deal with the pollutant's local effects, not its transformation into sulfates. Over the years, there were unsuccessful attempts to compel EPA to revise these NAAQS to address acid deposition, and to promulgate a separate NAAQS for sulfate particles. In connection with part of this litigation, the Second Circuit explicitly stated that the NAAQS "were not designed to protect against the deleterious effects of [sulfur oxides] associated with acid rain and dry deposition".[1] Moreover, even though the Clean Air Act has specific provisions seeking to constrain interstate pollution, the courts have consistently upheld EPA's position that the impact of transformed pollution, such as sulfates, need not be taken into account in evaluating whether the upwind pollution is excessive.

During the 1980s, Congress rejected various proposals to control acid rain. The failure of these initiatives was due in large part to regional feuding over how to allocate, among the various regions, the costs of addressing the problem.

After all these false starts, the Clean Air Act amendments of 1990 finally established the national sulfur dioxide trading program as a means of controlling acid rain. It was the first pollution trading program authorized by Congress, as well as the first with nationwide scope. Under the program, each allowance or permit authorizes the holder to emit one ton of sulfur dioxide in one year.

The program's implementation has proceeded in two stages. From January 1, 1995 through December 31, 1999, Phase I was in effect. It covered 261 sources at 110 electric utilities, which were considered the dirtiest in the nation. The goal of Phase I was to achieve a 3.5 million ton reduction in sulfur dioxide emissions. The initial allocation of permits was performed through a grandfathering scheme under which the covered sources obtained, for free, an amount equal to the product (in tons) of: (i) each source's baseline fossil fuel consumption (in British Thermal Units (BTUs)), which is taken to be the average consumption during 1985 through 1987, and (ii) an assumed emissions rate of 2.5 pounds of sulfur dioxide emissions per million BTUs of fuel consumption. The sources covered under Phase I all had actual emissions rates greater than or equal to 2.5 pounds of sulfur dioxide emissions per million BTUs of fuel consumption. Thus, their prior emissions were not fully grandfathered.

Phase II of the sulfur dioxide trading program, which began on January 1, 2000, subjects virtually all fossil-fueled electric generating plants, including new sources, to an annual nationwide cap on sulfur dioxide emissions. It is expected to lead to further reductions in sulfur dioxide emissions of 6.5 million tons. Although the Phase II allocation provisions contain numerous rules and bonus allocations, the underlying logic of the initial grandfathering is to give each existing source a number of allowances equal to the product (in tons) of: (i) the source's baseline fuel consumption, determined in the same manner as for Phase I, AND (ii) the lesser of 1.2 pounds of sulfur dioxide per million BTUs and its actual 1985 emissions rate (in pounds of sulfur dioxide per million BTUs).

A small portion of the permits to which sources would be entitled under this grandfathering formula – 2.8% – is withheld by EPA and auctioned annually. These allowances are obtained by withholding 2.8% of the permits to which each source would otherwise be entitled under the grandfathering formula discussed above. The Chicago Board of Trade conducts these auctions on behalf of EPA. The proceeds are distributed pro rata to sources from which the allocated allowances were withheld.

Following the initial allocation of allowances, the sulfur dioxide trading program authorizes their purchase and sale not only among covered pollution sources, but among anyone who chooses to participate in the market. The market is national and there are no geographic restrictions on trading. The requirements for a transfer to be effective are only ministerial. Brokers facilitate the functioning of the market by maintaining price information and by matching buyers with sellers.

In addition to these intra-source transfers, holders of allowances may "bank" them for use in a future year. Such banking is a form of intra-source trade. An allowance that is valid for use in a future year, however, may not be used before that year.

Shortcomings of the Trading Program
The sulfur dioxide trading program is poorly suited to guard against hot spots of acid rain. Most importantly, because trading occurs in a single national

market, it pays no attention to the impact of the location of the sulfur dioxide emissions on the production of acid rain.

The region that contributes most to the problem of acid rain in the United States is the Ohio River Valley. Prevailing winds generally carry sulfur dioxide emissions from this region to the Northeast, where acid deposition harms fragile lakes with limited buffering capacity. The sulfur dioxide trading program can exacerbate this problem by permitting sales from northeastern to midwestern polluters. Indeed, sulfur dioxide emissions in the Northeast produce acid rain over the Atlantic Ocean, which has a large buffering capacity that limits the extent of the harm. When northeastern emissions are sold to the Midwest, however, the acid rain falls over the Northeast, where the harm is far larger.

Some commentators suggest that, in practice, this problem is not serious. For example, an analysis of allowance transfer data suggests that, thus far, the trading program has not resulted in the development or exacerbation of hot spots (Ellerman et al., 2000, pp. 130–136; Swift, 2000, p. 954). The reason appears to be that the number of permits received by the dirty sources in the Midwest was smaller than their historic pattern of emissions. As a result, even if such sources became net purchasers of permits, their aggregate emissions would be less than they were before the implementation of the trading program. In essence, the argument of these supporters of the current trading program is that any adverse consequences of trading are dominated by the reduction in emissions produced by the grandfathering scheme. The fact remains, however, that it might be desirable to design a trading regime that does not produce adverse consequences.

Another study suggests that the Northeast will not suffer net harms as a result of Phase II trading (Burtraw & Mansur, 1999, p. 2). The study acknowledges, however, that "[t]rading leads to an increase in emissions in the Midwest and a decrease in the East and Northeast." It finds, however, that as a result of the geographic shift in emissions among sources in the Midwest, there is slight decrease in acid deposition in the Northeast (Burtraw & Manur, 1999, p. 2; Burtraw, 2000, pp. 143–149).

But regardless of what the aggregate impacts of trading might be, nothing in the sulfur dioxide trading program prevents the development or exacerbation of acid rain hot spots. More generally, the program does not allocate emissions between upwind and downwind sources in a way that minimizes their adverse consequences.

Efforts to Redesign the Trading Program
Concerns about the adverse environmental consequences of having trading take place in a single national market received sustained congressional attention

during the consideration of the 1990 amendments to the Clean Air Act. Moreover, almost immediately following the passage of the legislation, states and environmental groups sought to introduce constraints into the trading program as a way to prevent the development or exacerbation of hot spots.

Pre-Enactment. Even though the final acid rain legislation created a single national market for the trading of sulfur dioxide allowances, Congress did consider imposing various forms of trading constraints, in part to guard against the possibility of hot spots. A public policy study sponsored by Senators Heinz and Wirth in 1988 was the genesis of the trading scheme to control acid rain (Project 88, 1988, p. 32). The study, under the direction of Robert Stavins, envisioned that trading "would occur on a national or regional basis" (Project 88, 1988, p. 32). The report acknowledges that "[t]he source-receptor relationship must be considered, since reducing acid rain precursors in California, for example, will not reduce acid rain on the East Coast" (Project 88, 1988, p. 32).

On June 12, 1989, President Bush announced his proposal to control acid rain by means of a market in sulfur dioxide emission allowances. The proposal generally restricted trading during Phase I to intrastate transfers. Interstate trading was authorized only for intra-utility trades. During Phase II, "full interstate trading would be allowed" (White House Fact Sheet on the President's Clean Air Plan, 1989, p. 710).

The Administration's bill, transmitted to Congress on July 21, 1989, similarly authorized only intrastate or intra-utility trading during Phase I (Senate Committee on Environment and Public Works, 1993, p. 3972). With respect to Phase II, however, it modified (or perhaps clarified) the proposal announced a month earlier to provide that "allowances may be transferred among affected sources within each of the geographic regions of the country, as prescribed by regulation" (Senate Committee on Environment and Public Works, 1993, p. 3972). The bill did not explain how EPA would define these multi-state regions within which trading would be allowed, but commentators believed that there would be two regions.

Hearings before the Subcommittee on Environmental Protection of the Senate Committee on Environment and Public Works, conducted in October 1989, devoted considerable attention to the structure of the trading markets. For example, Senator Lieberman raised questions about unfettered national trading:

> If we allow an open market, free market of trading of permits here, is there any danger that the sources of the acid rain that are hitting New England will acquire the permits to continue to do so? In other words, that reductions, if they're imposed on a national basis, may actually occur in plants other than the ones in the Midwest that seem to be the source of our problem?[2]

Senator Lieberman thus was concerned that midwestern sources would buy additional permits, beyond their initial allocation, and thus exacerbate the acid rain problem in the Northeast.

Senator Heinz stated in response:

> I don't think so . . . for two reasons. The first is that the first wave of reductions is largely targeted to the area that I think you believe is responsible for the most acid rain in the Northeast and second, there are as I understand it, restrictions within the President's proposal that have the effect ensuring that you don't have those kind of bicoastal trades that you might be worried about.[3]

Senator Heinz appears to rely on the fact that the initial allocations to midwestern sources would be less than their historical emissions. His response fails to address Senator Lieberman's concern over the impact of subsequent trades on the distribution of emissions. Moreover, the division of the country into two trading zones, which would indeed have prohibited bicoastal trades as indicated by Senator Heinz, would probably have authorized the Midwest/Northeast trades that worried Senator Lieberman.

At the same hearing, David Hawkins of the National Resources Defense Council, testifying on behalf of the National Clean Air Coalition, expressed concern about the effects of the trading program on the West: "We think new sources of pollution in the western United States might be tempted if they were permitted to do so to purchase allowances from the eastern United States and build dirtier plants in the West."[4] He added:

> [T]he pristine air quality in the West is a global treasure. You cannot compensate for the damage to that air quality that would be caused by building a dirty plant in the West simply by getting emission reductions hundreds of miles away in another region of the country."[5]

Hawkins' statement reflects strong concern, consistent with a concave damage function, that one unit of sulfur dioxide produces greater harm in pristine western areas, which include many important national parks, than in industrialized areas.

The bill reported out of the Senate Committee on Environment and Public Works on November 16, 1989 direct[ed] that EPA regulations "shall permit transfers only within each of the two major geographic regions of the country as defined by such regulations" (Senate Committee on Environment and Public Works, 1993, p. 8205). The Report accompanying the bill explains: "Allowances transferred between the owners and operators of units that receive allowances or others who lawfully hold allowances under this title may only be transferred to units within the region in which the unit for which the allowances were originally issued is located" (Senate Committee on Environment and Public Works, 1993, p. 8645). It adds that "[t]he Administrator should give exclusive consideration to the environmental protection objectives of this title in defining such regions" (Senate Committee on Environment and Public Works, 1993, p. 8661). A plausible goal for EPA would therefore have been to define the regions in the manner least likely to lead to the violation of the NAAQS for sulfur oxides or to the formation of acid rain hot spots.

The bill passed by the Senate on April 3, 1990, provided for a single national market, with no geographic restrictions on trading. The debate on the Senate floor does not explain why the two-region restriction on trading was dropped.

On the House side, the bill reported out of the House Committee on Energy and Commerce on May 17, 1990, provided for two trading regions. It delegated to EPA the definition of "the two major geographic regions of the country" (Parker et al., 1991, p. 2041 & n. 90). Trading across regions was prohibited, with two exceptions. First, new sources could buy allowances without geographic restriction. Second, the bill permitted intra-firm trades among units in operation on the date of the enactment of the legislation. These restrictions on interregional trading, applicable during both Phase I and Phase II, were retained in the bill passed by the House on May 23, 1990, following only limited floor debate.

The conference bill followed the Senate model of interstate trading in a single national market with no geographic restrictions. The House passed the bill on October 26, 1990 and Senate on the following day, and it was signed by President Bush on November 15, 1990.

The decision to adopt a trading program with no geographic restrictions appears to be largely attributable to the fact that the emphasis throughout the debate was on the large decrease in aggregate sulfur dioxide emissions. For a sufficiently large reduction, the acid rain problem would be ameliorated across the board even if the emissions were not allocated between upwind and downwind sources in an optimal manner. As one commentator explained, "it was understood that the greater the overall size of the reduction [in overall emissions], the more indifferent society could be to the spatial impacts of trades . . ." (Kete, 1992, p. 83).

Post-Enactment. The concern over geographically unrestricted trading was not laid to rest with the passage of the legislation. Commenting on EPA's proposed regulations setting forth the details of the sulfur dioxide markets, several northeastern states criticized EPA for crafting a program that focused on the total sulfur dioxide emissions rather than on the existence of acid rain hot spots. They urged EPA to consider the effects of the structure of the trading program on the formation of acid rain. In addition, New York filed separate comments urging EPA to preclude emission permit trading between sources in different regions of the country until EPA could develop appropriate trading ratios. As a result, if one ton of sulfur dioxide had a greater impact on acid rain if emitted by the purchaser rather than the seller, the purchaser would have to buy more than one ton in permits in order to emit an additional ton, with the ratio being determined by the relative contributions of the two sources to the acid rain problem.

In adopting its final regulations, EPA rejected these criticisms, establishing a single national market, in which allowances for sulfur dioxide emissions traded on a one-to-one ratio, with no geographic constraints. The agency indicated that Congress had already made the policy choice on this matter: "Geographical limitations on allowance transfers were included in early Senate and House versions of the bill, but later rejected in favor of national transfers.[6] The regulations, moreover, bar the states from prohibiting or otherwise restricting the ability of sources within their jurisdiction from purchasing or selling permits.

The hypothetical concerns voiced by some states about the sale of in-state allowances to out-of-state sources soon became a reality. In 1992, the Wisconsin Power and Light Company sold 35,000 allowances to the Tennessee Valley Authority, and sought to keep the transaction secret. Similarly, in 1993, a New York utility – the Long Island Lighting Company (LILCO) – sold allowances to AMAX, a non-utility supplier of energy, that in turn intended to resell the allowances to midwestern utilities. LILCO also tried to keep the identity of the purchaser a secret, but New York's public service commission denied its request.

In March 1993, New York State and an environmental group filed an action in the D.C. Circuit complaining that the sulfur dioxide trading program improperly permitted such trades. The action sought to compel EPA to restrict trades that would lead to heightened acid deposition damage in environmentally sensitive areas of New York.

Moreover, bills introduced in the New York and Wisconsin legislatures sought to constrain or discourage the sale of in-state allowances to out-of-state sources. Wisconsin's bill, which was never enacted, would have required public disclosure of, and the approval of the state Public Service Commission for, any transaction of allowances by a Wisconsin utility. In New York, similar bills introduced first in 1993, and again in subsequent years also failed to be enacted.

In 1998, LILCO entered into a memorandum of understanding with New York State not to sell allowances to sources in upwind states, in certain nearby states, or in New York. The agreement also contained a provision purporting to prohibit any purchaser of LILCO allowances to sell them to polluters in states with which LILCO was not permitted to trade. Between 1994 and the date of this agreement, LILCO had sold 260,000 allowances, "the vast majority of them going to polluters in the upwind states" (Hernandez, 1998, p. B1).

In 2000, New York enacted a law designed to remove the incentive for sales of sulfur dioxide allowances to utilities located in upwind states.[7] Specifically, the law applies to sales of so-called "select SO_2 allowance credits," which are defined as sulfur dioxide permits under the national sulfur dioxide program "issued to generating sources located within the boundaries of the state of New York."[8] The law effectively allows the State Public Service Commission to deprive a New York seller of proceeds from the sale of such credits to a polluter located in an "acid precipitation source state" – generally states located upwind from New York.[9]

The statute provides that "[n]othing in this section shall discourage or prohibit allowance trades (such as for retirement purposes) that will have a beneficial impact on sensitive receptor areas in the state of New York."[10] It is unclear whether, under this provision, a sale to an "acid precipitation source state" might be exempted if the trade nonetheless would have a beneficial impact on acid deposition in environmentally sensitive areas of New York – perhaps by reducing in-state emissions causing worse harm.

The statute contains a safe harbor to sellers of allowances that include a restrictive covenant in the sale documents. To be effective, the covenant must restrict the use of the transferred allowances to an acid precipitation source state. This safe harbor suggests that the law's scope may extend to sales of an allowance originally issued to New York sources, even where the original New York holder sold the allowance to an out-of-state entity not located in an acid precipitation state, where this latter entity subsequently transfers the allowance to a source in a covered state. Absent the safe harbor, at the time of the second sale the New York seller would apparently become liable for the proceeds of the first sale.

Assessing the State Efforts to Constrain the Trading Market
Piecemeal efforts undertaken by the states are unlikely to be a desirable solution to the acid rain problem. Laws such as New York's may encourage a state-by-state patchwork approach to ameliorating the poor design of the national sulfur dioxide trading program. Such a piecemeal approach would undermine the market for sulfur dioxide allowances and potentially render the national program ineffective by subjecting it to multiple, and possibly inconsistent, state regulations. A national market subject to multiple limitations on trading may well not survive. Such approaches would also discourage sales that produce adverse effects in the state in which the seller is located even if there are greater beneficial effects elsewhere.

Moreover, the New York law does not necessarily promote even the state's own interest. Presumably not every sale of an allowance by a New York holder to a source in an "acid precipitation source state" would result in an increased acid rain problem in New York. Merely because some portion of a state is upwind from New York does not imply that emissions from every location in that state adversely affect New York. Thus, the law would constrain sales that, on balance, might be good for New York by reducing local pollution.

In addition to these policy flaws, the constitutionality of laws such as New York's is at best uncertain. The New York law may run afoul of the dormant Commerce Clause by unduly burdening the interstate market in sulfur dioxide allowances. New York might argue that its statute is necessary to protect in-state health and welfare. This argument, however, is somewhat compromised because the law overbroadly reaches transactions that might produce adverse effects on New York. Moreover, the New York law may be preempted by the provisions of the Clean Air Act establishing the market in sulfur dioxide emissions. The claim would be that New York's constraints on the transfer of allowances undermine Congress' goal of achieving pollution reduction on a nationwide scale as cost-effectively as possible, and impose unintended constraints on the ability of sources outside New York to purchase allowances. Indeed, in April 2002, a federal district court in New York invalidated the New York statute on the grounds that it was preempted by the federal Clean Air Act and unconstitutional under the Commerce Clause.

Regional NO_x Trading Programs and the Control of Tropospheric Ozone

Processes of Ozone Formation

Nitrogen oxides (NO_x) are the primary precursors of tropospheric, or ground-level, ozone. Ozone, in turn, is the primary ingredient in smog, which causes numerous human health risks, adversely affects plants and ecosystems, and contributes to acid deposition.

Tropospheric ozone is formed by the photolysis of nitrogen dioxide (NO_2). Other compounds, however, affect the rate at which ozone is produced. In particular, through a series of complex chemical reactions, certain volatile (or highly reactive) organic compounds (VOCs) help to convert nitrogen monoxide (NO) into nitrogen dioxide, thereby increasing the production of ozone.

The production of ozone, however, does not vary proportionately with the concentrations of NO_x and VOCs in the atmosphere. As a result, decreasing NO_x and/or VOC concentrations does not necessarily result in a proportionate decrease in ozone production. Indeed, for certain ratios of NO_x to VOC concentrations, a *decrease* in the concentration of NO_x actually will lead to an *increase* in ozone production. The reason for this facially counterintuitive result is the role of the hydroxyl radical (OH) in the atmospheric photochemistry that leads to the formation of ozone. Its presence is a prerequisite for the series of reactions that allows VOCs to accelerate the conversion of nitrogen monoxide into nitrogen dioxide. The hydroxyl radical, however, also reacts with nitrogen dioxide. Thus, at comparatively low VOC to NO_x concentrations (i.e. where NO_x is relatively abundant), the nitrogen dioxide "effectively competes with the VOCs for the [hydroxyl] radical" (National Research Council, 1991, pp. 167–168). This reaction decreases the ability of VOCs to convert nitrogen monoxide into nitrogen dioxide, and thus reduces the rate of production of ozone. As a result, if NO_x concentrations are lowered relative to VOC concentrations, "more of the [hydroxyl] radical pool is available to react with the VOCs, leading to greater formation of ozone" (National Research Council, 1991, p. 168).

A consequence of these complex interactions is that the best strategy for reducing ozone production depends on the local ratio of VOC to NO_x concentrations. At relatively low VOC to NO_x concentration ratios (of around 10 or less), reductions in NO_x concentrations may have little impact on, or even increase, ozone concentrations; the preferable strategy is to decrease VOC concentrations. At larger ratios (in excess of 20), reductions in NO_x will be far more effective than VOC reductions. For moderate ratios (between 10 and 20), there is no global "best strategy" – the control of NO_x emissions, VOCs emissions, or both, might work best, depending upon the particular circumstances (National Research Council, 1991, p. 353).

NO_x emissions not only affect the local concentrations of ozone, but also pose a regional pollution problem. Winds that carry chemicals great distances and mix atmospheric components can significantly augment the rate of ozone production.

In the United States, interstate pollution transport is generally a greater problem in the eastern portion of the country, as a result of a combination of prevailing weather patterns, topography, and a sizeable concentration of pollution sources east of the Continental Divide. In the case of NO_x and ozone, regional disputes have erupted between the South and Midwest – where many NO_x and ozone emissions originate – and the Northeast – where the impact of these emissions is often felt.

To date, the Clean Air Act's attempts to address the problem of long-distance ozone transport have been unsuccessful. The NAAQS for nitrogen dioxide and ozone are poorly suited to control interstate spillovers. They simply establish the minimum permissible levels of environmental quality at a location. They do not constrain the amount of the pollution at that location that can come from upwind sources. In fact, "a state might meet its ambient standards precisely because it exports a great deal of its pollution" (Revesz, 1996, p. 2350).

In the 1990 amendments to the Clean Air Act, Congress tried to address the problem of regional NO_x transport by imposing emissions limitations on certain sources. Unlike the case of sulfur dioxide, however, Congress did not create the possibility of trading. The resulting high costs imposed on polluting facilities by this command-and-control regime make it difficult for EPA to regulate at a level that would make a significant dent on the regional ozone problem.

As a result, there has been increasing momentum for a NO_x allowance trading regime in the Eastern United States, even though Congress has not specifically authorized such an approach. In the last few years, three independent regimes have been designed: the Ozone Transport Commission's program, EPA's optional state model program, and EPA's mandatory federal program. All three are modeled generally on the EPA's sulfur dioxide trading program. In particular, they create a single market with no restrictions on trading.

The preceding discussion of the chemistry of ozone production suggests problems with a single regional trading market. First, while NO_x emissions have a regional impact on ozone levels, they also have a substantial local impact. As a result, trades in nitrogen oxides emissions may lead to local ozone hot spots. For example, because of the higher concentration of VOCs in urban areas, trades that increase urban NO_x concentrations may lead to a net increase in the production of ozone. Second, because the best strategy for reducing ozone levels turns heavily on the ratio of VOCs to NO_x concentrations, trades between regions with different ratios can have negative effects at both the buyer's and seller's locations. For example, in rural areas reductions in NO_x leads to reductions in ozone formation, but in many urban areas, such reductions lead to increased ozone concentrations. Thus, a sale of a permit from such an urban area to a rural area has adverse effects at both locations.

The Ozone Transport Commission's Trading Program

To address the problem of regional ozone transport, the 1990 amendments to the Clean Air Act established the Ozone Transport Commission (OTC), which is composed of twelve Northeastern and mid-Atlantic states, and the District of Columbia. In September 1994, the members of OTC – with the exception of Virginia – signed a memorandum of understanding to reduce regional NO_x emissions in two stages – to 219,000 tons by 1999, and to 143,000 tons in 2003 (from 490,000 tons in 1990). The memorandum divided the OTC region into three zones defined by reference to geography and to the degree by which ambient air quality levels exceeded the NAAQS. It established different emission reduction requirements in each zone on the basis of the relative impact of emissions from different zones on regional air quality.

In 1996, OTC, in conjunction with EPA, developed a model rule that envisioned the implementation of a regional cap-and-trade NO_x emission allowance regime designed to achieve OTC's regional reduction goals. The model rule departed from the three-zone approach of OTC's memorandum of understanding by designing a unified regional market. It did not, however, formally adopt such a system, instead leaving this decision to the OTC members.

OTC's NO_x trading program is based upon tradeable emission allowances, each of which authorizes the holder to emit one ton of NOx. Each state receives a quota of emission allowances annually, set by reference to a 1990 baseline that is reduced proportionately to reflect the emission reduction goals. The states can choose how to distribute their allocations to their various sources. No source may emit more NO_x than authorized by its allowance holdings during the months of May through September – the period when ozone concentrations are most problematic.

Moreover, in an application of "regulatory tiering," a holder of an allowance may not use the allowance if to do so would result in a violation of some other federal or state limit on emissions (Tietenberg, 1991, p. 103). Participants are generally free to bank allowances for use in a future control period if they so choose. If too many allowances for a given annual control period are banked, however, a certain proportion of them become unusable.

OTC acknowledged the 1994 memorandum's division of the region into zones and noted that, "[o]ptimally, inter-zone trading would be established to encourage actual emission reductions where such reductions would do the most good for air quality" (Carlson, 1996, p. 17). It relied, however, on a simulation that demonstrated to OTC's satisfaction that there was "no discernible difference" in terms of environmental impacts between a scenario in which trading between zones was strictly prohibited and one in which trading was authorized region-wide with no trading ratios (Carlson, 1996, p. 17). OTC concluded that "institution of a trading ratio would appear to have no influence on where emissions would be reduced in the region" (Carlson, 1996, p. 17).[11] Accordingly, it decided "not to apply a trading ratio to discount allowances if they are traded across zones" (Carlson, 1996, p. 17).

EPA's Trading Programs
While EPA was helping OTC develop its NOx trading program, it also endeavored to establish a broader program, into which the OTC scheme could eventually be merged. It designed both a model program that would be implemented, on a voluntary basis, by the states in their State Implementation Plans (SIPs), and a mandatory, but more limited, federal program.

The State Model Program. The Clean Air Act requires each state to submit to EPA for its approval a SIP detailing how it intends to control sources within its jurisdiction so that the NAAQS are met within its borders. If, subsequent to such approval, EPA determines that a SIP has become inadequate "to attain or maintain" the NAAQS or "to mitigate adequately . . . interstate pollution transport", it can issue a "SIP call" requiring the state to correct the inadequacies in its SIP.[12] In October 1998, EPA issued a SIP call and promulgated a so-called NO_x SIP rule requiring upwind states to take action to ensure that the transport of ozone precursors – primarily NO_x – would not contribute significantly to nonattainment of the ozone NAAQS, or impede their maintenance, in downwind states. Various states and industry groups challenged EPA's authority to issue the SIP call and promulgate the NO_x SIP Rule, but in March 3, 2000, the D. C. Circuit upheld EPA's action. As a result of this challenge, however, EPA extended the initial date by which states must file SIPs complying with the NO_x SIP Rule to September 1, 2000.

The NO_x SIP Rule requires 22 Eastern states and the District of Columbia to reduce NOx emissions sufficiently to bring the majority of non-attainment areas into attainment with the NAAQS. In fashioning its remedial approach, EPA relied upon the work of the Ozone Transport Assessment Group (OTAG) – a working group comprised of representatives of states, EPA, and industry and environmental groups. OTAG had envisioned two possible trading programs. Under the first, states could adopt a NO_x emission permit cap-and-trade scheme. All states that elected this option would combine to form a single market in NO_x emissions.

Alternatively, states could elect to implement a trading regime without emission caps. Under such schemes, sources generally receive credits for emission reductions beyond those required by law, and are free to sell those credits to other sources. Trading would initially be intrastate, with the possibility of multistate arrangements evolving over time.

Although OTAG did not incorporate the notion of trading zones into its proposal, its final recommendation recognized the problem of hot spots and directed a workgroup to consider the possibility of trading zones:

> Subregional modeling and air quality analysis should be carefully evaluated to determine whether geographical constraints should be placed on emissions trading. Appropriate mechanisms, such as trading ratios or weights, could be developed if significant effects are expected (Gade & Kanerva, 2000, p. 91).

In the NO_x SIP Rule, EPA followed OTAG's first recommended track, announcing a model NO_x trading program that states could adopt as part of their revised SIPs. The model program is designed so as to allow for it to incorporate the preexisting OTC NO_x trading program. It applies to sources with high capacities, although certain other sources can opt into the program.

EPA's model trading program suggests a procedure states might use to allocate the allowances in their budgets, but like the OTC program leaves the decision to the individual states. Also, like its OTC counterpart, the EPA program allows for banking but places certain limits on the use of banked allowances.

In structuring its model NO_x trading program, EPA considered the possibility of imposing geographic restraints on trading. When EPA proposed the trading regime, it described a single-zone scheme, but invited comments on the issue. The agency raised the possibility of constructing subregions between which trading would either be barred, or allowed only subject to an interregional exchange ratio.

EPA received over 50 comments on the appropriateness of geographic constraints on trading. According to EPA,

> [t]he majority of commenters on this subject favored unrestricted trading within areas having a uniform level of control. Most commenters supporting unrestricted trading stated that restrictions would result in fewer cost-savings without achieving any additional environmental benefit and would increase the administrative burden of implementing the program. They expressed concern that discounts or other adjustments or restrictions would unnecessarily complicate the trading program, and therefore reduce its effectiveness.[13]

EPA decided to structure the model program as "a single jurisdiction trading program allowing all emissions to be traded on a one-for-one basis, without restrictions or limitations on trading allowances within the trading area."[14] In justifying its decision, EPA relied on a model predicting that "significant shifts in the location of emissions reductions would not occur with unrestricted trading compared to where the reductions would occur under command-and-control and intrastate only trading scenarios."[15]

The Federal Program. EPA also developed a mandatory federal NO_x trading program in response to petitions filed by eight states, under Section 126 of the Clean Air Act, complaining about an excessive upwind contribution to their failure to meet the NAAQS for ozone. EPA upheld portions of six of the eight petitions in May 1999. In a final rule issued in January 2000, EPA required that certain sources participate in a federal NO_x trading program to address the problems asserted in the petitions filed by the downwind states. This program formally applies to a limited number of large NO_x emitters in upwind states. In addition, certain other sources in those states may opt into the program.

The program generally follows the structure of the SIP Model NO_x trading program. Sources covered by the federal NO_x trading program that are located in states that choose to implement the model program can buy and sell allowances on that market. In authorizing this integration of the two programs, EPA relied on computer simulations showing that trading under the integrated programs "will not significantly change the location of emissions reductions."[16]

The Los Angeles Metropolitan Area's Local Smog Precursor Trading Program

The South Coast Air Quality Management District (SCAQMD) administers an urban smog trading program in the Los Angeles metropolitan area. The program is a part of the California SIP, and relies on a market in emission permits of two smog precursors: sulfur oxides and nitrogen oxides.

SCAQMD's trading regime consists of three distinct programs. The principal program, the Regional Clean Air Incentives Market (RECLAIM), is designed to achieve a regional 75% reduction in nitrogen oxide emissions, and a 60% reduction in sulfur oxide emissions, by 2003. All facilities within the SCAQMD's jurisdiction that emit at least four tons of either pollutant annually are subject to RECLAIM; other facilities can choose to participate in the program.

Each permit entitles the holder to emit one pound of a specific pollutant. Participating facilities receive annual emission permits in proportion to prior pollution levels. In order to meet the air quality improvement mandated by federal law, SCAQMD decreases the total number of RECLAIM permits, as well as the total number of permits allocated to each facility, by approximately five to eight percent each year.

The RECLAIM program establishes two trading zones – a coastal zone and an inland zone. Restrictions on trading between these zones apply only to new sources, sources that relocate, and sources that seek to increase emissions above their initial allocation of permits. Such sources may purchase permits only from sources in their zone. In contrast, inland facilities may purchase permits from either zone. These trading restrictions were predicated on the use of computer models: "In view of modeling results of the prevailing winds, trading was permitted from coastal sources to inland but not the other way around" (Kosobud, 2000, p. 24).

The second component of SCAQMD's trading regime is a mobile source emissions credit program. It is designed to address the complaint that the RECLAIM program, standing alone, did not incorporate automobile emissions, one of the largest sources of air pollution in the area. Under this program, which is voluntary, SCAQMD awards allowances to "licensed car scrappers" who purchase and destroy old cars based upon the predicted emissions of the cars had they not been destroyed (Chinn, 1999, p. 92; Drury, 1999, pp. 246–247). Mobile source emissions credits awarded for sulfur and nitrogen oxide reductions can be bought and sold on the RECLAIM market.

SCAQMD's third trading regime is a source credit program. This program, which is also voluntary, brings small pollution sources such as household and small business furnaces, which otherwise would be unregulated, into the trading fold. Owners of these small sources can earn area source credits by upgrading older, high emission equipment. Area source credits can be traded on the RECLAIM market.

Data from the first three years of the RECLAIM program reveal a "robust" trading market (Lents, 2000, p. 232). Through December 31, 1997, there had been 1200 trades, with a total value of $42 million. These trades represent the transfer of 244,000 tons of nitrogen and sulfur oxides.

Environmental justice groups have assailed SCAQMD's emission trading regime. These groups allege, consistent with the general environmental justice criticism of trading, that areas immediately proximate to pollution sources have not seen improvement, or have experienced deterioration, in air quality. According to these groups, the adversely affected areas tend to be economically disadvantaged and contain relatively higher percentages of ethnic and racial minorities.

Of particular concern is the fact that mobile source credits can be sold on the RECLAIM market. Environmental justice advocates find this arrangement problematic because automobile emissions are dispersed, so that their effects are distributed fairly homogeneously across the regulated area. In contrast, a small number of industrial polluters, primarily four oil companies, have purchased these allowances. These polluters are disproportionately located in minority areas. Indeed, one commentator notes that of the four companies that have purchased most of the emission credits, "three are located close together" in two communities that are heavily populated by Latinos (Drury, 1999, pp. 252–253).

In 1997, one local environmental group, Communities for a Better Environment (CBE) filed an administrative complaint with the EPA against SCAQMD and the California Air Resources Board, under Title VI of the Civil Rights Act of 1964. It claimed that the mobile source program exacerbates hot spots in Latino communities. Although there have been calls to halt the trading during the pendency of the litigation, the SCAQMD's various programs remain in place.

III. ADDRESSING THE PROBLEM OF AMBIENT STANDARD VIOLATIONS AND THE FORMATION OF HOT SPOTS

As we have explained, for local and regional pollutants, the location and extent of the damage caused by a unit of emissions vary according to the location of the emissions. Markets in emissions permits, such as those implemented in the regulatory programs discussed in Part II, however, treat emissions at different locations as interchangeable. As a result, such markets cannot ensure against the violation of ambient standards or the formation of hot spots.

In this part, we analyze the three leading proposals for constructing marketable permits schemes that avoid or at least mitigate this problem: subdividing the emissions market into zones and prohibiting interzonal trading; designing markets in units of environmental degradation, rather than units of emissions; establishing emissions offset markets: markets in which the trading in units of emissions does not take place on a one-to-one ratio.

For expositional ease, our primary focus in the remainder of the chapter is on the violation of ambient standards rather than on the formation of hot spots. The analysis we present, however, is equally relevant to both problems.

Emissions with Multiple Trading Zones

One way to reduce the potential negative effects of emissions trading is to divide a region into separate zones, create a separate permit market in each zone, and prohibit interzonal trades. The market price for permits in each zone would be determined by the supply and demand for permits within that zone. Market prices would therefore differ across zones.

In Part II, we discussed proposals, that ultimately were not implemented, to divide the nation into two trading zones in the national sulfur dioxide trading regime, and the possible division of the OTC NO_x trading program into three trading zones. The RECLAIM program in Los Angeles, in contrast, does feature two zones, with trading prohibited from the inland zone to the coastal zone.

The division of a regulated region into distinct trading zones may ameliorate the threat of ambient standard violations for two reasons. First, by precluding trades across zonal boundaries, a zoned market limits the number of permits that any source can acquire. Second, emissions from different sources in a smaller zone are likely to have a more similar impact – in terms both of location and magnitude – than emissions from different sources in a larger zone. Especially if the policy maker takes topography and the prevailing wind patterns into account in constructing zonal boundaries, emissions from within a trading zone may have substantially equivalent adverse effects. Under such circumstances, trading within the zone might not generate ambient standard violations.

A zoned emission permit regime, however, does not eliminate the possibility of ambient standard violations. No matter how much attention the policy maker devotes to constructing zonal boundaries in light of topography and wind patterns, emissions of local and regional pollutants from different locations are not equivalent. Rather, they remain spatially differentiated and will have somewhat different impacts – in terms of location and magnitude. As Alan Krupnick, Wallace Oates, and Eric Van De Verg explain:

> [T]he assumption under [emissions permit trading] that a unit of emissions from one source in a zone is precisely equivalent in its effects on air quality to a unit of emissions from any other source in the zone *may*, under certain circumstances, do serious violence to reality. The ambient effects of emissions do not depend solely on the geographical location of the source; they depend in important ways on such things as stack height and diameter, and on gas temperature and exit velocity (Krupnick et al., 1983, p. 244).

A further complication is raised by the fact that the ambient air quality levels in a zone depend not only on emissions from sources in that zone but also on transported pollution from upwind zones. Thus, for example, if an upwind source with a relatively short stack sells a permit to a source with a taller stack, the ambient air quality levels in the downwind zone will worsen.

Moreover, even if zoned regimes performed relatively well in terms of avoiding violations of ambient standards, one needs to worry about other complications. First, they reduce the size of each trading market (and increase the number of markets). Because a smaller market has fewer participants, market power is more likely to be concentrated in the hands of a relative few. These participants with market power – whether buyers or sellers – would have the incentive to engage in anticompetitive practices. In particular, sellers with power would be able effectively to set the price for allowances above the efficient market price, while buyers with power would do the opposite.

A non-global pollutant, by definition, causes different damage – in terms both of magnitude and location – depending on the location from which it is emitted. The factors giving rise to these spatially differentiated effects, such as topography, temperature, and wind patterns, can vary significantly across relatively small distances. As a result, to achieve a viable market that respects ambient standards, one of two conditions must hold: the regulated pollutant must have the characteristic that only relatively large shifts in its emission locations affect the spatial distribution of the harm, or there must be a sufficiently large concentration of potential market participants. Under the former condition, the trading zone can be relatively large; under the latter, the zone's small geographic size does not stand in the way of a robust market.

If neither of these conditions is met, the policy maker must make a tradeoff. It can assign sufficiently numerous sources to each zone, but heighten both the probability and magnitude of ambient standard violations. Alternatively, it can mitigate the probability and magnitude of ambient standard violations by assigning fewer sources to each zone, but heighten the probability of thin markets and of the resulting deleterious effects.

Another negative consequence of the division of a region into trading zones is that the lowest-cost reduction of pollution to the desired level (one of the primary promises of tradeable permit schemes) is unlikely to be realized. Indeed, dividing the regulated region into zones among which trading is barred can result in the preclusion of some cost-saving trades that would not lead to ambient standard violations.

More generally, a zoned scheme will not achieve the most cost-effective pollution reduction unless the policy maker allocates the "correct" number of permits to each of the various zones (Atkinson & Tietenberg, 1982, p. 104). That will not happen unless – as seems unlikely – the policy maker is privy to pollution-reduction cost information for polluters in each zone. Because trading across zonal boundaries is prohibited, subsequent trading cannot ameliorate the initial misallocation of permits. The policy maker could adjust the allocation of permits across zones over time as it gains more information about control costs, but the administrative costs of such an approach are likely to be substantial.

To address the shortcomings of zoned schemes, some commentators have suggested allowing trading across zonal boundaries under an exchange ratio that converts permits traded in one market into permits traded in another. Consider, for example, a situation in which permits traded within Zone I cause twice as much damage at a given location as permits traded within Zone II. A buyer in Zone I could then purchase permits for two units of emissions from a seller in Zone II for every unit of emissions that it wishes to discharge. But this relatively straightforward exchange works only because, in the example, the location of the damage is independent of the location of the emissions. If, instead, emissions in Zone I affected ambient air quality in both Zones I and II, but emissions in Zone II affected ambient air quality levels only in Zone II, trading at a fixed exchange rate would not guarantee the attainment of ambient standards.

Markets in Units of Environmental Degradation

To address the problem of ambient standard violations caused by trading, commentators have proposed schemes in which permits are denominated in units of environmental degradation, rather than in units of emissions. Under this approach – often referred to in the literature as an "ambient permit system" – the policy maker determines the ambient quality that it deems acceptable at numerous receptor points distributed over the region. For each receptor point, the policy maker issues permits of environmental degradation, so that the sum of the environmental degradation authorized by all the permits issued at a given receptor point is set by reference to the ambient standard at this point.

Each receptor point defines a separate market. A firm wanting to increase its emissions would have to determine, through the use of computer modeling, which receptor points are affected by its emissions and the extent of the environmental degradation at each such point. Then, it would need to obtain sufficient permits at each relevant market. As long as the policy maker chooses representative receptor points, the computer modeling accurately predicts the ambient impacts of emissions at different locations, and robust markets for the various environmental degradation permits develop, the trading will not interfere with the attainment of the ambient standards throughout the region.

Markets in units of environmental degradation have certain undesirable features, which in some cases may be sufficiently severe to undermine the vitality of the markets and compromise the benefits of trading. First, such schemes require the establishment and maintenance of permit markets at each of the receptor points. Thus, the system is likely to impose substantial market supervision costs, as well as costs on prospective market participants who must enter multiple markets.

Second, markets in units of environmental degradation will be thinner than single-zone emission markets. Indeed, polluters participate only in the markets at which their emissions cause damage; thus, not every polluter participates in every market. At some receptor points, the number of prospective market participants might not be sufficient to support an efficient market.

Third, these problems are exacerbated by the segmentation of the property rights. As we noted above, a firm that seeks to increase its emissions must obtain additional permits at each receptor point at which its emissions cause damage. If the firm is able to purchase permits only in some markets but not others, the purchased permits would be useless. Only a full portfolio of permits gives the firm the right to increase its emissions.

This emphasis on portfolios is likely to affect a firm's bidding strategy. For example, if a firm knew that obtaining permits in one market would be especially difficult, it might wait until it obtained those permits before attempting to purchase permits in other markets.

Prospective sellers face similar problems. While brokers presumably would work to match willing buyers to willing sellers, it is unlikely that the seller would find a buyer that needed the precise portfolio of permits that the seller was offering. Only in relatively rare instances would the seller be able to dispose of its permits in one transaction. In the more likely scenario, it would have to engage in multiple transactions, with the consequent rise in transaction costs.

The interlinking of the different markets, which results from the need of both buyers and sellers to trade portfolios of permits, gives rise to the potential that thin markets at some receptors will cause a domino effect. If the market for a permit to degrade the environment at one receptor point has an insufficient number of prospective market participants, markets at other points might also become thin if too many prospective participants rely on purchasing or selling permits on the first market to make transacting on the other markets worth their while.

Pollution Offset Markets

Given the problems of zoned emission markets and of markets in units of environmental degradation, environmental economists have searched for alternative ways to construct pollution markets. Alan Krupnick, Wallace Oates and Eric Van De Verg (1983, p. 233) have proposed a pollution offset market, which entails a single market in emission permits but in which trades are not effected on a one-to-one basis. Rather, "the parties exchange emission permits at ratios depending on the relative effects of the associated emissions on ambient air quality at receptors with potential to violate the standard . . . ." (McGartland, 1988, p. 37). Under a pollution offset market regime, a buyer need only purchase emission permits if the buyer's emissions would otherwise cause a violation of an ambient standard at a receptor point. A buyer requiring permits in order to increase its emissions, can purchase them from a particular seller only if the emissions of both the buyer and seller adversely impact ambient air quality at a common receptor point. The exchange rate is determined by reference to the relative rates by which emissions of the regulated pollutant by the buyer and seller contribute to ambient concentrations at this receptor point: it equals the ratio of the seller's contribution to that of the buyer.

A numerical example is illustrative. Let us say that an ambient standard of 10 $\mu g/m^3$ of pollutant P governs at receptor point ρ, and that emissions from firms A and B contribute to levels of P at ρ. In particular, let us say that the level of P at ρ increases by 3 $\mu g/m^3$ for every ton of P emitted annually by A, while the level of P increases by 1 $\mu g/m^3$ for every ton of P emitted annually by B. At present, A has 2 permits and B has 4; each permit entitles the holder to emit 1 ton of P annually. Thus, the ambient level of P at ρ precisely equals the ambient standard:

$$(2 \text{ tons}) \times (3 \ \mu g/m^3 \text{ per ton}) + (4 \text{ tons}) \times (1 \ \mu g/m^3 \text{ per ton}) = 10 \ \mu g/m^3.$$

Now say that A decides to purchase 1 permit from B. If the trade were conducted on a one-to-one basis, then A and B could each emit 3 tons of P per year, with the resulting annual average ambient level of P at ρ:

(3 tons) $\times$ (3 μg/m^3 per ton) + (3 tons) $\times$ (1 μg/m^3 per ton) = 12 μg/m^3.

This concentration exceeds the ambient limit of 10 μg/m^3. Instead, the trade must be accomplished using an exchange rate based on A's and B's relative contribution to the level of P at ρ. Here, the exchange rate is 1/3. Thus, the seller B reduces its emissions by 1 ton (since it is selling 1 permit), but the amount by which A is permitted to increase its emissions is restricted by the exchange rate:

$$(1/3) \times (1 \text{ ton/permit}) \times (1 \text{ permit}) = 1/3 \text{ ton.}$$

Thus, the additional permit obtained by A enables A to emit a total of 2-1/3 tons of P per year; B is permitted to emit 3 tons annually. The 10 μg/m^3 ambient level of P at ρ is then not violated:

$$(2\text{-}1/3 \text{ tons}) \times (3 \text{ } \mu\text{g/m}^3 \text{ per ton}) + (3 \text{ tons}) \times (1 \text{ } \mu\text{g/m}^3 \text{ per ton}) = 10 \text{ } \mu\text{g/m}^3.$$

The calculation of ratios is more complicated where a buyer and seller share more than one common receptor point, because the trade cannot result in an ambient standard violation at *any* receptor point. Thus, the trading ratio is determined by reference to the receptor at which the ratio of the buyer's to the seller's impact on ambient quality is lowest.

Offset markets have another layer of complexity: the treatment of "slack." Slack occurs when the pollution level at a receptor is below the ambient standard, so that this standard is not constraining. Under the logic of a pollution offset regime, a source can simply increase its emissions without purchasing any permit if these emissions affect only receptors at which the ambient standard is not constraining. This feature leads to a "first-come, first-served" allocation of the slack in which, by increasing its pollution, a source can claim a valuable economic right.

An offset market system is fraught with complexity, both for the market supervisor and for market participants. This complexity arises in large measure because the commodity subject to trade, while nominally an emissions permit, in fact conveys different rights to different holders at different times. The government must therefore maintain a record of the rights accompanying each permit. It also imposes higher transaction costs on prospective market participants. Also, the presence of slack gives rise to additional difficulties, again leading to higher administration and transaction costs.

IV. A PROPOSAL FOR A CONSTRAINED EMISSIONS PERMIT REGIME

In this part, we present our proposal for a constrained emission permit regime that respects ambient standards. Our regime relies upon the pre-clearance of transactions by a website containing information on the impacts of emissions on ambient air quality levels throughout a region. Section A sets forth the elements of our proposal. Section B explains why it is superior to the alternative means of structuring markets to avoid or mitigate the violation of ambient standards. Section C analyzes the impact of our approach on the structure of the emissions markets. Finally, Section D illustrates the operation of our proposal by means of a simulation of the air pollution dispersion model recommended by EPA.

Structure of the Proposal

The constrained trading regime that we propose consists of a single market in units of emissions, but in which a proposed trade is rejected if it leads to the violation of an ambient standard at any receptor point. The determination of whether a proposed trade is approved would be performed through computer modeling, using an atmospheric dispersion model that predicts the impact of emissions from each source in the region on ambient air quality levels at the various receptor points. The model calculates the impacts on ambient air quality levels of the increase in emissions by the prospective purchaser and the decrease by the prospective seller, and determines whether these changes cause a violation of an ambient standard.

Our proposed trading scheme consists of six steps, which are illustrated in Fig. 1. At Step 1, EPA, as the market supervisor, uses the computer model to verify that the initial allocation of permits did not result in the violation of any applicable ambient standards at any receptor point within the regulated region. This distribution of permits constitutes the 'base case' scenario. Thereafter, at Step 2, the website is ready to consider proposals for trades. We assume, as is currently the case in the sulfur dioxide emissions market, that brokers match prospective buyers with prospective sellers.

Under a simple emissions trading regime, once the buyer and seller agree on a price and quantity, the transaction is finalized, and the exchange of permits is registered with the governmentor other market supervisor. In contrast, under our proposal, the trade is approved and registered only if the computer model predicts no violations of the applicable ambient standards at any receptor point.

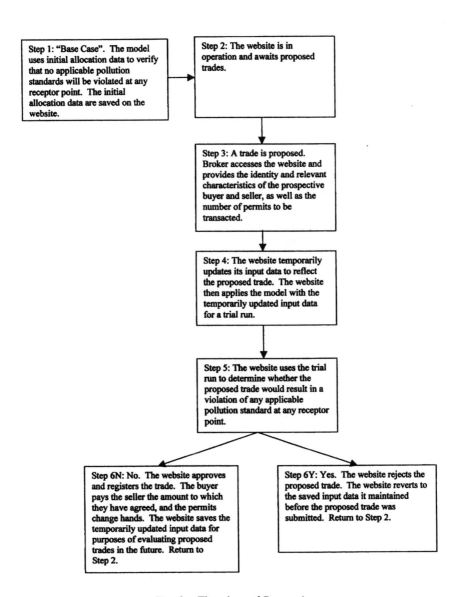

Fig. 1. Flowchart of Proposal.

The market supervisor facilitates this process by granting brokers (or the market participants themselves) access to its computer model over the internet. This verification takes place in Step 3 of Fig. 1. In particular, the website allows the broker to enter the sources' identities, the quantity of permits that would change hands, and the characteristics of the sources that have an impact on ambient air quality levels at any points (such as location, stack height, stack diameter, and speed and temperature of emissions).

The website program then prepares a trial run. As Step 4 in Fig. 1 indicates, for purposes of the trial run, the website modifies temporarily the saved input data by increasing the buyer's emissions, and decreasing the seller's emissions, to reflect the proposed trade.

In Step 5, the website uses the computer modeling program to predict how pollution will disperse, given the temporarily modified input data. It thereby determines whether the proposed trade will result in the violation of an ambient standard at any receptor point.

If the model indicates that the proposed trade will not result in a violation of any applicable pollution standard at any receptor point, then – as Step 6N indicates – the trade is approved and registered. The buyer pays the seller for the permits, and the permits change hands. Following approval of a trade, the website automatically updates the input data to incorporate the buyer's and seller's new emissions profiles. Thus, at the time that other prospective buyers and sellers approach the website with proposals for trades, the website's input data will reflect all the previously approved trades.

In contrast, if the computer simulation reveals that the proposed trade will result in a violation of an ambient standard at any receptor point, then – as Step 6Y indicates – the trade is not allowed to go forward. In the case of disapproved trades, the website does not update its input data. After the website finishes with Step 6, either by approving the trade in Step 6N or rejecting it in Step 6Y, the website returns to Step 2. There, it awaits the next proposal for a trade.

Comparison of the Alternative Trading Regimes

We turn now to a comparison of our proposal for a constrained single-zone emission regime with other pollution trading regimes. We perform this analysis by reference to four criteria: the likelihood that the trading will result in the attainment of the applicable ambient standards; the thickness of the markets; the transaction costs of market supervision and participation; and the treatment of slack. Table 1 presents our conclusions with respect to each of the criteria.

Table 1. Characteristics of Tradeable Pollution Permit Regime Design Structures.

	Traditional Single-Zone Markets	Multiple Zone Markets	Environmental Degradation Markets	Offset Markets	Ambient-Restricted Single-Zone Markets
Likely Attainability of Ambient Standards	Poor	Fair	Very Good	Very Good	Very Good
Thickness of Markets	Very Good	Poor	Fair	Good	Good
Ease of Implementation and Participation	Very Good	Fair	Poor	Poor	Good
Treatment of Slack	Very Good	Very Good	Fair	Poor	Very Good

Attainability of the Ambient Standards

Only three of the five trading regimes that we analyze – environmental degra-dation markets, offset market, and our proposal for constrained single-zone markets – provide for the attainment of ambient standards. As we have explained, of the remaining two regimes, multiple-zone markets perform somewhat better from this perspective than single-zone markets, but fall short of meeting the goal.

Even single- and multiple-zone markets can lead to the attainment of ambient standards if they are coupled with a second regulatory tier. This tier would impose use restrictions on certain permits in order to attain ambient standards. We believe that our approach is preferable, in that it provides a tangible means by which a prospective purchaser can ascertain, in advance of the purchase, whether the permits that it seeks would in fact allow it to increase its emission. In contrast, under a system of use restrictions, there is not a clear way of determining when an ambient violation takes place. Presumably such a violation would come to the attention of a regulator only after the violation had already occurred. Moreover, to the extent that more than one source contributes to the degradation of ambient air quality at the point of the violation, it is not clear how the use restrictions would be apportioned among the various contributing sources. A determination of this type would need to be made through some kind of administrative procedure, adding further delay to the resolution of the problem.

In contrast, our approach is quite specific as to what trades can and cannot go forward. Thus, it lowers the transaction costs and provides better assurances that the ambient standards will, in fact, be met.

Thickness of Markets

Thin markets can interfere with the goal of reducing pollution in a cost-effective manner. They can produce situations in which either buyers or sellers have market power and in which, as a result, trades do not take place at competitive prices. Multiple-zone emission markets fare most poorly in this regard because they permit trading only between buyers and sellers in the same zone, rather than among all the sources in the region.

Environmental degradation markets may fare somewhat less poorly. The market at each receptor point is open only to participants who affect ambient air quality level at that point. Unlike the zoned-market scheme, eligibility for trading on a market is determined not by location of the source, but by the location of the receptor points at which the source's emissions cause damage. Sources, however, typically trade in multiple markets, potentially ameliorating the problem of thin markets.

Pollution offset markets also permit only less-than-universal trading. Like environmental degradation markets, they preclude trades that would produce ambient violations at a particular point. Unlike environmental degradation markets, however, they allow trades between parties that do not affect ambient levels at the same point. Thus, the resulting markets are somewhat thicker.

From the perspective of market thickness, our proposal for constrained single-zone markets has comparable properties to pollution offset markets. It leads to somewhat thicker markets however, because any increase in emissions requires resort to the market, whereas under pollution offset markets, increases that do not produce any ambient violations can be appropriated as slack. Only a single-zone market with no trading constraints produces thicker markets.

Transaction Costs

Transaction costs, whether borne by the government as market organizers or by the trading parties, impede the effectiveness of market systems, both by consuming resources that could otherwise be used in more productive endeavors and by preventing trades that otherwise would be socially beneficial. Single-zone markets give rise to the lowest transaction costs because they require the establishment of, and participation in, a single market, and impose no trading constraints. The government needs only to keep track of who holds which permits at what times.

Environmental degradation market systems and offset markets have comparatively high transaction costs. The former requires the government to establish a market at each receptor point, and requires each source to purchase permits at each receptor point at which it has an adverse impact on ambient concentrations. The latter require the government and market participants to keep track of what rights are associated with each permit at all times. Moreover, the proposals for such trading schemes do not contain a low-cost mechanism for determining the appropriate trading ratios.

A multiple-zone market scheme does only slightly better. It too involves the establishment of multiple markets. It also imposes on the government the burden of determining how to allocate permits among the various zones. On the other hand, each source needs to buy permits in only one market.

The constrained trading regime that we propose adds some complexity to the standard single-zone market. We believe, however, that the website containing the atmospheric dispersion model provides a fast and straightforward way to determine whether ambient standards would be violated, thereby imposing relatively low transaction costs on the government and market participants.

Slack

Market-based schemes that contain slack give rise to conceptual and administrative problems. If slack is offered on a "first-come, first-served" basis, it results in inefficient expenditures by sources to capture the slack as well as in inefficient allocations of the slack among the sources. The government must also keep track of which sources have obtained the right to utilize slack. If the government allocates the slack in some other way, for example, by periodic auctions, it faces complicated design problems in determining at what rate and in what manner to make the slack available to the market.

As we discuss above, slack is endemic to pollution offset schemes and has received the attention of academic commentators. Slack is also a problem for environmental degradation markets, though the literature has not addressed the issue. Environmental degradation markets exhibit slack if the government's goal is to meet an ambient standard where some areas had ambient quality levels that are better than this standard. Then, a source could increase its emissions without purchasing any permits if this increase did not lead to the violation of an ambient standard at any receptor point.

In contrast, single-zone markets, multiple-zone markets, and the constrained single-zone markets that we propose do not have slack problems. Any source that wishes to increase its emissions needs to buy permits from another source.

In sum, our proposal performs well across the relevant criteria. It dominates multiple-zone markets, environmental degradation markets, and offset markets, doing better along some dimensions but no worse along any dimension. Our proposal does not perform as well as single-zone markets with respect to two of the criteria: thickness of markets and transaction costs. If one attaches sufficient importance to the attainment of ambient standards, however, our proposal is preferable.

Likely Impacts of the Proposal

This section analyzes some likely features of the market for permits under our proposal to require website approval as a precondition to a valid trade. We focus on the impact of our proposal on the order of trades, the ability of particular prospective buyers and sellers in fact to engage in trades, and the price of permits.

Importance of the Order of Trades

The order in which trades occur may have an impact on what trades will be permitted thereafter. Recall that the website modifies its saved input data when – and only when – a trade is approved and registered. Although the website would modify temporarily the saved data to reflect a proposed trade for purposes of a trial run, that modification would become permanent only if the trade were ultimately approved; if not, the saved input data would remain unchanged.

This structure has ramifications for prospective market participants. For example, it is possible that multiple pairs of buyers and sellers may contemplate trades at substantially the same time. Even if each of the trades would be accepted if it were the first to be presented to the website for approval, some trades might not be acceptable if presented later, after other trades have been registered. Conversely, a prior trade may render viable a subsequent trade that otherwise would have been impermissible.

The following example illustrates this feature. Let us say that A is contemplating purchasing 10 permits from B, and at the same time that C is contemplating purchasing 100 permits from D. Assume that emissions from A and C contribute substantially to high pollution levels – though not in excess of any applicable standards – at a receptor point D. In the case of a local pollutant, this effect could be attributable to the proximity of the sources, whereas in the case of a regional pollutant it might be caused by their relative stack heights and the prevailing wind patterns at and around their respective locations. Assume further that, given the existing distribution of permits, the website would approve either trade – separate trial runs of the computer model would reveal no violation of any applicable ambient standard.

Let us say that A and B carry out their trade before C and D do so. The shift of 10 permits from A to B is approved, and the website modifies – now permanently – its saved input data to reflect the transaction.

C and D now seek approval for their trade. Assume that the additional 10 permits purchased by A have increased the ambient concentration at receptor point D enough that the acquisition of another 100 permits by C would produce a violation of the ambient standard. In that case, the website would disapprove the proposed transaction between C and D, even though it would have approved this trade before A's transaction with B. Similarly, had C and D carried out their transaction first, A's transaction with B would have been precluded.

Ability of Particular Buyers and Sellers to Engage in Trades
The constrained emissions trading regime that we propose potentially precludes certain sellers from selling permits to certain buyers. We discuss in this section a common pattern of constraints on the permissibility of trades.

We consider a situation in which emissions from a group of sources cause harm at the same location. This location has the worst ambient air quality levels of the region throughout which trading is allowed and the ambient standard constrains further degradation. As already indicated, in the case of a local pollutant this location would be in the vicinity of the sources, whereas in the case of a regional pollutant it could be hundreds of miles away.

Our proposed trading regime generally would allow sources in this group to trade permits with one another. In such cases, the increase in the purchaser's emissions would be counteracted by a decrease in the seller's emissions. If the emissions of the two parties have similar impacts on ambient air quality at the point at which the standard is constraining, the transaction should not lead to the violation of these standards.

Similarly, our trading regime would generally allow sales from the concentrated group of polluters to sources elsewhere in the region. Such transactions would ease the pressure on ambient air quality levels at the location with the highest concentrations. A comparable increase in ambient concentrations at locations with better air quality would not be problematic.

In contrast, our trading regime would not allow sources in the concentrated group to purchase permits from outside the group. Increased net emissions from within the group would lead to the violation of the ambient standards. Thus, sources in this group can generally trade permits with one another, can sell permits to sources outside the group, but cannot purchase permits from sources outside the group.

This discussion highlights an important difference between our constrained single-zone trading regime and a multiple-zone regime. Under the latter, a group of sources located close together would presumably be placed in the same zone in which trading among the sources was authorized, and trading with sources outside the zone would be prohibited. Our scheme preserves the ability of sources within the group to trade with one another. But unlike the case of multiple-zone markets, it also allows sources in this group to sell to sources at other locations, thus reducing the barriers on permissible transactions.

Price of Permits

The preceding discussion shows how our proposal would inhibit the ability of certain polluters to sell permits to certain buyers. This feature implies that, at any given time, permits in the market might sell at different prices.

The price of permits, like prices of all goods, is determined by their supply and demand. If a potential purchaser affects ambient air quality levels at a receptor at which the ambient standard is constraining, the universe of potential sellers will be reduced. In particular, the potential purchaser will not be able to buy permits from sources that do not have an impact at the same location, or from sources for which the impact of a unit of emissions at the constraining location is smaller.

This reduced supply potentially leads to an increase in the price of the permits that the prospective purchaser can acquire. In particular, sellers that might have been able to supply permits at relatively low prices would not be permitted to participate in certain transactions.

In contrast, a potential purchaser that affects ambient levels in areas in which the quality is better than the ambient standard would not face similar constraints. Its market for permits would not be similarly truncated. The larger supply of potential sellers would reduce the market price.

One feature of our regime is that it would provide disincentives for new sources to locate in areas that contribute to ambient air quality levels at receptors in which the ambient standards are constraining – or, by extension, where these standards are close to being constraining. The higher market price for permits at such locations should encourage new sources to locate elsewhere. We regard this feature of our regime to be highly desirable.

V. REFINEMENTS OF THE PROPOSAL

In this part, we analyze several possible refinements to our proposed trading regime. The first two relate to the choice of and use of the air pollution dispersion model. Then, we consider the purchase of allowances by new sources and the possibility of having different ambient standards at different locations. The remaining refinements concern ways to incorporate into our proposal certain practices of existing trading regimes, such as auctions of allowances, retirement of allowances, purchases of allowances for investment purposes, and banking of allowances. All of these extensions could easily be incorporated into our simulation.

The Choice and Use of the Air Pollution Dispersion Model

Choice of Model
One objection that might be leveled against our proposal concerns the reliability of the air pollution dispersion model used to determine violations with the ambient standards. Such models are useful devices to predict what should happen to air pollutants based upon physical formulae and prior empirical data. It is possible, however, that a model does not accurately capture all the intricacies of weather patterns, topography, and chemistry that determine air pollution dispersion. Further, even if a model is otherwise designed

accurately to predict air dispersion, the model's predictions will be only as good
as the empirical data that are used as inputs.

Our proposal, however, does not rely on models and input data any more
than existing EPA programs and regulations do. For example, the Clean Air
Act requires states to develop SIPs indicating how they will control their existing
sources in order to meet the NAAQS. EPA mandates that states use
atmospheric modeling to demonstrate that their SIPs will lead to the attainment
of the NAAQS. The Act also requires that SIPs include a means of evaluating
modifications of existing sources and constructions of new sources to ensure
that such changes do not interfere with the attainment of the ambient standards.
Atmospheric models must be used to make these determinations. EPA
maintains a list of approved atmospheric dispersion models. For our simulation, we used one of the models that EPA recommends, the ISC3 Model, which
is the "work-horse" for EPA's compliance determinations. Thus, to the extent
that the ISC3 Model might lead to imprecise predictions, its use under our
trading scheme would not lead to more inaccurate results than those generated
under the current regime.

We recognize, however, that, as technology and science's understanding of
the physics of air and weather patterns and the chemistry of air pollutants
advance, more accurate atmospheric models will emerge. For example, two
newer models were presented at a recent EPA conference on air quality
modeling – AERMOD and CALPUFF. AERMOD, another Gaussian dispersion
model, is more advanced than ISC3. Accordingly, EPA has proposed that
AERMOD replace the ISC3 Model "for many air quality impact assessments."[17]
Moreover, EPA is proposing to recommend the use of CALPUFF "for refined
use in modeling long-range transport and dispersion to characterize reasonably
attributable impacts from one or a few sources for [prevention of significant
deterioration] Class I impacts."[18]

Another atmospheric dispersion model that might one day be incorporated
into our trading proposal is the Regional Air Pollution Information and
Simulation (RAINS) model. RAINS was developed by the Austrian-based
International Institute for Applied Systems Analysis to evaluate problems of
transboundary transport of air pollutants in Europe, and has been relied upon
in treaty negotiations (Institute for Applied Systems, 2000). In particular, RAINS
has been used to match areas that contribute sulfur emissions with the areas
and ecosystems that are harmed by those emissions, and to determine what
emission reductions would be required from what regions to protect adequately
the downwind areas and ecosystems.

The use of the ISC3 Model is not essential to our proposal. The website
could be programmed to rely upon any atmospheric dispersion model to make

predictions as to ambient pollutant concentrations at the receptor points. Thus, there is nothing in our proposal that would preclude the integration of AERMOD, CALPUFF, or RAINS (or of any more advanced, and more accurate, atmospheric dispersion model) in place of the ISC3 Model when EPA determines that other models are indeed preferable.

Refining the Use of the Model

In our simulation, we placed receptors homogeneously throughout the regulated region. Because our regime bars a trade only if it would result in the violation of an ambient standard at a receptor point, a trade might result in a violation at a location between receptor points but nonetheless be approved. While current models do not allow for the calculation of expected pollutant concentrations at all points within a region, which would involve the inclusion of an infinite number of receptor points, three modifications to our trading regime would ameliorate this potential problem.

One possibility is simply to place more receptors. For example, the ISC3 Model, as run on a DOS system, allows for the placement of up to 500 receptors and up to five receptor networks. Moreover, those numbers can be increased, depending upon the available memory capacity. Additional receptor points, however, may slow down the computations. Thus, there may be a tradeoff between more accurate predictions and predictions that are sufficiently fast to satisfy the market participants.

Second, receptors should be placed where violations of ambient standards are most likely to occur. For example, if measurements of air quality indicate that pollutant concentrations are especially high in a given area, the computer program should include additional receptors in that area. Similarly, additional receptors should be included on the windward side of a mountain range because substantial amounts of pollution may be trapped on the side of the range.

A third course of action would be to program the website automatically to include additional receptors in the immediate vicinity of an existing receptor if the model predicts that the ambient pollutant concentration at this receptor comes within a certain range of the applicable ambient standard. For example, if a standard run of the model for sulfur dioxide concentrations reveals that the predicted annual average ambient concentration at a given receptor will be 79 $\mu g/m^3$ – within 1 $\mu g/m^3$ of the applicable 80 $\mu g/m^3$ NAAQS – the website program could add, say, 20 additional receptors, at a relatively small radius around the existing receptor.

New Sources

The simulation in Part IV dealt only with transfers of allowances among existing sources. The treatment of purchases of allowances by new sources would proceed no differently. The website would obtain the location of the new source and temporarily update its emission data to consider whether the proposed emissions increase at that location, combined with the decrease of emissions at the seller's location, would contribute to the violation of an ambient standard. If the proposed transfer would produce a violation, the website would reject it. If not, the website would approve the transfer and save permanently the new emissions data.

Variable Ambient Standards

The simulation in Part IV also assumed, for expositional convenience, that uniform ambient standards would apply throughout the region. Despite the uniformity of the NAAQS, the Clean Air Act in fact imposes disuniform ambient standards throughout the country. In areas that have ambient air quality levels that are better than the NAAQS, the Prevention of Significant Deterioration (PSD) program imposes a more stringent ambient standard, defined by reference to a baseline plus an increment. In contrast, areas that have failed to comply with the NAAQS – the non-attainment areas – are subject to a less stringent requirement. Until some time in the future, such areas must make "reasonable further progress" toward attainment, rather than actually achieve attainment.[19]

The website could easily be programmed to apply different ambient pollution concentration constraints at different locations. For example, the website might reject trades that would result in annual average ambient concentrations of sulfur dioxide in excess of 80 :g/m^3 at some locations, but in excess of 60 :g/m^3 at other locations. The website could also be programmed to reject trades that would result in any increases in ambient concentration levels at certain receptor points.

Current Market Practices

We focus here on how our proposal would deal with various practices under the national sulfur dioxide trading program: auctions of allowances, retirement of allowances, purchases of allowances for investment purposes, and banking of allowances.

Auctions of Allowances

As we explained above, the sulfur dioxide program provides for an annual auction of allowances, in addition to the grandfathering of existing sources. The Chicago Board of Trade, operating on EPA's behalf, currently conducts the sulfur dioxide allowance auction. Interested parties submit bids indicating the number of allowances they seek to purchase and the price they are willing to pay. The Chicago Board of Trade determines the price that would lead to the sale of all the available allowances. All bids at or above that price are then accepted, with each successful bidder paying the amount it bid.

Consistent with the logic of our proposed trading regime, before accepting a bid the Chicago Board of Trade could verify that the transfer of the allowance to the successful bidder would not result in the violation of an ambient standard at any receptor point. In light of the current auction structure, it would make sense for bidders placing higher bids to receive priority in determining whether the bidder's purchase would result in an ambient standard violation. In other words, to the extent that two bidders' purchases taken together would result in an ambient standard violation but individually would not, the system would approve the higher bidder's bid and reject the lower bid.

Our approach would have the effect of lowering somewhat the total revenues from the auction. Indeed, because certain high bidders might not be able to purchase permits as a result, the Chicago Board of Trade would need to turn to lower bidders.

Retirement of Allowances

Environmentalists and environmental groups sometimes purchase allowances for the purpose of retiring them so as to reduce the total amount of pollution. Our proposal readily accommodates such transactions. The purchaser simply would indicate that it sought to retire them. The website would then decrease the seller's emissions accordingly, but there would be no corresponding increase in emissions at another location. As a result, the website would never prohibit a retirement transaction.

Purchases of Allowances for Investment Purposes

Investors might be interested in purchasing pollution allowances for investment purposes, intending to hold them until they can be sold at a profit. Also, brokers may purchase allowances to act as 'market-makers,' without intending to use the allowances or resell them immediately. As in the case of retirement transactions, the website would credit an emission reduction but not add a corresponding increase in emissions. Again, no such transaction would ever be rejected.

When the holder was ready to sell the allowances, however, it would have to receive permission from the website to proceed with the sale. As in the case of all other transactions, the website would determine whether the increase in emissions would result in violations of ambient standards. Unlike the case of transactions between contemporaneous emitters, there would be no corresponding decrease in emissions, because this decrease would already have been taken into account at the time the investor purchased the allowances.

Our proposal would complicate matters for allowance purchasers with investment motives. Prospective investors would have to take account of the possibility that allowances that they purchase may now be saleable only to a limited pool of buyers at some point in the future. In that case, the allowances would command a lower price than in a system of unconstrained trading – although, for the same reason, the investor may have bought them at a lower price as well. Moreover, investors might feel pressure to sell allowances earlier out of fear that a delay might preclude (or make less attractive) sales later. At the same time, however, if an investor finds that it cannot sell the allowances at an acceptable price because of a limited pool of buyers, it would have the option of holding them until a later date.

Banking of Allowances

Some entities obtain allowances with the intent of using them only in the future. Our proposed trading regime is compatible with such banking of allowances. It simply would require that the website maintain emission input data for multiple years into the future.

A source seeking to bank an allowance for future use would notify the website of the year in which it intended to use the allowance. The website would first test whether the proposed use of the allowance in the future year would result in the violation of an ambient standard in that year, given the existing emission input data for that year, as augmented by the proposed additional emissions. If the model predicted that an ambient standard would be violated, it would reject the proposed banking. If not, the transaction would be approved and the website would update the emissions profile for the future year. It would also modify the emissions date for the current year to reflect the resulting decrease in emissions.

We acknowledge that our trading regime would impose some limits on the ability of sources to bank their emissions. Moreover, sources would have an incentive to declare their intent to bank sooner, before too many other sources had done so. Such constraints, however, are necessary to ensure the attainment of the applicable ambient standards.

CONCLUSION

This chapter highlights how traditional tradeable pollution permit regimes fail to guard against the violation of ambient standards and the formation of hot spots. The existing regulatory programs that provide for such market transactions all exhibit these problems. Similarly, three alternative design structures advocated by commentators either provide incomplete solutions or give rise to other potentially serious shortcomings.

We propose a new trading regime that might increase the attractiveness of market-based schemes. Under this regime, the government maintains a website containing current emissions data for all the sources of a regulated pollutant. Using an atmospheric dispersion model, the website determines whether a proposed trade leads to a violation of an applicable ambient standard (or the formation of a hot spot), and disapproves trades causing such violations. We demonstrate, by means of a computer simulation, how our trading regime would function, and explain why it performs better than the existing alternatives.

NOTES

1. Environmental Defense Fund v. Thomas, 870 F.2d 892, 895 (2d Cir.), *cert. denied*, 493 U.S. 991 (1989).
2. S. Hrg. No. 101–331, pt. 5, at 13 (Oct. 3, 1989).
3. S. Hrg. No. 101–331, pt. 5, at 13 (Oct. 3, 1989).
4. S. Hrg. No. 101–331, pt. 5, at 38 (Oct. 3, 1989).
5. S. Hrg. No. 101–331, pt. 5, at 38 (Oct. 3, 1989).
6. Acid Rain Program: General Provisions and Permits, Allowance System, Continuous Emissions Monitoring, Excess Emissions and Administrative Appeals, 58 Fed. Reg. 3590, 3614–15 (1993).
7. Ass'y Bill 9090, 223d Leg. (N. Y. 1999) (enacted May 24, 2000) (to be codified in N. Y. PUB. SERV. LAW § 66-k, N. Y. ENVTL. CONSERV. LAW. § 19–0921(3), N. Y. STATE FIN. LAW § 99-G, N. Y. PUB. AUTH. LAW § 1854(10-a)).
8. Ass'y Bill § 2 (adding N. Y. Pub. Serv. Law § 66-k(1)(c)).
9. Ass'y Bill § 2 (adding N. Y. Pub. Serv. Law § 66-k(1)(d)).
10. Ass'y Bill § 2 (adding N. Y. Pub. Serv. Law § 66-k(5)).
11. Interview with Bill Gillespie, Senior Program Manager at the OTC (June 27, 2000).
12. 42 U.S.C. § 7410(k)(5) (1994).
13. NOx SIP Final Rule, 63 Fed. Reg. at 57,475.
14. NOx SIP Final Rule, 63 Fed. Reg. at 57,475.
15. NOx SIP Final Rule, 63 Fed. Reg. at 57,475.
16. 65 Fed. Reg. at 2693.
17. 65 Fed. Reg. at 21,507.
18. 65 Fed. Reg. at 21,508.
19. 42 U.S.C. § 7502(c)(2); 42 U.S.C. § 7501(1) (1994).

REFERENCES

Ackerman, B. A., & Stewart, R. B. (1985). Reforming Environmental Law. *Stanford Law Review*, 37, 1333–1365.

Atkinson, S. E., & Tietenberg, T. H. (1982). The Empirical Properties of Two Classes of Designs for Transferable Discharge Permit Markets. *Journal of Environmental Economics and Management*, 9, 101–21.

Burtraw, D. (2000). Appraisal of the SO_2 Cap-and-Trade Market. In: R.F. Kosobud et al. (Eds), *Emissions Trading: Environmental Law's New Approach*.

Burtraw, D., & Mansur, E. (1999). The Effects of Trading and Banking in the SO_2 Allowance Market. Resources for the Future Discussion Paper 99–25.

Carlson, L. J. (1996). NESCAUM/MARAMA NO_x Budget Model Rule. Available from: *http://www.epa.gov/airmarkets/otc/index.html* (visited October 1, 2001).

Chinn, L. N. (1999). Comment, Can the Market Be Fair and Efficient? An Environmental Justice Critique of Emissions Trading. *Ecology Law Quarterly*, 26, 80–125.

Dales, J. H. (1968). *Pollution, Property & Prices*.

Drury, R. T. et al. (1999). Pollution Trading and Environmental Injustice: Los Angeles' Failed Experiment in Air Quality Policy. *Duke Environmental Law and Policy Forum*, 9, 231–289.

Ellerman, A. D. et al. (2000). *Markets for Clean Air: The U.S. Acid Rain Program*.

Gade, M. A., & Kanerva, R. A. (2000). Emissions Trading Designs in the OTAG Process. In: R. F. Kosobud et al. (Eds), *Emissions Trading: Environmental Law's New Approach*.

Hernandez, R. (1998). LILCO Is to Stop Selling Credits to Upwind Polluters. *New York Times*, April 30, p. B1.

Institute for Applied Systems (2000). Cleaner Air for a Cleaner Future: Controlling Transboundary Air Pollution. *http://www.iiasa.ac.at/ Admin/INF/OPT/Summer98/feature.htm* (visited July 21, 2000).

Kete, N. (1992). The U.S. Acid Rain Control Allowance Trading System. In: *OECD, Climate Change: Designing a Tradeable Permit System*.

Kosobud, R. F. (2000). Emissions Trading Emerges from the Shadows. In: R. F. Kosobud et al. (Eds), *Emissions Trading: Environmental Law's New Approach*.

Krupnick, A. J. et al. (1983). On Marketable Air-Pollution Permits: The Case for a System of Pollution Offsets. *Journal of Environmental Economics and Management*, 10, 233–247.

Lents, J. M. (2000). The RECLAIM Program (Los Angeles' Market-Based Emissions Reduction Program) at Three Years. In: R. F. Kosobud et al. (Eds), *Emissions Trading: Environmental Law's New Approach*.

McGartland, A. (1988). A Comparison of Two Marketable Discharge Permits Systems. *Journal of Environmental Economics and Management*, 15, 35–44.

National Research Council (1991). *Rethinking the Ozone Problem in Urban and Regional Air Pollution*.

Parker, L. B. et al. (1991). Clean Air Act Allowance Trading. *Environmental Lawyer*, 21, 2021–2068.

Project 88 (1988). *Harnessing Market Forces to Protect Our Environment: Initiatives for the New President*.

Revesz, R. L. (1996). Federalism and Interstate Environmental Externalities. *University of Pennsylvania Law Review*, 144, 2341–2416.

Senate Committee on Environment and Public Works (1993). *A Legislative History of the Clean Air Act Amendments of 1990*.

Swift, B. (2000). Allowance Trading and SO2 Hot Spots – Good News from the Acid Rain Program. *Environment Reporter (BNA)*, 31, 954.

Tietenberg, T. (1991). Tradable Permits for Pollution Control When Location Matters. *Environmental and Resource Economics*, 5, 95–113.

White House Fact Sheet on the President's Clean Air Plan, I Pub. Papers: George Bush 710 (June 12, 1989).

COMMENTS ON PAPER BY JONATHON REMY NASH AND RICHARD L. REVESZ

Daniel H. Cole

This paper tackles an issue that is becoming increasingly important as tradeable quota instruments proliferate: the distribution of the costs and benefits of environmental regulation. No one should be surprised that this issue has arisen. Like most economists, the advocates of tradeable permits have had little to say about distributional issues. But distributional issues have become prominent, especially with the rise of the environmental justice movement. The question is whether it is possible to maintain the enhanced efficiency associated with tradeable quotas without sacrificing equity. Conversely, can we better account for distribution concerns in designing emissions trading programs without losing their efficiency advantages? More than anything else, the Nash and Revesz paper illustrates the difficulties inherent in efforts to combine equity and efficiency in a tradeable permit regime.

To cut to the chase, I think their idea of an web-based system for pre-approving trades of emissions allowances is clever, but not without problems. The first problem is that the cost of setting up the pre-approval system may not be worth the benefits. I am thinking here primarily of the acid rain emissions trading program under the CAA, which simply has not generated the kind of hot spots with which Nash and Revesz are concerned. This is not to say that the potential for hot spots does not exist, but to this point at least, distributional problems have not materialized. Consequently, I think their conclusion, that the national sulfur dioxide trading program is poorly designed, is unwarranted. To argue that just because it doesn't deal with feasible but, as

An Introduction to the Law and Economics of Environmental Policy: Issues in Institutional Design, Volume 20, pages 379–381.
© **2002 Published by Elsevier Science Ltd.**
ISBN: 0-7623-0888-5

yet, non-existent hot-spot problems is tantamount to arguing that all contracts are poorly designed because they inevitably fail to deal with all possible contingencies.

The case for a solution to hot-spot problems seems much stronger for other emissions trading programs, including the regional NO_x trading programs and Southern California's RECLAIM program. However, it's unclear whether their proposed solution would help with the hot spot issues in Los Angeles, mainly because those hot spots are due not just to the RECLAIM program. They talk about a situation in which just four polluters – all refineries located in poor parts of town – purchased most of the available mobile source credits. Nash and Revesz's pre-approval program would disable those four refineries from purchasing credits, but that would hardly resolve the hot-spot problems that exist around those refineries. Indeed, any real solution to that hot spot problem would seem to require some form of command-and-control regulations on top of, or instead of, the market-based program.

A more discrete problem concerns the authors' numerical example. The emissions from the two sources described in their example have differential effects on ambient concentration levels. But why should this be the case? A unit of P should have the same impact on ambient concentrations at a single receptor point, unless other factors, such as wind patterns, are significantly different. That may, realistically, be the case. Still, the authors should be clear about it. It's important because, if the only difference between the two sources is their location viz-à-viz the common receptor point, then the location of such receptor points becomes all important to the trading system, particularly to the design of the computer model for pre-approving trades.

Indeed, the very fact that their proffered pre-approval system for trades is model-based raises some important questions. Presumably the model would be based on estimated average conditions, including prevailing winds and so forth. It would simply be too difficult to design a model that covered all contingencies, right? So, what if we have a certain model, which is based on average conditions, but some unusual conditions obtain in a certain year? You could still get a hot spot resulting from a trade that was pre-approved. What is worse, a particular trade could create a hot spot that would not have occurred in the absence of the approval system. Imagine two sources A and B with a potential hot spot between them. Assume that the prevailing winds blow from source A to source B. All other things being equal, the model would reject trades of emissions allowances from B to A, but not from A to B because, when the prevailing winds blow, B's emissions blow away from the potential hot spot. However, in a certain year the winds may blow otherwise than the model suggests, in which case, we might prefer trades from B to A, rather than from A to B; but only trades from A to B will be approved by the model.

Generally speaking, a climate model cannot ensure ambient compliance any more than a climate model can tell you what effects a certain level of emissions reduction will have on, say, global warming.

Finally, Nash and Revesz seem to presume a robust trading market, with enough trading activity and profit-potential to lure in brokers, who are needed, in their model, to minimize the transaction costs associated with contracting for emissions rights. The market seemingly would have to be much larger and deeper than the Clean Air Act's sulfur dioxide emissions-trading market, for example. In the absence of a large enough market to support emissions-trading brokers, the transaction costs – particular the costs of identifying potential partners and determining which trades are likely to be pre-approved (or not) – could exceed the benefits from trading, in which case the market would fail.

Having said all this, I think Nash and Revesz's proposal is ingenious, and could potentially be quite useful in trying to achieve greater equity without sacrificing the efficiency advantages of emissions trading regimes. I think it would be a most interesting to implement their proposal experimentally, as part of the regional NO_x trading program, where I sense it may have the greatest utility.

SECTION C:
THE LAW AND ECONOMICS OF COMPLIANCE AND ENFORCEMENT

PART 1:
OPTIMAL MONITORING
INSTITUTIONS

MONITORING FOR LATENT LIABILITIES: WHEN IS IT NECESSARY AND WHO SHOULD DO IT?

Timothy Swanson and Robin Mason

ABSTRACT

The problem of intentional environmental harms is discussed as the combined consequence of limited liability laws and specific capital structures. When the conditions are "right", then the option of planned liquidation is optimal for certain types of firms. These firms are typified by capital structures that make it easy for them to engage in "hit and run" strategies on their industries, i.e. they operate with the intention of early exit. When this is the case it can be demonstrated that mandatory insurance forms of institutions are inadequate to the regulation of this particular problem. It is necessary to engage some manner of continuous monitoring institution, capable of observing and identifying when the capital structure of the firm begins to resemble those that would choose to engage in strategic liquidation. The paper concludes by stating that it might be possible to introduce private sector (bonding) institutions that perform this function, but it is more likely that a combination of public and private sector institutions will be able to best undertake the combined monitoring/insurance role that this requires. In short, the paper recommends that the public

An Introduction to the Law and Economics of Environmental Policy: Issues in Institutional Design, Volume 20, pages 387–411.

ISBN: 0-7623-0888-5

*sector should have the obligation to monitor/audit all limited liability
institutions for the existence of the pre-conditions for "looting" behaviour,
and then to charge these firms for limited liability in order to cover the
costs of the monitoring. In this way the costs of incorporation would
be raised sufficiently to render the worst forms of looting unlikely from
the outset.*

1. INTRODUCTION

One of the major causes of large-scale environmental damages is the system
of incentives for strategic liquidation inherent within a limited liability system.
In effect the intersection between the regulatory role of a civil liability
system and the exculpatory function of a limited liability regime makes
available a loophole to appropriately structured firms. These are firms that are
designed to operate on a "hit and run" basis and able to escape liabilities
that take time to accumulate within the environment. In this case strategic
liquidation is the option that is chosen from the outset as the most profitable
approach to the joint use of the firm's assets, and the environment's resources.

This theory of strategic bankruptcy was developed initially in the context of
the American savings and loan (S&L) crisis. The analysis there was undertaken
by Akerlof and Romer (1993) and they stylised the use of strategic bankruptcy
for the appropriation of S&L rents as "looting". In an earlier paper (Mason &
Swanson, 1996) we extended and generalised their analysis to the environmental
context, and identified the structure of firms that would make the "looting
strategy" viable. In this paper our objective is to set forth our conclusions
regarding the appropriate means for regulating for looting. Who can monitor
for this problem and when is it necessary?

Looting is a phenomenon with which most people are now familiar. It has
taken two very noticeable forms recently. In the U.S., the thrifts crisis has been
estimated to have generated societal losses in the amount of US$140 billion.
According to Akerlof and Romer, the S&L crisis resulted from the deliberate
relaxation of the regulatory regime for the savings and loan industry. The
regulators reduced reserves requirements and relaxed monitoring, and this then
produced the loophole through which the looting strategy was able to operate.

Another example of the same phenomenon of looting occurred on a much
more widespread basis in the form of the toxic waste disposal site problem.
This is now a worldwide problem occasioned by the almost complete absence
of any regulatory system regarding the disposal of toxic wastes prior to the
1980s. In this regulatory context, an obvious loophole existed whereby firms
were established for the single purpose of providing landfill sites for the disposal

of problematic wastes. These firms often existed with few assets other than the land on which the disposal occurred. After years of dumping, and before detection of any leakage, the firms would often liquidate its assets and abandon its disposal sites. The U.S. has expended more than US$100 billion on the cleanup operations associated with this problem (Menell, 1991), and the problem of "orphaned waste sites" is prevalent within the European Union as well.

This is the essence of looting: Operation of a firm in a context in which it is possible to incur benefits today while postponing the associated costliness into the future, with dissolution and liquidation occurring in the interim. The necessary conditions for looting are therefore: (a) the availability of limited liability; (b) the capacity to create deferred costliness; and (c) the capital structure that renders liquidation the optimal strategy to pursue. The environment is always going to be one of the areas that absorbs more than its share of the costs of this sort of behaviour. Environmental resources are natural locations for looting to occur, because of the level of difficulty and costliness involved in the monitoring of common resources. Given the institutional costliness of monitoring for these latent liabilities within the environmental context, the question of how to otherwise monitor for the phenomenon of looting must be considered.

In short, the problem of environmental damages is to some extent a problem of endogenously created risks, i.e. the creation of long term deferred liabilities expressly for the purpose of avoiding them. It is a case of "extreme moral hazard" in the face of the limited liability loophole afforded by society by their laws of incorporation. Limited liability laws constitute societally provided insurance for enterprises, in the absence of a premium. Some of those insured will see the benefits to be had from exploiting the underpriced policy, and this is when looting occurs. When this is the case, we are dealing with a situation that is at the boundary of the public/private interface. Although insurers are able to provide monitoring and rate adjustments for all sorts of situations (even including the "bail bonding" of suspected criminals), there is a clear additional role for the state to intervene to monitor for behaviour that is, firstly, generated by its own lax system of insurance (laws of corporation) and, secondly, generating only negative social value. Looting behaviour is no more an exclusive matter for private insurance markets than is any other form of theft, and it calls for the sorts of public intervention that these other forms of rent-seeking behaviour command. The public sector must be involved in monitoring for the pre-conditions to looting, and not just for the existence of looting behaviour itself.

This article recounts the conditions which make looting possible in Section 2. Then in Section 3 it specifies the conditions under which monitoring might be predicted to occur; that is, it answers the question: When is monitoring for this sort of problem necessary? Section 4 discusses some of our general ideas about the public/private interface in environmental problems, and how endogenously created risks must be regulated; that is, it answers the question: Who should be monitoring for looting, and how? Section 5 concludes.

2. THE PRE-CONDITIONS FOR LOOTING

This section will reprise our findings from our previous article on the choice of liquidation and accumulating liabilities. Those who wish to avoid the analytics may move directly to the summary of the section (Mason & Swanson, 1996). In that paper we developed the conditions under which looting will occur. The primary findings were that looting is the optimal strategy whenever the firm is structured so that its revenues are frontloaded and its costs backloaded, with the costliness of exit from the industry low relative to its early benefits. The costliness of exit is made up of three primary factors: (1) the opportunity costs of forgoing further involvement in the industry; (2) the costs of liquidating sunk capital; and (3) the costs of preventing future liabilities. A firm with a relatively thin capital structure or with an easily liquidated capital stock is the typical candidate for looting behaviour (Ringleb & Wiggins, 1990).

2.1. The Factors that Determine the Choice of Strategic Liquidation

In Section 2, we re-state our model of strategic liquidation, and then discuss its implications for monitoring in Sections 3 and 4.

In brief, the model of strategic liquidation concerns the firm's choice of initial capital stock (K_0) at cost ω, and the choice of the amount of capital to scrap in the next period (K_1) at price ϕ_1, taking these benefits as dividends (Δ), as opposed to continuing to use the capital in production in the enterprise, returning a future flow of revenue $f(K_0 - K_1)$. Given the conditions in which looting occurs, there must also be the prospect of future damages (D) occurring with probability P(E) where E is the amount of costly effort which the firm expends on preventive measures. The firm's choice concerns the amount of investment to undertake in precaution (harm avoidance) and the amount of capital to retain in the enterprise. The firm will be balancing the benefits from making a long-term commitment to the industry (by investing in harm avoidance and maintaining capital in the firm) against the benefits of undertaking a "hit and run" strategy (by not investing in harm avoidance and running down the capital stock).

The objective of the limited liability firm is therefore to maximise its value given these choices (as set forth below):

Assumption: The Objective Function Facing the Limited Liability Firm:

$$V^{LL} = \max_{\{K_0, K_1, \Delta_1, E\}} - \omega K_0 + \frac{\Delta_1}{\delta_1} + \frac{E[\Delta_2]}{\delta_1 \delta_2}$$

$$\Delta_1 + E \le f_1(K_0) + \phi_1(K_1)$$

where:

$$\Delta_2 = \max\left[0, f_2(K_0 - K_1) + \delta_2(f_1(K_0) + \phi_1(K_1) - \Delta_1 - E - D \right.$$

$$\left. + \frac{\sum\limits_{t=3}^{\infty} f_t(K_0 - K_1)}{\prod\limits_{t=3}^{\infty} \delta_1} \right] \text{ with probability } (E),$$

$$\Delta_2 = f_2(K_0 - K_1) + \delta_2(f_1(K_0) + \phi_1(K_1) - \Delta_1 - E$$

$$+ \frac{\sum\limits_{t=3}^{\infty} f_t(K_0 - K_1)}{\prod\limits_{t=3}^{\infty} \delta_1} \text{ with probability } (1 - p(E))$$

The meaning of the objective function set out above is simple. It states that the firm has its potential liability bounded at 0 by reason of the existence of the limited liability loophole afforded by the laws of incorporation. The enterprise is effectively fully insured for any costs that are incurred in excess of the assets existing within the enterprise during that period. The firm will take the existence of this implicit insurance policy into consideration in the course of choosing its capital structure and investment strategy. Here we assume a simple two-period choice problem for the firm, and the issue concerns the amount of capital to retain in the industry at the end of period two. A firm that is going to "hit and run" will remove its assets from the industry in the second period, and will commence its investment strategy with this endpoint in mind.

One set of commentators has discussed this problem as a case of *judgement proofness*: the impact of unpaid insurance on the choice of the level of capital stock with which to commence operations. Shavell (1986) argued intuitively that this free insurance would act as a subsidy, increasing the amount of capital employed within the industries with higher risks; in essence, he argued that the subsidy would generate excessive risk-taking by limited liability corporations. Others have shown, however, that the insurance subsidy has counteracting effects. In addition to the reduced prospect of feeling the full effects of the potential damages, there are also effectively reduced costs of care expenditures and effectively increased costs of capital. The overall effect of limited liability on ongoing enterprises is ambiguous, an empirical matter determined by the relative size of these various effects. (Beard, 1990; Posey, 1993; Mason & Swanson, 1996).

The problem that we address here is very different from the judgement proof problem. It concerns the range of most extreme responses to the availability of societally supplied insurance (the impact of which is the bounded liability within the objective function set forth above). In certain cases the firms involved will not simply have incentives to undertake more risk-preferring sorts of behaviour; they will in fact structure their corporations for the single purpose of planned liquidation. In these cases there are no exogenous risks. The risks created and avoided by the firm in this situation, and insured by society under limited liability, are entirely endogenous. The firm is creating them on a deferred basis for the purpose of receiving immediate benefits, while avoiding them through liquidation in the interim. This is precisely the nature of the problem that has existed in the context of the toxic waste disposal industry: extreme moral hazard (endogenous risk creation) in the face of unregulated and unpriced insurance provided by society.

It is possible to identify the nature of the industries which will produce this sort of looting behaviour. These firms will perceive the benefits from one period's production, scrapping and liquidation as being in excess of the benefits from ongoing involvement in the industry. This condition is stated more precisely below:

Proposition: Conditions under which Limited Liability Firms will Choose Liquidation

There are conditions under which planned liquidation is the optimal choice for a limited liability firm. In this case maximum value may be obtained by running the firm for one period with the plan of scrapping it in the second. Under such conditions, the firm would make no provision for future damages or liabilities. The conditions under which planned liquidation (or "looting") is the optimal choice are as follows:

$$\text{Max}_{\{K_0\}} \frac{f_i(K_0) + \phi_1(K_0)}{\delta_1} >$$

$$\text{Max}_{\{K_0, K_1, E\}} \frac{f_2(K_0-K_1) + \delta_2(f_1(K_0)-E)-p(E)D + \dfrac{\sum\limits_{t=3}^{\infty} f_t(K_0-K_1)}{\prod\limits_{t=3}^{\infty}\delta_t}}{\delta_1\delta_2}$$

This condition merely states that the value from operating the firm for one period and scrapping it in the second (the left-hand side) is greater than the value from operating the firm into the indefinite future (the right-hand side). The reasons for this would be: (a) the avoidance of the potential scale of the liabilities (D) from remaining in the industry; (b) the avoidance of the costs of taking precaution (E); and (c) the value acquired from scrapping the capital within the firm (ϕ). Where the expected future costs from remaining in the industry are greater than the expected future benefits, then it makes sense to have a very short-term objective with regard to the industry.

2.2. Summary: The Pre-Conditions to Looting

The pre-conditions to looting may be summarised as follows: the existence of highly mobile capital stocks (little sunk capital) in relatively low productivity industries. Where these conditions hold, the choices of the firm's capital structure renders future risks and liquidation prospects endogenous. The firm is then choosing to create a relatively thin capital structure made up of relatively liquid assets for the purpose of implementing a "hit and run" sort of strategy. The firm is operating with the single purpose of becoming bankrupt in the next period in order to avoid the losses it will create. This is what Akerlof and Romer termed "looting":

> ... the normal economics of maximising economic value is replaced by the topsy-turvy economics of maximising current extractable value, which tends to drive the firm's economic net worth deeply negative .... Because of [the] large disparity between what the owners can capture and the losses that they create, we refer to bankruptcy for profit as looting (Akerlof & Romer, pp. 2–3).

Before proceeding to the policy analysis of the problem of endogenous liquidation, we will first identify the precise set of conditions under which these problems will occur. The problem of regulating for this loophole comes down to the problem of regulating these sorts of industries. The first step in this enquiry must be the identification of the set of conditions which will define the problematic industries.

3. WHO MUST BE MONITORED?: CHARACTERISTIC CONDITIONS

Defining the problematic industries is not difficult. These are industries in which a "hit and run" sort of strategy is likely to be effective. They are therefore typified by modest amounts of relatively liquid capital stocks, operating in areas with high potential for societal damages. Chemical waste disposal now seems like an obvious case in point with the benefit of hindsight, but it is also possible to see how it might have been identified before the fact. The time required for chemical leachage into groundwater provides for the initial requirement, while the capacity to provide disposal facilities with nothing more than a short-term lease on waste land provided the remainder. These sorts of industries will be generated by the limited liability loophole, if they are not otherwise regulated.

Annex I gives an illustration of the precise conditions under which firms will elect various strategies. In particular, the annex provides a graphic portrayal of the relative parameter values under which firms will make choices between the socially optimal investment, risky investment, or planned liquidation. In Figs 1–4, the area above the indicated curve (and below the next) represents the set of parameter values for which that particular strategy will be the optimal choice for the given firm (possessed of the characteristics represented by those values). **Region I** represents those parameter values for which the firm will elect to pursue the socially optimal investment strategy (which objective function is represented by (V^{SO}). **Region II** represents the set of parameter values for which the firm will elect to pursue an excessively risky investment strategy (V^{JP}) (i.e. greater levels of social risk than would exist under the socially optimal strategy). **Region III** represents the set of parameter values for which the firm will elect to pursue the strategy of planned liquidation, i.e. looting (V^L).

The parameter values that determine the incentives for looting concern: (a) the productivity of capital investments in the industry; (b) the relatively sunk nature of the capital investment in the industry; and (c) the costs of precaution. We turn now to discuss these various factors, and their respective meanings.

3.1. The Importance of Capital Investment Productivity

Capital productivity measures the returns that can be generated from a stock of capital. Its relevance is two-fold. First, it is a major determinant of the returns to be acquired from remaining within an industry (as opposed to liquidating early). Secondly, it measures the extent to which firms will invest in fixed, rather than mobile forms of assets (e.g. labour); this in turn influences the relative attractiveness of the judgement proof strategy (see Wiggins & Ringleb, 1992). We will identify three regions of interest regarding firm structures, the third of which corresponds to the conditions that we are interested in regulating.

A firm that operates in region I will make its choices in a socially optimal fashion despite the availability of limited liability ($V^{SO} > V^L$). This is most likely to be the case when capital productivity is high (other things being equal). In this case, the industry will be capital intensive, generating high returns on that capital. This results in the vesting of the firm within the industry, since its return from maintaining its current capital stock operational in the industry is far greater than the return from liquidating to avoid prospective liabilities. This corporation will therefore be willing to provide for the future – even substantial liabilities – in order to remain in operation. An example of such an industry would be marine-based or pipeline petroleum transport. Companies in this industry achieve high levels of productivity and returns only by operating at the scale that requires massive capital investments (e.g. oil pipelines and tankers). Equally, these capital investments are capable of generating substantial future liabilities – the Exxon Valdez incident is indicative of the scale of the liabilities that potentially can flow from these investments. The scale of the required investment, most of it relatively immobile (especially in the case of the pipeline) means that the firm has a long-term vested interest in the industry. So, the firms' responses to the realisation of these liabilities has not been to exit. Instead, all of the firms in the business of large scale oil extraction and transport have formed a syndicate to spread their risks across the entire industry. This insurance agreement is indicative of their commitment to the industry.

A firm that operates within region II ($V^{SO} > 0$, but $V^{JP} > V^{SO}$ and $V^{JP} > V^L$) is a socially beneficial concern with a structure that presents the temptation for taking greater than socially optimal risks but not looting; Firms operating in region II suffer from the "judgment proof problem", i.e. they recognise that their potential risks are bounded by zero below and hence do not make provision for liabilities that would exceed the value of their assets. Examples of such firms would be land-based hazardous chemical transporters. Unlike the large pipelines and marine transport operators, these firms are able to operate at relatively small scales. A single tanker truck would be capable of initialising a firm. Given the low scale of investment, the amount of capital available to cover future liabilities is relatively limited. For this reason the firm has reduced incentives for taking precautions against the largest-scale damages that the nature of the firm might inflict on society. In this context, it is important that the firm is made to carry liability insurance in excess of the value of its assets, and this will be an effective regulatory strategy.

Firms that operate in region III (in which $V^{SO} < 0$, $V^L > 0$) undertake activities that have negative net social worth. The combination of the technology of the industry and the institution of limited liability, however, generates incentives for shareholders to pursue the activity. The firms in such an industry would be characterised by relatively small amounts of capital investment in the firm. The classic example of such a 'region III' enterprise was the toxic waste disposal business in the U.S. as it was sometimes operated prior to the adoption of the Resource Conservation and Recovery Act (RCRA). At that time it was frequently the case (due to the lack of regulation) that toxic waste disposal sites were operated with little or no capital investment. All that was required was a lease on land, and little or any other form of capital investment. The operator of the dump site depended upon the waste containers and local geology to ensure that any liability arising from the dumping of waste occurred well into the future. This behaviour and the liabilities it generated almost certainly rendered these operations of negative social value; but the deferral of liabilities and the availability of liquidation in the interim made the operations financially attractive. This is a clear example of strategic liquidation or "environmental looting": the operations existed in these forms solely because of the availability of limited liability. It resulted in an estimated 10,000 orphaned waste sites and clean-up costs in excess of US$100 billion to date (see Menell, 1991).

3.2. The Importance of Sunk Capital Investments

Capital is sunk when there is a cost implied in the liquidation of any part of the capital stock. Clearly, it is not simply the amount of capital invested in a firm that matters when determining the degree of commitment to the industry, it is also the extent to which that capital can be mobilised overnight. Even firms with extremely large capital investments in given industries will be able to pursue suboptimal strategies, if that capital can be removed to other industries relatively quickly.

One of the more interesting attempts at mobilising capital was undertaken by large firms with operations within the asbestos industry, who attempted to create subsidiaries that were wholly concerned with the asbestos operations and then evacuate their assets from those subsidiaries. Of course this ex post approach to creating mobility in capital is problematic to the courts, and so the firms experienced very limited success in their attempts at creating mobility. But this does indicate the importance of maintaining judical capacities for disallowing these sorts of mobility-enhancing strategies (Roe, M., 1986).

It also indicates that it is important to look not just at the capital investment levels of a firm in order to determine its potential for anti-social behavious but also at the degree to which that capital is immobile ("sunk"). An excellent example of relatively sunk capital is tradename-based capital. Tradenames are capital items that are representative of the stock of goodwill attaching to the reputation of a firm in its current operations. These are highly important assets for many firms, often representing a substantial proportion of firm value as indicated by brandname motivated mergers; recently, accounting principles have been revamped in order to allow for the incorporation of such assets within the financial assets of most firms. Such assets are highly illiquid, in that they usually are valuable only within the context of the firm's current line of operations; there is some but not substantial value in British Airways marketing its own line of food products, for example. Tradenames are highly valuable capital assets, but usually only in connection with the current set of operations and only if there is a future in that business. For this reason it would not be anticipated that firms which invest in the development of a tradename would also engage in significant amounts of looting; for these firms, "their name is their bond".

3.3. The Effectiveness of Care Expenditures

The final factor of importance in determining which incentives will inhere with regard to liquidation is the effectiveness of expenditures on damage avoidance. A highly effective form of investment in precaution (damage avoidance) will create the incentives for the firm to exercise this option, and will simultaneously remove the option to loot. The crucial point is that it is not just expenditure on pollution that is indicative of commitment to the industry, but rather it is expediture on pollution elimination rather than deferral. In the context of toxic waste disposal, it would have been very difficult for firms to mitigate fully the future costliness of these wastes simply by means of storage (Swanson & Vighi, 1996). Basic expenditures on waste storage in this industry are only effective in isolating and deferring, but not reducing, the ultimate accumulation of these chemicals within these media. Most of the actions taken by toxic waste firms in the past (e.g. plastic or clay lining of pits) were of this nature. These had virtually no mitigative effect, but merely the effect of deferring the arrival of those damages. Thus, these expenditures on pollution control were effective only in creating a *larger loophole* by extending the gap between benefit and cost incurrence. If, on the other hand, there had been a readily available low-cost method for treating these materials in a manner that actually reduced their ultimate costliness, then these same firms would have had the incentive to undertake these measures rather than storage in order to continue in the business. Therefore, monitoring for commitment to an industry does not require evidence only of expenditure on pollution control, but evidence of pollution elimination.

3.4. Summary: When is Regulation for Looting Necessary?

The discussion in this section has indicated the conditions under which firms might be expected to pursue the looting strategy. The initial issue concerns the conditions under which the strategy of planned liquidation is feasible. There are two preconditions. First, it is necessary that the industry is capable of making access to limited liability laws (usually the case). Second, it is necessary that the industry has the technological option of making use of a resource today, but having the liability accumulate at some point in the future (by reason of accumulation within some environmental medium). When these two necessary conditions exist, then the looting strategy is technically feasible.

When is it profitable? The key facets determining profitability, and hence determining the identity of the firms and industries that would pursue a looting strategy are: relatively *low* long-term productivity of capital stocks (and hence low opportunity costs of exit from the industry); relatively *high* mobility of capital (and hence low explicit costs of exit from the industry); and, relatively *high* costs of precaution (rendering the choice of liquidation relatively preferable). When these conditions exist, then society must be careful to create some method of monitoring for looting problems. Now we turn to the question of how this monitoring might be implemented.

4. HOW SHOULD LOOTING BE REGULATED?

The problem of looting is an interesting one because once it has been identified that it exists, it is too late to regulate. This is why the regulation of looting is essentially a matter of monitoring rather than enforcement. Regulation in this area must take the form of identifying the situations in which the pre-conditions for looting exist, rather than the identification of looting itself.

So, the fundamental issues concerning looting are: Who should monitor for the pre-conditions to looting? What should be the response if these pre-conditions are found to exist?

The first point to note is that the problem of looting is one of a set of problems for which the pre-condition is the existence of the limited liability institution. The elimination of this institution would eliminate a necessary condition for both the looting problem and the judgment proof problem. The first policy question concerns whether the benefits from retaining the former institution justify the costliness of the problems that it entails (Hansmann & Kraakman, 1991). Taking the continued existence of limited liability forms of enterprise as a given, then the policy questions concern the creation of new forms of regulatory institutions in order to monitor for the existence of the other pre-conditions for looting.

The second point to note here is that looting is a distinct phenomenon from excessive risk-taking (i.e. the judgment proof problem) and requires independent regulation. Most forms of environmental damages (toxic wastes, chemical accidents, etc.) have previously been analysed as issues concerning excessive risk-taking by limited liability corporations, and the ultimate responsibility for uncompensated risks that they create (Shavell, 1986). The analysis here indicates that there is a wholly distinct problem, based upon a strategy that involves little or no risk-taking. In the case of the judgment proof problem (as indicated above) the solution is to require liability insurance coverage in order to cover these excessive risks. In the case of the looting problem, it is necessary to consider whether these institutions are of the form required to resolve this separate problem.

What sorts of regulatory policies does the looting problem require? The key to solving looting problems will be the development of ex ante institutions that either: (a) remove the incentives to loot; or (b) monitor and identify when the conditions for looting are in existence. We will now consider these alternative policies in turn.

The first policy to consider is the requirement of mandatory insurance over a lengthy time horizon (e.g. 30–50 years), that must be provided in advance of incorporation and extend beyond liquidation within industries capable of long-term damages. Such an approach would substitute immediate (but smaller) payments in lieu of deferred (but greater) damages, and hence remove the necessary condition of non-simultaneity. This is part of the approach adopted in the U.S., which has a 30 year financial responsibility undertaking (usually in the form of insurance) as a prerequisite to the licensing of waste facilities under the Resource Conservation and Recovery Act. This approach should be effective in addressing the judgment-proof problem, so the question is whether it can be used to mitigate the looting problem as well.

The problem with exclusive reliance upon insurance-based solutions is that private insurers (using only the instruments that they traditionally use) are unable to eliminate the loophole, irrespective of the amount they charge for a premium. Historically, private insurers have operated via two primary instruments: rate adjusting and co-insurance. Taking the latter first, it is not possible to co-insure an insured no longer in existence, so this instrument is unavailable in the context of endogenous liquidation. In the case of rate adjustment, it is simply not possible (by definition) to make the rate charged high enough to cover the costs that may be generated by this loophole. To be effective, the insurance premium would have to be so high as to remove all incentives for liquidation from the industry; however, given that planned liquidation can result in the near certainty of damages occurring and extremely high damages (in certain industries dealing with chemicals or other technologies with pervasive effect), this would imply the application of a virtually infinite premium to such industries. Anything less would leave some residual loophole to be exploited by those engaging in the most extreme forms of moral hazard. In short, if mandatory insurance is the chosen regulatory strategy, then looting may continue to occur through the choice of extreme capital structures, irrespective of the parameters used to set the terms of the insurance policy. Therefore, the traditional instruments of the insurer are incapable of addressing the full problem of looting by themselves; one is unavailable and the other has necessarily limited effect. This does not mean that mandatory insurance should not be required (as it addresses the judgment proof problem) but only that it does not resolve this additional problem.

This indicates that other, more intrusive forms of instruments are necessary for the regulation of this problem. This is one reason why there is a necessary role for further *monitoring* in this arena. Insurance firms are capable of engaging in the casual monitoring of and rate-adjusting for a wide range of activities, but when more intrusive forms of monitoring are required for their effectivenss, then it is important to recognise the limits of the insurer's traditional approach to monitoring.

What sort of monitoring is required? The monitoring that is required will look to the existence of the pre-conditions for looting, i.e. (a) thin capital structure; (b) liquid or mobile capital; and/or (c) little effective expenditure on long-term damage limitation. In addition, these monitoring functions must be undertaken relatively continuously, as the capital structure could be re-shaped fairly quickly between audits (and looting elected) if the monitoring had long gaps in-between.

One possibility is that the monitor for such pre-conditions might be a financial institution, under a mandatory bonding regime for example. In this case the financial institution would operate as the guarantor of the firm, and then would have the incentive to monitor the capital structure of that firm in order to protect its bond. Financial institutions would then be replacing the regulated firm's level of commitment with their own, and they would be operating in a realm within which they are relatively specialised (i.e. the monitoring of the financial status and structure of operating firms). So, an independent bonding agent that is required of all firms operating in hazardous industries might make sense as a mechanism for monitoring for the pre-conditions for looting.

There are two important reasons why this monitoring function might be retained in the public sector rather than devolved to financial institutions in the private. The first is the level of intrusiveness that this sort of monitoring requires. It could be argued that the financial institution might be captured by the regulated firm (as has been argued in the context of some independent auditors of failed enterprises generally). It could also be argued that the public sector has built-in checks on the reasonableness of such intrusive forms of monitoring. These checks allow for the maintenance of an arbiter over the reasonableness of the monitoring, i.e. the judiciary, and it provides a more general right to intervene in the public interest.

The second reason that public intervention might be desirable is that the activities described here more closely approximate theft than legitimate risk-creating activity. The area with which we are concerned with is that region in which the activity itself is of negative societal value (Region III above) but in which the limited liability loophole renders the activity profitable. As in the case of other forms of theft, there is a public good function to be served by means of monitoring and deterring these forms of rent-seeking behaviour, and so the public sector might want to control for the effort that is placed into monitoring for looting activities. It might also be argued the this loophole should be regulated by the state in order to internalise the costs of limited liability to the public sector.

To some extent, this division of responsibilities between the private and public sectors must remain somewhat arbitrary, as insurers should be allowed to insure companies' activities in any event and this insurance should extend beyond the point of liquidation (i.e. should flow from the fact of holding insurance at the time that the activities giving rise to future liabilities take place). However, insurers cannot regulate for looting behaviour, and they should not attempt to adjust their rates to regulate for it (because this will deny affordable insurance to legitimate enterprises). Since it is probably not possible to segregate between those liquidations which were endogenous and those which occurred in response to exogenous risks, the only other bases upon which to separate out between private and public functions will be necessarily arbitrary.

One reasonable basis for separation would be time. Looting can only occur in circumstances in which the costliness of activities are separated from their benefits, and deferred into the future. It is reasonable to expect insurers to be able to foresee some forms of deferred costliness, but not all. They are not privy to all of the information within the industry, only the hazard rate of associated industries. One basis for separating out between private and public functions would be to require implicitly that insurers be aware of the deferred risks of enterprises over a given time period, e.g. ten or twenty years after cessation of activities. This would mean that insurers would be given an explicit and limited time horizon over which they must be prepared to manage risks. They would then focus on the problem of monitoring and adjusting for the reasonably anticipated medium term risks of the enterprise. This would seem to be a manageable problem for an industry managing risks ex post on a foundation of historical data. It is a problem that may be solved based upon experiences with associated industries and reasonable inferences based upon the associated technologies.

Firms which are going to choose to liquidate will not insure voluntarily once the full costs of liquidated firms are included in the premium. This is an avoidable cost in the context of a looting strategy. Therefore, the only means by which private insurance is able to perform a function in this area is through the requirement of mandatory insurance for incorporated status. This is the first pillar of the recommended policy for dealing with looting behaviour: Liability insurance with a life extending some (say, 10–20) years beyond the event of liquidation should be a requirement for the granting of limited liability status.

It must once again be emphasised that this requirement of mandatory insurance does nothing to prevent looting forms of behaviour. It merely establishes the division of responsibilities for monitoring for these risks. Insurers under such a regime will now include a premium within their policies which will account for anticipated liquidations (even endogenous ones), but opportunists will continue to see that activities that either: (a) create higher than anticipated costs and benefits; or (b) defer costs beyond the policies coverage will generate opportunities for profitable strategic liquidation. In effect, the insurer's premium will establish the parameters around which the new looting forms of behaviour will take shape. It will be important for the insurer to monitor for planned liquidation and adjust for those which are clearly risky, otherwise its market will be constrained by excessive premiums. This is a manageable task so long as the time horizon is maintained short enough that the insurer is able to foresee near-term risks from historical data and reasonable inferences from associated technologies. The time horizon for long tail risks managed by insurance firms must be maintained short enough so that this is a manageable task given this sort of information base and this range of instruments.

The remainder of the tail of risk incidence could then be managed publicly. This allows a wider range of instruments for information acquisition, monitoring and management for those industries which are susceptible to looting. In brief, the public sector would then manage the residual long-term risks as follows: (1) it should use the conditions identified above as the basis for identifying those industries which are most vulnerable to looting strategies; (2) it should then use the wider powers of the state to require continuing information on the crucial parameters of the firms; and (3) it should also use those powers to require firms to alter those parameters, when it becomes apparent that looting is an optimal strategy.

This means that the state would be monitoring firms in industries with capital structures that render "hit and run" sorts of strategies available. They could use the wider powers of the state to require information on both the activities being undertaken and the capital structures that are being maintained. If it becomes apparent at some point that the potential damages are far in excess of the capital being maintained, then the state should be able to recognise the potential for looting as an optimal strategy. Then once again the wider powers of the state could make it feasible for the requirement of further capital injections into these firms, in order to maintain some sort of parity of the firm's assets with the damages of which they are capable.

This has been the essence of the banking regulation strategy over the past half century. After the banking collapse of the depression era, the states saw that the entire economy was at risk from bank liquidations and intervened to manage for this risk. They did so by means of ongoing monitoring of operations (independent bank audits) and mandated maintenance of capital assets (reserve requirements). These mechanisms were effective in preventing another collapse of the banking system, until the 1980s when they were relaxed in the savings and loan industry, almost immediately unleashing the looting epidemic that resulted. The prevention of looting necessitates broader sets of powers than are available to the insurance industry: the monitoring of the firm's capital structure and the power to require minimum levels of capital. These are more appropriately functions of the public sector.

Therefore, the recommended policy for looting is a combination of private and public sector mechanisms. There should be mandatory liability insurance for incorporation status, applicable for a fixed term of years beyond the event of liquidation. This will leave a residual amount of endogenous liquidation, which is best managed by the state. The state should manage this by, firstly, recognising the structural conditions under which it occurs and monitoring those industries which exhibit those characteristics. Secondly, it should not only monitor but also mandate levels of capital required for continuance within the monitored industries. If a firm moves from an adequate capital structure towards an inadequate one, it should be required to immediately transfer new assets into the ongoing enterprise or cease operations.

This of course sounds like a lot of public intervention into a wide range of possibly very small enterprises (because most of the most suspect ones would be relatively tiny in size). In fact, the undertakings by the state will be much more limited than might be expected, precisely because of the scale of the monitoring required. This is because the costliness of such public monitoring should be passed onto the industries involved, in the form of an additional premium for engagement in the enterprise. Then these firms with relatively slight capital structures but relatively powerful technologies would be responsible for paying both the mandatory insurance premium (covering the medium tail risks) and the government monitoring supplement and risk premium (covering the long tail risks) in order to procure incorporation status. These charges are going to be excessive for firms with relatively small capital requirements, and they will then elect other forms of organisational structure. They will proceed as partnerships or sole entrepreneurs in those areas in which the costs of incorporation exceed its benefits.

Of course this is precisely what is required. At present the social insurance of limited liability is available to all those for which this benefit is important (e.g. those requiring large pools of capital) and those for which this benefit is less so. For the latter group, the uncharged for social insurance becomes an opportunity to make use of societal assets while putting little or any of their own on the line. State involvement to monitor and manage capital structures of small corporations is simply the means by which those involved are required to have some of their own assets involved in the enterprise. If they are willing to put their own assets on the line in the first instance (as partnerships and sole owners), then no further governmental involvement is required. In effect, the social insurance subsidy of limited liability should be channelled toward those enterprises requiring it for their existence (primarily the large pools of capital), while the others should be channelled toward other forms of organisation by means of charging them the full cost of their insurance.

5. CONCLUSION

We have discussed the appropriate form that policies must take to address one very specific form of long tail risk problem: the problem of endogenous liquidation in the face of deferred liabilities. This may seem to be a very peculiar part of the overall problem, but it probably is not. Other forms of long tail risks (e.g. the problems of insuring products liabilities where the scale of the risks are unknown) may be successfully managed through insurance futures markets, where the pessimistic insurers reinsure with the more optimistic ones. The problem of long tail risks that cannot be managed completely through any form of private insurance market is the problem of endogenous liquidation. This simply requires too broad an information base and intervention capacity for private insurers to manage it successfully. The problem of endogenous liquidation requires continuous monitoring for the conditions which make looting profitable (thin capital structures, liquid capital, deferred damages), not monitoring for the existence of damages. Once the damages are occurring (in the context of looting), it is too late; the firms are already liquidated.

Of course there is a role for the private insurance market, but this role must be demarcated in a fashion that will allow them to perform their tasks well based on historical data and reasonable inferences. This implies the necessity of truncating the long tail of liability, at some point some years after liquidation of the firm. The insurer is then responsible for monitoring for and managing the risks that occur within this near term time horizon. The purpose of such long tail risk truncation is to make clear the role for the state. The state must then implement a monitoring system for those industries exhibiting characteristics that render liquidation profitable, and it can use its wider powers for the acquisition of broader information on the capital structure of the firm and to manage that same capital structure. This division between private and public sector places the insurance industry in its usual role as risk manager, while the state is performing its customary function as the policeman of anti-social (negative net worth) behaviour.

More importantly, the mere adoption of this system should have the effect of rendering itself unnecessary. The charges for these insurance mechanisms, private and public, should lie squarely with the firms requiring them. All firms which choose to incorporate should be made to procure madatory liability insurance of a fixed term after liquidation. Those firms whose capital structures warrant public scrutiny should be subject to a separate charge, representative of the public costs of monitoring and management, should they elect to incorporate. The conditions which warrant public scrutiny are those firms with limited capital structures but dealing in dangerous substances or powerful technologies. Clearly, these firms should be charged the full cost of the social insurance to which they are subscribing when they incorporate, since they represent substantial societal risks while placing few assets of their own on the line. They should also be subjected to a user-fee assessed system of continuous monitoring and capital management. However, once this policy of full cost social insurance is adopted, the relative benefits of incorporation status will be vastly diminished for all firms other than those with requirements of large pools of capital. The smaller enterprises will instead elect to pursue their activities as partnerships and sole enterprises. Then the costliness of the system will be very small, as these large corporations will seldom if ever require the monitoring system that the smaller ones would. In short, the mere creation of a system that is capable of passing through the full cost of social insurance to those responsible for those costs will render that system largely unused.

This is another example of how private markets are unable to operate well in the context of poorly constructed public frameworks. The social insurance subsidy of limited liability creates a framework that renders the private marketing of insurance untenable, because it makes it possible for extreme moral hazard to be profitable. Endogenously created risks cannot be managed by the instruments available to private insurers. For these to be managed the public sector must act to either close this loophole explicitly (by limiting the availability of incorporation status) or implicity (by charging the full cost of the insurance to the users). Then the private insurance market will be able to return to the task of pooling and managing the remaining risks to society.

REFERENCES

Akerlof, G., & Romer, P. (1993). Looting: The Economic Underworld of bankruptcy for Profit. *Brookings Papers on Economic Activity, 2,* 1–60.

Beard, T. (1990). Bankruptcy and Care Choice. *Rand Journal of Economics, 21*(4), 626–634.

Cooter, R. (1991). Economic Theories of Legal Liability. *Journal of Economic Perspectives, 5*(3), 11–30.

Hansmann, H., & Kraakman, R. (1991). Toward Unlimited Shareholder Liability for Corporate Torts. *Yale Law Journal, 100*(7), 1897–1934.

Mason, R., & Swanson, T. (1996). Long Term Liability and the Choice of Liquidation. *Geneva Papers on Risk and Insurance, 79,* 204–224.

Posey, L. (1993). Limited Liability and Incentive when Firms Can Inflict Damages Greater than Net Worth. *International Review of Law and Economics, 13*(3), 325–330.

Ringleb, A., & Wiggins, S. (1990). Liability and Large-Scale, Long Term Hazards. *Journal of Political Economy, 98*(3), 574–595.

Roe, M. (1986). Corporate Strategic Reaction to Mass Tort. *Virginia Law Review, 72*(1), 1–59.

Shavell, S. (1986). The Judgement Proof Problem. *International Review of Law and Economics, 6*(1), 45–58.

Swanson, T., & Vighi, M. (Eds) (1998). *The Regulation of Chemical Accumulation in the Environment.* Cambridge: Cambridge University Press.

Wiggins, S., & Ringleb, A. (1992). Adverse Selection and Long Term Hazards. *Journal of Legal Studies, 21*(1), 189–215.

ANNEX I: SIMULATION OF THE OPTIMAL CHOICE OF STRATEGY

A Simulation to Demonstrate the Choice of Liquidation

Section 2 gave the theoretical conditions under which looting will be the most profitable strategy for limited liability firms. In this section, the model that has been developed is simulated in order to answer the question: what types of industries will pursue which strategy? The three strategies under consideration are: (1) the social optimum (V^{SO}); (2) the judgment proof strategy (V^{JP}); (3) the looting strategy (V^{L}). The various strategies represent the values (V) that the firm will receive from pursuing a strategy reprentative of: (a) optimal social investment; (b) excessively risky investment (compared to the social optimum); or, (c) short run "hit and run" sorts of investment.

The following assumptions on particularly simple functional forms will be used to generate some graphs that demonstrate the choice between the social optimum (SO), risky investment (JP), and strategic liquidation (L):

Assumption 1: The production function is stationary and Cobb-Douglas
Assumption 2: The capital scrapping function is linear
Assumption 3: The probability of damage occurence function is iso-elastic

Figures 1–4 depict the solutions graphically, showing their dependence on the underlying parameters a (the marginal productivity of capital), b (the extent to which capital is sunk, i.e. the fraction of the capital's value that can be recouped by shareholders on scrapping), and ε (the elasticity of the damage probability function, which measures the firm's ability to influence the likelihood of the damage occurring). Essentially, the greater is a, the more profitable is the investment of assets in the specified industry. The greater is b, then the more liquid (less sunk) is the capital invested in the particular industry. The greater is ε, then the greater is the capacity for investment in precaution to prevent damages from occurring. We want to investigate how these various factors determine the relative profitability of choosing between the different investment strategies.

Figures 1–3 show the variations of the value functions V^{SO}, V^{L} and V^{JP} with changes in the three parameters. Figure 4 shows the values of a and b for which V^{SO} is greater than V^{L}, and vice versa. There are two other important parameters in the model. D, the size of the damage, is set equal to 5; this value is smaller than period-2 flow revenues in the socially optimal case, but nevertheless represents a large liability (in the examples of this section, it is

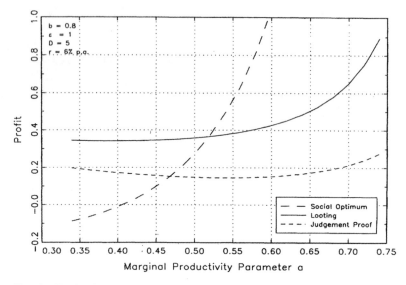

Fig. 1. Profits from Alternative Stategies with Changes in Model Parameters.

approximately 25–100% of the assets of the firm managed by the social planner, depending on the parameter values used). r, the discount rate, is set equal to 78%; with an annual interest rate of 6%, this means that each period lasts about ten years, and that therefore the liability occurs after twenty years. Both D and r are set, therefore, to satisfy Ringleb and Wiggins' condition that the liability be large and long-term for strategic behaviour to be worthwhile.

Figure 1 shows the dependence of V^{SO}, V^L and V^{JP} on the marginal productivity parameter a (with the parameters b and ε set equal to 0.8 and 1 respectively). V^{SO} and V^L both increase with a, as is to be expected. For the values of b and ε chosen, V^L lies above V^{JP} for all values of a: a shareholder protected by limited liability will prefer to loot than choose the JP strategy. V^L is positive for all a (it can never be negative), but V^{SO} is less than zero for a less than about 0.41. For marginal productivity in this region, the social costs of the firm's activities outweigh the social benefits, and the planner would not choose to operate the firm. Shareholders avoid part of the costs of their activities, however, and the firm has positive value under the JP and L strategies.

Figure 2 carries out the same analysis, but with respect to the sunk capital parameter b. Strategies SO and JP do not involve scrapping of any capital (K_1 equals zero in both, and, due to condition (ref: damlim), the social planner need not scrap capital in the second period to pay for the damage); the profit from

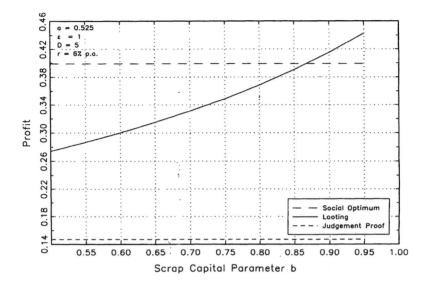

Fig. 2. Profits from Alternative Strategies with Change in Model Parameters.

these strategies is therefore constant with respect to changes in b. V^L increases with b, however – profits from looting, in which all capital is scrapped after one period, increase as the value loss due to scrapping decreases. Figure 2 shows that for b sufficiently large (greater than 0.86, with $a = 0.525$ and $\varepsilon = 1$, that is, less than 15% of the value of capital lost on its scrapping), the profit from looting exceeds that from socially optimal behaviour; for b lower than this (i.e. higher levels of sunk capital), V^L drops below V^{SO}. (With the parameter values used, V^{JP} lies below the returns from the other two strategies.)

Figure 3 shows how returns vary with the elasticity of the damage probability function ε; a and b are fixed at 0.5 and 0.8 in this figure. V^L is independent of ε, since the looting strategy involves no care expenditure. The signs of both V^{SO} and V^L depend on the parameter value. For low ε, both are negative; for higher ε, both are positive; and for still higher ε, both exceed the return to looting. This analysis echoes that of Fig. 1 – for low parameter values (i.e. low productivity of capital or highly ineffective care expenditures), production has negative social value, and the social planner would not choose to operate the firm. For higher values, V^{SO} becomes positive, and may exceed the returns from either looting (or the judgement proof strategy), for high enough a and ε.

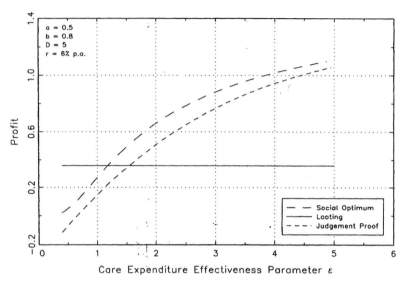

Fig. 3. Profits from Alternative Strategies with Changes in Model Parameters.

The figures indicate that the parameter values **define three regions**. The relevance of these three regions lies in their policy implications, to be discussed in the next section. In the **first region (I)**, socially optimal behaviour is the most attractive strategy, i.e. V^{SO} is greater than both V^L and V^{JP}. In Fig. 1, this region lies to the right of $a = 0.52$; in Fig. 2, it is to the left of $b = 0.87$; and in fig. 3, it is to the right of $\varepsilon = 1.2$. The **second region (II)** is defined by V^{SO} > 0, but less than V^L (or V^{JP}, whichever is the greater). In this region, the firm has positive social value, but shareholders will choose looting over socially optimal behaviour. Region II in Fig. 1 lies between the values of 0.41 and 0.52 for a; in fig. 2, it is defined by any value of b less than 0.87; and in fig. 3, it lies between $\varepsilon = 0.4$ and $\varepsilon = 1.2$. Finally, region III has V^{SO} < 0 – the firm has negative social value. In Fig. 1, this region lies to the left of $a = 0.41$; Fig. 2, as drawn, does not have a region III (although with other parameter values, it might); region III lies to the left of $\varepsilon = 0.4$ in Fig. 3.

Figure 4 shows the pairs of (a, b) parameter values for which socially optimal behaviour yields the greatest return, and the values for which looting is best for shareholders. ($\varepsilon = 1$, $D = 5$ and $r = 6\%$ per annum in this figure.) The region to the north-west of the upward-sloping line is the set of values in which $V^{SO} > V^L$: this region corresponds to highly productive (high a) and highly sunk (low b) capital (as Figs 1 and 2 suggest). Conversely, the region to the

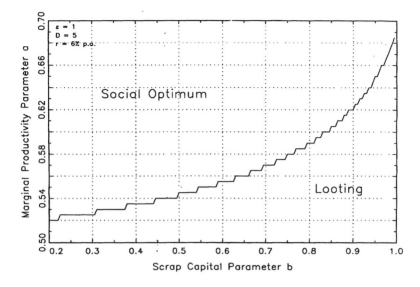

Fig. 4. Region in Which Looting is Profitable.

south-east favours looting – marginal productivity is low, as is the degree to which capital is sunk. It is clear from the plot that regions II and III occupy a significant part of the total space: V^L is greater than V^{SO} for many parameter values, and the figure suggests that looting will hardly be an exception. Moreover, the frontier marked on the figure will shift as ε, D and r change. Higher D and r – larger and longer-term damages – or lower ε – less effective care expenditure – will cause the frontier to move to the north-west, so that the region in which looting is optimal increases in range (and vice versa).

The above discussion makes clear that the looting strategy becomes more likely to be chosen as: (1) the amount of sunk capital increases; (2) the productivity of capital in the industry decreases; and (3) the costs of taking precaution increases.

COMMENTS ON PAPER BY TIMOTHY SWANSON AND ROBIN MASON

Anthony Heyes

It is well understood that most of the optimistic predictions about the ability of liability-based regimes to solve environmental problems falter when firms may – at the critical moment – be 'judgement proof'. As such, the potential for judgement proofness – the inability of a firm to pay, either in part or in whole, environmental damages imposed by it – is something that economists need to take seriously when thinking about legal and regulatory design. Tim Swanson and Robin Mason contribute substantially to this thinking in the current paper.

The authors adopt the memorable terminology of Akerlof and Romer's model of 'looting' society. They seek to characterise the circumstances in which firms may use antisocially low environmental standards to 'loot' society.

A firm can loot by operating in a way that generates future environmental risks, then either exiting the scene altogether or else restructuring in such a way that when the damage come to light their assets have been structured in such a way that there's little left to be seized. An interesting central tenet of this paper is that firms will rearrange their financial assets in such a way as to *make* themselves judgement proof. This is in contrast to most contributors (including Shavell's original analysis) in which the seizable assets of the firm are assumed to be exogenously determined – albeit too small to cover some liability scenarios. Ringleb and Wiggins, in a well-known paper in the *Journal of Political Economy* in 1990 made the same suggestion – and provided compelling empirical evidence of what they referred to as 'strategic subsidiarisation' in

An Introduction to the Law and Economics of Environmental Policy: Issues in Institutional Design, Volume 20, pages 413–415.
© 2002 Published by Elsevier Science Ltd.
ISBN: 0-7623-0888-5

certain sectors in the U.S. Strategic subsidiarisation refers to the situation in which a firm hives-off some of its more dangerous or high-risk activities (those activities most likely to generate future liability – such as the reactor cleaning activities of a nuclear power building and maintenance corporation) into a separately incorporated subsidiary with low capitalisation. In so doing the parent corporation is able to (or hopes that it is able to) protect its own assets from seizure in the event of a major claim against the subsidiary. The case law remains ambiguous as to the extent to which judges in different jurisdiction would be able and willing to access the assets of the parent – 'pierce the corporate veil' – in these circumstances. Even if parent companies don't subsidiarize, it is natural that activities that have the potential to impose substantial and deferred damage are ones that will, other things being equal, come to be dominated by: (a) shallow pocketed and/or (b) fly-by-night firms.

This paper explores some of the subtleties that arise when thinking about situations where the opportunity to loot arises. More importantly, it points to contexts in which looting behavior will be particularly prevalent, and ways in which policy can be adjusted to discourage looting. They argues that necessary conditions for looting to feature in equilibrium are threefold: (a) the availability of limited liability (b) the capacity to create deferred costliness and (c) a structure that renders liquidation the optimal strategy of players.

Conditions (a) and (b) should be apparent, (c) gives us plenty to think about. What sorts of things make liquidation less attractive to a firm – and hence looting a less attractive ploy? If a lot of capital is sunk in an industry that means that it cannot be recovered if the firm decides to exit, and this provides an incentive to want to stay. If the market for the activity in question is not particularly competitive – such that those in it are able to extract a nice flow of operating profit – that provides another push in the direction of not wanting to follow a looting strategy.[1] If the firm in question is also engaged in other markets, and is able to make profits in those, then that is something else that would make bankruptcy costly. It is also interesting to think about how a degree of tolerance in the operation of the liability regime may make looting less attractive, and what implications that may have for legislation and legal practice in this field.

Of course, in some cases the difficulties that arise from trying to tinker with the legal regime in such a way as to restore the incentives for private action may get so cumbersome – and throw up so many secondary problems – that we might just want to scrap private sector involvement altogether.

This is a nice piece of work, offering a number of insights into an important topic, and pointing towards a number of avenues for further work.

NOTE

1. In most contexts economists trumpet the benefits of perfect competition. Here competition is likely to be destructive. Under perfect competition supernormal profits are zero and the incentive to stay in the market is, therefore, also zero. This points to ways in which the work of the environmental regulator might interact with the work of competition (antitrust) regulators. A firm operating in an industry where a cozy cartel is tolerated or encouraged (e.g. through a program of licenses) is much less likely to want to give up its engagement in the activity.

FINANCIAL RESPONSIBILITY FOR ENVIRONMENTAL OBLIGATIONS: ARE BONDING AND ASSURANCE RULES FULFILLING THEIR PROMISE?

James Boyd

ABSTRACT

Financial assurance rules, also known as financial responsibility or bonding requirements, foster cost internalization by requiring potential polluters to demonstrate the financial resources necessary to compensate for environmental damage that may arise in the future. Accordingly, assurance is an important complement to liability rules, restoration obligations, and other regulatory compliance requirements. The paper reviews the need for assurance, given the prevalence of abandoned environmental obligations, and assesses the implementation of assurance rules in the United States. From the standpoint of both legal effectiveness and economic efficiency, assurance rules can be improved. On the whole, however, cost recovery, deterrence, and enforcement are significantly improved by the presence of existing assurance regulations.

An Introduction to the Law and Economics of Environmental Policy: Issues in Institutional Design, Volume 20, pages 417–485.
Copyright © 2002 by Elsevier Science Ltd.
All rights of reproduction in any form reserved.
ISBN: 0-7623-0888-5

1. INTRODUCTION

A bedrock principle of environmental law and regulation is that pollution costs should be borne by their creators. U.S. environmental laws and regulations give this principle form by making polluters liable for property, health, and natural resource damages and unperformed resource reclamation obligations. Unfortunately, many environmental obligations, despite being well defined in theory and in law, are not always met in practice. Bankruptcy, corporate dissolution, and outright abandonment are a disturbingly common means by which polluters avoid responsibility for environmental costs.[1]

Financial assurance rules, also known as financial responsibility or bonding requirements, address this policy problem. Assurance rules require potential polluters to demonstrate – before the fact – financial resources adequate to correct and compensate for environmental damage that may arise in the future. Accordingly, assurance acts as an important complement to liability rules, restoration obligations, and other compliance requirements.[2] A benefit of assurance rules is that they can harness the expertise and scrutiny of private, third-party financial providers. The insurers, sureties, and banks who provide the financial products used to demonstrate compliance, for their own commercial reasons, train an extra, self-interested set of eyes on the financial and environmental risks posed by potential polluters. In this way, assurance rules can yield a flexible, market-based approach to compliance and monitoring.

Financial assurance is demanded of a wide variety of U.S. commercial operations, including municipal landfills, ships carrying oil or hazardous cargo, hazardous waste treatment facilities, offshore oil and gas installations, underground gas tanks, wells, nuclear power stations, and mines. Firms needing assurance can purchase it in the form of insurance, surety obligations, bank letters of credit, and deposit certificates. Alternatively, firms can establish trust funds or escrow accounts dedicated to future obligations. Most programs allow wealthy and financially stable firms to comply via demonstration of an adequate domestic asset base and high-quality bond rating. A wealthy financial parent can in some cases guarantee the obligations of a subsidiary or affiliate via an indemnity agreement.

This study provides an overview of financial assurance policies based on a review of the rules' implementation in the U.S. Relatively little analysis of the rules' practical implementation exists.[3] The goal is not an exhaustive review of specific regulatory programs, but rather a synthetic overview of the many issues common to environmental assurance programs. From the standpoint of both economic efficiency and legal effectiveness assurance rules can be improved. Assurance programs raise a set of design issues, including the level of assurance to be required, the financial mechanisms to be allowed, the conditions under which bonds are released, and the interaction of assurance rules with other areas of law; most importantly, bankruptcy law. This report illustrates those issues and identifies a set of correctable weaknesses present in some assurance programs. For instance, in some regulatory contexts inappropriately low levels of assurance are required; in others, the mechanisms used to demonstrate responsibility undermine the goal of cost internalization.

Despite a set of criticisms regarding the details of policy, this report should be read as a spirited defense of financial assurance's desirability as a regulatory tool. Absent assurance, too many firms can and do abandon obligations. As will be evident from the cases and data cited in this report, the evasion of environmental liabilities and cost internalization by defunct or insolvent firms is relatively common. On average, 60,000 U.S. firms declare bankruptcy each year and an untold number cease or abandon operations without even entering legal bankruptcy proceedings.[4] Clearly, not all of these firms leave unfunded environmental obligations behind them; but many do. Mandatory assurance addresses the insolvency problem in a direct way, and thereby strengthens the effectiveness of environmental regulation and law.

1.1. The Underlying Problem – Unperformed Obligations and Non-Recoverable Liabilities

Conceptually, polluter cost internalization is nearly unassailable as a guiding principle for environmental regulation. Cost internalization by responsible parties yields the most equitable means of victim compensation, the alternatives being no compensation or compensation provided by public funds. Polluter cost internalization also promotes deterrence, risk reduction, and innovations to reduce environmental harm.[5] Accordingly, with few exceptions, most U.S. environmental laws make polluters liable for damages caused by commercial activities that injure the public health or that cause property or natural resource damage.

Unfortunately, cost internalization's importance in law and regulation is not always matched by its achievement in practice. Even the most unassailable legal obligation can quickly evaporate when presented to a bankrupt, dissolved, or absent polluter. Consider first the implications of bankruptcy. Generally speaking, debtors are protected from creditors by the "automatic stay" provision of the U.S. bankruptcy code.[6] This means that both private and public environmental claims can be discharged in bankruptcy.[7] In other words, environmental costs are only partially recoverable once bankruptcy occurs, if they are recoverable at all.[8] To compound the problem, firms may purposefully increase the likelihood of bankruptcy by divesting themselves of capturable assets in order to externalize costs. In industries where liability costs are potentially significant, firms' business organization and capital investment and retention decisions may be influenced by the desire to externalize liabilities. For instance, firms may avoid retained earnings, choosing to not vertically or horizontally integrate, or shelter assets overseas.[9]

Environmental cost recovery can also be defeated if a polluter has legally dissolved prior to the realization of liabilities or performance of obligations. There are limits to this strategy. A liable firm that is simply sold does not automatically escape liability since those liabilities will be transferred to the purchasing firm.[10] If assets are sold piecemeal or simply retired over time, however, environmental costs can more effectively be externalized. This possibility is enhanced by the nature of many environmental risks and obligations, which often materialize only over a period of years or decades.[11] Dissolution can be a rational, if socially irresponsible, way to avoid future obligations. Irrespective of the precise strategy used to avoid liability and reclamation obligations, the lack of a solvent defendant defeats the ability of victims or governments to collect compensation. Insolvency also undermines the law's ability to deter environmental injuries in the first place.

1.2. The Scale and Scope of Unrecovered Environmental Costs

Non-recoverable environmental obligations are more than a theoretical possibility. Over the past decades untold numbers of environmentally damaging operations have been abandoned or have avoided liability via bankruptcy. There is no central repository of statistics regarding the scale of unrecovered environmental obligations but figures from a range of environmental programs illustrate the significance of these costs.

Underground Storage Tanks.
Leaking USTs pose a significant risk to the nation's groundwater supplies. There are currently an estimated 190,000 abandoned petroleum underground storage tanks in the U.S.[12] According to the EPA, "these USTs pose a challenge in that the owner is either disinclined or financially unable to comply, or is often difficult to locate." In addition, billions of dollars in public funds have been expended to clean up USTs that were not abandoned, but where owners and operators would have been unable to bear remediation costs themselves.[13]

Oil and Gas Wells

Unplugged oil and gas wells can pollute both ground and surface water. Many states have programs that have identified thousands of abandoned oil and gas wells. State well-plugging funds have spent $70M to plug and abandon approximately 13,000 orphan wells. There are an estimated 57,000 remaining orphan wells nationwide.[14] With an average plugging cost of $5,400, the cost to state agencies for plugging these orphan sites will be an additional $560M.

Oil Spills

Beginning with the 1972 Clean Water Act, and now under the Oil Pollution Act, the U.S. has maintained a public fund for the cleanup of oil spills associated with offshore accidents and onshore accidents contributing to surface water pollution. A goal of the fund is to recover public expenditures on oil spill response from responsible parties. According to one study, however, the current fund has recovered only 19% of its expenditures from responsible parties.[15] Accordingly, the remaining percentage corresponds to costs externalized by polluters.

Landfills and Other Disposal Facilities

A recent inventory by the state of Texas located 4200 abandoned landfills in that state alone.[16] A nationwide study of permitted, operating hazardous waste landfills in 1984 and 1985 identified 54 owned by bankrupt firms.[17] A more recent EPA study of medium-sized municipal solid waste disposal firms found that, of 40 firms studied, 37 had estimated financial assurance obligations exceeding their net worth.[18] As recently as 1999 a Canadian company, exploiting exemptions in waste disposal regulations, was able to abandon a site in Tacoma Washington leaving $4.3m in uncompensated cleanup costs.[19]

Hardrock Mining

The Bureau of Land Management has identified 900 environmentally hazardous abandoned mine sites on BLM managed lands.[20] A 1986 GAO study found that of a sample of BLM mine sites surveyed, 39% had not been reclaimed.[21] One non-governmental study estimates a total of 557,000 abandoned mine sites nationwide, with an estimated cleanup cost of $32 to $72 billion.[22] Sixty-seven abandoned mines are on the EPA's Superfund National Priority List. The Agency estimates that it will cost approximately $20 billion dollars to clean up mine sites currently on the NPL.[23] In terms of mine bankruptcies, a study of mining operations found 26 large-scale Western hardrock mines in bankruptcy as of 1999.[24] The Summitville mine in Colorado, abandoned in 1993, alone has an estimated cleanup cost of $150 to $180m.[25] A 1999 National Research

Council report identified site abandonment and unfunded obligations as a significant regulatory issue for the industry.[26]

Coal Mining

The federal government's Abandoned Mine Land (AML) program estimates that there are $7.9 billion worth of high priority coal related AML problems, including health, safety, and environmental problems.[27] A study of coal mining sites in Pennsylvania found more than 22,000 mining acres with forfeited mining bonds and that 67% of all acres covered by bond requirements had not been reclaimed.[28] A Congressional Hearing in 1986 identified poor reclamation rates from a variety of other states, including reclamation rates of only 7%, 19%, and 13% in Indiana, Kentucky, and Tennessee, respectively.[29] A recent actuarial study placed a lower bound of $1b on Pennsylvania's long-term mine drainage costs, associated primarily with abandoned mines.[30]

National Priority List (NPL) Sites

Many Superfund sites were polluted by parties that no longer exist or that have declared bankruptcy.[31] EPA refers to these parties' contribution to contamination as "orphan shares." One EPA study estimated that the cost of orphan shares associated with NPL sites will range from $150 to $420 million each year.[32] The EPA's current orphan share compensation program has spent $175 million at 98 sites where responsible parties were willing to negotiate long-term cleanup settlements.[33] It should be noted that these expenditures represent only a lower bound on non-recoverable NPL costs, since orphan share contributions are strictly limited to 25% of remedy and removal costs.[34] The lion's share of orphan shares are allocated to viable responsible parties under principles of joint and several liability. Also, these numbers only refer to orphan shares at the 1300 NPL sites, which represent only a fraction of polluted sites nationwide.[35]

It should be emphasized that many of the unrecovered environmental obligations indicated above are due to the failure of past, rather than current regulatory programs. As described below, a variety of regulatory programs have been developed in recent years to minimize the environmental and financial problems created by bankrupt or unidentifiable polluters. The scale of problems indicated above suggest that these new programs will fill an important gap in environmental regulation. However, as will also be described below, current programs have by no means eliminated the externalization of significant environmental costs by polluters.

One conclusion to be drawn from the above statistics is that it is not only notorious catastrophes, such as oil tanker spills, that signal the need for financial responsibility. Individually smaller risks, such as tank leaks at filling stations, or unplugged wells, can in aggregate create even greater externalized costs due to large numbers of operations and the shallower pockets of firms responsible for them. Finally, it is important to realize that large companies, not only small ones, can externalize costs via bankruptcy. A current example is the chemical manufacturer W. R. Grace, which has recently filed for bankruptcy due primarily to asbestos-related liability claims. The effect of the firm's bankruptcy on its multi-million dollar environmental cleanup liabilities remains to be seen.[36]

1.3. The Benefits of Assurance

Liability rules and reclamation obligations lead to polluter cost internalization only in theory. In practice, liability, many administrative requirements, and any other after-the-fact penalties or obligations suffer from an important weakness: since the financial damages or obligations arise only after environmental damage has occurred, polluters can escape cost internalization via prior dissolution or bankruptcy. Financial assurance rules counter this weakness.

In concrete terms, financial responsibility ensures that the expected costs of environmental risks appear on a firm's balance sheets and in its business calculations. If new investments imply possible future environmental costs, financial responsibility increases the relevance of these costs to the firm's decision making. When firms self-insure, they must possess demonstrable wealth and financial stability. Firms with fewer resources often cannot self-insure and must therefore acquire rights to financial assets from third parties, such as banks and insurers. Third-party assurance providers are obviously concerned that their capital will be consumed by their clients' future liabilities. As a result, they have a strong incentive to monitor the environmental safety of firms they underwrite.[37] Capital providers can also base the cost of capital or premiums on observable attributes of the firms to whom they provide assurance. For example, more favorable premiums can be offered to firms with meaningful risk management and safety programs. In the extreme, financial coverage may be denied altogether to firms that fail to demonstrate acceptable levels of safety. In these ways, the capital markets that arise to satisfy demand for financial responsibility generate incentives to reduce environmental risks.[38]

Financial assurance can also foster timely, relatively low-cost public access to compensation. This can be beneficial when a swift response allows for the minimization of damages. When assurance is held by a public trustee, such as a state regulatory agency, it minimizes the public transaction costs associated with collecting compensation. Even when liability is firmly established, the possibility of appeal, delay, and uncertainties associated with penalty collection can complicate the actual transfer of funds from defendants to victims and resource trustees. Some financial assurance instruments, such as letters of credit, allow almost instant access by regulators to reserved funds. This shifts the burden of proof from the government to the plaintiff. Instead of the government having to prove that compensation is due and seek the funds, the burden falls to the polluter to demonstrate that they are not liable.[39]

Assurance is a time-tested concept. Its application is neither new nor confined to environmental problems.[40] Mandatory automobile insurance and minimum capital requirements for banks share similar motivations: namely, the desire for victim compensation and the deterrence of inappropriate risk-taking.[41] Bail and construction bonds, like environmental bonds, guarantee performance of a future action by making a solvent third party liable for the costs of a performance failure. In terms of their environmental application, assurance has been advocated for decades as a complement to environmental law and regulation.[42] The academic literature on tort law has long identified defendant insolvency as a source of inefficiency associated with the use of liability rules.[43]

1.4. Alternatives to Assurance

Perhaps the strongest motivation for assurance requirements arises from contemplation of the alternatives. Since environmental costs never simply vanish on their own, someone must pay. The question is, who? Two principal alternatives exist: the externalization of costs to society and the extension of environmental costs to polluters' business partners. As argued above, the externalization of environmental costs to society is highly undesirable because it undermines deterrence and the ability to compensate victims. The extension of liability to business partners is a more complex case. But it, too, highlights the desirability of assurance.

The law routinely extends liability to the business partners of insolvent or absent defendants. Retailers and distributors can be liable for injuries due to defects in products they sold but did not manufacture and employers can be liable for damages caused by independent contractors employed by them.[44] The motivation for extending liability is the same as that for assurance: deterrence and compensation are served by an internalization of costs. Firms exposed to their business partners' liability will more closely monitor those partners' safety. Business partners also provide a source of compensation. In the environmental context, joint and several liability extends liability in this way and for these purposes. Under CERCLA, an acquiring firm takes on the liabilities attached to property owned by the seller.[45] Liability is also extended from operators of disposal facilities to the original generators of waste.[46] And liability can be applied without reference to fault or the liable firm's proportional contribution to the damage.

Assurance is preferable to extended liability for a variety of reasons. First, the extension of liability does not guarantee cost internalization since there may be no applicable business partners from whom to seek compensation, or if there are they may themselves be insolvent. Second, as the history of CERCLA has shown, joint and several liability entails significant transaction costs associated with *ex ante* contracting between mutually liable firms and the resolution of *ex post* claims for contribution among jointly liable defendants.[47] Finally, extended liability can distort production decisions, such as investments in capital and the pattern of transactions between contracting parties.[48]

2. WHEN IS ASSURANCE REQUIRED?

Although some assurance rules have existed for decades under U.S. law, in the past decade their implementation has become much more widespread.[49] Assurance regulations are now associated with many of the U.S.' most important environmental laws. Financial assurance is required under the Oil Pollution Act (OPA), the Comprehensive Environmental Response, Compensation, and Liability Act (CERCLA), the Resource Conservation and Recovery Act (RCRA), the Outer Continental Shelf Lands Act (OCSLA), the Federal Land Policy and Management Act (FLPMA), Atomic Energy Act (AEA), Toxic Substances Control Act (TSCA), Safe Drinking Water Act (SDWA), and the Surface Mining Control and Reclamation Act (SMCRA). Not all enterprises regulated under these laws are subject to assurance requirements, but financial assurance is required for vessels carrying oil or hazardous substances, underground petroleum storage tanks, solid and hazardous waste landfills, many types of industrial, oil, and gas wells, offshore oil-drilling facilities and pipelines, nuclear power plants and disposal facilities, and coal and mineral mining operations.

2.1. Federal Assurance Regulations

Assurance rules differ somewhat depending on their precise application, but always feature descriptions of implementation schedules, types of facilities to which the rules apply, financial instruments with which compliance can be achieved, and enforcement procedures. This section provides a brief overview of the types of facilities and obligations governed by U.S. federal assurance rules. Section 3.4 describes the variety of financial mechanisms firms can use to demonstrate financial responsibility.

2.1.1. Vessels Carrying Oil and Hazardous Cargo

A financial assurance rule, authorized by both OPA and CERCLA, governs waterborne vessels that carry oil or hazardous substances.[50] Before the passage of OPA and CERCLA, financial responsibility was required for vessels carrying oil and hazardous cargo under the Federal Water Pollution Control Act.[51] The current rules apply to a wider range of vessels and facilities, cover a wider range of damages, and require higher levels of coverage than earlier rules.[52] Full implementation of these rules has occurred only recently.[53] Deadlines for compliance, which depended on the type and size of vessel, occurred between 1994 and 1997.[54] The vessel rule applies to tank vessels of any size, foreign-flag vessels of any size, and mobile offshore oil- and gas-drilling units.[55] Some smaller commercial vessels, such as barges not carrying oil or hazardous substances, are excluded from the regulations. Mandatory assurance amounts are based on the type of cargo, type of vessel, and the vessel's tonnage. For a large vessel, assurance requirements can range into the tens of millions of dollars.

2.1.2. Offshore Oil Facilities

Another assurance rule authorized by OPA governs offshore facilities used for oil exploration, drilling, production, or transport.[56] Notice of the offshore facilities rule was given in 1997 and finalized in 1998.[57] Compliance for all regulated facilities had to be demonstrated by 1999. Prior to OPA, financial responsibility was required for offshore facilities under OCSLA, and for oil pipelines under the Trans-Alaska Pipeline Act.[58] The offshore facility rule applies to facilities "in, on, or under" navigable waters. Covered facilities include platforms, terminals, refineries, and pipelines used for oil exploration, drilling, and production.[59] Onshore oil facilities are not covered. Assurance amounts are based on calculations of "worst-case" discharge volumes from the facilities and can go as high as $150m.[60]

2.1.3. Underground Petroleum Storage Tanks

RCRA requires financial responsibility for the owners and operators of underground petroleum storage tanks (USTs), such as those used at gas stations.[61] The rules were codified in 1988, but compliance deadlines for certain operators were extended until 1998. UST owners and operators must demonstrate the ability to perform corrective action to restore a contaminated site and compensate third parties for property damage or injury arising from a leaking tank. The amount of financial assurance that must be demonstrated can be significant. For example, most gas stations are required to carry $1 million in insurance coverage.

2.1.4. Solid Waste Landfills and Hazardous Waste TSDFs

RCRA also requires financial assurance for solid-waste (non-hazardous) landfills and hazardous waste treatment, storage, and disposal facilities (TSDFs).[62] The final municipal landfill compliance deadlines were in 1997. Facilities must provide financial guarantees designed to assure the internalization of costs associated with the closure of these facilities and their long-term maintenance.[63] Closure requirements include the capping of landfills and long-term monitoring of groundwater impacts. Hazardous facilities must also demonstrate liability coverage to compensate third parties suffering bodily injury or property damage resulting from an accident.[64] Coverage amounts for a typical site run into the millions of dollars.

2.1.5. Wells

To protect drinking water quality, the Safe Drinking Water Act of 1974 established rules for the regulation of "underground injection control" (UIC) wells. Operators of Class I, II, and III wells are required to demonstrate financial responsibility for their eventual plugging and abandonment.[65] These wells include wells used to dispose of hazardous waste, wells used to dispose of fluids associated with production of natural gas and oil, and wells used to inject fluids for the extraction of minerals.[66] Unplugged wells can lead to migration of contaminants into aquifers, saltwater intrusion into a freshwater aquifer, and surface soil contamination. In addition to plugging, requirements can include re-vegetation, erosion control, and removal of tanks and lines. Bond amounts vary greatly depending on the well type.[67] There is no assurance required for 3rd-party liability.

2.1.6. Coal and Hardrock Mines

Coal mining is regulated at the federal level by the Surface Coal Mining and Reclamation Act of 1977 (SMCRA). SMCRA governs mining impacts including both surface effects, such as strip mine reclamation, and subsurface effects, such as damaged water quality from mine drainage.[68] Prior to the Act's passage, states had regulatory authority and often required bonds, though these bond amounts were often inadequate.[69] SMCRA increased bond amounts for site reclamation, including revegetation, backfilling, grading, and mine drainage controls. Bond amounts are based on acreage and vary depending upon the type of mining activity and site characteristics.[70]

Assurance is also required for hardrock mining operations. Hardrock mining continues to be regulated primarily by state law and state bond policies vary.[71] However, federal law requires hardrock bonds when mining occurs on federal lands.[72] Mining on lands administered by the Bureau of Land Management and

U.S. Forest Service is subject to those agencies' respective rules.[73] Like coal mine bonds, hardrock bonds are based on acreage and site characteristics.

2.1.7. PCB Storage Facilities

Under the Toxic Substances Control Act, commercial PCB storage facilities must demonstrate financial assurance for costs associated with the closure of storage facilities, including final disposal, decontamination, and monitoring costs.[74]

2.1.8 Nuclear Reactors, Fuel Processing and Enrichment Plants, and Radioactive Disposal Facilities

The Atomic Energy Act requires financial assurance for the costs associated with nuclear power plant decommissioning and for the closure of radioactive waste disposal facilities.[75] Minimum amounts for plant decommissioning are in excess of $100m. Bonds are also required for the closure of uranium and thorium mill sites.[76] Assurance is also required for liabilities arising due to nuclear accidents. The Price-Anderson Act, while limiting the industry's liability, also requires coverage for reactors, reprocessing facilities, and fuel enrichment facilities.[77] The private insurance requirement is currently $200m for reactor units.[78]

2.2. The States' Role in Assurance Regulation

State laws sometimes complement and expand upon federal assurance regulations. States also often implement the assurance rules mandated by federal law. For these reasons, it is most appropriate to think of assurance regulations as emerging from a combination of state and federal rules and enforcement.

A comprehensive survey of state financial assurance requirements is beyond the scope of this paper. However, it is worth noting that individual states can have assurance requirements that in some cases exceed those under federal law. For example, California recently passed a law requiring oil-carrying vessels to demonstrate $1 billion in coverage for oil pollution damages.[79] The law also requires marine terminals, fueling facilities, and barges to demonstrate assurance coverage. Alaska law mandates financial responsibility for oil terminals, pipelines, tank vessels, and barges with coverage levels higher than under federal law.[80] In addition, a new Alaska law extends financial responsibility to vessels other than tankers, including cruise ships, and railroad tank cars carrying oil.[81] Similarly, Washington State requires oil vessel coverage in excess of federal requirements and extends the requirements to a broader range of facilities.[82]

In other cases, states require assurance for operations or situations not required under federal law. Again, a comprehensive review is beyond the scope of this study, but Michigan, for example, requires holders of sand dune mining permits to provide assurance for the reclamation and re-vegetation of sand dune areas.[83] Several states require bonds to cover closure costs for scrap tire disposal facilities.[84] Texas requires transporters of medical waste to demonstrate insurance for automobile and pollution liability.[85] Several states require financial responsibility for the closure of agricultural operations producing animal waste.[86] And North Carolina established financial responsibility requirements for dry-cleaning operations.[87]

States are often responsible for the implementation of assurance regulations, even when assurance is required by federal law. This is true, for example, under RCRA. In general, UST, landfill, and TSDF assurance programs are operated by the states, subject to federal oversight and approval.[88] Under the SDWA, the federal government regulates wells only if states do not administer their own programs.[89] In the hardrock mining context, states have their own mine bonding regulations, but states must come to agreement with the federal government over bonding criteria for mines on federal land.[90] Similarly, Under SMCRA, the Department of Interior's, Office of Surface Mining Reclamation and Enforcement (OSM) enforces the rules until individual States achieve what is called "primacy," or independent enforcement authority approved by OSM.[91]

3. DEMONSTRATING FINANCIAL RESPONSIBILITY

Financial responsibility can be demonstrated in a variety of ways. All of the assurance rules described above allow a choice compliance mechanisms. This section describes the variety of mechanisms in more detail. Before doing so, it is useful to note some basic distinctions: between insurance and performance bonds, and between self-assurance and assurance that is purchased from third parties.

3.1. "Assurance as Insurance" vs. "Assurance as a Bond"

There are two basic types of environmental costs that require assurance: uncertain environmental liabilities – typically associated with remedial site-cleanups, property damage, or health impacts – and more defined environmental obligations, such as site restoration, land reclamation, or long-term water treatment obligations.

The distinction is subtle, but important. Assurance for uncertain environmental costs is best thought of as mandatory insurance. An important characteristic of insurance is that, by forcing cost internalization, it creates an incentive to reduce uncertain environmental risks via improved technology or management. In contrast, when obligations are fully known *ex ante*, there is no need for insurance, *per se*. Instead, what is needed is a guarantee that the known obligation will be performed. Typically, bonds are used to guarantee performance of a known, future obligation.

Consider an example: landfill closures involve both relatively certain obligations and relatively uncertain hazards. Known obligations include the need to re-vegetate, cap, and monitor the site. These obligations tend to be guaranteed via bonds. Uncertain risks from the landfill include future groundwater contamination, health impacts, and damage to neighboring property. These uncertain liabilities tend to be assured via insurance coverage. To be clear, the motivation for assurance in the bonding context is nearly identical to the motivation for assurance in the insurance context. In both, assurance guarantees that funds will be available in the future to internalize costs.

The difference, though, has practical implications for the instruments used to demonstrate assurance. First, bond agreements typically assume that the principal bears ultimate responsibility for the loss. In other words the bond provider pays only if the principal is unable to do so, due to insolvency or abandonment.[92] Consequently, bond pricing is primarily a function of the principal's bankruptcy risk and bonds tend to be priced as a simple percentage of their face value.[93] Insurance products are different since insurers typically pay the claims of both solvent and insolvent clients. This means that insurance is priced to reflect a greater likelihood and range of possible claims. Consequently, insurance is usually priced with much greater sensitivity to the risks presented by the insured.

A bright line between assurance as insurance and assurance as a bond should not always be drawn. Moreover, the distinction should not be applied to the suppliers of these forms of assurance, since surety bonds are often sold by insurance companies.

3.2. Self-Demonstrated vs. Purchased Assurance

All assurance programs allow firms to purchase assurance from a 3rd party. Insurance, bonds, bank certificates, and letters of credit can be purchased from private financial providers, including insurers, sureties, and lenders. Some programs allow firms to self-demonstrate assurance as an alternative to purchased assurance. Self-demonstration is essentially a demonstration of profitability and stability. In theory, wealthy, stable firms can be counted on to internalize their future costs, without the involvement of 3rd party capital providers.[94]

There are clear differences between purchased and self-demonstrated assurance. The most important difference is in the government's monitoring role. Self-demonstration requires the government to monitor the firm's financial condition over time. For instance, asset ratios, profitability indicators, and bond ratings may be used to pass a self-demonstration test. Accordingly, regulators must regularly audit this kind of financial data to determine their accuracy and adequacy. Note, however, that corporate financial auditing is not a traditional strength of environmental regulators. In contrast, purchased assurance is relatively easy to monitor.[95] Two basic things must be verified: first, the existence of a valid assurance contract with a 3rd party provider and second, the financial strength of that provider. The financial strength of capital providers is easier to monitor since oversight is usually already in place. The Treasury Department, for example, keeps an up-to-date list of government-approved sureties. In contrast, self-demonstration requires verification of changeable, complex, and often subjective financial data.

Another difference is that purchased assurance inevitably directs the attention of private financial providers to the risks presented by the potential polluter. After all, it is in the commercial interest of private financial providers to accurately analyze and minimize risk. This virtue is not harnessed when firms self-demonstrate assurance.

Some assurance mechanisms blur the distinction between purchased and self-demonstrated mechanisms. Trust funds, for example, are funded by the firm itself and thus are not technically purchased. However, when appropriately designed they involve an independent trustee and funds releasable only on the approval of the regulator. Accordingly, trust funds need not suffer from the weaknesses of self-demonstration. Another mechanism that blurs the distinction is captive insurance. Captive insurance is insurance provided by the firm itself or by a collection of similarly-regulated firms. Like purchased insurance, captives typically charge risk-sensitive premiums. Because they are not independent firms, however, they present many of the same monitoring problems as self-demonstrated assurance.[96]

3.3. Publicly Subsidized Assurance

In some instances environmental assurance is provided by public funds. For example, most states under RCRA's underground storage tank rules set up state guarantee funds to help owners comply with RCRA's financial responsibility provisions. Funds were financed via taxes on gasoline sales or retail deliveries, not by UST owner-operators themselves.[97] In a limited set of cases, publicly-funded remediation is a defensible public policy.[98] In general, however, public financing of pollution costs is undesirable. Public funds are usually funded via taxes that do not distinguish on the basis of a firm's safety record, technology, or ability to manage risks effectively. For this reason, coverage costs do not reflect risk and thus fail to create an incentive for risk reduction. One particularly troubling aspect of publicly-operated assurance funds is that they undermine private markets for assurance. Public assurance funds tend to be cheaper and easier to qualify for than privately-purchased insurance. Private insurance is likely to be better monitored and more accurately priced, however, due to private providers' incentives to minimize their own risks and collect premiums that will cover the costs they are insuring.[99] Most states have already phased out the publicly financed UST guarantee funds, or are in the process of doing so.

3.4. Mechanisms

This section provides more specific descriptions of the financial products, mechanisms, or tests firms can use to demonstrate assurance. Assurance programs allow firms to choose from a variety of the mechanisms described below and combine mechanisms in order to meet their full assurance obligation.[100] Under most programs, firms do not all choose the same type of mechanisms in order to comply.[101]

3.4.1. Insurance
Insurance policies are generally purchased from independent insurance providers. For a premium the insurer promises to compensate the purchaser for claims covered in the insurance contract. Contracts are of two basic forms, "claims made" and "occurrence." Claims made policies provide coverage for claims presented to the insured and reported to the insurer during the coverage period. Claims falling outside the coverage period, even if caused by acts during the coverage period, will not be covered.[102] Accordingly, care must be taken with claims made policies so that they provide assurance over long time horizons. In contrast, occurrence policies cover claims arising even after the policy period has ended, providing the cause of the claim occurred during the policy period. Insurers like to avoid occurrence coverage, as a way to reduce the scale and enhance the predictability of their exposures. From the standpoint of public policy, however, occurrence coverage addresses the goals of assurance better than claims made coverage.

Another concern associated with insurance is that the policy may feature "exclusions" which weaken coverage.[103] For this reason, care must be taken to verify that policies fully cover the kinds of claims subject to assurance requirements.

3.4.2. Letters of Credit and Surety Bonds

Letters of credit are purchased from banks.[104] They require the bank to pay a third party beneficiary, in this case the government, under certain specified circumstances, such as the failure of the purchaser to perform certain obligations. Banks may require collateral or deposits before providing a letter of credit, depending on the purchaser's financial health. Letters of credit are typically priced as a small fraction of their face value and are granted for annual terms. Typically, letters of credit are automatically extended at the completion of a one year term, subject to the purchaser's continued good credit and adherence to contract terms. The instrument can be altered only with the agreement of the purchaser, the provider, and the beneficiary. The credit provider does not generally pay out on claims. Rather, the purchaser indemnifies the bank, making the bank liable only if the purchaser defaults. Designed properly, beneficiaries can draw on the letter of credit if its term is not extended and if a replacement form of assurance is not put in place.

Surety bonds are similar to letters of credit, though usually purchased from an insurance company. Sureties usually pay out on claims only in the event of default by the purchaser.[105] Under most programs surety companies must be certified by the U.S. Treasury Department in order to qualify as an acceptable source of assurance.[106] Bonds, like letter of credit, cannot be cancelled without prior notice being given to the regulator and the government is the beneficiary of the bond in the event of default by the principal.

"Blanket bonds" are a special form of bond, allowable as assurance for oil and gas wells, where a relatively large number of sites are covered by a single bond. With proof of past good behavior and passage of financial tests, well operators can bond a large number of wells for a relatively small fraction of the assurance they would have to demonstrate if they bonded the wells individually.[107] Since, almost by definition, the assurance amount is less than the firm's obligations, blanket bonds do not guarantee full cost recovery.

3.4.3. Cash Accounts and Certificates of Deposit

Cash accounts and certificates of deposit are a particularly iron-clad form of assurance. They place cash or some other form of interest-bearing security into accounts that are made payable to or assigned to the regulatory authority.[108] In the event of default, the accounts may be liquidated by the regulator for the payment of covered obligations. Key safeguards for the use of these instruments are that the public authority be made the sole beneficiary, the accounts are managed by independent financial institutions, and that terms can be changed only with the approval of regulators. Assets remaining after the fulfillment of obligations are revert to the control of the firm.

3.4.4. Trust Funds

Trust funds are vehicles for the collection of monies dedicated to a specific purpose. So-called third party trust funds are administered by an independent trustee who is in charge of collecting, investing, and disbursing funds.[109] Money is typically paid in over some period of time. This means that trust funds may not be fully funded at the time of a claim. Accordingly, shorter-term pay-in periods are preferable for assurance. The regulator should be the sole beneficiary of any such trust fund. The trust agreement, administered by the trustee, specifies the conditions under which trust monies are paid out. After obligations are fulfilled trust assets are returned to the firm. It is essential that regulators monitor payments into the trust.

Less desirable are 1st party trusts, where trust funds remain in the custody of the principal. Because there is no independent trustee, 1st party trusts should allow the regulator to make direct inquiry into the trust's status. Also, the principal's ability to alter the trust's terms or access its funds must be restricted.

3.4.5 Self-demonstration

Self-demonstration, or a "financial test," is a mechanism that allows companies with relatively deep pockets to satisfy coverage requirements by demonstrating sufficient financial strength.[110] For example, rules may require that the firm's working capital and net worth both be greater than the coverage requirement. Some require or allow a bond rating test. Usually, a combination of tests must be passed.[111] There may also be a domestic assets tests to foster cost recovery. For example, working capital may be defined as the value of current assets located in the United States minus current worldwide liabilities; and net worth may be defined as the value of all assets located in the United States minus all worldwide liabilities.[112] Ideally, when using the financial test, firms must make annual reports that are independently audited according to generally accepted accounting practices and that are consistent with the numbers used in the firm's audited financial statements for Securities and Exchange Commission reporting.[113] And changes in a firm's financial status should also be reported.

3.4.6 Corporate Guarantee

A financial guaranty, or indemnity agreement, allows another firm, such as a parent corporation, to satisfy the coverage requirement. Financial guarantors must themselves pass the corporate financial test and agree to guarantee the liabilities of the firm seeking assurance. The requirements are identical to those for self-demonstrators, including the domestic assets requirement. Some programs require that an indemnity agreement be with a single firm that is either a corporate parent or an affiliate.[114]

As financial responsibility instruments, self-demonstration and indemnity are popular with the regulated community, because no third party must be involved and compensated. A common refrain in the regulated community is that the financial tests should be made less stringent, thus allowing a larger number of firms to qualify. However, these instruments are less desirable from a regulatory standpoint. They require more administrative oversight than insurance and sureties, and they provide less of a guarantee that costs will be recoverable in the future. Accordingly, some programs have resisted changes favoring the more widespread use of self-demonstration.[115]

4. THE POLITICS AND COST OF ASSURANCE

The regulated community typically opposes new or strengthened assurance rules.[116] New assurance rules produce dire predictions of significantly higher insurance rates, the withdrawal of insurers and sureties from markets, and the demise of businesses unable to meet the assurance requirements.[117] The response to OPA vessel assurance rules is illustrative of the alarm with which some in the private sector received new assurance rules. The law was predicted by some to increase the cost of insurance by seven to nine times relative to pre-OPA rates, if insurance was to be available at all. Even more dire predictions included the possibility of a total halt in maritime trade[118] and the collapse of worldwide vessel insurance markets.[119] RCRA's UST regulations were met with similar fear and opposition, one U.S. Representative vowing that he would not "just sit around and watch the small businesses be legislated out of business by the Federal Government."[120] More recently, changes in hardrock mining rules have prompted opposition based on their impact on small mining operations.[121] Should these fears call into question assurance's social desirability?

First, it should be noted that much opposition can be attributed to an underlying distrust and fear of expanded liability, rather than fear of assurance requirements themselves. Over the last few decades the widespread adoption of assurance rules has occurred alongside a broad expansion of liability for environmental damages under U.S. law. For example, the adoption of strict, joint and several, and retroactive liability rules have vastly expanded the conditions under which polluters are liable. Also, federal enforcement is not a potential polluter's only concern. In addition to the federal government, private citizens, states, and localities can sue to recover environmental damages. Another source of concern to many is that OPA, CERCLA, and other statutes have expanded liability to include damages to natural resources, as distinct from damages to private property or human health.[122] Natural resource damages (NRDs) can be difficult to value and methods used to calculate NRDs are controversial.[123] By definition, NRDs involve damages to ecosystem services or resources that are not "marketed" and for which there is no observable price. This means that NRDs are unpredictable and highly sensitive to the valuation methodologies employed by the government and courts.[124]

All of these factors have generated fears in the regulated community of large, unpredictable, and uninsurable obligations. This is true even when liability is capped, as it is in some cases.[125] One way the expansion of liability can be opposed is via opposition to assurance, since for many firms assurance requirements are the way in which bottom-line liabilities are actually defined. There is an important corollary to this statement: opposition to assurance can be reduced by reducing the uncertainty of liability standards and the methodologies used to value damages.

4.1. Cost Creation vs. Cost Redistribution

Another way to explain opposition to assurance is to draw a distinction between created and redistributed regulatory costs. As with any regulation, assurance comes at a cost. And costs generate opposition. It is important, however, to distinguish between costs that are merely "redistributed" by assurance and new, "true" social costs. First consider the way in which assurance redistributes costs. Most obviously, assurance can raise a regulated firm's costs by forcing the internalization of otherwise avoided obligations – that being the very point of assurance. From the perspective of a regulated firm, newly-internalized costs are very real and can be expected to reduce profitability. Accordingly, it is not surprising that assurance rules generate opposition. From the social perspective, however, costs newly-internalized by polluters are redistributed, not new, costs. Without assurance society bears the cost. Assurance simply redistributes those costs to the polluter. Thus, from a social welfare standpoint, redistributed costs do not count as a true cost of assurance.

However, assurance can create real costs. For instance, assurance products must be purchased, contracts signed, paperwork administered, and compliance and coverage conflicts litigated. Also, regulators must monitor compliance and enforce the rules – tasks that create administrative costs. These costs are true social costs, since they are costs that would not be present, absent assurance regulation. Note that a benefit-cost analysis of assurance should weigh only this latter type of cost against the benefits of assurance.

In light of this distinction, political opposition to assurance should be placed in its proper perspective. As described in Section 2, environmental costs redistributed by assurance can be quite large, since many firms' otherwise avoided obligations are large. Society should embrace this redistribution, however, since it represents a fairer and more efficient allocation of financial responsibility for environmental harm. Of more appropriate concern are costs associated with administration and compliance. But the evidence suggests that these costs are relatively low. In environmental market after environmental market assurance is readily available at reasonable rates. This is a strong indication that assurance's social costs are not overly significant.[126]

4.2. Availability and Affordability

The history of assurance implementation speaks for itself. Assurance does not bankrupt whole industries and it does not mean the end of small business. In every regulatory context to date, private financial markets have developed to provide the insurance, bonds, and other financial instruments necessary to demonstrate assurance, and they provide them at reasonable cost.[127] Consider the market for vessel assurance required by OPA. Despite fears, a host of financial assurance products are currently available at rates that have been easily absorbed by the regulated community. None of the worst-case predictions – bankruptcies, failure of the insurance market – came to pass and fears were exaggerated.[128] According to the Coast Guard, which administers the program, traditional vessel insurers "confirmed that [they] had no hard and fast information to support their testimony in July 1994 that the cost of commercial [assurance] would greatly exceed the cost of [prior] coverage" provided by the insurers.[129] New specialty providers have come into existence and are currently providing coverage at affordable rates.[130] To date, there have been no complaints regarding these new providers' ability to offer coverage.[131]

The government has conducted its own analyses of financial assurance compliance costs under the vessel and offshore facility programs. According to the Coast Guard, combined annual premiums for vessel coverage were $70 million in 1996, two years after the program went into effect. This number is significantly lower than the pre-implementation worst-case compliance cost estimate of $450 million per year.[132] Coverage rates vary by the type of vessel and the cargo carried, but at the low end small, dry cargo vessels can get millions of dollars in coverage for a $1,000 annual premium.[133] As for the offshore facility program, administered by the Department of Interior (DOI) the industry-wide annual cost of coverage is estimated at only $6.3 million.[134] Moreover, DOI does "not agree with the comment that the costs of complying with this regulation threaten the viability of many small businesses, because our estimated annual compliance cost is only $14,000 per business."[135]

Assurance under other programs is also readily available. According to a government study of hazardous waste facilities: "Every Subtitle C permit official interviewed, regardless of whether their state allowed the financial test, stated that no financially viable facility in the state was unable to obtain a valid financial assurance mechanism."[136] An estimate of assurance costs for non-hazardous waste landfills placed them at only 2 to 3% of total annual land-fill costs.[137] According to the General Accounting Office, mining bonds too are widely available.[138]

Assurance rates are a particularly good indicator of availability and affordability. The costs associated with specific assurance products are difficult to summarize. However, a 1994 government study of environmental bond prices revealed a price of approximately 1–1.5% of the bond's face value. More specifically, the 1994 rates for non-collateralized bonds covering environmental obligations were as listed in Table 1.[139]

The same report suggested that larger firms with good environmental records could obtain bonds at rates less than 1%.[140] Annual rates ranging from 1 to 3% of the coverage are reported by a range of sources.[141] Bonds used to guarantee safe nuclear facility closure exhibit a similar range of costs.[142] Offshore facility rates are even lower. According to the government, "90% of the 200 designated applicants will demonstrate an average of $35 million in financial responsibility using insurance or a surety that costs $35,000."[143] Annual premiums for $10 million in OSFR coverage average $10,000. These figures imply annual rates of only one tenth of 1% of the coverage's face value. Finally, UST owners can insure a tank for $400 a year, less than it costs to insure a car.[144]

Table 1. Environmental Bond Rates.

Level or layer of coverage	Bond rate
First $100,000	$25 per $1,000 in coverage
Next $100,000	$15 per $1,000 in coverage
Next $2,000,000	$10 per $1,000 in coverage
Next $2,500,000	$7.50 per $1,000 in coverage

In conclusion, opposition to assurance, based on fears of mass disruption to business, are unwarranted. Opposition is best explained as a reaction to the re-distribution of costs to responsible parties and as a lobbying tactic to reduce the stringency of regulatory requirements. Claims that assurance mechanisms will be unavailable and that insurance and bond markets will dry up should be viewed in the same context. In the words of one commentator, "frequently the assertion of bond unavailability has been used as an attempt to ratchet reclamation standards downward and to reduce periods of operator/surety responsibility. It has also led to the use of inadequate bond amounts in some states."[145]

4.3. An Important Exception: Assurance Availability and Retroactive Liability

In 1994, the U.S. GAO issued a report on the availability of environmental insurance products. A principal finding was that "the majority of companies operating treatment, storage, and disposal facilities in 1991 that attempted to obtain pollution insurance found that it was difficult to obtain"[146] and that 44% of surveyed firms attempting to obtain insurance between 1982 and 1991 were denied coverage at least once.[147] These conclusions are clearly at odds with the argument that coverage is easily available and affordable. In large part, the discrepancy reflects short-term difficulties in the adjustment of insurance markets to assurance. Subsequent technological changes have improved the safety of facilities (a desirable consequence of assurance regulations) and the insurance industry today has an improved ability to predict exposures and tailor products to specific risks. Another explanation for the discrepancy is that the U.S. environmental insurance market in the 1980s and early 1990s was hobbled by uncertainties and costs arising from retroactive, unanticipated liabilities.

Environmental laws passed in the 1970s and 1980s significantly strengthened regulatory requirements and expanded the scope of polluters' liability. CERCLA, for example, imposed liability on firms retroactively. In one stroke, firms were liable for damages due to pre-existing conditions, conditions that may not have created liability prior to CERCLA's passage. It is important to emphasize that financial assurance rules foster prospective deterrence, but they do little to promote the cleanup of existing environmental problems. Firms with wealth adequate to absorb existing risks are already "financially responsible." Firms without adequate wealth have no incentive to demand – and capital providers have no incentive to supply – coverage for existing, known liabilities. For this reason, financial responsibility rules should not be applied to retroactive liabilities.[148] In fact, the failure of regulation to account for the interaction between financial assurance rules and retroactive liability largely accounts for the insurance availability problems observed in the U.S. in the last decade. An explanation for why insurance was unavailable or unaffordable is that insurers were afraid of exposing their own assets to retroactive liability when underwriting future liabilities.[149]

Consider the experience with UST assurance rules and liability. When RCRA mandated financial responsibility for UST owners the law did not distinguish between financial responsibility for future risks and responsibility for the cleanup of already-existing contamination. Because many USTs had already leaked, the immediate effect of assurance requirements was to require insurance for environmental damages that already existed. Because many owners were small businesses unable to afford the cleanup of their sites, the UST requirements led to the publicly-financed assurance funds described in Section 3.3. But as these funds are phased out, sites are remediated, and new technologies are installed, USTs are increasingly insurable by private markets.[150] The EPA lists 13 major insurers and 97 agents and brokers as current providers of UST financial responsibility coverage.[151] The lesson to be drawn from the UST example is that public financing can be a desirable short-term financial mechanism for pre-existing, retroactive liabilities. As long as they are strictly limited in duration, public funds foster the transition to a workable and afford-able system of prospective financial responsibility provided by third-party private-sector providers.[152] Markets for financial assurance coverage may at first be problematic, but over time they adapt to new environmental technologies and risks, resulting in greater availability and lower prices.

4.4. The Politics of Small Business Regulation

A significant political barrier to assurance arises from its disproportionate impact on small businesses. This is unavoidable, of course, since small firms – by definition – are in particular need of financial responsibility regulation. In general, small firms are less wealthy and are thus more likely to become insolvent in the face of large environmental obligations. Small firms may also be monitored less effectively than larger firms. But clearly, it is harder and more costly for small firms to demonstrate financial assurance. For large firms, com-pliance with financial responsibility may involve little more than the preparation of audited financial statements. Small firms, by definition, cannot self-insure and so must pay for the involvement of a third-party insurance or capital provider. Also, small firms may be required to participate in risk assessments, paperwork, and a set of transactions with which they are unfamiliar.

In general, regulating small business is not politically popular. Small-business regulatory relief bills are a common Congressional offering.[153]

A particular issue for agencies proposing assurance rules that apply to small businesses is the Regulatory Flexibility Act (RFA), which requires agencies to evaluate, offer flexible compliance alternatives, and minimize the impact of regulations on small business.[154] The RFA can be thought of as a procedural safeguard to ensure that small firms are not overly burdened by regulation. It can also be viewed as a warning to agencies targeting small firms for regulation. From a policy standpoint, and accepting the desirability of objective regulatory impact analysis, the "smallness" of firms should not be used as a barrier to assurance regulation. After all, firms' smallness lies at the very root of the policy problem addressed by assurance.[155]

5. DESIGN AND IMPLEMENTATION: THE SCOPE OF ASSURANCE RULES

Assurance is an elegantly simple concept: firms must provide a financial or contractual demonstration of their ability to meet environmental obligations. This simplicity obscures a set of important design issues, however. These issues can be grouped into two basic categories. First what is the appropriate *scope* of assurance requirements? Second, how can the *security* of the assurance mechanism be guaranteed?

Issues of scope relate to the liabilities and obligations that are covered by assurance, and the dollar value of coverage or bonding that must be demonstrated. There is a tension between the desire to maximize the scope of assurance, in order to maximize deterrence and compensation, and the desire to minimize compliance costs by minimizing assurance requirements.[156] Issues of security relate to the collection of obligations in the future, given the financial mechanisms used to comply with the assurance rule. One way for responsible parties to reduce costs and their own financial risks is to reduce the security of the instruments they purchase or provide as assurance. A major challenge created by financial assurance rules is that they require regulators to monitor and ensure the mechanisms' security over long periods of time.

5.1. How Much Coverage is Enough Coverage?

To internalize costs and get environmental obligations performed assurance rules need to guarantee firms' ability to internalize costs in the future. So how high should coverage requirements be? The answer is, just high enough to guarantee the performance of the required obligation or internalization of future liabilities. Coverage requirements higher than these levels are wasteful, because they tie up capital (which always has an opportunity cost) but yield no additional social benefit. Coverage requirements lower than these levels are undesirable because they do not guarantee cost internalization, and thus yield inadequate deterrence and compensation.

If it is known that a future restoration obligation will cost a firm C, then the appropriate level of assurance is C. Requiring less raises the possibility that the firm will fail to internalize the full cost.[157] In other situations, the prescription is less clear. For instance, a firm's landfill may not leak, may leak a little, or may leak a lot. If a range of possible future costs can arise, what is the optimal level of assurance? If the possibilities range from zero cost to some higher bound C^U the appropriate level of assurance is the upper bound C^U. Call this the "maximum realistic environmental cost." Unless there is assurance for the maximum realistic cost there will be situations in which firms fail to fully compensate victims and, as a consequence take insufficient care to avoid those costs.[158] In practice, assurance rules always mandate coverage up to some finite dollar value, even if there is no real upper limit to the possible damages arising from an operation.

5.2. How are Required Assurance Levels Actually Determined?

In practice, firms and regulators rarely know with certainty what environmental costs will eventually be. Even the cost of a certain obligation, such as the capping, restoration, and monitoring of a landfill, can be difficult to estimate with precision. Over a period of decades will climate and biological variables allow for successful re-vegetation. Will the site's hydrology and geology prove stable? Will the site be subject to encroachment? As environmental conditions go, these are fairly predictable concerns. Even so, cost estimates are subject to error.

At the other extreme, liabilities associated with pollution events are even harder to predict. The environmental cost of a vessel grounding, for instance, may be very high or relatively low depending on the cargo, location, and weather conditions associated with the spill. In other words, while it may be clear that we should require coverage up to the maximum realistic obligation C^U, how do we know what C^U is?

Given these uncertainties, the determination of required assurance amounts can be problematic. A variety of methods is used to determine coverage requirements. In some cases, coverage requirements are determined on a case-by-case basis, taking into account the specific risks posed by an operation. In others, greater procedural formality is imposed via established estimation methodologies. For example, some states require hazardous waste treatment, storage, and disposal facilities to prepare, based on a routinized methodology, an estimate of costs required to close the facility.[159] This methodology typically involves the use of standard software and "worksheets" associated with specific cost categories. Even so, the characteristics of particular facilities, and hence closure cost estimates, can vary widely. To compound the challenge, it is common for cost estimates to change dramatically over time.[160] Bond amounts must be adjusted for cost inflation and changes in a site's environmental conditions.[161]

Accordingly, estimation of required coverage amounts places a significant burden on the regulator to audit the quality of the numbers and estimation methodology. Under some regulatory programs, a relatively fixed "schedule" of requirements is imposed across a whole industry. An example of this are the OPA and CERCLA coverage requirements for vessels carrying oil and hazardous cargo. Under these rules, coverage requirements are simply a function of the vessel's size, type, and cargo (oil vs. hazardous substances) and can be more easily calculated and verified.[162] As another example, offshore facility assurance requirements are based on the facility's location and the volume of a "worst-case" oil discharge.[163]

In general, however, agencies may have difficulty determining appropriate assurance levels.[164] Recent cases highlight the procedural challenge. For example, in *Leventis, et al., v. South Carolina DHEC, et al.*, the Sierra Club successfully argued that the State environmental agency failed to adequately determine and require adequate cleanup, closure, and restoration assurance amounts for a hazardous waste disposal facility.[165] The case's history is largely one of motion and counter-motion to determine appropriate levels of financial assurance. In 1989, the South Carolina DHEC issued a draft determination requiring $30m in third party insurance coverage for property and bodily injury and a $114m trust fund for cleanup, closure, and restoration costs. In 1992, those requirements were raised to $33m and $132m, respectively. A later administrative decision revised the requirements downward slightly. In turn, the Sierra Club appealed to the DHEC Board. The Board agreed in part, raising the trust fund component to $133m, with part to be satisfied via a corporate guarantee. At that point, the landfill owner and Sierra Club both sought judicial review, challenging various aspects of the decision. Based on the State agency's failure to honor procedural safeguards relating to public comment, the court found in favor of the higher assurance amounts.[166]

One way in which an agency's assurance requirements (in particular, their amounts) may be challenged is via the National Environmental Policy Act (NEPA). This is particularly true for operations taking place on federal lands, as in the mining and forestry contexts. Primarily a procedural statute, NEPA requires agencies to consider the full environmental consequences of allowing a project to proceed.[167] NEPA cannot be used to require assurance, per se. It can be used to force analysis and identification of restoration requirements that in turn would demand assurance.[168]

Also, federal and state agencies can be compelled to promulgate assurance requirements, as a matter of administrative law, if assurance is found to be short of legal requirements.[169] In general, the cost estimates that determine assurance requirements under many programs should be taken with a grain of salt and considered as good candidates for regular review by both regulators and environmental advocates.

5.3. The Need to Audit Self-Estimated Assurance Requirements

While regulators can perform cost estimation themselves, estimation is costly and time-consuming. In some cases, firms are asked to develop their own environmental cost estimates as a basis for their own assurance obligations. Absent adequate oversight, these estimates may prove to be too low. After all, low-balling estimates of future environmental obligations is a good way for firms to minimize the costs of assurance. A low estimate translates into lower coverage requirements and, consequently, lower compliance costs. Accordingly, audits, ideally conducted by certified third-parties, are imperative to ensure that adequate assurance is put in place. Note that a virtue of fixed assurance schedules is that they minimize this auditing burden.[170]

Absent a meaningful audit procedure, it is inadvisable to allow firms to estimate their own obligations.[171] In fact, there is evidence that firms routinely under-estimate obligations in the course of complying with assurance regulations. One recent EPA study found that 89 of 100 facilities submitting landfill cost estimates underestimated their closure costs and thus posted inadequate levels of assurance. Moreover, the total amount of the under-estimates was significant, estimated at $450m, just for those 89 sites.[172] Because the effectiveness of assurance rules hinges in large part on having enough assurance, and because the level of assurance is often based on cost estimates, verification of estimates should be an important regulatory priority.

5.4. Are Coverage Levels Adequate?

Not always. The best test of whether or not coverage levels are adequate is the degree to which firms' environmental obligations are met over a span of decades. Because many assurance rules are relatively recent, and cover obligations that arise over a period of decades, it is difficult to draw firm conclusions regarding the adequacy of coverage levels, for example under RCRA's waste disposal assurance rules. To be sure, isolated examples suggest that coverage amounts may be inadequate.[173] But longer-term, overall patterns of cost recovery have yet to be established.

Mining bond levels are an exception. Mining bonds have been required for decades and there is ample evidence that mining bond levels have been, and in many cases remain, inadequate. The Surface Coal Mining and Reclamation Act (SMCRA) of 1977 was enacted largely in response to the coal mining industry's poor record of surface mine reclamation. Over the last two and a half decades, SMCRA's bonding requirements have improved, though not completely solved, the problem of un-reclaimed coal mining sites and their associated environmental problems. The adequacy of required bond levels has been an ongoing issue. A General Accounting Office study and House Hearing in 1986 highlighted the problem. For example, as of 1986, and since the passage of SMCRA, 67% of all acres covered by bond requirements in Pennsylvania had not been reclaimed.[174] In West Virginia, 30% of disturbed lands had gone un-reclaimed despite the presence of bonds.[175] The problem was due largely to the inadequacy of the bond amounts. For example, in Pennsylvania average per-acre reclamation costs were $6200 over the period, while average bond amounts were only $730.[176] GAO testimony suggested that States were uncritically accepting reclamation cost estimates from mine operators, resulting in inadequate bond amounts.[177] More recent studies have also been critical of SMCRA bond implementation.[178] A study of Pennsylvania's coal bonding program suggests that the under-bonding problem continues in that state.[179] One problem is that bonding programs have failed to adequately anticipate problems associated with long-term acid mine drainage (AMD).[180]

Bond levels for hardrock mining on Western lands are also inadequate in many cases.[181] A 1997 EPA Inspector General's report found "strong agreement" among EPA officials that "financial assurance limits now in place at mines are, in large part, inadequate."[182] The report also found that only two of eight states studied required full bonding for the estimated costs of addressing toxic contamination."[183] A 1987 General Accounting Office study focused on bonds for mining on Forest Service-administered lands found federal bond procedures to be lacking.[184] The report cites Forest Service studies documenting poor management of bond programs. One finding is of particular importance: namely, that reclamation standards, which determine bond amounts and the criteria for the release of bonds were "not well documented" and are "generally subjective and difficult to measure."[185] This emphasizes the importance of standardized, audited reclamation cost estimates and performance standards. Other studies have emphasized the importance of extending bonding requirements to even the smallest mine operations, some of which are exempt under current rules.[186]

Another concern relating to the adequacy of bond amounts arises from the use of trust funds as an assurance mechanism. If a trust fund is fully funded at its inception, then coverage will be adequate provided that the required coverage amount is adequate. Some programs, however, allow firms to pay funds into a trust fund over time.[187] Note that if a firm becomes insolvent before a trust is fully funded then the actual amount of available coverage will be inadequate. And, in fact, incompletely funded trusts are relatively common.[188]

5.5. Does Assurance Lead to Confiscation?

Some have raised a concern that bonds and other forms of assurance may aid the government's ability to confiscate private property.[189] Put differently, if the government is the beneficiary of a bond, what is to guarantee that the bond will be released to a firm upon satisfaction of its obligations? Recall that bond agreements include a set of performance criteria. If those obligations are fulfilled the bond is released – at least in theory.

Assuming a bond agreement is well-specified *ex ante* and governments are subject to independent judicial oversight, there is little reason to fear confiscation. First, clear restoration criteria, and a firm's success in achieving those criteria, are interpretable by courts.[190] Second, liability for the environmental damage must be established before bond funds can be forfeited.[191]

Finally, bonds funds cannot be used to cover liabilities not specified in the bond agreement. A good example is *Long v. City of Midway*, a construction bond case, where tort claimants not explicitly covered by a bond sought construction bond funds as a source of compensation.[192] The plaintiffs' effort was rejected on the grounds that "if tort claimants are permitted to share in the amount of the bond equally with claimants for labor and material, such claimants can never be certain they will be paid, because a great many tort claims for personal injuries and injury to property would materially reduce or amount to perhaps, in some instances, more than the penalty of the bond."[193] Empirically, there is little evidence that environmental bonds are used for claims not specified in the bond.[194]

However, it is important to note that many bonds are so-called "penal bonds." Penal bonds authorize the forfeiture of an entire bond amount for failure to perform as agreed. As a result, even though the performance failure may have a relatively small cost, a larger bond sum can be collected by the government.[195] This is by design, however, and is agreed upon mutually by the parties before the fact. Accordingly, penal bond collections represent less a worrisome form of confiscation, and more a penalty used to motivate compliance with performance standards.

5.6. Should Liability Be Limited to the Coverage Requirement?

Assurance requirements, even if based on sound estimation procedures, may be exceeded by the eventual costs of reclamation or liability. If so, is the firm's liability limited to the assured amount? In practice, it may be, since the firm may have no other funds available to cover environmental claims.[196] Legally, however, a firm's liability is not generally limited by the amount of required assurance.[197] That is, a firm is liable for any environmental damages it causes, irrespective of the amount of required assurance. There are exceptions, however. Under OPA and CERCLA, oil and hazardous waste vessel and offshore facility liability is capped at a statutory limit that is equal to the facilities' financial assurance requirements.[198] Nuclear facility liability is also limited, and equal to the amount of mandatory insurance coverage.[199]

From a public policy standpoint, the choice of liability limits reflects a trade-off. On one hand, truncated damage awards reduce uncertainty. Reduced uncertainty can be expected to reduce the costs of assurance (above and beyond the cost reductions implied by the limitation itself) and thus may promote the development of markets for third-party assurance products. Also, from a regulated firm's standpoint, liability limits discipline the government's pursuit of claims the polluter may feel are unsubstantiated. Accordingly, liability limits may ameliorate political opposition to financial assurance requirements. On the other hand, these benefits to the regulated community must be weighed against the obvious drawback of capped liability: namely, that environmental costs above the cap will be uncompensated by responsible parties.

6. DESIGN AND IMPLEMENTATION: THE SECURITY OF ASSURANCE MECHANISMS

Assurance rules must ultimately be judged on the basis of their ability to deliver compensation when environmental obligations come due. Thus, it is important to understand the ways in which the effectiveness, or security, of assurance can be thwarted. In some cases, firms may overtly fail to comply with coverage requirements. In other cases, third party providers of assurance may themselves be unable to deliver on obligations, due to their own insolvency. The financial mechanisms used to demonstrate compliance may be flawed, by design or lax regulatory oversight. In this regard, self-demonstrated financial assurance is a particularly problematic compliance mechanism. Finally, regulators may fail to administer assurance instruments effectively, allowing funds to be released prematurely.

6.1. Compliance Evasion

A virtue of financial assurance rules is that they create an incentive for third party assurance providers to monitor the environmental safety and performance of the firms whose obligations they guarantee or underwrite. This can relieve some of the enforcement burden on regulatory agencies. An enforcement burden that is not relieved, however, is the need to ensure that firms comply with the assurance requirements themselves.[200] Like any regulation, assurance requirements require penalties and monitoring to promote compliance.[201]

Non-compliance has been defended with a variety of novel arguments, most of which fail. In *U.S. v. Ekco Housewares, Inc.*, for instance, Ekco failed to comply with RCRA's hazardous waste financial assurance requirements and a consent order requiring assurance.[202] The firm argued, unsuccessfully, that it was excused from assurance requirements since the facility in question accepted no new waste after 1984.[203] The defendant also filed a liability insurance policy as proof of assurance, knowing that it contained exclusions rendering it unacceptable as an assurance mechanism, and backdated the instrument in an attempt to conceal its failure to comply over a period of years. Finally, the firm argued that the $4,600,000 penalty imposed for these violations was unreasonably high.[204] The Court of Appeals ultimately reduced the penalty somewhat, but retained most of it, concluding that "the deterrence message sent by the district court's penalty was one sorely needed" given "Ekco's apparent view that financial responsibility requirements take a far-distant seat to its other RCRA obligations." Another example of non-compliance includes a firm's argument that payments into a state UST trust fund constituted funds applicable to compliance with financial assurance requirements. In that case, the court held that RCRA's UST assurance rules required the firm to secure its own assurance.[205]

Another case worthy of note, one testing the federal government's ability to "overfile" a state enforcement action, centered around Power Engineering Company's failure to provide financial assurance for a hazardous waste treatment facility.[206] The case's history involved numerous RCRA violations associated with a metal refinishing plant and the defendant's failure to comply with several regulatory orders. The federal government initiated an action when Colorado failed to require financial assurance for the facility's closure. Assurance enforcement was urgent since, as the Court noted, the defendant had "recently engaged in a pattern or debt reduction and asset forfeiture . . . [and] threatened bankruptcy or abandonment of the facility if the federal or state government continues seeking the facility's compliance with applicable hazardous waste regulations."[207] Based on the federal government's motion, the district court required the defendant to provide $3.5m in financial assurance.[208] The defendant subsequently appealed, arguing that the federal government does not have the authority to override a completed state enforcement action under RCRA. The firm's appeal was based in large part on another RCRA financial assurance case, *Harmon Industries, Inc. v. Browner.* In that case, the 8th Circuit held that the federal government can initiate an enforcement action only if the state fails to initiate *any* enforcement action, or if the federal government completely withdraws the state's authorization to implement RCRA.[209] Power Engineering's appeal failed, however, upon the 10th Circuit's refusal to decide the "overfile" issue and upon the Supreme Court's refusal to hear the case. Upon its return to District Court, Power Engineering was required to comply with the financial assurance requirements originally imposed on it. The District Court also explicitly rejected the 8th Circuit's argument in *Harmon* limiting federal enforcement authority under RCRA.[210] The case is important because it affirms the federal government's ability to force compliance with assurance rules, and other RCRA provisions, despite pre-existing, and potentially inadequate, state enforcement actions.

6.2. Evasion via Bankruptcy?

Assurance rules reduce the risk that firms with environmental obligations will be insolvent when the obligations come due. In some cases, however, assurance is imposed, or greater amounts must be posted, while a firm is already in bankruptcy.[211] This creates a clash between assurance requirements and bankruptcy law. For instance, environmental cleanup costs, once a firm is in bankruptcy, may be a dischargeable "claim" under the bankruptcy code.[212] With the bankruptcy code as a shield, firms have attempted to evade assurance requirements by claiming that assurance-related expenditures are dischargeable obligations.

In general, however, courts have held that assurance costs, including the required posting of bonds or increased bond amounts to cover reclamation costs, are not "money judgments" under the bankruptcy code and fit within the "police and regulatory powers" exception to the automatic stay.[213] Consider the decision In re Industrial Salvage, Inc., which involved cleanup and closure orders for a set of landfills in Illinois.[214] As Industrial Salvage filed for bankruptcy the Illinois Pollution Control Board required the facility's closure, revoked the owner's development permit, and required it to post financial assurances for closure of the facility. Industrial Salvage filed a petition for the discharge of debts, and in particular claimed that the facility's closure and assurance costs should be discharged in bankruptcy. They argued that the order to post financial assurances constituted a dischargeable claim since the state could collect on the bonds in the event of non-performance. The Court disagreed, however, finding that the "obligations under the Board's order for closure and post-closure care of the three landfills were not discharged as a claim in their Chapter 11 bankruptcy proceedings."[215]

Another decision supportive of assurance in the bankruptcy context is *Penn Terra, Ltd. v. Department of Environmental Resources*.[216] The bankrupt Penn Terra was subsequently asked to expend funds under Pennsylvania's SMCRA law to reclaim lands it had previously mined. The 3rd Circuit reversed a district court ruling that the reclamation request was a money judgment and thus dischargeable. In its ruling, the Circuit court argued that the DER's attempt to remedy future harm, rather than past damages, did not constitute a money judgment, but rather was an exercise of the state's police powers.[217] Accordingly, while the precise limits of the police and regulatory powers exception remain somewhat murky, closure and reclamation obligations, such as those associated with assurance, are not easily dischargeable in bankruptcy.

6.3. Assurance Provider Insolvency

Insurers, banks issuing letters of credit, and sureties issuing bonds can themselves become insolvent, thus threatening the availability of assurance funds. Unfortunately, there is no insurance against an assuror's financial failure.[218] Regulations typically guard against the possibility of assuror insolvency by requiring U.S. Treasury certification of bond issuers, "secure" ratings for insurers, or, at a minimum, some form of licensing for financial institutions providing assurance.[219] Nevertheless, provider bankruptcies are relatively common. Eight U.S. insurance companies failed in 1998, ten in 1999, and 16 in 2000.[220] Between 1982 and 1986 ten to fifteen sureties serving the surface mine bond reclamation market become insolvent, leaving a total of $36M in bonds unfunded.[221] According to the EPA, between 1984 and 1990 the average annual number of insolvencies among property/casualty insurers was 32 out of 3800, or an average annual failure rate of 0.85%.[222] Over the same period, the average annual failure rate for FDIC-insured banks was 1.14% and U.S. Treasury-approved sureties were de-listed at an annual rate of 0.95%.[223]

A particular concern when assurors fail is that their former customers must acquire assurance elsewhere and on fairly short notice. For financially healthy customers this is not typically a problem. When firms in need of assurance are experiencing financial difficulties of their own, however, replacement can prove difficult. In some cases, new assurance may not be available. Recent problems with an important assurance provider, Frontier Insurance Company, are illustrative.[224] Due to financial weakness, the U.S. Treasury in 2000 removed Frontier's qualification to issue federal bonds. As a result, Frontier customers had to find providers in order to remain in compliance with their assurance requirements. Most were able to. But two large customers, landfill operator Safety-Kleen Corporation, and mining company AEI Industries were unable to quickly replace their environmental bonds.

When an assurance provider fails suddenly, and a firm with assurance obligations is in financial distress, regulators face a dilemma.[225] Technically, non-compliance with assurance regulations is grounds for an injunctive action, including facility closure. This kind of penalty can be a powerful compliance motivator if a firm is financially healthy. When a firm is near bankruptcy, however, facility closure yields no real environmental benefit, since closure starves the firm of cash flow that could be used to finance obligations, improve the firm's ability to find alternative bonds, and avoid insolvency.

In light of the dilemma, consider the difficulties faced by the states and EPA in motivating Safety-Kleen to replace their bonds. Safety-Kleen filed for bankruptcy in 2000, raising questions about a large number of closure obligations associated with its operations.[226] Safety-Kleen and the EPA entered into a Consent Agreement requiring regular financial reports, reports on the firm's attempts to find alternative assurance, and independent environmental audits of sites formerly covered by Frontier bonds.[227] The agreement also specified a set of deadlines for bond replacement. Unfortunately, these deadlines are not particularly credible. Three deadlines had already passed as of August 2001, without compliance and, according to Safety-Kleen itself, "there can be no assurance that the Company will be able to replace Frontier on a schedule acceptable to the EPA and the states."[228] Without any meaningful threat except facility closures, the EPA's hand is weak. Compounding Safety-Kleen's problems, another one of its assurance providers, Reliance Group Holdings, Inc, filed for bankruptcy protection in June 2001.[229]

Frontier's weakness caused difficulty for at least one other large bond holder, AEI Resources, Inc.[230] AEI held $680m worth of Frontier bonds and relied heavily on debt financing prior to Frontier's failure. In turn, the withdrawal of

Frontier bonds led Moody's to downgrade the firm's debt to a Caa2 rating.[231] With such poorly-rated debt and a lack of collateral, sureties have not been willing to supply AEI with replacement bonds.[232]

Safety-Kleen and AEI Resources are large firms. Even so, the weakness of a single surety created a significant barrier to compliance for both firms and a significant financial crisis for AEI. While assuror failures remain an infrequent occurrence, Frontier's failure underscores the importance of regulatory oversight and the screening and monitoring of assurance providers' financial health.

6.4. Defenses, Exclusions, and Cancellation – The Importance of Instrument Language

For assurance to be effective the financial instruments used to demonstrate it should not contain defenses or exclusions that might hamper the government's ability to collect obligations. It is also important that the instruments not be easily withdrawn by providers in the event that large environmental costs develop. In most situations, insurers and insureds voluntarily agree on cancellation terms and coverage exclusions. For instance, non-payment of premiums is typically grounds for cancellation. Exclusions may be included to reduce the insurer's risk exposure and correspondingly, the customer's cost of coverage. These voluntary coverage limitations are inappropriate for the purposes of environmental assurance, however. Coverage limitations, while potentially desirable for the customer and insurance provider, undermine the ability to recover costs and ensure future environmental obligations.

6.4.1. Defenses

It is common for assurance rules to require that assurance instruments adhere to a format with terms established by regulation. As an example, consider the OPA and CERCLA vessel and offshore facility assurance rules. Under the rules, allowable assurance instruments must include an "acknowledgment of direct action."[233] This acknowledgment states that "the insurer [or surety] consents to be sued directly with respect to any claim."[234] The direct action provision is designed to foster resolution of claims and access to compensation. In practice, direct action allows cost recovery independent of a defendant's bankruptcy status.[235] The direct action requirement also eliminates a set of defenses that are typically available to insurers, such as fraud or misrepresentations by the insured.[236] In a typical insurance agreement, fraud and misrepresentation are grounds for a denial of coverage.[237] OPA and CERCLA remove this possibility, as do some state laws.[238] All of the third-party financial assurance mechanisms authorized under the statutes require an acknowledgment that the guarantor agrees to direct action.[239] The only defense available to a guarantor is that the loss was caused by the "willful misconduct" of the owner or operator.[240] The motivation for the direct action provisions is sound. Both cost recovery and deterrence are served by the limitation on policy defenses.[241]

6.4.2 Exclusions

Not all assurance rules feature such a clear-cut limitation on defenses available to an insurer.[242] Most programs, however, guard against the use of policy "exclusions." Exclusions, if allowed, are features of an insurance contract designed to limit the exposure of an assurance provider to certain kinds of risks. Exclusions are problematic for an environmental assurance program, however.[243] Most obviously, they may directly exclude coverage for costs that are intended to be assured.[244] Even if an exclusion is not ultimately honored, exclusions complicate interpretation of the insurance contract, which can open the door to costly and time-consuming litigation.[245]

Because exclusions can so directly undermine the effectiveness of assurance, many state programs rely on the use of "boilerplate" endorsements that must accompany instruments used to demonstrate coverage.[246] These endorsements require the insurer to acknowledge the scope of coverages required by regulation and rule out any exclusions that would limit that coverage.[247]

In general, contract law offers protections against the use of exclusions that are not voluntarily agreed to by the insured or by the beneficiaries of assurance. Misrepresentations of an insurance contract by an insurer, for example claiming coverage when coverage was in fact excluded, are not tolerated.[248] When bonds are issued to satisfy a customer's regulatory obligations, the coverage mandated by the regulations defines the bond provider's obligation. In cases where the regulatory requirement and the bond's language are in conflict, courts tend to favor the regulatory definition of coverage.[249] Courts also accord little credence to a surety's claim of misunderstanding a surety agreement.[250]

6.4.3. Cancellation

The cancellation of coverage prior to the satisfaction of claims and obligations is also a concern. Accordingly, assurance instruments, at a minimum, are required to carry cancellation clauses that require notification prior to cancellation. Consider RCRA's hazardous waste facility closure rules. The rules require advance notification of cancellation, whether the instrument is a bond, letter of credit, or insurance policy.[251] Cancellation of an insurance policy is prohibited unless alternative coverage is acquired, or unless the insured fails to pay premiums.[252] Letters of credit must be automatically renewed, absent a cancellation notice.[253]

In the case of the OPA and CERCLA vessel and offshore facilities rules, the Coast Guard or Minerals Management Service must be notified at least 30 days prior to the cancellation of coverage. Moreover, the instruments must specify that "termination of the instrument will not affect the liability of the instrument issuer for claims arising from an incident . . . that occurred on or before the effective date of termination."[254] And with respect to litigation, guarantor liabilities survive well past coverage termination.[255] Because assurance can be difficult to purchase once environmental or financial difficulties arise, cancellation restrictions are an important component of any assurance program.[256]

6.4.4. Claims-Made Policies

One way in which insurers limit exposure to environmental risks is via the use of "claims-made" policies. Under claims-made coverage, coverage is limited to claims made against the insured during the period of insurance. Claims made after the insurance expires or is withdrawn are not covered. In contrast, "occurrence" coverage covers claims resulting from events during the coverage period, even if the claim is brought after coverage is withdrawn.[257] Claims-made policies can complicate cost recovery since they place time pressure on regulators to discover pollution and initiate cost recovery actions.[258] For this reason, some assurance programs place restrictions on the use of claims-made insurance policies. For example, if a claims-made policy is used, regulations may require that the coverage period be extended beyond the policy's cancellation date.[259]

6.4.5. Arrangements Worthy of Special Attention

The regulator's administrative problems are multiplied when different mechanisms and providers are used in combination. This is typically allowed, so long as the assorted coverages equal the aggregate requirements.[260] In some cases, however, there are restrictions on the number of providers. Under the OPA/CERCLA vessel rule, for example, no more than 4 insurers or 10 sureties can be used to satisfy a firm's coverage requirement.[261] The offshore facility rules place a limit on the number of insurers (either 4 or 5, depending on the facility's location). Also, contribution percentages, in insurance parlance, must be "vertical," not "horizontal."[262] Vertical contributions associate a specific fraction of liability to a provider, irrespective of the dollar value of the claim. Horizontal contributions delineate provider liability as a function of the total dollar claim.[263] Horizontal layering of coverage by different providers is prohibited under the rules, apparently because of administrative difficulties associated with that type of contract.[264]

Increased attention should also be given to the use of so-called "captive" insurance plans. A captive is an insurance company formed to insure the risks of a parent company or set of affiliated companies. Captives do not supply insurance to the general market. While captives are entirely appropriate as a risk-reduction tool for firms, they are inappropriate as a demonstration of financial assurance. The key characteristic of a captive insurer is that its financial strength is tied to the parent company's financial strength. Thus, unlike a third party insurer, a captive insurer's ability to absorb claims is weakest when its strength is most needed: upon the insolvency of the parent.[265] Some, but not all, assurance programs prohibit the use of captives as an assurance instrument.[266] A problem for regulators is that identification of captive policies can be difficult. Policies do not necessarily specify the insurer's structure.

6.5. Monitoring, Administration, and Record-Keeping

Assurance instruments must be monitored by regulators in a variety of ways. First, the initial establishment of an approved mechanism must be verified, usually by inspection of a coverage contract from an approved assurance provider. The issues highlighted in Section 6.4 also highlight the need for regulatory oversight of the insurance, bond, and other instruments used to demonstrate assurance. But, as important, the ongoing of validity of assurance contracts must be verified.

Regulatory rules themselves can help simplify they regulator's task. For example, requiring letters of credit to automatically renew relieves the regulator of one burden: the need to verify annual renewals. But sound bookkeeping and monitoring of instruments is crucial in order to ensure that the contracts will be valid and provide funds in the future. A particular problem is the release of assurance funds – letters of credit, certificates of deposit, and trust funds – by providers without regulatory approval.[267] Again, regulations can help address the problem, in this case by requiring the state agency to be the sole beneficiary of a bond, letter of credit, certificate of deposit, or trust fund.[268] Changes in bank accounts or trust agreements can occur over time, providers themselves can merge or restructure, and computer records need to be updated to reflect changes in the instruments.[269] At a minimum regulatory rules and administrative procedures need to place emphasis on basic record-keeping to facilitate the legal and financial maintenance of assurance instruments.[270] The fact that regulators are typically not accountants, insurance experts, or contract lawyers complicates the task.

Another key element of the regulator's burden is the decision to release assurance funds after a firm's reclamation, closure, post-closure, and other obligations are met. This requires scientific and engineering expertise, rather than financial acumen. But the administrative challenge is clear. The quality of restoration and site closure efforts can be difficult to assess.[271] Public involvement in these determinations can help, but cannot be relied upon in all circumstances.[272] Firms also have the right to challenge an agency's determination of whether or not bonds should be released. Litigation over these issues is common in some cases and adds to administrative costs.[273]

6.6. Problems with Self-Demonstration and Corporate Guarantees

Self-demonstrated assurance and corporate guarantees allow firms to pass a set of accounting tests as a substitute for purchased assurance. When a firm self-demonstrates, its own financial status is used to fulfill the tests. When a corporate guarantee is used, the corporate parent or affiliate's financial status is used. Almost all financial assurance programs allow self-demonstration and corporate guarantees as forms of compliance.[274] To the regulated community, self-demonstration is the cheapest, and thus most desirable, form of compliance since no coverage need be purchased or dedicated funds set aside.[275] Accordingly, agencies and legislatures may be pressured to relax self-demonstration standards in order to allow more firms to comply in this nearly-costless fashion.[276] Self-demonstration is desirable when used by the wealthiest, most environmentally responsible, and most financially stable firms because it avoids the cost of purchased assurance.[277] Unfortunately, it can be surprisingly difficult to distinguish between those firms and their less stable and scrupulous counterparts.

The problem with self-demonstration and guarantees, in a nutshell, is that there exists no financial assets dedicated to environmental obligations.[278] In recognition of self-demonstration's dangers, regulations feature a set of safeguards designed to ensure the firm's ability to absorb future costs. Under RCRA's hazardous waste facility rule, for example, firms must pass one of two tests: a bond rating test or a set of financial ratio tests based on "total liabilities to net worth," "sum of net income plus depreciation, depletion, and amortization to total liabilities," and "current assets to current liabilities." In addition, there is a tangible net worth test, a domestic assets test, and a "net working capital and tangible net worth to estimated closure and post-closure costs" ratio test.[279] This daunting set of accounting challenges means that many firms cannot self-demonstrate.[280]

The regulator's task is equally daunting. Interpretation, verification, and monitoring of the financial tests over time requires either significant in-house accounting expertise or reliance on 3rd-party audits. Regulations typically require independent accounting reports, but this is not an iron-clad safeguard. Accounting fraud is relatively common, and most common among smaller firms and firms in financial distress – precisely the kind of firm and situation that can pose the most serious assurance problems.[281] Unfortunately, the occurrence of financial reporting fraud is not eliminated by independent audits, even those by so-called "Big Six" firms.[282] Moreover, accounting standards for environmental liabilities and other obligations are not adequately standardized.[283] There tends to be great variability in the way environmental obligations are recognized for accounting purposes.[284] Also, it can be very difficult to fully assess the degree to which a firm's assets are obligated to other liens or creditors.[285] From a book-keeping standpoint alone it is very difficult to assess all of the environmental obligations attached to a single firm. Firms often operate multiple facilities with multiple obligations in multiple jurisdictions. Accordingly, "adding up" all these obligations and accounting for them properly is central to the goal of assessing a firm's ability to internalize costs years in the future.[286] In sum, environmental assurance accounting is a problem not only for regulators untrained in its subtleties, but for accountants themselves.

Another serious concern is that a firm's financial status can quickly deteriorate. When this happens, the regulator may not even be notified of a financial crisis for a period of many months. Consider a firm that experiences a loss of revenue or an increase in costs leaving it unable to pass the financial test criteria. RCRA hazardous waste rules only require notification "within 90 days after the end of the fiscal year for which the year-end financial data show that the owner or operator no longer meets the requirements."[287] The firm then has an additional 120 days in which to find alternative, 3rd-party assurance. Since financial conditions can deteriorate early in a firm's fiscal year, notification may not occur for an extended period of time.

As an example of both the rapidity with which a firm's financial fortunes can turn and the subjective, and inappropriate, use of accounting data and techniques, consider the case of Dow Corning. Between 1994 and 1995 Dow Corning went from a AA bond rating to bankruptcy, largely due to breast implant litigation costs.[288] As a result, the firm no longer qualified for self-demonstration for a hazardous waste disposal facility in Michigan. Nevertheless, the firm submitted a claim of self-demonstration based on dubious accounting techniques and un-audited data that was ultimately inconsistent with audited financial reports. In effect, the firm claimed that its balance sheet, for the purposes of assurance, improved as a result of its bankruptcy filing.[289] In that short period the firm went from compliance to non-compliance and left the site without an adequate assurance of the firm's ability to provide closure, post-closure, and liability obligations. Moreover, any firm finding itself in this situation faces the challenge of finding alternative assurance at the very time – a bankruptcy filing – when providers will be most reluctant to offer it.[290]

Another problem with self-demonstration is that it involves no specific financial asset to which a regulator can lay claim in the event obligations are not performed.[291] As discussed above, trust funds, insurance policies, letters of credit, bonds, and cash deposits, may not always be easily converted into compensation. Nevertheless, these instruments are more likely to yield liquid sources of compensation.[292] This is particularly true if, as is ideal, the regulating agency is made the sole beneficiary of the instrument. Purchased coverage will also tend to be viewed by courts as specifically dedicated to reclamation or liability obligations, and thus more likely to be recoverable by regulatory agencies.[293] The assets claimed by a self-demonstrating firm are much more ephemeral. Assets are not specifically dedicated to assurance in a legally binding way and must therefore be sought in competition with other creditors, if those assets are in place and have value at all once obligations come due.

7. CONCLUSION

Unfulfilled environmental obligations, whether due to abandonment or insolvency, are disturbingly common. Cost recovery, deterrence, and enforcement is improved directly by the presence of financial assurance requirements. Assurance is desirable in theory because it helps assign costs to the parties best able to plan for and reduce them – potential polluters themselves. Assurance is desirable in practice because it achieves its goals at relatively low cost and without significant commercial disruption – contrary to fearful rhetoric that typically accompanies the imposition of new assurance requirements. It is particularly desirable when viewed in relation to the alternatives: costs abandoned to the public or imposed after-the-fact on offending firms' commercial partners. Compared to these alternatives, assurance leads potential polluters to a transparent, in-advance appreciation of future environmental obligations. Assurance's value as a deterrent is enhanced further when firms have to purchase assurance from third parties, since coverage rates and availability will be determined by the customer's environmental track record and expectations of future environmental performance. The breadth of operations and risks covered by current rules is an additional testament to assurance's practicality. Markets for assurance coverage provide a wide variety of financial instruments that can be tailored to the needs of individual firms, facilities, and regulatory requirements.

If there is to be a criticism of assurance requirements it may be that they do not go far enough. It is clear, for example, that many mining bonds have not been sufficient to ensure adequate reclamation. In other programs, more experience with cost recovery over longer periods of time is needed to judge whether the scope of assurance requirements is adequate. The security of particular assurance instruments is also worthy of ongoing scrutiny. Self-demonstrated assurance, claims-made insurance policies, captive insurance arrangements, and trust funds with lengthy pay-in periods may hamper cost recovery, particularly the recovery of costs that arise only after a period of decades. Also, state assurance programs could benefit from centralized administration and record-keeping and the creation of databases to foster cross-state comparison of firms' financial statements, aggregate environmental obligations, assurance coverages, and reclamation performance. As it stands, most state programs operate independently of one another, both within and across state boundaries.

Finally, it should be noted that many of the most significant environmental obligations guaranteed by assurance mechanisms have yet to come due. Long-tailed hazards associated with landfills, for example, will not reveal themselves for decades. Accordingly, the legal and financial security provided by current assurance rules will be tested in earnest only in years to come. Ongoing analysis should be trained on the various allowed mechanisms' ability to internalize costs over the long run. In turn, regulation should be prepared to respond to any weaknesses that are revealed, through the elimination of particular mechanisms, greater mandatory coverage amounts, improved auditing, and assurance mechanisms with more ironclad contractual foundations.

NOTES

1. See Section 1.2 infra.
2. Liability rules create future obligations associated with damage to property, human health, and natural resources. Restoration obligations create a future liability for failure to perform necessary reclamation or restoration. In addition, assurance rules promote compliance with immediate regulatory requirements such as monitoring, control, and reporting standards. Assurance does this by fostering the internalization of administrative penalties used to motivate such operational standards.

While liability and restoration obligations feature most prominently in the following analysis, it should be emphasized that the deterrent effect of – and thus the value of assurance to – any type of penalty is blunted by insolvency or abandonment. For a particularly dramatic example, see In re Gary Lazar and Divine Grace Lazar, U.S. Bankr. Cent. D. CA, Case No. LA 92-39039 SB, October 24, 1996 (administrative fines totaling hundreds of millions of dollars, associated with violations of gas station operating standards, most failing to receive priority in bankruptcy).

3. See, however, EPA Office of Inspector General, RCRA Financial Assurance for Closure and Post-Closure, March 30, 2001.

4. American Bankruptcy Institute statistics for annual business bankruptcy filings, 1980–2000. Available at http://www.abiworld.org/stats/1980annual.html.

5. An important exception is the cost internalization achieved by so-called retroactive liability. Since retroactive liability, by definition, is not anticipated by potential defendants it does not promote deterrence. See 4.3 infra.

6. 11 U.S.C. § 362(a).

7. Bankruptcy may be forced by environmental obligations themselves or by conditions unrelated to those obligations. In either case, environmental obligations can be discharged.

8. See Section 6.2 infra. In general, environmental claims do not enjoy any special priority over other creditor claims. There is an important exception, however. In some cases governments can employ the "police and regulatory power exception" to the automatic stay. The exception states that the automatic stay does not apply to the "commencement or continuation of an action or proceeding by a governmental unit to enforce such governmental unit's police or regulatory power," 11 U.S.C. § 362(b)(4). In some cases, this exception can improve the government's ability to recover funds from a bankrupt polluter, though it is no guarantee of full recovery. See Richard L. Epling, *Impact of Environmental Law on Bankruptcy Cases*, 26 Wake Forest Law Review, 69, 1991.

9. To investigate the impact of liability on firm scale, Ringleb and Wiggins (1990) explored the rate of small firm incorporation as a function of the riskiness of a given industry. Their evidence suggests that liability has a direct impact on enterprise scale. They compared the number of small firms in 1967 – a period before the routine use of strict liability for tort claims – to the number of such firms in 1980, when the use of strict liability was routine and expected. Their analysis suggests that the incentive to avoid liability led to a 20% increase in the number of small corporations in the U.S. economy between the two periods. For a description of offshore financial havens, or "asset protection trusts" see "Salting it Away," *The Economist*, Oct. 5, 1991, at 32.

10. Whether or not liability is inherited normally hinges on a determination of the degree to which there is a continuation of the seller's business. See Ray v. Alad Corp. 19 Cal. 3d 22 (1977) (136 Cal.Rptr. 574, 560 P.2d 3), which held that, in appropriate circumstances, the successor to the manufacturer of a defective product may be held liable for damages caused by the product at a time after the successor acquired the manufacturer. Specifically, the purchaser assumes liability if: (1) there is an express or implied agreement of assumption, (2) the transaction amounts to a merger or consolidation of the two corporations, (3) the purchasing corporation is a mere continuation of the seller, or (4) the transfer of assets to the purchaser is for the fraudulent purpose of escaping liability for the seller's debts.

11. The fact that exit can create inefficiencies through risk-externalization is discussed extensively in Hansmann and Kraakman, Toward Unlimited Shareholder Liability for Corporate Torts, 100 Yale Law Journal 1879, 1991 who argue that, "[a factor creating] inefficient incentives under limited liability is the shareholder's option to liquidate the corporation and distribute its assets before tort liability attaches. Since products and manufacturing processes often create long-term hazards that become visible

only after many years, firms can – and often do – liquidate long before they can be sued by their tort victims."

12. This includes 38,000 registered but abandoned tanks and 152,000 unregistered and abandoned tanks. U.S. EPA, Report to Congress on a Compliance Plan for the Underground Storage Tank Program. EPA 510-R-00-001, June 2000, at 11–12.

13. Congressional Research Service, Report for Congress, Leaking Underground Storage Tank Cleanup Issues, updated February 17, 1999. Beginning in 1987, the federal government began collection for the so-called Leaking Underground Storage Tank (LUST) Fund. $1.6 billion was collected for the Trust Fund before the taxing authority expired in December 1995. Congress reinstated the LUST tax in the Taxpayer Relief Act of 1997 (P. L. 105-34). As of December 31, 1998, the Trust Fund balance was $1.25 billion. In addition, forty-seven states established financial assurance funds. For 1997, the total balance of state funds was approximately $1.34 billion, annual revenues were $1.31 billion, and outstanding claims against the funds were $2.31 billion, Vermont Department of Environmental Conservation, Waste Management Division, Summary of State Fund Survey Results, June 1997.

14. See Thomas, supra note 14, at 2. Kentucky alone has 12,000 wells waiting for plugging by the state.

15. The analysis was based on Congressional documents and financial statements obtained from the Coast Guard under the Freedom of Information Act. See Brent Walth, "Spill Laws Fail to Halt Seepage of Public Cash," *The Oregonian*, February 27, 2000. Records show that Oil Spill Liability Trust Fund has paid out $262 million for oil spills since 1990 and has been reimbursed $49 million, or about 19%. The Coast Guard claims a significantly higher recovery rate (60%) based on recoveries associated with closed cases.

16. www.tnrcc.state.tx.us/oprd/wasteplan/swinvent.html.

17. Docket materials in support of the April 10, 1998 Financial Assurance Mechanisms for Corporate Owners and Operators of Municipal Solid Waste Landfill Facilities; Final Rule, Issue Paper: Assessment of First party Trust Funds, at 7 (citing ICF Incorporated, Preliminary Results of Case Studies of Bankrupt TSDFs, June 1985.)

18. 63 Federal Register 17706, 17731, April 10, 1998, Financial Assurance Mechanisms for Corporate Owners and Operators of Municipal Solid Waste Landfill Facilities. (Hereafter "Federal Register 1998.")

19. Andrew Ballard, Financial Assurance, Closure Changes Urged by Washington State Regulator, *Environment Reporter*, April 27, 2001, at 807.

20. U.S. Department of Interior, Bureau of Land Management, Abandoned Mine Land Inventory and Remediation: A Status Report to the Director, November 1996.

21. General Accounting Office, Public Lands: Interior Should Ensure Against Abuses From Hardrock Mining, GAO/RCED-86-48, 1986, at 24.

22. Lyon, J. S. et al., 1993, "Burden of Gilt," Mineral Policy Center.

23. Office of Inspector General, Audit Report, EPA Can Do More To Help Minimize Hardrock Mining Liabilities, E1DMF6-08-0016-7100223, June 11, 1997, at 2.

24. The study defined large-scale mines as those with bond obligations greater than $250,000. James Kuipers, Hardrock Reclamation Bonding Practices in the Western United States, National Wildlife Federation, February 2000.

25. See www.epa.gov/unix0008/superfund/sites/sville.html.

26. National Research Council, Hardrock Mining on Federal Lands, National Academy Press, 1999. ("The Committee observed instances of recently abandoned but

un-reclaimed exploration and mining sites that had not been covered by any financial assurance . . .. The Committee also found that long-term water treatment and monitoring at mine sites generally does not carry financial assurance at either the state or federal level . . .. Based on the Committee's findings, inadequate protection of the public and the environment caused by current financial assurance procedures is a gap in the regulatory programs" at 65.)

27. See Office of Surface Mining, Reclamation, and Enforcement 1999, http://www.osmre.gov/aml/remain/zintroun.htm.

28. Cited in U.S. Government Printing Office, Adequacy of Bonds to Ensure Reclamation of Surface Mines, Hearing Before a Subcommittee of the Committee on Government Operations, House of Representatives, 99th Congress, June 26, 1986, at 4.

29. U.S. GPO, 1986, at 148.

30. Actuarial Study of the Pennsylvania Coal Mining Reclamation Bonding Program, Milliman & Robertson, Inc., July 16, 1993, at 13. As a concrete example of the inability to collect funds necessary for mine discharge treatment, consider Glacial Minerals, a mining company that went bankrupt in the early 1990s. The firm left 28 mine sites with post-mining discharges in western Pennsylvania. Bond recoveries associated with the firm's sites have allowed for water treatment at only three sites. Testimony of John Hanger, Hearing on "Current and proposed Bonding Requirements on Coal Mining," Before the Pennsylvania Environmental Resources and Energy Committee, December 14, 1999. Also see Commonwealth of PA, DEQ Fact Sheet: Reed and Strattanville Mine Reclamation Projects, at http://www.dep.state.pa.us/dep/deputate/minres/BAMR/Strattanville/FS2386.pdf.

31. According to The Superfund Progress Report: 1980 – 1997, U.S. EPA 540-R-98-044 October 1998, "at almost every Superfund site, some parties responsible for contamination cannot be found, have gone out of business, or are no longer financially able to contribute to cleanup efforts."

32. U.S., EPA, OSWER, Mixed Funding Evaluation Report. The Potential Costs of Orphan Shares, September 1998.

33. Statement of Lois Schiffer, Assistant Attorney General, Environment and Natural Resources Division U.S. Department of Justice, before the Superfund, Waste Control, & Risk Assessment Subcommittee of the Environment and Public Works Committee, United States Senate, March 21, 2000.

34. Interim Guidance on Orphan Share Compensation for Settlors of Remedial Design/Remedial Action and Non-Time-Critical Removals, June 3, 1996.

35. Most states have developed cleanup programs to deal with an estimated 30,000 sites unable to qualify for the NPL program. Congressional Research Service, Report for Congress, Superfund and States: The State Role and Other Issues, October 16, 1997.

36. The firm has cleanup liabilities in the tens of millions of dollars. "W. R. Grace Files for Bankruptcy Protection, Citing Huge Increases in Asbestos Litigation," Environment Reporter, April 6, 2001, at 640.

37. This was also the rationale behind the "lender liability" provisions of CERCLA. These provisions have been clarified, and narrowed, to apply only to lenders actively participating in management of an enterprise (CERCLA, Section 101(20), P. L. 104–208, September 30, 1996). Environmental advocates urged broader application of lender liability in order to foster the detection and remediation of polluted sites.

38. See generally Goran Skogh, Insurance and the Institutional Economics of Financial Intermediation, *The Geneva Papers on Risk and Insurance*, 1991 (describing

the benefits from monitoring that come when intermediate financial guarantors expose their assets to the liability claims of the firms they underwrite).

39. The corollary, of course, is that the transaction costs borne by regulated firms will increase. Whether or not this improves overall welfare is a more complex issue.

40. Bond agreements can be found in the Old Testament, as in Genesis 43:9 ("I will be surety for him; of my hand you shall require him. If I do not bring him back to you and set him before you, then let me bear the blame for ever") and Proverbs 20:16 ("Take a man's garment when he has given surety for a stranger . . .").

41. For instance, the 1988 Basle Accord is an international agreement setting minimum capital requirements for banks in order to prevent bank failures. Bank insolvency creates a compensation problem because it means depositors cannot be paid. It creates a deterrence problem because the possibility of insolvency can create incentives for excessive risk-taking, in this case excessive risk in the granting of loans. See Robert Merton, *A Functional Perspective of Financial Intermediation, Financial Management* 24, 1995, pp 23–41 (identifying three ways for banks to reduce their risk exposures: hedging, insuring with others, and possession of an adequate capital cushion).

42. Peter Bohm and Clifford Russell, Comparative Analysis of Alternative Policy Instruments, in Allen v. Kneese and James L. Sweeney, eds., *Handbook of Natural Resource and Energy Economics*, Volume 1, Elsevier, 1985 and Peter Bohm, Deposit-Refund Systems: Theory and Applications to Environmental, Conservation, and Consumer Policy, Resources for the Future, by the Johns Hopkins University Press, 1981.

43. Because insolvency truncates the expected penalties borne by potential defendants it also undermines the motivation to take precaution against risk. For analyses that explored or employ this reasoning, see Alan Schwartz, Products Liability, Corporate Structure, and Bankruptcy: Toxic Substances and the Remote-Risk Relationship, 14 J. Legal Stud. 689, 1985; Steven Shavell, The Judgment-Proof Problem, 6 Int'l Rev. L. & Econ., 45, 1986; William Landes and Richard Posner, The Economic Structure of Tort Law, 1987; Lewis Kornhauser and Richard Revesz, Apportioning Damages Among Potentially Insolvent Actors, 19 J. Legal Stud. 617, 1990; and James Boyd and Daniel Ingberman, Noncompensatory Damages and Potential Insolvency, 23 J. Legal Stud. 895, 1994.

44. For the liability of retailers and distributors see Section 402A of the Restatement of Torts (Second) and Products Liability: Manufacturer and dealer or distributor as joint or concurrent tortfeasors, 97 ALR 2d 811. A recent case to this effect is Pepper v. Star Equipment, Ltd. 484 NW2d 156, CCH Prod. Liab. Rep. 13162, in which a defendant distributor in a products liability action was not allowed to seek contribution from a manufacturer in the midst of Chapter 7 bankruptcy.

For the liability of employers for injuries caused by independent contractors see Sections 416 and 427 of the Restatement of Torts (Second) which states that when the contractor's activities are likely to entail significant or inherent risk, the employer of the contractor is liable for the contractor's failure to exercise reasonable precaution, even if the employer had required that precaution in contract.

45. See United States v. Kayser-Roth Corp., 910 F.2d 1032 (1st Cir. 1990).

46. 42 U.S.C. § 9607 (1988).

47. For discussion of the transaction costs associated with joint and several liability under CERCLA see Lloyd Dixon, The Transaction Costs Generated by Superfund's Liability Approach, in Richard Revesz and Richard Stewart (Eds), Analyzing Superfund: Economics, Science, and the Law, Resources for the Future, 1995.

48. See James Boyd and Daniel Ingberman, The Search for Deep Pockets: Is Extended Liability Expensive Liability?, J. Law Econ. & Org., 1997 and James Boyd and Daniel Ingberman, The Extension of Liability Through Chains of Ownership, Contract, and Supply, in Anthony Heyes ed., *The Law and Economics of the Environment*, 2001.

49. California required bonds for oil well plugging as early as 1931. See Thomas, supra note 14 at 2.

50. 33 U.S.C § 2702; 42 U.S.C § 9607(a)(1). The rules are codified at 33 CFR, Part 138.

51. FWPCA, Section 311, 33 U.S.C 1321 (1970).

52. For instance, the Clean Water Act § 311(f) limited liability to $150 per vessel ton. The corresponding limit under OPA is $1,200 per gross ton. Moreover, before OPA there were traditional admiralty shipowner liability protections that limited the application of liability to negligent parties and situations in which plaintiffs were "physically impacted or touched by the oil."

53. The rule was finalized in 1996, Financial Responsibility for Water Pollution (Vessels), codified at 33 CFR 138; Final Rule, 61 FR 9274, March 7, 1996, and 61 FR 9263, March 7, 1996.

54. 59 FR 34212–34213. 33 CFR 138.15.

55. 33 CFR 138.12.

56. OPA § 1016. The offshore facility financial responsibility rules are codified at 30 CFR, Part 253.

57. 62 FR 14052, March 25, 1997 (notice of proposed rulemaking); and the final rule, codified at 30 CFR, Part 253, Oil Spill Financial Responsibility for Offshore Facilities, 63 FR 42699, August 11, 1998.

58. See 30 CFR 250,251, 256, 281, 282 (mandatory bond coverage for Outer Continental Shelf lessees). The Outer Continental Shelf Lands Act had a $35 million FAR for certain oil and natural gas facilities. OPA increased the required amounts (to as much as $150 million) for some facilities.

59. 30 CFR 253.3.

60. 30 CFR 253.13.

61. RCRA's subtitle I covers UST facilities. The UST financial responsibility rules are codified at 40 CFR 280, and see 53 FR 43370, October 26, 1988.

62. RCRA's Subtitles C and D govern hazardous and solid waste disposal facilities, respectively. The RCRA C financial responsibility rules are codified at 40 CFR 264 and 265 ("subpart H"). The RCRA D financial responsibility rules are codified at 40 CFR 258 ("subpart G").

63. For the Subtitle C requirements see 40 CFR 264/265.144 and 264/265.145. For Subtitle D see 40 CFR 258.72 and 258.73.

64. Coverage requirements may be for both "sudden" and "non-sudden" accidental occurrences. 40 CFR 264/265.147.

65. Codified at 40 CFR 144.28(d), 40 CFR 144.52(a)(7), and 40 CFR 144.60–144.70.

66. Injection wells are "bored, drilled or driven shafts or dug holes whose depth is greater than the largest surface dimension into which fluids . . . are emplaced. That is, any hole that is deeper than it is wide and through which fluids can enter the ground water is an injection well." 40 CFR 144.3.

67. Oil and gas wells are typically regulated by individual states. Bond amounts vary state-to-state. For instance, a single well bond for a well 500 feet deep

or less is $500 in Kentucky, but $100,000 in Alaska. See Thomas, supra note 14, at 2.

68. 30 CFR 800. For an overview see James McElfish, Environmental Regulation of Coal Mining: SMCRA's Second Decade, Environmental Law Institute, 1990. The Mineral Leasing Act also requires bonds for compliance with approved mining and exploration plans on public lands. 43 CFR 3474.1.

69. Inadequate bond amounts were one reason for the Act's passage. See McElfish, supra note 68 at 91, citing H. R. Rep No. 128, 95th Cong., 1st Sess. 57–58, reprinted in 1977 U.S. Code Cong. & Admin. News 595–596.

70. To illustrate, Pennsylvania requires minimum per-acre bond amounts that range from $1000 to $5000, depending upon site characteristics. http://www.dep.state.pa.us/dep/deputate/minres/bmr/bonding/bondingrpt021000a.htm

71. See generally, Kuipers supra note 24.

72. The Federal Land Policy and Management Act (FLPMA) directs the Secretary of the Interior to prevent unnecessary or undue degradation of the public lands. Financial assurance is considered part of this charge. See 43 U.S.C.1732(b).

73. BLM mining rules are codified at 43 CFR 3809. USFS reclamation rules are codified at 36 CFR 228.

74. Codified at 40 CFR 761, Subpart D.

75. Plant decommissioning assurance rules are codified at 10 CFR 50.33(k) and 50.75; disposal assurance at 10 CFR 61.62.

76. 10 CFR Part 40, Appendix A.

77. 42 U.S.C. § 2210.

78. See U.S. Nuclear Regulatory Commission, The Price-Anderson Act – Crossing the Bridge to the Next Century: A Report to Congress, 1998.

79. California Government Code § 8670.37.53. The law went into effect on January 1, 2000.

80. Some oil terminals and pipelines must demonstrate $50 million in coverage. Tank vessels and barges must demonstrate up to $100 million. Alaska Stat. 46.04.040 (Supp. 1994).

81. Alaska Stat. 46.04.055, as of June 2000.

82. The coverage requirement for oil-carrying vessels is $500 million. Washington Rev. Code Ann. 88.40.020(2)(a). Coverage is also required for onshore facilities that could discharge oil to navigable waters or adjoining shorelines. Washington Rev. Code Ann. 88.40.025.

83. MCL 324.63712.

84. For example, Michigan, MCL 324.16903(1)(j); Ohio, OAC 3745-27-15(B)(1); and Texas, TAC, Title 30, Part 1, Chapter 37, Subchapter M.

85. TAC, Title 30, Part 1, Chapter 37, Subchapter U.

86. Kansas requires financial responsibility for large-scale swine facilities, K. A. R. 28-18a-23. Illinois requires financial responsibility for the closure of waste lagoons used in livestock production, 35 IAC § 506.601.

87. G. S. 143-215.104F (f). These rules have not been fully implemented. Facilities were required to obtain liability insurance of no less than $1 million or provide regulators with a surety bond or deposit of securities in the amount of $1 million. These requirements may be waved if the operation is unable to comply and is found to be uninsurable.

88. 42 U.S.C. § 6926(b), § 6943, § 6991(c). The EPA delegates implementation via a state authorization process. Federal approval of state programs places a floor on

standards and ensures consistency, while providing for some flexibility in terms of program details. Individual states can implement stronger standards, 42 U.S.C. §6929.

89. So-called "direct implementation" states are those in which the U.S. EPA administers the UIC program. As an example, Class II wells are federally administered in New York, Pennsylvania, Florida, Kentucky, Tennessee, Michigan, and Montana. www.epa.gov/r5water/uic/ffrdooc2.htm.

90. See Kuipers, supra note 24, at I-7.

91. http://ciir.cs.umass.edu/ua/October1995/priority/pfile-7.html

92. Even in the absence of an express written indemnity agreement, common law indemnity would favor the surety against the principal. See Lawrence Moelmann and John Harris (Eds), The Law of Performance Bonds, American Bar Association, 1999, at 6 (and also for more on the difference between performance bonds and insurance and a legal overview of performance bonds generally).

93. Moelmann and Harris, supra note 92, at 5, (referring to the relative simplicity of bond pricing, "this is a monumental difference from casualty underwriting, where the loss experience of the given insured can result in a premium that is several multiples of what an insured with a better record might pay").

94. But see Section 6.6 infra.

95. Section 6.4 and 6.5 infra discuss the need to monitor purchased assurance.

96. See discussion in Section 6.4.6 infra.

97. Because retail gasoline is a highly competitive business, these taxes are simply passed along to the consumer. So while the industry is taxed, the tax liability falls primarily on consumers.

98. Subsidized assurance can be justified if it is used to finance so-called retroactive liabilities created by a change in regulation. During a period of legal transition, public financing promotes the timely remediation of existing pollution and compliance with the prospective, deterrent aims of the law. See James Boyd and Howard Kunreuther, Retroactive Liability or the Public Purse?, 11 *Journal of Regulatory Economics*, 79, 1997.

99. See U.S. EPA, State Funds in Transition: Models for Underground Storage Tank Assurance Funds, Office of Underground Storage Tanks, 1996 (updated 1998). [www.epa.gov/swerust1/states/statefnd.htm]   ("In   1996,   commercial   pollution liability insurance (which meets the federal financial responsibility requirements) is readily available and generally affordable, especially for 'good' tanks meeting all technical requirements. Growth of this insurance market has not been constrained by a lack of supply, but rather by a lack of demand due to competition from state assurance funds"), at 4. Also see "Financial Responsibility Long Term Study," State of California, State Water Resources Control Board," January 1995, 94-2CWP. (The State UST fund "is a hindrance to insurance providers"), at 5. Available at http://www.swrcb.ca.gov/~cwphome/ustcf/resource/finrelts.htm

100. Typically, different mechanisms can be used in combination, with the aggregate coverage equaling the liability limit. For example, self-insurance can be used to cover the deductible included in an insurance policy. 63 FR 42704, August 11, 1998.

101. For examples, see "Distribution of Subtitle C Facilities Among Financial Assurance Mechanisms. Docket materials in support of the April 10, 1998 "Financial Assurance Mechanisms for Corporate Owners and Operators of Municipal Solid Waste Landfill Facilities; Final Rule" Issue Paper: "Effects of the Financial Test on the Surety Industry." at 7 (TSDF assurance); Review of Hard Rock Mining Reclamation Bond Requirements, Legislative Request #98L-36, December 4, 1997, appendix (hardrock

mining bonds in Montana); U.S. Coast Guard data, available at http://www.cofr.npfc.gov (water borne vessels).

102. See discussion, Section 6.4.4 infra.

103. See discussion, Section 6.4 infra.

104. Credit issuers must be those whose operations are "regulated and examined by a Federal or State agency." 40 CFR 258.74(c).

105. Schmitt v. Insurance Co. of North America (1991) 230 Cal.App.3d 245, 257. Typically, though, either the principal or the surety may be sued on a bond, and the entire liability may be collected from either the principal or the surety. This character-istic of surety bonds is also tempered by FAR "direct action" requirements, described below.

106. See 30 CFR 253.31 (vessels); 33 CFR 138.80(b)(2) (offshore facilities); 43 CFR 3809.555(a) (hardrock mines); 40 CFR 258.74(b)(subtitle D), 40 CFR 264.143(b)(1) (Subtitle C), 40 CFR 280.98(a) (Subtitle I). "The surety company issuing the bond must, at a minimum, be among those listed as acceptable sureties on Federal bonds in Circular 570 of the U.S. Department of the Treasury."

107. Federal Financial Responsibility Demonstrations for Owners and Operators of Class II Injection Wells, (EPA 570/9-84-007). Federal blanket bond coverage is accepted only if the operator: (1) has a spotless past record of plugged and abandoned wells; (2) has at least one oil field or lease with an estimated remaining economic life exceeding 5 years; (3) has been in the oil business for more than 5 years; (4) is producing from more than one production field; (5) operates more than 10 injection wells; and (6) can pass a financial test.

108. Under the hardrock mining assurance rules, cash must be deposited and placed in a Federal depository account by the BLM, 43 CFR 3809.555(b).

109. Only regulated trustees are acceptable. "The Trustee must be an entity which has the authority to act as a trustee and whose trust operations are regulated and exam-ined by a Federal or State agency." 40 CFR 258.74(a) (Subtitle D municipal landfill regulations); 40 CFR 264.143(a)(1) (Subtitle C), 40 CFR 280.102(a) (Subtitle I). Trustees may be required to "discharge his duties with respect to the trust fund solely ion the interest of the beneficiaries and with the care, skill, prudence, and diligence under the circumstances then prevailing which persons of prudence, acting in a like capacity, and familiar with such matters, would use in the conduct of an enterprise of a like character and with like aims." 40 CFR 280.103(b).

110. USTs, 40 CFR 280.95; TSDFs 40 CFR 264.143(f); surface mines 30 CFR 800.23.

111. Take the Subtitle C assurance test: firms must pass one of two test, each featuring a set of subtests. Test 1 requires the firm to pass a domestic assets test, a net worth test, a net working capital to closure cost ratio, and two of three tests relating to asset/liability ratios. 40 CFR 264.143(f)(1).

112. See the rules governing vessels carrying oil and hazardous substances, 33 CFR § 138.80(b)(3); 40 CFR 258.74(e).

113. As under offshore facilities assurance rules, 30 CFR § 253.21-.28. RCRA land-fill rules allow discrepancies, but only when accompanied by a special report providing explanation. 40 CFR 258.74(e)(2)(B). Audited reports are always required, 40 CFR 264.143(f)(3)(ii).

114. For landfills see 40 CFR 258.74(g)(1); TSDFs 40 CFR 264.143(f)(10); and USTS 40 CFR280.96(a). In the case of offshore facilities rules, this restriction is the outgrowth of difficulties that arose in an earlier FAR program administered by DOI. See 63 FR

42705, August 11, 1998 ("When the USCG first started operating the OCSLA OSFR program in the late 1970s, more than one indemnitor was allowed for any one OSFR demonstration. However, this proved to be unworkable because the failure of any one of the indemnitors could and did cause the failure of the whole package of OSFR evidence" and "If the designated applicant and the indemnitor share non-OSFR business objectives, then the potential for disputes over who will pay a claim should be minimized. Likewise, the corporate affiliate requirement should maximize the potential for timely settlement").

115. See 61 FR 9270, 1996. "The Coast Guard does not consider self-insurance and financial guarantees to be ironclad methods of evidencing financial responsibility. Assets can be dissipated without the Coast Guard's knowledge, and continuous monitoring of a self-insured entity's asset base is not feasible . . . Accordingly, the Coast Guard believes that any amendment to the financial guarantor provision that reduces the protections afforded by that provision is inconsistent with the concept of financial responsibility."

116. Such as higher required bond levels.

117. See Jason Shogren, Joseph Herriges, and Ramu Govindasamy, "Limits to Environmental Bonds," 8 Ecological Economics, 109–133, 1993 for a theoretical analysis suggesting that bonds and insurance may not be readily and cost-effectively supplied by financial markets.

118. See Deadline Near for Compliance with U.S. Oil Spill Liability Rules, Oil and Gas Journal, August 1, 1994, at 14.

119. Testimony of Chris Horrocks, International Chamber of Shipping, Hearing Before the Subcommittee on Coast Guard and Maritime Transportation of the Committee on Transportation and Infrastructure, House of Representative, June 26, 1996 (hereafter, "1996 House Hearing"), at 44.

120. Representative Richard Ray, Nov. 18, 1987, Hearing before the House Committee on Small Business Subcommittee on Energy and Agriculture, Y4.Sm1/2:S.hrg.101–690. A front page article in the New York Times fanned the flames with the headline "Fuel-Leak Rules May Hasten End of Mom and Pop Service Stations," that included an estimate by the American Petroleum Institute that the rules would force the closure of 25% of the nation's service stations, The New York Times, June 19, 1989, at A1.

121. An economist for the Small Business Association concluded that "the regulated [hardrock] mining industries operate at the edge of profitability and that the rule would oust small businesses from the industry." Memorandum of Points and Authorities, National Mining Association v. Babbitt, U.S. District Court, D.C., No. 00–2998, January 3, 2001, at 29.

122. Section 107 of CERCLA establishes natural resource damage liability and authorizes federal trustees to recover damages for assessing and correcting natural resource injuries, 42 U.S.C 9607(f)(1). OPA's Section 1002 establishes liability for "injury to, destruction of, loss of, or loss of use of natural resources." 33 USC 2702(b)(2)(A).

123. See Testimony of Richard Hobbie, Water Quality Insurance Syndicate and American Institute of Marine Underwriters, 1996 House Hearing, supra note 119 at 41. "The major uncertainty to the continuation of the [financial responsibility] program is the natural resource damage assessment problem and those regulations, the lack of standards. Should our fears prove true, we may find that no insurers are going to be in a position to issue guarantees . . .. The dangers posed by potentially excessive and arbitrary assessments present the most serious threat to our ability to continue to insure liabilities under these federal pollution statutes."

124. The contingent valuation method is particularly controversial, but its role in damage assessment has been over-emphasized. See testimony of Douglas Hall, NOAA, Subcommittee on Water Resources and Environment, House of Representatives, July 11, 1995. "There have only been six contingent valuation studies completed to date, and only one in which the Federal Government was involved in litigation." Restoration or replacement, rather than monetized damage estimates, is the preferred damage calculation method for NRDs. See James Boyd, Financial Assurance Rules and Natural Resource Damage Liability: A Working Marriage? Resources for the Future, DP01–11, 2001.

125. OPA and CERCLA, for instance, limit liability for vessel spills 0.33 CFR § 138.80 and offshore facilities 30 CFR 253.13. This is not enough to counter the fears of some potentially responsible firms. According to one shipping industry representative, "there is fundamental concern about the exposure under OPA 1990 to potentially unlimited liability. We know, of course, that the act retains the principle of limitation. We know that there is legal dispute about whether, in fact, legal limitation would be breached in real life." Testimony of Chris Horrocks, 1996 House Hearing, supra note 119 at 44.

126. But see Section 6 infra, for a discussion of costs associated with the administration of assurance regulation.

127. There have been short-term shortages of assurance products in some industries. See 56 Federal Register 31602, Mining Claims Under the General Mining Laws. ("The traditional surety bond is no longer available. This lack of availability was clearly documented in the 1988 General Accounting Office Report, GAO/PEMD-88-17, Surface Mining: Cost and Availability of Reclamation Bonds .... The report found that surety bonds were much harder to obtain than when the existing regulations were promulgated, because of tightening of requirements in the surety industry during the 1980s, and that even when obtainable they required large amounts of collateral. The report concluded that small and mid-sized coal operators face a liquidity crisis when forced to use high cost alternatives to surety bonds or to offer large amounts of collateral to obtain a surety bond"), at 31604.

128. Consider an illustrative exchange between Congressman Sherwood Boehlert and Richard Hobbie, an insurance industry representative, during 1995 hearings relating to the fear of bankruptcies in the PRP vessel community (from 1995 House Hearing, note 171 *supra*): *Congressman Boehlert:* "Do you have any examples of [firms] that have already gone out of business?" *Mr. Hobbie:* "The escalation of costs so far in OPA have been within a context that the maritime industry has been able to sustain. I would suggest that there used to be a larger number of small tow- and push-boat companies all throughout the south intracoastal waterways. Many of those are no longer with us. The larger operators have purchased many of them. If I may, we have had a number of companies who have ceased transporting black oil – that would be Ingram Barge Lines, Bouchard Transportation of New York, and Canal Barge Lines in New Orleans – because of the insurance costs and the liabilities, so I think there would be a direct example where OPA has caused people to change the business pattern." *Congressman Boehlert:* "But no examples of anybody being forced out of business? I'm being intentional in my pursuit of this because so often we hear these horror stories up here and we are all alarmed and we can't proceed with anything because the bottom is going to fall out, and then when we ask to see where the bottom has fallen out no one can quite show us where that bottom has fallen out ...." Hearings before the Subcommittee on Water

Resources and Environment of the Committee on Transportation and Infrastructure, House of Representatives, July 11, 1995.

129. Statement of Daniel Sheehan, Director, National Pollution Funds Center, U.S.CG, 1996 House Hearing, supra note 119.

130. The traditional vessel insurance market is currently experiencing a period of health, at least on the loss side, which is translating into lower premiums. According to one insurance company document, "Excess oil pollution cover is again available from market underwriters for the 1999/2000 policy year. As a result of the excellent claims experience and the over capacity in the insurance market it has again been possible to achieve significant reductions in the rating structure." See http://www.nepia.com/Circulars/excess_oil.htm (accessed July 28, 2000).

131. "Traditional providers of COFR guarantees declined to provide coverage under the OPA 90 regime, necessitating the emergence of new guarantors. However, since the regulatory program became effective in December 1994, there has not been a single incidence where a guarantor has not met the expectations of the program. The new mix of guarantors has been as reliable as the old mix." Testimony of James Loy, U.S.CG, Subcommittees on Coast Guard and Maritime Transportation and Water Resources and Environment, House of Representatives, March 24, 1999.

132. Statement of Daniel Sheehan, Director, National Pollution Funds Center, USCG, 1996 House Hearing, supra note 119.

133. According to one company's advertisements, small dry cargo vessel operators can get up to $70 million in COFR coverage for $1,000 a year. See www.american-club.com/cir2-98.htm (accessed July 28, 2000).

134. 63 FR 42709, August 11, 1998.

135. These figures are DOI's estimates for small facilities (those requiring only $10 million in annual coverage). The total includes $10,000 in estimated annual premium costs and $4,000 annual administrative costs. 63 FR 42708, August 11, 1998.

136. Docket materials in support of the April 10, 1998 "Financial Assurance Mechanisms for Corporate Owners and Operators of Municipal Solid Waste Landfill Facilities; Final Rule" Issue Paper: "Market Effects of the Financial Test," at 7. The report also notes that In some cases, firms have been unable to obtain financial assurance. However, in every case, the problem was not the availability of financial assurance mechanisms, but the financial strength of the company," at 7.

137. See Federal Register 1998, supra note 18, at 17722.

138. U.S. GAO, Federal Land Management: Financial Guarantees Encourage Reclamation of National Forest System Lands, GAO/RCED-87-157, August, 1987. ("We did not identify any cases where the costs associated with posting a financial guarantee prevented operators from mining") at 1; (Neither Forest Service officials nor representatives of mining associations that we spoke with could cite an instance where mine operators decided not to mine because of the cost of obtaining a financial guarantee") at 6.

139. Office of Solid Waste, U.S. EPA, Subtitle C and D Corporate Financial Test Issue Paper: Performance of the Financial Test as a Predictor of Bankruptcy, April 30, 1996, at 5.

140. Ibid., at 5.

141. Interviews with Michigan Department of Environmental Quality financial assurance program administrators. Also see ICF Memorandum to Betsy Tam, EPA Office of Solid Waste, January 25, 1988 (cited in Office of Solid Waste, U.S. EPA, Subtitle C

and D Corporate Financial Test Analysis Issue Paper: Market Effects of the Financial Test, December 9, 1997, at 2), which reports an annual 1.5% of face value cost of environmental letters of credit and surety bonds. See also, McElfish, supra note 68 (citing a representative of the Surety Association of America, placing the cost of surface mining reclamation bonds at 1.25%), at 86; Kuipers, supra note 24 (hardrock mining bonds costing 1 to 3.5% annually), at I-12; and C. George Miller, "Use of Financial Surety for Environmental Purposes," paper prepared for the International Council on Metals and the Environment, 1998 (citing annual costs of mining letters of credit and surety bonds as between.37 1.5% of face value) at 5. Available online at http://206. 191.21.210/icme/finsurety.htm

142. A Nuclear Regulatory Commission study of decommissioning bonds found rates from 3% to less than 1% of the bonds' face value. Cited in U.S. EPA, Issue Paper, Assessment of Trust Fund/Surety Combination, docket materials in support of Financial Assurance Mechanisms for Corporate Owners and Operators of Municipal Solid Waste Landfill Facilities; Final Rule, April 10, 1998, at 5.

143. *Id.* In addition, the government estimates that each regulated firm bears $4,000 in annual administrative costs associated with compliance.

144. U.S. EPA, State Funds in Transition: Models for Underground Storage Tank Assurance Funds, Office of Underground Storage Tanks, 1996 (updated 1998). [www.epa.gov/swerust1/states/statefnd.htm]. ("Premiums have also come down since 1989, when some of these commercial programs began. Then, the average premium was approximately $1000 per tank (for good tanks). Today that average has been reduced to roughly $400 per tank. For a double-walled tank and piping system, the cost could drop to $200 per tank"), at 5.

145. McElfish, supra note 68, at 90.

146. See General Accounting Office, Hazardous Waste: An Update on the Cost and Availability of Pollution Insurance, GAO/PEMD-94-16, April 1994, at 3.

147. *Id.*, at 23.

148. See Boyd and Kunreuther, supra note 98. Public funds, by absolving firms of historic liabilities, allow for remediation of existing contamination without reducing firms' wealth. Firms left with greater wealth have a greater incentive to take efficient prospective risk reduction measures, assuming that they are prospectively liable and have to demonstrate privately-provided financial responsibility.

149. This is not to suggest that voluntary insurance for retroactive liabilities will not be provided or demanded. In fact, insurers have played an important role in the detection of risks and minimization of retroactive liabilities, particularly in the lender liability context. It is important to note, however, that these markets function best when they are non-compulsory.

150. See note 144 supra.

151. U.S. Environmental Protection Agency, List of Known Insurance Providers for Underground Storage Tanks, Office of Solid Waste and Emergency Response, EPA 510-B-00-004, January 2000.

152. As noted in Section 3.3, public financing is an undesirable form of prospective financial responsibility. By subsidizing private environmental costs, public assurance funds undermine deterrence.

153. See the Small Business Paperwork Reduction Act Amendments (HR 3310 & S. 1867), 1998, which would have prohibited federal agencies from fining small businesses for first time violations or for not complying with paperwork requirements, as long as

the company complied within six months of notice of the violation. Or the Small Business Liability Protection Act (H. R. 1831), 2001 a bill that provides Superfund liability relief for small businesses and other small contributors.

154. 5 U.S.C. 601, et seq. See also, the 1996 Small Business Regulatory Enforcement Fairness Act (which allows small businesses to challenge an agency in court for failure to comply with the RFA), 5 U.S.C. 801, et seq.

155. In at least one instance, an agency's assurance rules were overturned for failure to abide by RFA requirements. Revised hardrock mining bond rules were overturned in 1998 by as U.S. District Court, Northwest Mining Association v. Babbitt, F.Supp.2d 9, 1998 U.S. Dist.

156. It is always in the interest of a regulated firm to minimize its assurance requirements. Lower levels of assurance imply less cost internalization in the future and lower assurance coverage costs in the present. As an example, see Office of Inspector General, Audit Report, EPA Can Do More To Help Minimize Hardrock Mining Liabilities, E1DMF6-08-0016-7100223, June 11, 1997, at 11 (citing instances of mine owners converting land from federal land to private land in order to minimize bond requirements, where state bond requirements are less than federal requirements).

157. For the moment, we set aside issues raised by the time value of money. Clearly, what is important is that the firm has reserved C for use at the future time it is required. This can mean that an amount less than C is set aside today, with knowledge that that amount will grow if invested properly over time.

158. Note that the firm need not set aside this full amount. All it need do is purchase insurance adequate to cover the full amount.

159. See U.S. EPA, Region IV, Evaluating Cost Estimates for Closure and Post-Closure Care of RCRA Hazardous Waste Management Units, 1996.

160. Consider one example: bonds required for the Zortman-Landusky hardrock mine. Per-acre bond rates at the site increased from $750, to $8700, to $12,500, to $37,000 over a period from 1982 to 1998. See Kuipers, supra note 24.

161. Many assurance requirements have a fixed value over a period of decades. With the passage of time, fixed amounts may become significantly inadequate simply due to inflation. Some wells bonded in the 1940s and 1950s may still be operating under coverage amounts required 50 years ago. In some states, old well bonds are "grandfathered," meaning that wells with pre-existing bonds do not have to post updated bond amounts. As a consequence, many wells may be significantly under-protected. (Conversation with Dave Davis, Michigan DEQ, Aug 1, 2000.)

162. See 33 CFR § 138.80(f)(3).

163. As a rule of thumb, the worst-case discharge is approximately equal to four times the estimated uncontrolled first-day discharge. 63 FR 42707, August 11, 1998. The only exempted facilities are those with an estimated worst-case oil discharge of 1,000 barrels or less. Depending on location and potential discharge volume, coverage requirements range from $10m to $150m for individual facilities.

164. See U.S. EPA, Office of Inspector General, Audit Report, RCRA Financial Assurance for Closure and Post-Closure, March 30, 2001, at ii ("state officials have expressed concerns that the cost estimates are difficult to review.")

165. 340 S. C. 118, 530 SE2d 643, 2000 WL 502520 (S. C. App., Refiled April 4, 2000).

166. ("Sierra Club contends DHEC failed to issue proper notice and provide opportunity for adequate public comment. We agree.")

167. 42 U.S.C. 4321–4347.

168. See Interior Board of Land Appeals, IBLA 97–339, National Wildlife Federation, et al., September 23, 1998. ("We believe the proper course of action at the time the ROD issued in March 1997 would have been for BLM, an agency operating under a mandate to protect the public lands from unnecessary or undue degradation, to require the posting of a sufficient bond to protect against the uncertainties relating to groundwater quality identified in the FEIS, with the possibility of reducing that bond if further studies clarified those uncertainties"), at 360. ("The lack of information and BLM's failure to require a bond in light of the uncertainties created by that lack of information is what convinced the Board to grant a partial stay in this case"), at 366.

169. See *Pennsylvania Federation of Sportsmen's Clubs, et al. v. Com. of Pa. Dept. of Env. Resources 1868* C. D.1981, which sought higher coal mine bonding rates. The petition resulted in a 1988 Consent Decree requiring modifications to the state's bonding program, including higher bond rates if indicated by forfeitures and incomplete reclamation.

Also, see Trustees for Alaska v Gorsuch, 835 P 2d 1239 (Alaska 1992) wherein Trustees for Alaska challenged a surface coal mining permit issued by the Alaska Department of Natural Resources, claiming that DNR violated Alaska's mining laws by approving a bond amount which inadequately reflected the costs of reclamation over the life of the permit. The court held that DNR should "recalculate" the bonds so that they are "sufficient to assure the completion of the reclamation plan by [DNR] in the event of forfeiture," as under AS 27.21.160(a).

170. On the other hand, a weakness of fixed schedules is that they may fail to account for differences in the specific risks being assured.

171. See Kuipers, supra note 24, at 4 for a critique of Arizona and Nevada's hardrock mining regulations, in part on the basis of their willingness to allow companies to estimate their own reclamation costs.

172. Study cited in U.S. EPA, Office of Inspector General, Audit Report, RCRA Financial Assurance for Closure and Post-Closure, March 30, 2001, at 46.

173. For example, the first major post-OPA vessel oil spill created injuries valued at $90 million. The vessel was only required to post $10 million in assurance coverage, however. Brent Walth, Spill Laws Fail to Halt Seepage of Public Cash, The Oregonian, February 27, 2000. According to Walth, seven vessel spills since 1990 resulted in damages exceeding assurance requirements in seven vessel spills since 1990 (reporting on a statement from Daniel Sheehan, Director, National Pollution Funds Center, U.S.CG). Also see U.S. EPA Region V, UIC Permitting Guidance, Technical Support Document, Financial Responsibility for Class II Injection Wells, at http://www.epa.gov/r5water/uic/r5_02.htm, which suggests that coverage amounts for certain wells are not likely to be adequate ("The present coverage for blanket bonds in Michigan is $50,000 and in Indiana is $30,000. This is generally less than the Federal guideline of 10 times the cost to plug and abandon an injection well").

174. U.S. Government Printing Office, 1986. Adequacy of Bonds to Ensure Reclamation of Surface Mines. Hearing Before a Subcommittee of the Committee on Government Operations, House of Representatives, 99th Congress, 2nd Session, 26 June, at 5.

175. *Id.*

176. *Id.* In West Virginia, the average reclamation cost was $2500 per-acre and average bond $1100 per-acre.

177. ("If you read OSM oversight reports, the comment that was made by OSM was that the State was accepting what the operator submitted as the estimated bond amount with no independent verification or mathematical calculations by the State regulatory authority ... There isn't any written or formal criteria.") *Id*, at 71.

178. McElfish, supra note 68 ("SMCRA's bonding provisions have not been effectively implemented in all states. Bond amounts are often set based on faulty assumptions or under systems that have not accurately projected the need for reclamation funds. Some forfeited mine sites still remain un-reclaimed or have been reclaimed to lower than statutory standards because their bonds were insufficient for full reclamation"), at 85.

179. Assessment of Pennsylvania's Bonding Program for Primacy Coal Mining Permits, Office of Mineral Resources Management, Bureau of Mining and Reclamation, February 2000. The analysis derives reclamation costs for sites that forfeited bonds ranging from $5500 to 20,000 per-acre, while bond rates range from only $1000 to 5000 per acre, at 5, and 20–23.

180. Actuarial Study of the Pennsylvania Coal Mining Reclamation Bonding Program, Milliman & Robertson, Inc., July 16, 1993. Also see McElfish, supra note 68 at 92. ("... current bond-setting methodologies incorporate assumptions that do not consider all factors affecting reclamation costs, and thus result in bonds inadequate to cover all costs. For example, bond forfeiture sites frequently have water pollution problems, yet bond-setting methodologies overlook these costs.")

181. See Kuipers, supra note 24 ("the financial failure of numerous mining companies has exposed shortcomings in both bond methods and bond amounts. American taxpayers are faced with significant liability for mines left un-reclaimed, shifting the economic burden from the companies that profited from the mines and leaving environmental disasters behind for the public to clean up"), at 1. Also citing widely varying bond amounts, depending on the state program (average per-acre bond amounts in Alaska $2600 vs. $15,000 in Montana).

182. Office of Inspector General, Audit Report, EPA Can Do More To Help Minimize Hardrock Mining Liabilities, E1DMF6-08-0016-7100223, June 11, 1997, at 8. ("Federal and state land management agencies' authorities to require environmental performance standards and financial assurances at hardrock mines varied, leaving critical gaps in bonding requirements. Unreasonably low bond ceilings did not allow adequate financial assurance coverage for hardrock mining on some state and private lands. As a result, EPA may become liable for the considerable costs of cleaning up mines abandoned by the companies that operated them"), at v.

183. *Id.*, at 9.

184. U.S. GAO, Federal Land Management: Financial Guarantees Encourage Reclamation of National Forest System Lands, GAO/RCED-87-157, August, 1987.

185. *Id.*, at 5.

186. See National Research Council, Hardrock Mining on Federal Lands, National Academy Press, 1999 ("Financial assurance should be required for reclamation of disturbances to the environment caused by all mining activities beyond those classified as casual use, even if the area disturbed is less than 5 acres"), at 8. And U.S. Department of Interior, Office of the Inspector General, Hardrock Mining Site Reclamation, Bureau of Land Management (92-I-636), 1992 (recommending that all operators post financial guarantees, commensurate with the size and type of operation in question).

187. RCRA's hazardous waste disposal rules, for example, allow trust funds to be funded over the term of the facility operating permit, or the remaining life of the facility, whichever is shorter. 40 CFR 264.143(a)(3).

188. See U.S. EPA, Office of Inspector General, Audit Report, RCRA Financial Assurance for Closure and Post-Closure, March 30, 2001, at 21("In our Subtitle C sample, there were a significant number of facilities that went out of business or into bankruptcy with partially funded trust funds").

189. For a theoretical exploration of this concern see Jason Shogren, Joseph Herriges, and Ramu Govindasamy, Limits to Environmental Bonds, 8 Ecological Economics, 109–133, 1993.

190. United States v. Shumway, U.S. Court of Appeals for 9th Cir. (December 28, 1999), wherein the court rejected the U.S. Forest Service's attempt to increase required bond amounts for a hardrock mine operation. The Court found the bond amount to have been raised arbitrarily. More specifically, the court cited evidence that environmental problems had not become more serious over time and that existing site conditions were acceptable, thus calling into question the need for increased bond levels ("Based on our review of the evidence before the trial court, there is an issue of fact as to whether or not the government properly increase the bond amount.")

191. See *C & K Coal Co. v Commonwealth of Penn., Dept. of Environmental Resources*, Docket No. 91–138-E (Consolidated), 1992 Pa Envirn LEXIS 128 (Pa EHB September 30, 1992), where the state was found to have improperly denied a bond release due to its failure to establish liability for damages (". . . Since DER did not sustain its burden of proving there was a hydrogeologic connection between the discharge [emanating in the right-of-way of a public road and running along the boundary of the permitted area] and appellant's permitted area, DER's order to appellant directing it to treat the discharge was an abuse of DER's discretion. Likewise, as the only reason for DER's denial of the appellant's application for bond release was this discharge, DER's denial of bond release was an abuse of its discretion.")

192. 311 S.E.2d 508 (Ga. Ct. App. 1983)

193. *Id.* (citing *John L. Roper Lumber Co. v. Lawson*, 143 S. E. 847 (N. C. 1928), at 850) "If actions for a tort like the present or personal injuries are contemplated, this should be fully and clearly provided for by the surety bond in reasonably clear language. The remedy of plaintiffs is against the contractors."

194. See Moelmann and Harris, supra note 92, who reviewed surety contracts in the environmental field to assess whether bonds were reinterpreted to cover tort claimants ("In researching this field, previously thought to be a 'hot topic,' at no point was a performance bond surety castigated or found liable for any damages beyond those which are reasonably foreseeable or within the realm of a normal recovery under surety or contract law"), at 176.

195. See *American Druggists Ins. Co. v. Comm. of Kentucky Department of Natural Resources and Environmental Protection et al.*, No. 83-CA-807-MR, slip op. (Ky. Ct. App., November 11, 1983) (clarifying the nature of penal versus performance bonds and finding that failure to perform all reclamation requirements resulted in total bond forfeiture). Also see *Morcoal Co. v. Comm. of Pennsylvania*, 459 A.2d 1303 (Pa Commw Ct 1983) (ruling that mining reclamation bonds are intended to be penal and that the state Department of Environmental Resources was not required to prove precise damages in order to forfeit the bonds).

196. The assured amount is a minimum, guaranteed amount of money available for compensation.

197. See Regulatory History 48 FR 32932 (July 19, 1983), Final Rule, Bond and Insurance Requirements, Discussion of Comments and Rules Adopted ("The operator does have the underlying obligation to fully reclaim disturbed lands. A regulatory authority, in having reclamation performed on which the operator has defaulted in his obligation, may incur costs in excess of the forfeited amount. To make clear that the regulatory authority may recover that excess amount from the operator, the suggested addition is made to Sec. 800.50 in paragraph (d)(1))".

198. There are limits to the liability limitation. Specifically, there is no liability limit if a release is determined to be caused by "gross negligence or willful misconduct of, or the violation of any applicable Federal safety, construction, or operating regulation by, the responsible party" or if the incident is not reported in a timely fashion. 33 U.S.C § 2704(c)(1). But note that the liability of guarantors (the third parties guaranteeing coverage) is always strictly limited to amounts specified in the assurance contract, which in no case would be greater than the coverage requirement. 42 U.S.C § 9608(d).

199. See Section 2.1.8.

200. According to the EPA, 19% of hazardous waste facilities studied were not in compliance with financial assurance requirements. U.S. EPA, Office of Inspector General, Audit Report, RCRA Financial Assurance for Closure and Post-Closure, March 30, 2001, at 24.

201. For a set of cases involving penalties for failure to comply with financial assurance regulations see In the Matter of Marley Cooling Tower Co., No. RCRA-09-88-008, 1989 RCRA LEXIS 22 (Nov. 30, 1989) ($7000 penalty for failing to update financial assurances and in failing to demonstrate financial responsibility for third-party claims); In the Matter of Landfill, Inc., Appeal No. 86-8, 1990 RCRA LEXIS 65 (Nov. 30, 1990) (financial assurance penalty of $1900); In re Frit Indus., No. RCRA-VI-415-H, 1985 RCRA LEXIS 4 (Aug. 5, 1985) (financial assurance penalty of $1200). In the Matter of Harmon Electronics, No. RCRA-VII-91-H-0037, 1994 RCRA LEXIS 52 (Dec. 12, 1994) ($251,875 for four years of noncompliance); In the Matter of Standard Tank Cleaning Corp., No. II-RCRA-88-0110, 1991 RCRA LEXIS 47 (Mar. 21, 1991) ($145,313 for six years of noncompliance), aff'd, Appeal No. 91–2 (July 19, 1991).

202. 62 F.3d 806, 809, 812 (6th Cir. 1995).

203. The argument was based on a flawed reading of cases related to RCRA's "loss of interim status" (LOIS) amendment. The facility is in fact subject to assurance regulations until final closure is certified, even though it never obtained interim status by filing for a permit.

204. *U.S. v. Ekco Housewares, Inc.*, 853 F. Supp 975 (N. D.Ohio 1994).

205. In the Matter of B&R Oil Company, Inc., Respondent, United States EPA, before the Administrator. Administrative Law Judge, issued September 4, 1997 ("payment into the state tank fund constitutes a legal obligation separate and apart from respondent's obligation to comply with the Federal regulations. . .").

206. *United States v. Power Engineering Co.*, no. 97-B-1654 (D. Colo. Nov. 24, 2000).

207. *United States v. Power Engineering Co.*, no. 98–1273 (D. Colo., Sept. 8, 1999), at 8. And see United States v. Power Engineering Co., 10 F.Supp2d 1145, 1165 (D. Colo. 1998) at 1157, 1163, and 1165.

208. *United States v. Power Engineering Co.*, 10 F.Supp 2d 1145, 1165 (D. Colo. 1998).

209. 191 F.3d 894 (8th Cir. 1999).

210. *United States v. Power Engineering Co.*, no. 97-B-1654 (D. Colo. Nov. 24, 2000), at 15 ("With all due respect, I conclude that the *Harmon* decision incorrectly interprets the RCRA").

211. Of the cases referenced in note 200 supra, "financial difficulties and bankruptcies were significant contributing factors to facility non-compliance," at 24.

212. See note 8 supra. For general guidance on the conditions which discharge environmental costs and penalties see U.S. EPA, EPA Participation in Bankruptcy Cases, September 30, 1997, memorandum, available at http://es.epa.gov/oeca/osre/ 970930–1.pdf. An illustrative case exploring the issues is In Re Chateaugay Corp., 944 F 2d 997 (2nd Cir 1991) (finding that an injunction encountered in an environmental case that does no more than impose an obligation entirely as an alternative to a payment right is dischargeable). But see also, Ohio v. Kovacs 469 U.S. 274, 105 S Ct 705 (1985) (Dischargeability is limited to situations where a clean-up order is converted into an obligation to pay money. Regulatory orders that demand performance and which cannot be satisfied solely via a monetary payment are not dischargeable in bankruptcy). See also In re Commonwealth Oil Refining Co., 805 F.2d 1175 (5th cir. 1986) (a RCRA compliance order is not stayed by bankruptcy code even though compliance involved expenditure of money.)

213. See *Commonwealth of PA, Dept. of Environ. Resources v. Peggs Run Coal Co.*, 55 PA Commw 312, 423 A 2d 765 (Pa Commw Ct 1980)(DER injunction, including bond requirement, was a "proceeding to enforce its police or regulatory power and as such is exempted from the stay provisions of Section 362 of the Bankruptcy Code)".

214. 196 Bankr. 784, 702 (Bankr. S. D.Ill. June 6, 1996) (In the court's reasoning, the ability to collect on the bonds is not akin to a claim: "Environmental cleanup orders, in particular, often require an expenditure of money in order to clean up immediate and ongoing pollution, and the government may exercise its regulatory powers and force compliance with its laws even though a debtor must spend money to comply ... an obligation does not become a 'claim' merely because it requires the expenditure of money"), at 5.

215. *Id.*, at 4.

216. 733 F.2d 267 (3rd Cir. 1984).

217. *Id.*, at 278.

218. For example, the Federal Deposit Insurance Corporation (FDIC) does not insure letters of credit issued to governments, such as those that would be used as an environmental guarantee. Similarly, most states have an insurance guaranty fund to protect policyholders in the event of an insurer's insolvency. However, most enabling statutes include a "net worth exclusion" which eliminates governments as recipients of these funds. See Michigan, MCL 500.7925(3); and Illinois, 215 ILCS 5/534.3(b)(iv). Accordingly, government attempts to access such funds in environmental guarantee cases have not been successful. See Attorney General ex rel Department of Natural Resources v. Michigan Property and Casualty Guaranty Association, Court of Appeals of Michigan, 218 Mich. App. 342; 533 N.W.2d 700, 1996.

219. See notes 100,103, and 106 supra. Trust funds can be vulnerable to the insolvency of a financial institution acting as trustee. Some regulations require trustees to be only those regulated or regularly examined by a Federal or State agency, see 40 CFR 264.143.

220. See "American Insurance: Bungee Jump," The Economist, September 16, 2000, at 84.

221. McElfish, supra note 68, at 89 (citing Office of Surface Mine Reclamation and Enforcement, Record of Surety Insolvencies, August 1988, unpublished).

222. U.S. EPA, Issue Paper: "Assessment of Financial Assurance Risk of Subtitles C and D Corporate Financial Test and Third-Party Financial Assurance Mechanisms," in docket materials in support of "Financial Assurance Mechanisms for Corporate Owners and Operators of Municipal Solid Waste Landfill Facilities; Final Rule," April 10, 1998, at 7.

223. *Id.* at 6. Being de-listed is not equivalent to being insolvent, though a surety's financial health is the main determinant of whether or not it is listed as an acceptable government bond provider.

224. Frontier was a major supplier of environmental bonds. For example, of 198 solid waste landfills in Michigan in 2000 35 had closure bonds issued by Frontier, or 18% of the total.

225. According to an EPA official, "requiring the company to close its treatment, storage, and other services was not in the best interest of the environment". Quoted in Pat Phibbs, Safety-Kleen, EPA Agree on Deadline for Obtaining Insurance for Facilities, Environment Reporter, October 20, 2000, at 2200–1.

226. In re Safety-Kleen Corp., Bankr. D. Del. No. 00–2303, October 17, 2000. Safety-Kleen and its subsidiaries operate approximately 30% of the waste management facilities in the U.S. Approximately 50% of its financial assurance was provided by Frontier. It is important to note that Frontier bonds, while not acceptable due to Frontier's financial weakness, remain in place, with Safety-Kleen continuing to pay the premiums. See 10-Q Report for Safety-Kleen Corporation, SEC file 1-08368, February 28, 2001, at 9.

227. 10-Q Report for Safety-Kleen Corporation, SEC file 1-08368, February 28, 2001, at 9-10. Safety-Kleen was in financial difficulty for a variety of reasons, most unrelated to the withdrawal of the Frontier bonds.

228. *Id.*, at 9.

229. *The Wall Street Journal*, Reliance Files for Chapter 11 Protection, June 13, 2001, at A3.

230. AEI is the fourth largest producer of coal for energy production in the U.S. (Corp. website.)

231. Moody's Downgrades AEI Debt, Coal Outlook, July 31, 2000, at 1.

232. Ken Ward, Addingtons' Coal Company in Trouble, Downgrade of Reclamation Bond Provider Gets the Blame, The Charleston Gazette, July 7, 2000.

233. 33 U.S.C § 2716; 42 U.S.C § 9608(c)(1–2).

234. Appendix B to 33 CFR, Part 138. Also see 30 CFR 253.41(a)(4).

235. The offshore facilities rule, for instance, allows direct action against guarantors as long as insolvency is simply "claimed" by the responsible party. In the government's reasoning, "Establishing a regulatory process that might require a lengthy insolvency determination procedure before compensation could begin would be totally inconsistent with [OPA's objectives]." 63 FR 42707, August 11, 1998.

236. 61 CFR 9270. "No standard marine liability insurance policy of which the Coast Guard is aware meets [the direct action] requirement."

237. For instance, there is an admiralty rule that any evidence of a material misrepresentation cancels insurance coverage. This rule is generally respected in U.S. jurisdictions. See *Port Lynch, Inc.* v. *New England International Assurety, Inc.*, 754 F.Supp 816, 1992 AMC 225 (W. D. Wash, 1991), upholding the standard. In contrast, however, see *Albany Insurance Co.* v. *Anh Thi Kieu*, 927 F.2d 882, 1991 AMC 2211

(5th Cir.), at 890, holding that state law should govern the question of what voids coverage and that misrepresentations did not void coverage since insured did not intend to deceive the insurer.

238. 42 U.S.C § 9608(c)(1). "The guarantor may invoke all rights and defenses which would be available to the owner or operator under this subchapter. The guarantor may also invoke the defense that the incident was caused by the willful misconduct of the owner or operator, but the guarantor may not invoke any other defense that the guarantor might have been entitled to invoke in a proceeding brought by the owner or operator against him." 61 FR 9268. "A guarantor agrees to waive all other defenses, including nonpayment of premium." For a state law example, see Alaska Statute 46.04.040(e).

239. 33 CFR 138.80(d)(1). "Any evidence of financial responsibility submitted under this part must contain an acknowledgment by the insurer or other guarantor that an action in court by a claimant for costs and damage claims arising under the provisions of the Acts may be brought directly against the insurer or other guarantor."

240. 30 CFR 253.41(a)(4); 33 CFR 138.80(d). "There is no evidence that fraud and misrepresentation have been a problem in the current OSFR program." 63 FR 42707, August 11, 1998. The meaning of the "willful misconduct" standard has been previously addressed by U.S. courts. See *The Tug Ocean Prince, Inc.* v. *United States*, 584 F.2d 1151, 1978 AMC 1787 (2d. Cir 1978), cert. denied 440 U.S. 959 (1979): willful misconduct or gross negligence being equivalent to the equally vague "egregious conduct making an accident likely to happen."

241. In the words of the Minerals Management Service, which administers the offshore facilities assurance program, "Allowing such a defense is inconsistent with two objectives of the OSFR program: Ensure that claims for oil-spill damages and cleanup costs are paid promptly; and make responsible parties or their guarantors pay claims rather than the Oil Spill Liability Trust Fund. Limiting the types of defenses guarantors may use to avoid payment of claims is consistent with and furthers the achievement of these objectives. Furthermore, there is no evidence that fraud and misrepresentation have been a problem in the current OSFR program," 63 FR 42707, August 11, 1998.

242. But note, similar to the lack of insurer defenses under direct action provisions, that case law denies sureties a defense based on malfeasance by the bond purchaser. In general, fraud by the principal does not discharge the surety's obligations unless the obligee (the party to whom performance is owed) was involved in the fraud. *Rachman Bag Co. v,. Liberty Mutual Insurance Co.*, 46 F.3d 230,237 (2nd Cir. 1995).

243. From an assurance standpoint, the most problematic of all exclusions would be one that relieves an insurer of its coverage obligations in the event of a customer's insolvency. Assurance rules tend to explicitly prohibit this specific exclusion. For example, 280.97(b)(2)(a).

244. See State of California, State Water Resources Control Board, Financial Responsibility Long Term Study," January, 1995, 94-2CWP, describing difficulties associated with exclusions: "First, the products offered have many pre-insurance requirements and numerous policy exclusions so that the coverage desired is often not the coverage offered. Second, the policy coverage offered often does not match necessarily the type of coverage legally required," at 6.

245. "In spite of insurance certificates which provide a warrant that policies conform with regulations, policy terms and exclusions may make it difficult for states to obtain closure and post-closure funds from insurance policies without litigation," U.S. EPA,

Office of Inspector General, Audit Report, RCRA Financial Assurance for Closure and Post-Closure, March 30, 2001, at 18.

246. See Texas' assurance regulations 30 TAC §37.641 (2)(e) and certification that "the wording of this [overage] endorsement is identical to the wording specified in 30 TAC §37.641."

247. For example, the State of Michigan's hazardous waste management facility assurance program requires one of two endorsements. The first if for policies that are "pre-accepted" as limiting exclusions. Insurers without pre-accepted policies must sign an endorsement which includes the following declaration: "No condition, provision, stipulation, limitation, or exclusion contained in the Policy, or any other endorsement thereon, or any violation thereof, shall relieve the insurer from liability or from payment of any claim, within the stated limits of liability in this Endorsement, for bodily injury and property damage to a third party caused by a sudden and accidental occurrence."

248. See *Advanced Environmental Technology Corp. v. Brown*, 4th Cir., No. 99–2228, October 2, 2000 (insurance agent found liable for having "negligently misrepresented" coverage provided to a waste removal subcontractor, knowing an exclusion was for coverage sought by the insured).

249. Where a bond is required by law but does not conform to the regulatory requirement the bond is typically interpreted to provide the protections envisioned by regulation, 17 Am. Jur. 2d, Contractors' Bonds §8. Also see Davis v. Moore, 7 Ill App 2d 519, 130 NE 2d 117 (Ill Ct App 1955), "[T]his court holds that the statutory requirements of an appeal bond are a part of such bond, whether fully recited therein or not, that it is not error for a court to decree a reformation of a bond to conform to the statute (although it may not be necessary), and that judgment may be entered on an appeal bond according to the provisions of the statute, regardless of any error in the form of the bond."

250. See *U.S. v. Country Kettle, Inc.*, 738 F.Supp 1358, 1360 (D.Kan. 1990).

251. Bonds and letters of credit require at least 120 days notice prior to cancellation. 40 CFR 264.143(b)(8), 40 CFR 264.143(c)(8), 40 CFR 264.143(d)(5).

252. 40 CFR 264.143(e)(6),(8),(10). Failure to pay premiums is considered a violation of assurance regulations and accordingly, can lead to monetary or injunctive penalties.

253. 40 CFR 264.143(d)(5).

254. 30 CFR 253.41(a)(2).

255. "OPA makes guarantors subject to liability for claims made up to 6 years after an oil-spill discharge occurs." 63 FR 42704, August 11, 1998.

256. See 44 FR 14902, MARCH 13, 1979("This restriction [against cancellation of the bond] is based on the first principle of surety law, i.e. the surety undertakes the obligation to stand in the shoes of the principal, and his obligation may not be rescinded or terminated without the consent of the party to whom the duty is owed.")

257. For more on the distinction between claims-made and occurrence coverage see Chris Mattison and Edward Widmann, Environmental Insurance: An Introduction for the Environmental Attorney and Risk Manager, 30 ELR 10365, 2000.

258. Central Illinois Public Service Company v. American Empire Surplus Lines Insurance Company, 267 Ill. App. 3d 1043 (1994) (denying coverage on a claims-made policy, due to lack of a third party demand necessary to constitute a valid "claim," even though pollution had been discovered and the regulator was notified of the occurrence).

259. See RCRA's UST assurance rules, 40 CFR 280.97(e). When a claims-made policy is used the insurer must include an endorsement stating that "The insurance covers claims otherwise covered by the policy that are reported to the ["Insurer" or "Group"] within six months of the effective date of cancellation or non-renewal of the policy except where the new or renewed policy has the same retroactive date or a retroactive date earlier than that of the prior policy, and which arise out of any covered occurrence that commenced after the policy retroactive date, if applicable, and prior to such policy renewal or termination date." Also see 40 CFR 258.74(d)(6), 40 CFR 264.143(e)(8).

Some states make further requirements. Texas, for example, require firms using claims-made policies to place in escrow funds sufficient to pay an additional year of premiums for renewal of a policy by the state on notice of the termination of coverage. Texas Code §37.6031(f).

260. For example, self-insurance can be used to cover the deductible included in an insurance policy. 63 FR 42704, August 11, 1998.

261. 33 CFR 138.80(c)(1).

262. 30 CFR 253.29(c)(4); 33 CFR § 138.80(c)(1)(j). The offshore facilities rule, however, establishes specific horizontal layers that can be served by different guarantors. Multiple guarantors cannot cover intermediate, horizontal sub-layers.

263. For example, insurer A is liable for claims up to $1 million, insurer B is liable for claims from $1 million to $2 million, etc.

264. Problems have been indicated by the Mineral Management Service: "The reason we placed a limit on the number of insurance certificates and the amounts in the [coverage] layers is that in the past we received insurance certificates that did not add up to the total amount of coverage indicated. We found that insurance certificate problems likely increase with the number of certificates. Many times the problem was associated with 'horizontal' layering, which is the allocation of risk within an insurance sub-layer. Verifying that the total amount of the certificate was properly allocated among participating insurers is a burdensome process . . . 63 FR 42704, August 11, 1998.

265. U.S. EPA, Office of Inspector General, Audit Report, RCRA Financial Assurance for Closure and Post-Closure, March 30, 2001, ("For example, a significant portion of the assets of one captive, established by a large waste management firm, was represented by a note receivable from the parent company"), at 12; ("captive insurance policies in our sample do not meet the intent or requirements of RCRA financial assurance regulations"), at 26.

266. A Virginia law, passed in 2000, prohibits reliance on captive insurers, approved surplus line insurers, and risk retention groups as a means of assuring closure and post-closure costs. HB1022, passed January 24, 2000.

267. U.S. EPA, Office of Inspector General, Audit Report, RCRA Financial Assurance for Closure and Post-Closure, March 30, 2001 ("We were given examples during our audit where banks had released funds from trust funds to Subtitle C facility owners without the required approval"), at 21.

268. See Financial Responsibility Long Term Study, State of California, State Water Resources Control Board," January 1995, 94–2CWP ("The Fund has not directed owners or operators to send an original of these mechanisms to us even though the Fund is the designated payee . . . . The Fund, as the payee, should obtain the original document designating the SWRCB as the payee"), at 10.

269. Review of Hardrock Mining Reclamation Bond Requirements, Legislative Request #98L-36, Legislative Audit Division, State of Montana, December 4, 1997

("During the course of our review, we identified several potential control weaknesses which affect the department's ability to effectively manage performance bonds . . .. File documentation does not necessarily reconcile with computer system information. We noted instances of bonds without department signatures"), document available at leg.state.mt.us/audit/download/98L-36.pdf

270. See testimony from the General Accounting Office on mining bond collection problems, Adequacy of Bonds to Ensure Reclamation of Surface Mines. Hearing Before a Subcommittee of the Committee on Government Operations, House of Representatives, 99th Congress, 2nd Session, 26 June 1986. ("I spoke to the Director of the State regulatory authority. She indicated that the problem in Oklahoma was the 'paper' on which some of those bonds were written. In essence, the bond paper was bad. Once the bonds are written off on a legal technicality, you are not going to get any money"), and ("Some of these bonds – I think four of them, had letters of credit amounting to about $425,000 which were allowed to expire. Therefore the money is not going to be available to reclaim the sites"), at 70.

271. See Kuipers, supra note 24 ("The measurement of success can be highly subjective and is often dependent upon the interpretation of specialists hired by the mining company") at I-16.

272. Review of Hardrock Mining Reclamation Bond Requirements, Legislative Request #98L-36, Legislative Audit Division, State of Montana, December 4, 1997 ("The department relies on public comment and scrutiny as a [bond release] control measure"), at 6.

273. Adequacy of Bonds to Ensure Reclamation of Surface Mines. Hearing Before a Subcommittee of the Committee on Government Operations, House of Representatives, 99th Congress, 2nd Session, 26 June, 1986. (discussing problems with inappropriate bond release and stating that 66% of mined Pennsylvania acres were appealed to an Environmental Hearing Board on the basis of conflicts over release. In all cases, the Board eventually sided with state, but Hearings took on average 16 months for resolution), at 4.

274. Self-demonstration is allowed under the OPA/CERCLA vessel and offshore facility rules, all of the RCRA programs (Subtitles, C, D, and I), SMCRA, and many state hardrock mining programs.

275. Firms unable use self-demonstration are particularly aware of this advantage. According to the testimony of a firm unable to comply with the self-demonstration criteria, "The market is now divided into those who can self-insure and do not have to pay the additional premium cost, and those who cannot and must assume this enormous expense." The Federal Requirements for Vessels to Obtain Evidence of Financial Responsibility for Oil Spill Liability Under the Oil Pollution Act of 1990, Hearing before the Subcommittee on Coast Guard and Maritime Transportation of the Committee on Transportation and Infrastructure, House of Representatives, 104th Cong., June 26, 1996, at 33.

276. As an example of the tendency to reduce the criteria necessary for self-demonstration, consider Michigan's UST assurance rules which state, in part, that "the amount of the financial responsibility requirements required under the provisions of this subpart shall be reduced to the amount required by the federal government upon passage by the federal government of a reduction in the financial requirements of this part." R 29.2161(f), amending Section 280.90. Also see, Minerals Management Service Press Release, May 4, 1995, OCS Policy Committee Passes Recommendations on Oil Pollution

Act Financial Responsibility Requirements (#50033), reporting on an advisory committee's approval of a resolution to seek "additional mechanisms for qualifying as a self-insurer" so that "that the costs of demonstrating OSFR do not cause serious economic harm to responsible parties." Available at http://www.mms.gov/ooc/press/1995/50035.txt

277. See Federal Register 1998, supra note 18 ("The financial test allows a company to avoid incurring the expenses associated with the existing financial assurance requirements which provide for demonstrating financial assurance through the use of third-party financial instruments, such as a trust fund, letter of credit, surety bond, or insurance policy"), at 17708. An EPA analysis of its self-demonstration rules for municipal landfills concluded that self-demonstration, by eliminating third-party assurance costs for qualifying firms, would save approximately $77 million annually. *Id.*, at 17719.

278. Disturbingly, and perhaps not coincidentally, Nevada's hardrock mining program, which as of 2000 featured 13 mines in foreclosure or bankruptcy, also features a particularly high rate of self-bonding (approximately 50% of Nevada's hardrock mine reclamation bonds are in the form of self-bonds). Kuipers, supra note 24, at II-44.

279. The financial tests are not arbitrary. For example, the EPA compared, using retrospective analysis, the ability of different tests to predict future bankruptcy. For example, firms with less than $10m in tangible net worth went bankrupt four times more frequently than firms with tangible net worth greater than $10m. Federal Register, Vol. 59, no. 196, October 12, 1994, at 51524. Also, see Federal Register 1998, supra note 18 ("An analysis of bond ratings showed that bond ratings have been a good indicator of firm defaults, and that few firms with investment grade ratings have in fact gone bankrupt"), at 17709; justifying the use of debt-to-equity ratio profitability ratios as an alternative to bond ratings, "The Agency selected these two specific financial ratios with their associated thresholds based on their ability to differentiate between viable and bankrupt firms," *Id.* at 17709.

280. Self-demonstration tests differ slightly under the various programs. For example, see note X supra.

281. See Mark Beasley, Joseph Carcello, and Dana Hermanson, Fraudulent Financial Reporting: 1987–1997, An Analysis of U.S. Public Companies, Committee of Sponsoring Organizations of the Treadway Commission, 1999 ("Relative to public registrants, companies committing financial statement fraud were relatively small") ("Pressures of financial strain or distress may have provided incentives for fraudulent activities for some fraud companies"), at 2.

282. *Id.*, at 3. Fifty-six percent of the sample fraud companies were audited by a Big Eight/Six auditor during the fraud period, and 44% were audited by non-Big Eight/Six auditors.

283. See Federal Register 1998 supra note18, at 17717 ("The financial analysis of firms with net worth between $1 million and $10 million show that environmental obligations may not be universally recognized. When EPA examined the liabilities, net worth and estimated financial assurance amounts for forty firms with net worth between $1 and $10 million, it found that many of these firms had estimated financial assurance obligations that exceeded their net worth (thirty-seven) and their reported liabilities (thirty-five). In the instances of firms with financial assurance obligations that exceed their liabilities, this strongly implies that they are not recognizing these obligations as liabilities, particularly because liabilities also include money owed to creditors such as banks. This inconsistent reporting of landfill closure obligations has been reported by the Financial Accounting Standards Board.")

284. For discussion of environmental obligation accounting standards see Financial Accounting Standards Board, Exposure Draft, Proposed Statement of Financial Accounting Standards, Accounting for Certain Liabilities Related to Closure or Removal of Long-Lived Assets, No. 158-B, February 7, 1996. Given the subjectivity of standards another concern is that audits may favor the interests of the audit's purchaser. See Comment Response Document for Financial Test and Corporate Guarantee for Private Owners or Operators of Municipal Solid Waste Landfill Facilities, October 12, 1994 Proposed Rule (59 FR 51523), ("Compliance with the proposed financial test relies on the opinion of an independent certified public accountant. The experience of [The Michigan Department of Natural Resources] is that even independent certifications are slanted to the benefit of the owner/operator to the maximum extent allowed by law"), at 111.

285. In a bankruptcy filing creditors compete to recover money owed to them. Environmental agencies are not typically guaranteed any priority in this competition. For this reason, some assurance rules require self-demonstrating firms to base asset calculations only on their unencumbered assets (those with no other claim attached to them). As under the offshore facilities rule, 30 CFR § 253.26; 63 FR 42703, August 11, 1998.

286. In theory, this problem is addressed by a requirement that all costs being assured are revealed. "Requiring that the owner or operator include all of the costs it is assuring through a financial test when it calculates its obligations prevents an owner or operator from using the same assets to assure different obligations under different programs." 63 Federal Register 1998 supra note18, at 17712.

287. 40 CFR 264.143(f)(6).

288. See "The People v. America Inc," The Economist, March 24, 2001, at 71.

289. See Correspondence, Waste Management Division, Michigan Department of Environmental Quality, to the Dow Corning Corporation, October 19, 1995 [on file with author] ("In making the demonstration, the company relied upon the bankruptcy filing as a basis to exclude certain liabilities, receivables, and special charges for the breast implant litigation. The MDEQ cannot accept the bankruptcy filing as a basis to exclude the amounts attributed to the breast implant litigation . . . . The bankruptcy filing cannot be used as a basis to improve Dow Corning Corporation's ability to pass a financial test that it previously failed"). The data submitted to MDEQ was un-audited and in conflict with subsequent, audited data. According to MDEQ "The August 2, 1995 letter from the independent accountant, price Waterhouse LLP, noted many significant deviations from the un-audited financial statements."

290. This problem has been previously noted. See section X supra.

291. In the words of the Michigan Department of Natural Resources, commenting on the RCRA D financial test, "A financial test does no provide a state or the U.S. EPA access to funds to complete closure, post-closure, or corrective action should the financially responsible corporation refuse to take the needed actions. . . .. The only recourse to a state or the U.S. EPA would be a lengthy and costly lawsuit with the owner or operator." Comment Response Document for Financial Test and Corporate Guarantee for Private Owners or Operators of Municipal Solid Waste Landfill Facilities, October 12, 1994 Proposed Rule (59 FR 51523).

292. This distinction is acknowledged by the EPA. Third-party mechanisms "provide easier access to funds to fulfill financial obligations. A State may, therefore, decide that it has facilities with poor compliance histories that do not make them a good candidate for the financial test in order to eliminate potential delays in obtaining closure,

post-closure or corrective action. Similarly, States may decide to forego altogether adoption of the financial tests." Federal Register 1998 supra note 18, at 17726.
    293. See Section 6.4.2 supra.

ACKNOWLEDGMENT

This research was supported by a grant from the Michigan Great Lakes Protection Fund. The views expressed are solely those of the author.

COMMENTS ON PAPER BY JAMES BOYD

Donatella Porrini

The paper "Financial Responsibility for Environmental Obligations: Are Bonding and Assurance Rules Fulfilling Their Promise?" written by James Boyd presents a very interesting and complete description of financial responsibility as an environmental policy instrument. It describes very well and in depth the topic of financial responsibility: the theoretical characteristics and the practical applications.

With the expression "financial responsibility" all the instruments are considered that require polluters to demonstrate *ex ante* financial resources adequate to correct and compensate for environmental damage that may arise. In practise, financial responsibility implies that the operations of hazardous plants and other business are only authorised if firms show proof that future liability claims will be covered by insurance companies products or other financial sources.

The starting point of the paper is the importance of financial responsibility as an instrument to solve the environmental damages internalisation problem. In fact a firm strict liable is not stimulated to choose an efficient level of care if there is the possibility to go into bankruptcy, following the judgement proof theorem.[1] On an economic point of view, this is a problem of internalisation in the sense that it causes the fact that some of the losses of the victims will go unclaimed under conventional strict liability and in some cases firms facing considerable liability risks can reduce their capital[2] and use judgement proofness as a evasion strategy.[3]

An Introduction to the Law and Economics of Environmental Policy: Issues in Institutional Design, Volume 20, pages 487–490.
© 2002 Published by Elsevier Science Ltd.
ISBN: 0-7623-0888-5

The internalisation problem arises when the financial resources of the firms are not sufficient to cover the size of the damage caused by an environmental accident: the author outlines that this is not only a characteristic of the so-called catastrophic accidents. As it happens in U.S. the problem can very easily arise in case of reduced size firm operating in risky production activity. So very often financial responsibility can be a useful and necessary instrument of internalisation.

Other advantages correlated with the general one of internalisation can derive from the application of this instrument. First of all, given the contractual relationship between the financial institutions and the firms, there exists a strong incentive for the insurance companies to monitor that the firm is taking an efficient level of preventive measures. Secondly, the firm itself is stimulated to take care because financial responsibility ensures that the expected costs of environmental risks appear on the firm's balance sheet and its business calculation.

Another aspect very well addressed by the paper is the existence of alternative instruments to financial responsibility used with the same target to solve the judgement proof problem. The most important one is the so called "lenders liability". Lenders liability, applied mainly in the CERCLA system by the courts, is a kind of *ex post* instrument of internalisation that considers the financiers of the firm responsible for the residual part of the environmental damage not covered by the firm for judgement proof problems.

The comparison between financial responsibility and lenders liability gives the opportunity of considering the advantage of *ex ante* in the respect of *ex post* instruments in terms of informational problems. The question is if *ex ante* monitoring stimulated through a system of financial responsibility is more effective than *ex post* incentive to monitor deriving from a lenders liability system.

We can find in the economic literature some interesting contributions that demonstrate the efficiency of the lenders liability solution to face the problem of internalisation in the case of judgement proof firms even in the presence of informational problems, such as moral hazard.[4] We can affirm that most of the results of this literature can be applied to support the efficiency of the financial responsibility solution, that presents the same advantage from the relationship between the firms and the financiers with less informational problems. A part of the fact that we can find other contributions, even if not so many as for lenders liability, that have demonstrated specifically that financial responsibility can be a fully efficient solution.[5]

The second part of the paper deals with the analysis of the application of the financial responsibility instrument in the U.S. experience: it presents an overview of financial assurance policies, a review of the rules of implementation in U.S., an examination of the ways to demonstrate financial responsibility (mechanism of implementation), and the politics and the costs of this instrument.

From this part of the paper, we can derive that financial responsibility has been widely applied in the U.S. under the CERCLA system and others, moreover this instrument has been also developed in its characteristics and practical application. And this is in contrast with the European experience where lenders liability and financial responsibility and other form of mandatory insurance are not very much applied in practise.

The difference between U.S. experience and Europe experience appears even more evident if we have look to the recent "White Paper on Environmental Liability"[6] that aims to design a new liability system for the European Community countries to face the problems connected with the assignment of liability in case of environmental damages.

In the *White Paper* in paragraph 4.9. about "Financial Security" the importance of the financial responsibility instrument is affirmed: "Insurability is important to ensure that the goals of an environmental liability regime are reached. Strict liability has been found to prompt spin-offs or delegation of risky production activities from larger firms to smaller ones in the hope of circumventing liability. These smaller firms, which often lack the resources to have risk-management systems as effective as their larger counterparts, often become responsible for a higher share of damage than their size would indicate. When they cause damage, they are also less likely to have the financial resources to pay for redressing it. Insurance availability reduces the risks to which companies are exposed (by transferring part of them to insurers). They should therefore also be less inclined to try to circumvent liability".

But the European legislator seems to not believe that insurance companies can be ready to face the problem to compensate environmental risks, and in particular natural resource damage. The insurance market, considered one of the possible ways of having financial security, is seen to be relatively undeveloped, despite the clear progress made in parts of the financial markets specialising in this area. For this reason "*the EC regime should not impose an obligation to have financial security*, in order to allow the necessary flexibility as long as experience with the new regime still has to be gathered. The provision of financial security by the insurance and banking sectors for the risks resulting from the regime should take place on a voluntary basis".

Given that the Commission intends to continue discussions about these problems in order to stimulate the further development of specific financial guarantee instruments, the contribution of James Boyd appears to be particularly useful to make the U.S. experience a basis for the future development of the European environmental liability system.

NOTES

1. Shavell (1986).
2. Heyes (1996).
3. Van't Veld, Rausser, Simon (1997).
4. Cfr. Beard (1990); Pichford (1995); Heyes (1996); Boyd e Ingberman (1997); Boyer, Laffont (1997).
5. Feess, Hege (2000); Feess, Hege (2001).
6. European Commission, Directorate-General for the Environment, "White Paper on Environmental Liability", COM(2000) 66 final, 9 February 2000.

REFERENCES

Beard, R. (1990). Bankruptcy and Care Choice. *RAND Journal of Economics, 21*(4), 627–634.
Boyd, J., & Ingberman, D. (1997). The Search of Deep Pocket: Is "Extended Liability" Expensive Liability? *Journal of Law, Economics,and Organization, 13*(1), 233–258.
Boyer, M., & Laffont, J. J. (1997). Environmental Risk and Bank Liability. *European Economic Review, 41*, 1427–1459.
Feess, E., & Hege, U. (2000). Environmental Harm, and Financial Responsibility. *Geneva Papers of Risk and Insurance: Issue and Practice, 25*, 220–234.
Feess, E., & Hege, U. (2002). Safety Monitoring, Capital Structure, and 'Financial Responsibility'. *International Review of Law and Economics*, (forthcoming).
Heyes, A. (1996). Lender Penalty for Environmental Damage and the Equilibrium Cost of Capital. *Economica, 63*, 311–323.
Pitchford, R. (1995). How Liable Should a Lender Be? *American Economic Review, 85*, 1171–1186.
Shavell, S. (1986). The Judgement Proof Problem. *International Review of Law and Economics*, June, 6, 45–58.
Van't Veld, K., Rausser, G., & Simon, L. (1997). The Judgment Proof Opportunity. *FEEM Nota di Lavoro*, (83).

PART 2:
OPTIMAL ENFORCEMENT

POLLUTION AND PENALTIES

Anthony Ogus and Carolyn Abbot

ABSTRACT

This paper investigates the economic implications of applying different sanctions, notably criminal penalties, the suspension or revocation of licences and administrative fines to environmental regulatory contraventions. Using familiar economics of law enforcement models, we predict that the almost exclusive reliance by British environment agencies on criminal justice sanctions leads to under-deterrence. We argue that the agencies should be given powers to levy administrative financial charges from offenders without the procedures and onus of proof with which the criminal process protects defendants, but which also inhibits prosecution. The German system of Ordnungswidrigkeit provides an excellent model for this purpose.

INTRODUCTION

The nature of regulatory penalties and appropriate enforcement policies might appear to be very familiar issues, particularly in the context of environmental protection, and they have received much discussion in the law and economics literature (e.g. Heyes, 2000). Nevertheless developments are occurring which suggest good reasons for looking at these matters afresh.

While governments and environmental lawyers have reiterated traditional concerns that the enforcement of command-and-control regulatory systems is insufficiently effective (e.g. Bugler, 1972; House of Commons Select Committee

An Introduction to the Law and Economics of Environmental Policy: Issues in Institutional Design, Volume 20, pages 493–516.

on the Environment, Transport and the Regions, 1999–2000, 34 I-II) they
have also explored and encouraged regimes which, because they focus on
management structures, incentive devices and forms of self-regulation, seem no
longer to rely on conventional sanctions and therefore orthodox deterrence
theory (Gunningham et al., 1998; Balch, 1980). Central to this paper is a third
area of debate. In the United Kingdom, the imposition of penal sanctions has
historically been reserved for the ordinary courts and always subject to the
normal processes of criminal justice. Elsewhere in the common law world,
particularly in the USA and Australia (Australian Law Reform Commission
2000) there has been a growing tendency to grant powers to the public
enforcement agencies to impose "civil" or "administrative" penalties, and
without the procedures and protection of the criminal law. However, to the best
of our knowledge, there has been little attempt to rationalise the use of these
powers, nor to relate them to more traditional deterrence theory.

Our main purpose in this paper is to explore the deterrence dimension to
the use of administrative penalties.[1] We do so within the context of U.K.
environmental regulation policy[2] by contrasting current reliance on the criminal justice system with what might be achieved by greater use of administrative
penalties. To complete the picture, we also consider other sanctions – notably
the revocation and suspension of licences. We begin with a description of the
enforcement powers and practices of the Environment Agency of England
and Wales. Drawing on the law and economics literature, there follows an
evaluation of the current reliance on the criminal justice system and of the use
of the sanction of suspension or revocation of a licence. We then turn to
alternative enforcement strategies, in particular the use of administrative
penalties.

ENFORCEMENT POWERS AND PRACTICE:
THE ENVIRONMENT AGENCY IN
ENGLAND AND WALES

General

One feature of most modern systems of pollution control is that the regulatory
authority has strong powers of enforcement. Whilst the Environment Agency,
in many instances, adopts a co-operative approach to enforcement, using
persuasion, warnings and other informal, non-statutory measures to ensure
compliance, it has recourse to a number of formal mechanisms which vary from
function to function. These powers of enforcement fall into two broad
categories. The first covers what are termed, administrative enforcement

mechanisms; these include the variation, suspension or revocation of a licence,[3] the issuing of enforcement, prohibition or works notices, and, if appropriate, the seeking of an injunction through the courts. The second involves criminal justice proceedings and contains the power to initiate prosecutions and to serve formal cautions[4] and warnings[5] where a criminal offence has been committed.

At the heart of the enforcement process lies a number of criminal offences which provide for prosecution through the criminal courts. These offences fall into several different categories two of which are important for our purposes: the general offences which can be committed regardless of whether a licence has been breached or, indeed, whether a licence is required;[6] and those specifically based on non-compliance with a condition of a licence or operating without an appropriate licence.[7] Offences in the first category are often couched in terms of 'cause', 'knowingly cause' or 'knowingly permit' pollution of the environment, the precise meaning of which has been subject to much judicial deliberation. In general, they are treated as offences of strict liability therefore requiring no *mens rea* in the sense of intention or negligence.[8] Environmental offences are triable either in a Magistrates' Court or the Crown Court, the latter option being invoked where the Agency is seeking an unlimited fine and/or imprisonment.

As regards administrative mechanisms, water discharge consents, integrated pollution control (IPC) authorisations and waste management licences can be revoked entirely[9] and indeed, a waste management licence can be partially revoked, thereby leaving parts of the licence in force.[10] Alternatively, the Agency can suspend an IPC authorisation and waste management licence, in the former case by serving a prohibition notice on the operator.[11] These administrative penalties of revocation and suspension have potentially large financial consequences for operators, and yet they are subject to administrative procedures, without the protection offered to offenders by the criminal process. Admittedly, the administrative decision is subject to an appeal to the Secretary of State, but this is not an adjudicative procedure and will not necessarily involve an oral hearing.[12]

The Agency has considerable discretion when exercising the powers of prosecution, revocation and suspension but this is constrained by its internal Enforcement and Prosecution Policy (Environment Agency, 1998). A policy statement, which sets out the general principles which the Agency intends to follow in relation to enforcement and prosecution, is used in conjunction with associated functional guidelines. The policy is divided into two sections, the first dealing with principles of enforcement, and the second with prosecution.

The stated purpose of enforcement is to ensure that action is taken to protect the environment or to secure compliance with the regulatory system and whilst it is stressed that the Agency expects voluntary compliance, enforcement powers will be used where necessary.[13] 'Firm but fair regulation' is the guiding policy in relation to enforcement. Underlying this policy are four principles: proportionality in applying the law and securing compliance, consistency of approach, targeting of enforcement action and transparency about how the Agency operates and what those regulated may expect.[14] The first principle is designed to ensure that any action taken by the Agency to secure compliance is proportionate to the associated environmental risks and the seriousness of any potential legal breach. Whilst consistency requires that a similar approach be taken in similar circumstances to achieve similar ends it does not mean simple uniformity, as regulatory officials need to take into account variables such as the scale of environmental impact, the attitudes and actions of management and the history of previous incidents or breaches. Under the principle of transparency, the regulated have a right to understand their obligations and what they should expect from the Agency.[15] Finally, the concept of targeting provides that those activities which give rise to serious environmental damage, uncontrollable risks or deliberate crime are made the primary target of inspection.[16]

Prosecution is expected to be taken in relation to, *inter alia*, incidents or breaches which have significant environmental consequences, operating without the relevant licence, excessive or persistent breaches of statutory requirements and reckless disregard for management or quality standards.[17] Public interest factors that should be considered when deciding whether or not to prosecute include the environmental effect of the offence, the foreseeability of the offence, the history of offending and the attitude of the offender.[18]

Further functional guidance on the implementation of the Agency's Enforcement and Prosecution Policy has been provided (Environment Agency, September, 1999). This relates enforcement to the Common Incident Classification Scheme (CICS) which is a national system for recording incidents, assessing a response level and categorising incidents.[19] With regard to environmental protection and water management, an incident which has an actual environmental impact can fall into one of three categories: "major" (Category 1), "significant" (Category 2) or "minor" (Category 3); and there is a fourth category to cover incidents where there is no environmental impact.[20] The occurrence of a Category 1 incident should generally lead to prosecution, whereas Category 2 incidents might result in prosecution or formal caution

It can therefore be seen that the enforcement powers of the Agency are broad and varied, ranging from statutory administrative tools such as suspension and revocation notices to the formal criminal tool of prosecution. This wide range and the issued guidance should enable the Agency to adopt an enforcement policy consistent with the public interest goals of the relevant legislation. To explore whether such a policy has in fact emerged, we now turn to the Agency's practice.

Use of Criminal Justice System

As we have seen, the Agency's enforcement process is underpinned by a number of criminal offences, the majority of which relate to specific Agency functions such as water, waste and integrated pollution control. According to the Agency, the initiation of a prosecution should punish wrongdoing, avoid a recurrence and deter others from committing an offence (Environment Agency, 1998, para. 21). However, the use of the criminal justice system in achieving such aims is, in practice, dominated by two key factors: the low number of prosecutions brought by the Agency and the failure of the courts to impose fines that reflect the nature of the offence.

An examination of enforcement statistics indicates that the number of substantiated pollution incidents is much higher than the number of cases prosecuted. In 1999–2000, 17,592 waste incidents were investigated by the Agency, leading to the conclusion of a mere 342 prosecutions (Environment Agency, 1999–2000, p. 131). As regards water discharge consents, 246 prosecutions resulted from the investigation of 14,417 substantiated pollution incidents (Environment Agency, 1999–2000, p. 129). Prosecutions relating to Integrated Pollution Control are slightly higher, with 22 resulting from 599 substantiated pollution incidents (Environment Agency, 1999–2000, p. 130).[21]

Such evidence of the Agency's failure to prosecute is amplified by what has been gleaned from a recent confidential internal review, that prosecutions are taken for less than a quarter of the worst environmental incidents (Environmental Data Services, 2001a).[22] The review also indicates that the Agency is failing to comply with its Enforcement and Prosecution Policy, the guidance for which prescribes enforcement action according to the nature and extent of environmental harm resulting from the offence. Although the normal response to Category 1 incidents is prosecution, only 23% of such incidents, where the offender was identified, led to such action being taken and in 17% of cases, no action was taken at all.[23] With regard to Category 2 incidents, prosecution or formal caution are the possible responses. In only 27% of cases where the offender was identified was either action taken and in 30% of cases, no action was taken at all.[24] There was also extensive non-compliance with the recommended response to Category 3 incidents, that being a warning letter or in some areas, such as waste, a formal caution. For the 5 regions which provided data, warning letters or notices were issued in 4% to 29% of cases, with prosecutions or formal cautions being the response to another 2% to 10% of incidents.

Alongside the failure to prosecute, there are problems concerning the level of fines imposed by the courts. It is alleged that judges and magistrates have very little experience of dealing with these offences, and have therefore failed to impose fines that reflect the impact that business has had on the environment and human health.[25] Despite the fact that the overall level of fines has increased since the Agency was created in 1996,[26] there is evidence that the amounts imposed are low relative to the profitability of the contravening activity.[27] Of all the companies that the Agency prosecuted in 1999, only 32 businesses were fined over £10,000. The average fine for prosecuted businesses and individuals was £6,800 and £1,000 respectively.[28] In some cases, such as illegal waste disposal, the level of fines is actually falling (House of Commons Written Answers, 30 November, 2000, Cols. 768–773W). The Parliamentary Select Committee on Environment, Transport and the Regions (1999–2000 34 I-II, para. 95), in its report on the Agency, expressed concern at the consistently low level of court fines for environmental offences. This perception was shared by the respondents to the consultation paper including, not surprisingly, the Environment Agency itself which has also publicly drawn attention to what it regards as a failing on the part of the criminal courts.[29]

In recognition of the problem, the Home Secretary issued a direction to the Sentencing Advisory Panel to produce sentencing guidelines to the Court of Appeal for a group of environmental offences.[30] The Panel's advice was published in March 2000. As the appropriate sentence will depend on the particular circumstances of the case, the Panel recommends that the court, in exercising its discretion, take into account a number of aggravating and mitigating factors.[31] In setting the level of fine, the Panel stresses that it should not be cheaper to offend than to prevent commission of an offence. It recommends that the court considers the culpability of the offender and the extent of the damage which has actually occurred or has been risked. In addition, the fine should reflect the means of the individual or company concerned.

By way of contrast it should be noted that the Agency is generally successful in those prosecutions which it does bring before the criminal courts, securing a conviction in over 95% of the prosecutions brought under waste, water and integrated pollution control. Nevertheless, although such figures might appear to be impressive, they are apt to mislead. To put them in their appropriate context, it is necessary to refer to the Agency's Enforcement and Prosecution Policy (Environment Agency, 1998, para. 21) which places great emphasis on the evidence being sufficient to justify a prosecution:

A prosecution will not be commenced or continued by the Agency unless it is satisfied that there is sufficient, admissible and reliable evidence that the offence has been committed and there is a realistic prospect of conviction. If the case does not pass this evidential test, it will not go ahead, no matter how important or serious it may be.

Whilst the adoption of such a high threshold for bringing a prosecution is therefore likely to lead to a positive success rate, it also contributes to the low level of prosecutions brought by the Agency.

A final aspect merits attention. The Agency recognises that adverse publicity has a significant impact on the behaviour of potential offenders and may be more important than the other consequences of prosecution (de Prez, 2000). This is evidenced by the Agency's strategy of 'naming and shaming' operators with poor compliance records, their environmental performance being assessed by the level of court fine awarded against the company.[32] Furthermore, the enhanced use of public registers, containing *inter alia* data on enforcement, enables the public to measure the extent of non-compliance.

To summarize this section, the Environment Agency adopts a cautious approach to use of the criminal process, only prosecuting when there is considerable confidence that the evidence will sustain a conviction, and not always in cases satisfying this condition. Although the consequence is a very high rate of convictions, the number of cases prosecuted is significantly low. For their part, the courts are reluctant to impose large fines.

Use of Administrative Powers of Revocation and Suspension

Having examined the Agency's use of the criminal justice system in pursuing compliance, we now turn to the tools of revocation and suspension, and the extent to which the Agency has recourse to such administrative powers. Evidence pertaining to such matters is far less readily available than that relating to prosecution. It would seem that the power of revocation is used extremely sparingly. In February 2001, it was reported that only six waste management licences had been revoked by the Agency since its establishment in 1996 (House of Commons Written Answers, 26 February, 2001, Col. 314W). In relation to water discharge consents, it has been suggested that only 4% of such consents are revoked (Bell & McGillivray, 2000, p. 580) again, a relatively low number considering that, for example in 1999–2000, there were approximately 14,400 substantiated pollution incidents. As for IPC, one empirical study carried out in 1993, three years after the IPC system came into force, found that the power to revoke had not been used although, at this time, the system was monitored and enforced by Her Majesty's Inspectorate of Pollution whose enforcement practices differ from those of the Environment Agency (Allott, 1994). Secondary sources have revealed only three instances in which an IPC authorisation has been revoked (Environmental Data Services, 2000; HMIP Bulletin Issues 34 and 350).

However, it would appear that, in some instances, the Agency is slightly more willing to exercise its powers of suspension. Under the original IPC system, the Agency served a total of 30 prohibition notices between 1997 and 2000 (Environment Agency, 1999–2000, p. 130). It would also seem that waste management licences are suspended more often than they are revoked. Nevertheless, there is still a marked reluctance on the part of the Agency to exercise its suspension powers. This is illustrated by its reaction to operators' non-compliance with the technical competence provisions.[33] The Waste Management Licensing Regulations 1994[34] made it a statutory requirement for managers of thousands of waste sites to obtain certificates of technical competence (COTCs) by August 1999. In derogating from its original policy of serving, in August 1999, suspension notices on all sites which lacked a technically competent manager, the Agency announced that due to high levels of non-compliance it would only serve notices on those operators who had failed to make "substantial ongoing" progress and where a manager had achieved all the technical units of the relevant COTC and was making "substantial progress", six months grace would be given with suspension notices not taking effect until February 2000. In September 1999, out of the 8,000 waste managers registered with the awarding body WAMITAB[35], only 1,400 were awarded the relevant COTC by the August 1999 deadline (Environmental Data Services, 1999). Only approximately 22 suspension notices were served, many of which gave operators a further six months to comply (Environmental Data Services, 2001b). Relatively few led to the closure of operational sites.

Of course, suspension or revocation is not an option where an offender is unlawfully operating without a licence. Subject to this, we should attempt to understand the Agency's approach to these sanctions in the light of its published Enforcement and Prosecution Policy (Environment Agency, 1998). This makes it clear that recourse to the administrative action of revocation and suspension, should, in most instances, be a last resort. Revocation of an IPC authorisation will normally be considered only in cases where other enforcement measures have been used exhaustively to the point where the Agency is satisfied that the operator is unable to carry on the process in accordance with the conditions of the authorisation (Environment Agency, September, 1999, para. 7.1.2). A prohibition notice (equivalent to suspension) will only be served where there is an imminent risk of serious pollution of the environment (Environment Agency, September, 1999, paras. 7.1.5 and 7.1.6). Revocation or suspension of a waste management licence should only occur where a Category 1 incident has resulted from or is about to be caused by activities to which a licence relates, or where continuation of the licensed activities would cause a Category 2 incident and the serving of an enforcement notice or modification of the licence conditions is inappropriate. As for water discharge consents, the Environment Agency has informed us that the Agency will only revoke a consent if it is notified by the consent holder that the discharge has, for example, ceased. Otherwise, the Agency will not revoke consents without the permission of the consent holder.

In summary, the above statistics would suggest that revocation notices are served in only the most serious of cases and the power of suspension is exercised with extreme caution. The exercise of revocation and suspension powers can prevent the occurrence of environmental harm and its use can clearly be rationalised on that basis, but it can obviously also serve a deterrence function. As such it should be compared with the prosecution of criminal offences. The latter is necessarily more formal, but the penalties are generally, and paradoxically, less severe. Unsurprisingly, the Agency exercises its powers of prosecution more readily than those of suspension and revocation. From a deterrence perspective, it might seem to make sense to proceed with these sanctions only when the criminal process does not appear to achieve compliance.[36] But there is no compelling evidence that this is Agency policy; all we know is that where a licence has been suspended due to the imminent risk of further damage occurring, it typically pursues the case through the criminal courts, if a criminal offence has been committed.

OPTIMAL COMPLIANCE AND DETERRENCE THEORY

General

To evaluate the different aspects of enforcement policy described in the previous pages, we can use models derived from the standard literature on the economics of law enforcement.[37] The starting point is to recognise that since regulation is not self-enforcing and that the securing of compliance by a public agency involves (substantial) resources, perfect compliance is neither possible nor desirable. Rather, the goal of the system should be optimal compliance, that is, the point at which the marginal social benefits accruing from compliance are equivalent to the marginal costs incurred in securing that level of compliance (Stigler, 1970). If we identify the social benefits of compliance as the reduction of unprevented damage costs, the losses which would not have been incurred if there had been compliance (*UDC*), and the costs as comprising mainly those incurred in administrating the system of enforcement (*EC*), then the principal economic goal of enforcement policy is to minimise *UDC* + *EC*.

To understand the relationship between these two sets of costs, and therefore to explore the means of achieving the minimisation goal, we need a theory to explain how the level of compliance relates to enforcement activity. We adopt a simplified version of the familiar Becker (1968) deterrence model, assuming that individuals and firms comply with regulatory obligations if the benefits derived from contravention are exceeded by the costs. This condition can be expressed as:

$$U < pD$$

where *U* is the profit to the offender from the contravening act(ivity), *p* is the probability of apprehension by a public agency and *D* the costs to the offender resulting from such apprehension.

Some aspects require clarification before we proceed further. First, the *p* and *D* variables reflect the potential offender's subjective perception of the probability of apprehension and of the associated level of costs respectively, rather than their objective values; and the accuracy of such perceptions will be a function of the offender's information costs. It follows, too, that *pD* should be weighted to reflect the degree of risk aversion (if any) towards the consequences (Polinsky et al., 1979). Secondly, since *p* refers to apprehension by the public agency and not, more narrowly, to a formal determination of liability (or guilt) by the court or agency with power to impose a penalty, *D* covers a far wider range of costs than any formal sanction. It thus includes the "hassle" costs of pressure by an agency to comply, legal and other defence expenditures and any stigma or loss of reputation resulting from the apprehension and subsequent events.

There are, of course, various strategies which can be employed by an enforcement agency: it can select a process leading to a particular sanction, for example a criminal penalty or revocation of a licence; and it can determine how far to proceed within any such process, for example, merely issuing a warning or instituting formal procedures. Different probabilities and associated costs attach to these possibilities, whether they are sequential or alternatives.[38] It thus becomes preferable to rewrite the condition for compliance as:

$$U < pD_1 + pD_2 + pD_3 + pD_4 + pD_5 + \ldots pD_n$$

where each element in the right-hand side of the inequality represents the probability and the associated costs of a different predictable event in the enforcement process (Ayres & Braithwaite, 1992, p. 36).

The Criminal Justice Process

We begin by exploring the implications of this model for the criminal justice process which, as we have seen, in a previous section is assumed to be the principal or ultimate instrument of enforcement in many of the regimes of environmental protection in Britain. The steps taken towards this outcome may be illustrated by a pyramid (Fig. 1), with the percentage of apprehended contraventions indicated on the horizontal axis (Ayres et al., 1992, p. 35).[39]

Here we observe the procedures of increasing severity which may be undertaken by the Environment Agency once a contravention has been identified. The logic is that the agency proceeds through these stages until it is satisfied that it has secured a particular offender's compliance.

In general, the costs (*D*) to the offending firm increase with each higher stage. At the same time, viewed *ex ante*, the probability (*p*) that that stage will be reached is reduced. At the apex we have a formal conviction and (generally) the imposition of a formal sanction. For most contraventions, the courts have power to impose a large financial penalty, but the average penalty falls well short of this: as we have seen above, the current figure is just under £7,000. We have also seen that the proportion of cases proceeding to the top of the pyramid is very small, between 2% and 10% of known contraventions. Given the presumably large number of undetected offences, the proportion of all contraventions reaching the prosecution stage must be significantly smaller.

Fig. 1. Criminal Justice Enforcement.

With such a relatively low value to the *pD* of a formal penalty following conviction, and – we are entitled to assume – a significant value to *U*, one or more of the following contingencies must occur if the compliance condition is to be met:

- The potential offender is highly risk averse;
- The potential offender's subjective perception of the formal sanction likely to be imposed is very high;
- The potential offender significantly over-estimates the probability of a conviction and formal sanction;
- Significant adverse consequences attach to enforcement procedures which fall short of a criminal conviction.

Empirical evidence goes some way to supporting these hypotheses. For example, in his study of enforcement of water pollution laws, Hawkins (1983; 1984, pp. 149–153) found that officials were wont to "bluff" firms into believing in the heavy hand of the law being applied to offenders;[40] and other research reveals that enforcement agencies are able to badger firms and thus impose quite substantial costs in no way related to prosecutions and formal sanctions (de Prez, 2000; Blowers, 1984, p. 306).

As regards the costs (*D*) arising from procedures which fall short of a conviction, we know that adverse publicity is considered to be an important corollary to prosecution.[41] It is reasonable to assume that reputation is to some extent affected by an enforcement agency's involvement at an earlier stage. A more interesting possibility for reconciling non-prosecution with the deterrence condition emerges from an influential paper published by Fenn and Veljanovski (1988).[42] They drew on empirical material but also developed a new dimension to the theoretical model. Given the high costs to the agency of observing acts of contravention, it is argued that there is potential for a mutually profitable bargain between agency and offender. Just as in plea-bargaining, a prosecuting authority can "purchase" a guilty plea by reducing the proposed penalty, so the enforcement agency may secure a commitment to future compliance by an agreement not to prosecute a past offence. Of course, for the offender to be motivated to adhere to the agreement, there must still be a credible threat to prosecute on the occasion of the second bite of the cherry. But since the firm in question will be aware that it is now "under observation", the probability of being apprehended is significantly increased – compared to its first contravention. And hence a higher perceived value to the *pD* of a formal conviction might be plausible.

The Fenn-Veljanovski model may be persuasive but it is still dependent on there being a plausible threat of a sufficient penal sanction being ultimately imposed. As we have seen, this is problematic and helps to explain the current disquiet about the enforcement policy of the Environment Agency. To proceed further, we need to address two related issues: Why are prosecutions so rare? Are there alternative approaches which might secure adequate deterrence at lower cost?

An explanation of low rates of prosecution much in vogue in the 1970s was that the relevant agency was the subject of capture or, at least, was pusillanimous.[43] Later studies reveal this to be simplistic, suggesting there are strong instrumental reasons for an agency's reluctance to prosecute (Hutter, 1997). Perhaps the most important explanation is the very high cost to the enforcement agency of marshalling sufficient evidence to obtain a conviction, given existing principles of criminal procedure including, notably, the burden of proof, restrictive rules of evidence and (in very serious cases) requirement of a jury trial (Rowan-Robinson et al., 1990, pp. 255–263). Why is the preparation of a criminal prosecution so onerous? The conventional reason is that it is to protect the innocent from wrongful conviction (Williams, 1958, pp. 154–158)[44] – in economic terms, to reduce error costs (Posner, 1973). Nevertheless, the more precautions we take to reduce these wrongful conviction costs (*WCC*), the more we generate increases to both *EC* (enforcement costs) and *UDC* (unprevented damage costs), the latter because we will fail to convict, and thus also to deter, some contraventions. It follows that there is, in theory, an optimal level of procedural safeguards where the marginal benefit (reduction in *WCC*) is approximately equal to the marginal cost (increase in *EC* and *UDC*).[45]

The costs of false convictions are generally assumed to be significantly higher than those of false acquittals. Blackstone (1836) famously argued, if only intuitively, that it was "better that ten guilty persons escape, than one innocent suffer". The "beyond reasonable doubt" burden of proof is an example of a rule which reduces *WCC* but increases *UDC*. Hylton and Khanna (2001) have recently estimated that for to this be, on economic grounds, preferable to the "balance of probabilities" standard, *WCC* must be, in the given area, at least 2.64 times *UDC*.[46] However, this and other procedural rules serve to increase *UDC* not only by acquitting guilty parties, but also by failing to deter others, because of the reduction in the probability of conviction. Add to this the very significant additions to administrative costs resulting from the stricter rules and the weighting necessarily attributed to *WCC* by the current criminal process must be much larger than the 2.64 mentioned.

Of course, assessment of the costs in this area is particularly difficult, not the least because of the assumed large negative externalities associated with lack of confidence in the judicial system as well as a high degree of subjectivity attaching to the hurt arising from wrongful conviction. Using American empirical criminal justice data on some of the cost variables, Hylton and Khanna (2001) feel able to conclude that the higher standard of proof cannot be justified by the reduction in error costs rationale.[47] We remain agnostic of this judgement *insofar as it applies to mainstream criminal offences.* It may be that the traditional procedural safeguards associated with the criminal justice system approximate to that optimal level, once account is taken of the fact that criminal conviction for such offences often leads to imprisonment and generally a high level of social stigma. But intuition suggests that *in a regulatory context,* where firms are mostly the defendants, the costs of a wrongful conviction should not be exaggerated. Imprisonment is not, in practice, an option and stigma may not be serious. If that intuition is correct, the traditional criminal justice safeguards are not optimal.[48]

Suspension or Revocation of Licence

We have seen that the Environment Agency has another important instrument of enforcement, the suspension or revocation of a licence. Since this has the effect of depriving the firm of the lawful right to engage in the activity causing the pollution, it can be treated as incapacitating the offender, in the same way that imprisonment incapacitates an individual in relation to mainstream crime (cf. Shavell, 1987). As such, it may obviously serve a preventative function, and thus is particularly appropriate where the social harm arising from the contravention is very large and/or there are problems in deterring the conduct by ex post sanctions (Shavell, 1993, pp. 261–262). But the very fact that it is a very heavy penalty, foreclosing profit-making activities and perhaps individual livelihoods, means that it can also play a significant deterrent role. Indeed, in some circumstances, it may be more effective than a conventional financial penalty, since the effect of the latter is dependent on the wealth of the offender. Imprisonment is not a feasible option for an insolvent firm and potential bankruptcy thus impedes deterrence (Coffee, 1981, p. 390).

It is, therefore unsurprising that Ayres and Braithwaite (1992, p. 35), in their model enforcement pyramid (Fig. 2), place the suspension and revocation of a licence at the apex, to be invoked if, but only if, the conventional criminal penalty fails to deter.

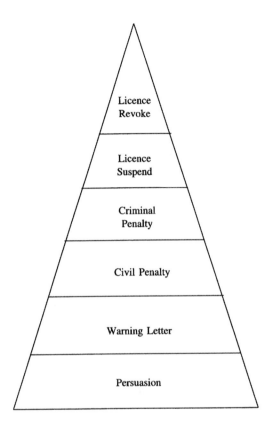

Fig. 2.   Ayres and Braithwaite Enforcement.

The rationality of such an approach is evident. The ultimate sanction, loss of the licence, is, as Ayres and Braithwaite (1992, p. 36) observe, "such a drastic one . . . that it is politically impossible and morally unacceptable to use it with any but the most extraordinary offences". A situation in which the ordinary processes of criminal justice have failed to secure compliance, presumably because U is high and/or the D of the penal sanction is low, may be considered "extraordinary" for this purpose. With the loss of a licence, D necessarily exceeds U, and the fact that there has been a previous conviction should mean that the value of p is relatively high. The sanction will, presumably, be rarely invoked, but the threat should be sufficient to render it an effective deterrent.

Ayres and Braithwaite (1992) are nevertheless rightly sceptical of the appropriateness of loss of the licence as a sanction if it is the only deterrence option, or if it is not consequent on the criminal justice process. The problem here is that the understandable and presumably widely known reluctance to use the device robs the threat of its credibility (Ayres & Braithwaite, 1992, p. 36). In these circumstances, the perceived low value of p will lead most potential offenders to apply a very high discount to D.

This argument can be forcefully applied to the arrangements in the United Kingdom, even though suspension or revocation of the licence is not the only sanction but ranks alongside the criminal process, described above. The historically-founded[49] decision to characterise the withdrawal of a licence as an administrative act, exercisable by the regulatory agencies – and not as a "last resort" judicial sanction – might seem at first blush to compensate for what we have seen to be problematic deterrence features of the criminal process. Sadly it exacerbates, rather than alleviates, the problem. The Environment Agency can threaten to invoke loss of a licence as an alternative to the criminal process, but clearly there must be evidence that it is willing to initiate proceedings if the threat is to be credible and, as we have seen,[50] it is known to be very reluctant to do this.

The present arrangements must be treated as even more unsatisfactory when account is also taken of the procedural aspects. Given the high financial and non-financial costs associated with the loss of a licence, we can assume that errors of the agency involving the false imposition of this sanction lead to high wrongful conviction costs (**WCC**), even though the stigma of a criminal conviction is absent. As such one would expect there to be, in analogy with the criminal process, a system of procedures – and an increased administrative expenditure – designed to reduce the errors. Such procedures do not exist and, although appeals or an application for judicial review may be brought against decisions, agency accountability in this area is considered to be low.[51]

Administrative Penalties

The preceding application of deterrence theory suggests how the enforcement policy of the Environment Agency might be improved. A substantial, but not crushing, financial penalty imposed administratively would not involve the complexities and attendant costs of the criminal justice process. And the need for the added protection of the criminal procedural rules would be reduced because, it is assumed, there would also be a decrease in wrongful conviction costs (**WCC**), given that the stigma of a criminal conviction would be avoided.

At the same time, because the costs associated with imposition (D) would be lower, the probability of imposition (p) would have to be increased for the deterrence value of the enforcement policy to be maintained (Polinsky et al., 1979). There should, however, be no problem in achieving this if the power to impose the penalty is conferred on the enforcement agency and the restrictive burden of proof, and other procedural rules inhibiting prosecution, do not apply (Posner, 1973, p. 416).[52]

Such an approach is implicit in the "civil penalty" tier of the Ayres-Braithwaite (1992) pyramid (Fig. 2). Recent discussion reveals this device is still seen as somewhat anachronistic in the common law world (Australian Law Reform Commission 2000).[53] Common lawyers are notoriously blinkered when it comes to observing ideas in rival legal cultures and it is amazing that in the now plentiful literature on administrative penalties there is so little by way of reference to civilian systems where there has been much experience of the phenomenon.[54]

Here we focus on Germany which has, perhaps, the most coherent and comprehensive system of regulatory enforcement.[55] Its origin can be traced to the end of the 19th century and the major expansion in that period of state intervention. Theorists then argued that non-compliance with administrative regulation should be distinct from traditional criminal processes (Goldschmitd, 1969). However the concept of "administrative offences" (*Ordnungswidrigkeiten*) was introduced only in 1949.[56] A general legislative framework for dealing with them was established in 1952,[57] the current system dating from 1968.[58]

Its main features can be briefly summarised. In comparison with the criminal law, notions of blameworthiness are largely absent: the legislation uses the term "*verwerfbar*" (objectionable), rather than *schuldig* (morally guilty) to characterise liability, which is effectively strict. Criminal procedures, especially the rules of evidence are relaxed. The financial sanctions employed (*Geldbuße*) are conceptually distinct from criminal fines (*Geldstrafe*) and there is formal adjudication in an administrative, and not criminal, tribunal only if there is an appeal against the agency administrative penalty.

An enforcement pyramid devised from the German system appears in Fig. 3. There are, of course, similarities – we can assume that the probabilities and the cost to an offender of the fine order being formally confirmed by the tribunal at the apex are broadly equivalent to those in the U.K. criminal justice model. The key difference is that, whereas the German agency can issue an administrative financial penalty, at the equivalent stage the British agency can only issue an abatement order. In short, with significantly reduced administrative costs, it is to be assumed that the German system is able to

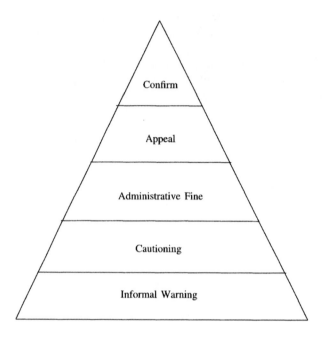

Fig, 3.   German 'Ordnungswidrigkeit' Enforcement.

secure an increased *pD* figure and presumptively thus minimise *UDC* (unprevented damage costs) and *EC* (enforcement costs) more effectively than its traditional common law counterpart.

CONCLUSIONS

The enforcement policy of the Environment Agency of England and Wales reveals a cautious approach to the prosecution of criminal offences, often because of the problems of securing a conviction, and an even greater reluctance to suspend or revoke a licence, except where this is deemed necessary to prevent further environmental harm. We have seen that, viewed from a deterrence perspective, the policy has generated concern, not the least because of the relatively low fines typically imposed by the courts.

Our theoretical analysis suggests that the concern is justified. The ex ante cost to the potential offender of contravening environmental regulation is assumed to be too low, given the small probability of a substantial imposition.

To resolve the problem, we argue that the Agency should be given powers to levy administrative financial charges from offenders without the procedures and onus of proof with which the criminal process protects defendants, but which also inhibits prosecution. The German system of *Ordnungswidrigkeit* provides an excellent model for this purpose.

NOTES

1. Administrative penalties are at the core of the U.S. environmental protection system. The EPA has the power, in many instances to impose administrative penalties on violators without recourse to the criminal or civil courts. For, under Section 113 of the Clean Air Act 1970 (as amended in 1990), the EPA Administrator has the power to issue an administrative penalty order where a person has violated an applicable implementation plan or permit (up to $25,000 per day limited to a total penalty of $200,000). In addition, field citations up to $5,000 for lesser infractions can be imposed by officers or employees designated by the Administrator.

2. We focus on the regulatory regimes of waste management, water quality and integrated pollution control (IPC). Although the latter has been superseded by a new system of pollution prevention and control (PPC), enforcement statistics relate to IPC, hence the continuing reference to that regime.

3. The pollution control regimes in the U.K. also refer to licences as permits, authorisations and consents.

4. A formal caution is a written acceptance by the offender that an offence has been committed. This power can only be used where a prosecution could properly be brought.

5. A warning can consist of either a site warning by the investigating officer or a warning letter.

6. See, e.g. Section 33(1)(c) Environmental Protection Act 1990 (EPA, 1990) and Section 85(1) and (3) Water Resources Act 1991 (WRA, 1991).

7. See, e.g. Section 6(1) EPA 1990 and Section 33(1)(a) and (b) EPA 1990.

8. See, e.g. the House of Lords decision in *Empress Car Co. (Abertillery)Ltd.v National Rivers Authority* [1998] 1 All ER 481 on Section 85 of the WRA 1991.

9. See Schedule 10 paragraph 7(3) WRA 1991, Section 12 EPA 1990 and Section 38 EPA 1990 respectively.

10. Section 38(3) EPA 1990.

11. Sections 14 and 38 EPA 1990.

12. See Sections 15 and 43 of the EPA 1990 and Section 91 WRA 1991. With regard to IPC and waste management, it is interesting to note that the bringing of an appeal against the serving of a suspension or prohibition notice does not suspend the operation of that notice. This can be contrasted with an appeal against revocation of a licence where the notice is suspended until the appeal is determined.

13. Environment Agency, 1998, paras. 4 and 6.

14. Environment Agency, 1998, para. 9.

15. Environment Agency, 1998, para. 14.

16. Environment Agency, 1998, para. 16.

17. Environment Agency, 1998, para. 28.

18. Environment Agency, 1998, para. 21.

19. An incident is defined (Environment Agency, 1998, para. 3.2) as a "specific event which comes to the attention of the Agency, is of concern to the Agency, and which may have an environmental and/or operational impact."

20. The same categorisation scheme can be used in relation to incidents or other events which have a potential environmental impact.

21. See further Mehta and Hawkins (1998)

22. The findings are based on reviews carried out to a standard specification by the Agency's 8 regions.

23. With decisions pending in 17% of cases.

24. Figures are based on a 10% sample.

25. This section focuses on the level of fines imposed by the courts. Although a number of other penalties, including prison sentences, can be imposed by the courts, Home Office statistics in 1997–1998 indicate that where individuals or companies are convicted of offences under Sections 23 and 33 of the EPA 1990 and Section 85 of the WRA 1991, they are invariably fined.

26. For example, the average level of fine increased from £2,786 in 1998 to £3,500 in 1999 (£4,750 if Milford Haven is included). Spotlight on Business Environmental Performance 1999 (Environment Agency, 1999) p. 8.

27. See, for example, evidence given to the Select Committee on Environment, Transport and the Regions by the Environment Agency (HC Session, 1999–2000, 34 I-II at 34 I para. 95): "Despite the large penalty in the Sea Empress case, fines are generally small in relation to the turnover or profits of companies, although there is no limit on what can be imposed by the crown court. A fine of £7,000 for one million gallons of largely untreated sewage discharged in a marina on a Bank Holiday, and a £20,000 fine for a company which saved £180,000 by illegally disposing of waste send the wrong signals to Board rooms. Bigger penalties from the courts are needed."

28. With regard to the average fine for businesses, this is £9,000 if Milford Haven is included.

29. See, for example, the Agency's *Water Pollution Incidents in England and Wales 1998* (1999, para. 8.4) where the Agency comments that "the general level of fines imposed for water pollution offences remains relatively low".

30. Offences considered by Panel are listed in Annex A to the Panel's report. For a useful summary of the Report see Papworth 2000.

31. Sentencing Advisory Panel (March, 2000) paras. 6–12.

32. See "Hall of Shame" on Environment Agency website (*www.environment-agency.gov.uk*). According to the recent Select Committee report on the Agency (1999–2000 34 I-II, para. 88–94), the 'naming and shaming' of companies has an important role to play in deterring non-compliance. However, the Committee was critical of the basis on which the tables were compiled, stressing the need for fair and consistent publication of environmental data.

33. Under Section 74 of the EPA, the operator must ensure that the licensed activities are in the hands of a technically competent person.

34. SI 1994/1056.

35. Waste Management Industry Training and Advisory Board.

36. See Ayres and Braithwaite 1992, p. 35; Fig. 2 and accompanying text.

37. For a survey with bibliography, see Polinsky and Shavell (2000).

38. As was indicated by the discussant, if the events are sequential a game-theoretic analysis might provide richer insights. See, for example, Harrington (1988).

39 For a slightly different version see Hutter (1988) pp. 149–153.

40. See also Hawkins (1984), pp. 149–153.

41. Above n. 32 and accompanying text.

42. A somewhat similar model, leading to similar conclusions, is used in Heyes and Rickman (1999).

43. See Rowan-Robinson et al. (1990, p. 8) and the studies there cited.

44. In an interesting recent paper, it is justified as an effort rather to constrain the costs associated with abuses of prosecutorial or government authority: Hylton and Khanna (2001) available at http://papers.ssrn.com/paper.taf?abstract_id=265795

45. To similar effect Posner (1973) pp. 410–417; Kaplow (1994), p. 349–362 and for commentary, Hylton and Khanna (2001).

46. Assuming the "beyond reasonable doubt" test to imply a probability of accuracy of 95%.

47. Hence their alternative explanation; above n. 43.

48. Posner (1973) reaches the same conclusion. See also Garoupa and Gomez-Pomar (2000), available at http: //papers.ssrn.com/paper.taf?abstract_id=260212

49. See, for example, Section 7(4) Control of Pollution Act 1974 which provided for the revocation of waste disposal licences by the waste regulation authority and section 43 Water Resources Act 1963 which gave a river authority the power to revoke water abstraction licences.

50. See Section on "Use of Administrative Powers of Revocation and Suspension". See to similar effect, on consumer credit licensing, Scott and Black (2000) p. 450.

51. Criticised on this ground in Rowan-Robinson et al. (1990) pp. 251–252.

52. Posner (1973) p. 416 suggests, nevertheless, that since the agency (unlike a court) might weigh the benefits from a successful prosecution more heavily than the costs of punishing the innocent, constraints on the agency's power through, for example, judicial review are important.

53. See also: Funk (1993); Kerrigan et al. (1993); Gillooly and Wallace-Bruce (1994).

54. Conseil d'Etat, *Les pouvoirs de l'administration dans le domaine des sanctions*, (Documentation Française, 1995); European Commission, *The system of administrative and penal sanctions in the Member States of the European Communities* (1994, Office for Official Publications of the European Communities).

55. The following two paragraphs draw on the German national report in European Commission, above n. 54.

56. *Gesetz zur Vereinfachung des Wirtschaftsstrafrechts.*

57. *Gesetz über Ordnungswidrigkeiten*, 25 March 1952.

58. *Gesetz über Ordnungswidrigkeiten*, 24 May 1968, as amended on 19 February 1987.

ACKNOWLEDGMENTS

We are grateful for the comments at the Workshop of the discussant and of other participants. We have been able to incorporate some of the suggestions in this revised version.

REFERENCES

Allott, K. (1994). *Integrated Pollution Control: The First Three Years*. London: ENDS.

Australian Law Reform Commission (December 2000). Civil and Administrative Penalties. Consultation Paper.

Ayres, I., & Braithwaite, J. (1992). *Responsive Regulation: Transcending the Deregulation Debate*. Oxford: Oxford University Press.

Balch, G. I. (1980). The Stick, the Carrot, and Other Strategies: A Theoretical Analysis of Government Intervention. *Law and Policy Quarterly*, 2, 35–60.

Becker, G. (1968). Crime and Punishment: An Economic Approach. *Journal of Political Economy*, 76, 169–217.

Bell, S. & McGillivray, D. (2000) (5th ed.). *Environmental Law*. London: Blackstones.

Blackstone, W. (1836) *Commentaries*, Vol. IV. (19th ed.).

Blowers, A. (1984). *Something in the Air: Corporate Power in the Environment*. London: Harper and Row.

Bugler, J. (1972). *Polluting Britain*. Harmondsworth: Penguin.

Coffee, J. (1981). 'No Soul to Damn; No Body to Kick': An Unscandalised Inquiry into the Problem of Corporate Punishment. *Michigan Law Review*, 79, 386–459.

Conseil d'Etat (1995). Les pouvoirs de l'administration dans le domaine des sanctions. *Documentation Francais*.

de Prez, P. (2000). Beyond Judicial Sanctions: The Negative Impact of Conviction for Environmental Offences. *Environmental Law Review*, 2, 11–23.

Environmental Data Services (1999). Waste Competence Rules Force Few Site Closures For Time Being. *ENDS Report 297*, 18.

Environmental Data Services (2000). Petrus Oils Denouncement. *ENDS Report 305*, 18.

Environmental Data Services (2001a). Agency Flounders with Prosecution Policy. *ENDS Report 315*, 3–4.

Environmental Data Services (2001b). Agency Faces Enforcement Headache on Waste Competence. *ENDS Report 314*, 19–20.

Environment Agency (1998). *Enforcement and Prosecution Policy*. Available to view online at: www.environment-agency.gov.uk

Environment Agency (1999). *Water Pollution Incidents in England and Wales 1998*.

Environment Agency (September 1999). *Guidance for the Enforcement and Prosecution Policy*. Available to view online at: www.environment-agency.gov.uk

Environment Agency Annual Report and Accounts (1999–2000).

European Commission (1994). *The system of administrative and penal sanctions in the Member States of the European Communities*. Office for Official Publications of the European Communities.

Fenn, P., & Veljanovski, C. (1988). A Positive Economic Theory of Regulatory Enforcement. *Economic Journal*, 98, 1055–1070.

Funk, W. (1993). Close Enough for Government Work? Using Informal Procedures for Imposing Administrative Penalties. *Seton Hall Law Review*, 24, 1–69.

Garoupa, N., & Gomez-Pomar, F. (2000). Punish Once or Twice: a theory of the use of criminal sanctions in addition to regulatory penalties. Harvard Law and Economics Discussion Paper *No 308*, available at: http://papers.ssrn.com/paper.taf?abstract_id=260212

Gesetz uber Ordnungswidrigkeiten (24 May 1968), as amended on 19 February 1987.

Gesetz uber Ordnungswidrigkeiten (25 March 1952). Gesetz zur Vereinfachung des Wirtschaftsstrafrechts.

Gillooly, M., & Wallace-Bruce, N. L. (1994). Civil Penalties in Australian Legislation. *University of Tasmania Law Review, 13*, 269–293.

Goldschmidt, J. (1969). Das Verwaltungsstrafrecht. *Scientia* (reprinting of original 1902 Berlin edition).

Gunningham, N., Grabosky, P., & Sinclair, D. (1998). *Smart Regulation: Designing Environmental Policy*. Oxford: Clarendon Press.

Harrington, W. (1988). Enforcement Leverage When Penalties are Restricted. *Journal of Public Economics, 37*, 29–53.

Hawkins, K. (1983). Bargain and Bluff: Compliance Strategy and Deterrence in the Enforcement of Regulation. *Law and Policy Quarterly, 5*, 35–73.

Hawkins, K. (1984). *Environment and Enforcement: regulation and the social dehmition of pollution*. Oxford: Oxford University Press.

Heyes, A. (2000). Implementing Environmental Regulation: Enforcement and Compliance. *Journal of Regulatory Economics, 17*, 107–129.

Heyes, A. G., & Rickman, N. (1999). Regulatory Dealing – Revisiting the Harrington Paradox. *Journal of Public Economics, 72*, 361–378.

HMIP Bulletin Issues 34 and 35.

House of Commons Select Committee on the Environment, Transport and the Regions, HC Session (1999–2000), 34 I-II.

House of Commons Written Answers. (30 November 2000), Cols 768-773W.

House of Commons Written Answers, (26th February 2001), Col 314W.

Hutter, B. (1988). *The Reasonable Arm of the Law*. Oxford: Oxford University Press.

Hutter, B. (1997). *Compliance, Regulation and Environment*. Oxford: Oxford University Press.

Hylton, K. N. & Khanna, V. S. (2001). *Toward an Economic Theory of ProDefendant Criminal Procedure*. Available at: http://papers.ssrn.com/paper.taf?abstract_id=265795

Kaplow, L. (1994). The Value of Accuracy in Adjudication: An Economic Analysis. *Journal of Legal Studies, 23*, 307–401.

Kerrigan, L. J. et al. (1993). Project: The Decriminalisation of Administrative Law Penalties. *Administrative Law Review, 45*, 367–434.

Mehta, A., & Hawkins, K. (1998). Integrated Pollution Control and Its Impact: Perspectives from Industry. *Journal of Environmental Law, 10*(1), 61–77.

Papworth, N. (2000). Environmental Offences: Views from the Sentencing Advisory Panel. *Environmental Liability, 3*, 91–96.

Polinsky, A. M., & Shavell, S. (1979). The Optimal Tradeoff between the Probability and Magnitude of Fines. *American Economic Review, 69*, 880–891.

Polinsky, A. M., & Shavell, S. (2000). Public Enforcement of Law. In: B. Bouckaert & G. De Geest (Eds), *Encyclopedia of Law and Economics* (Vol. 5, pp. 307–344). Cheltenham: Elgar.

Posner, R. (1973). An Economic Approach to Legal Procedure and Judicial Administration. *Journal of Legal Studies, 2*, 399–458.

Rowan-Robinson, J., Watchman, P., & Barker, C. (1990). *Crime and Regulation: A Study of the Enforcement of Regulatory Codes*. Edinburgh: T & T Clark.

Scott, C., & Black, J. (2000) *Cranston's Consumers and the Law* (3rd ed.). London: Butterworths.

Sentencing Advisory Panel (March 1st 2000). *Environmental Offences: The Panel's Advice to the Court of Appeal*.

Shavell, S. (1987). A Model of Optimal Incapacitation. *American Economic Review, 77*, 107–110.

Shavell, S. (1993). The Optimal Structure of Law Enforcement. *Journal of Law and Economics, 36*, 255–287.

Stigler, G. (1970). The Optimum Enforcement of Laws. *Journal of Political Economy, 78*, 526–536.
Williams, G. (1958). *The Proof of Guilt: a study of the English criminal trial* (2nd ed.). London: Stevens and Sons.

COMMENT ON PAPER BY ANTHONY OGUS AND CAROLYN ABBOT

Richard L. Revesz

Anthony Ogus and Carolyn Abbot present a lucid and informative analysis of the choice between civil and criminal environmental enforcement. The picture that emerges from their theoretical model and empirical data is troubling: under almost any plausible scenario, there appears to be under-enforcement of the environmental laws in England and Wales, principally because the penalties for non-compliance are far lower than the economic benefit that accrues to violators.

This comment addresses three issues. First, the role that stigma plays in the enforcement model deserves some careful scrutiny. As Ogus and Abbot explain, the stigma that attaches to environmental violators can have important deterrent effects. As a result, it can mitigate the effects of economic penalties that are too low (particularly penalties that are smaller than the economic benefit received by the violator) and prosecutions that are too infrequent. But stigma is a function of the context surrounding the imposition of liability. Ogus and Abbot would like to move from criminal to civil enforcement, and from low penalties to higher penalties. What effect are these two changes likely to have on the deterrent effect of stigma? It might be, for example, that stigma is greater when the violation is criminal rather than civil: The public might view criminal violations as morally more blameworthy. On the other hand, it might be that stigma increases with the size of the financial penalty: Regardless of

An Introduction to the Law and Economics of Environmental Policy: Issues in Institutional Design, Volume 20, pages 517–518.
Copyright © 2002 by Elsevier Science Ltd.
All rights of reproduction in any form reserved.
ISBN: 0-7623-0888-5

whether the violation is civil or criminal, the public might not perceive the conduct to be particularly egregious if the penalty is trivial. As a result, a complete theory of environmental enforcement would pay close attention to the impact of policy changes on the deterrence effects of stigma.

Second, the enforcement model analyzed by Ogus and Abbot consists of a single-period maximization problem that focuses on the polluter. Richer insights might be obtained from a game-theoretic multi-period model that focuses on both the regulator and the polluter. Such a model would be particularly fruitful given the pyramid of possible enforcement actions discussed by the authors. In each period the enforcer would undertake a particular action, and the polluter would decide on the extent of its compliance. Then, in light of an analysis of the benefits of further sanctions weighed against the resulting enforcement costs, the regulator would decide whether to proceed further. Game-theoretic models of this type provide far richer insights than single period, single-party, maximization models.

Third, there may be a desirable role for criminal sanctions beyond that contemplated by Ogus and Abbot. Enforcement can be significantly eased if regulated firms are required to report on their discharges. In the United States, several important regulatory statutes impose such obligations. Self-reporting can eliminate the need for cumbersome governmental testing of a plant's emissions in order to determine whether it complies with the law's requirements. But, of course, an enforcement scheme that relies on self-reporting will work well only if firms report truthfully, even when their emissions violate the applicable standards. Criminal penalties against corporate officers are a desirable tool to ensure truthful reporting. Then, a firm that truthfully reports a violation would be subject to only civil penalties, but a firm that reports fraudulently would place its officers in criminal jeopardy.

EIGHT THINGS ABOUT ENFORCEMENT THAT SEEM OBVIOUS BUT MAY NOT BE

Anthony G. Heyes

ABSTRACT

Enforcement of any rule or regulation is where 'the rubber hits the road'. Many economists and policy analysts have been guilty of proposing and promoting legal and regulatory instruments having given scant or no regard to the problems that might surround their implementation. Having said this, a significant and rapidly growing economics literature (both theoretical and empirical) has sought to think through some of the practical issues of implementation. Many of its conclusions for policy are not always obvious, and sometimes downright surprising.

1. 8 CONVENTIONAL WISDOMS[1]

1.1. CW 1: We Have a Good Idea What Fraction of Firms Comply With the Major Environmental Laws

It is obvious that enforcement issues matter in designing and appraising any regulatory regime. Cost-benefit evaluation of a particular piece of regulation which implicitly (or explicitly) assumes full compliance is likely to be misleading if 'slippage' occurs during implementation – particularly if that

An Introduction to the Law and Economics of Environmental Policy: Issues in Institutional Design, Volume 20, pages 519–537.
ISBN: 0-7623-0888-5

slippage is substantial. It is also right and proper that enforcement institutions and practices themselves come under cost-benefit scrutiny.

Of course, true compliance rates with regulatory requirements are often, by their nature, difficult to know with certainty. Published government statistics need to be interpreted with care. 'Compliant' is almost always the default categorization such that a polluting source being deemed compliant (for the purposes of recording) means only that the agency has failed to demonstrate non-compliance. This can be a very different thing – a 1979 report by the Government Accounting Office estimated that only 3% of sources designated fully-compliant with air quality regulations were actually compliant.[2]

The default designation being 'compliant' means that – apparently paradoxically – improvements in the accuracy of the enforcement agency's inspection technology can, other things being equal, be expected to lead to a decrease in population compliance. Econometricians have, in recent years, made efforts to deal with the issues of non-detection in evaluating compliance data. Feinstein (1989), in a frequently-cited example, uses data from over 1000 NRC inspections to (jointly) estimate the occurrence of violations, inspections and abnormality at nuclear power plants. Feinstein pays particular attention to the econometric problems arising from non-detection and is able to construct variations in 'propensity to detect' at the individual-inspector level.

These are approaches that need to be developed further and used more widely – official data could be adjusted to take account of detection probabilities (adjusted figures could be reported alongside raw figures, with a brief outline of the adjustment methodology used).

A well-known study by the White House Council on Environmental Quality – and reported in Russell (1990) – estimated the following compliance with air pollution limits by industrial sources: percentage of sources in violation 65; percentage of time sources that were in violation – 11; excess emissions as percentage of standards – 10. In the U.K. published compliance rates with many key water quality standards are significantly below 100%, and true compliance rates can be expected to be even lower.

1.2. CW 2: More Firms Would Comply With a Rule if the Penalty for Non-Compliance was Higher

Perhaps the most strongly held – and frequently invoked – conventional wisdom in the field of enforcement in general, and environmental enforcement in particular, is that the problem of under-compliance can always and everywhere be reduced simply by raising penalties. When enforcement agencies, academics and other policy commentators oppose increases in penalty levels – or when the penalty that an agency levies or seeks to have levied upon a firm is less than maximal – they are deemed to have 'gone soft on' pollution or to have been captured by regulated interests.

The fundamental model of rational compliance is due to Becker (1968). Penalties for illegality are treated as any other cost of doing business and polluters act to minimize the sum of expected compliance costs plus expected penalties. If we assume that the compliance decision is 'binary' – or yes/no – such as might be the case if a regulation requires a firm to install a well-defined and indivisible item of abatement equipment. Under standard simplifying assumptions, firm i will choose to comply if and only if its cost of compliance c_i is no greater than the expected penalty from non-compliance. If enforcement is by means of random inspection (which occurs with probability α) and lump-sum fine D, then firm i complies if and only if $c_i < \alpha.D$. If c_i is distributed across the population according to the distribution function $f(c)$ (with associated cumulative $F(c)$) then the rate of compliance across the population, which we will denote Λ throughout the paper, will be $F(\alpha.D)$. This yields the 'obvious' comparative statics: $\partial\Lambda/\partial\alpha = D.F' > 0$, $\partial\Lambda/\partial D = \alpha.F' > 0$. Increasing either the rate of inspection or the size of penalty increases population compliance, with the size of that increase depending upon the distribution of firms by cost type.

Whilst the assumption of binary compliance makes things tractable, it is clear that most real compliance decisions are likely to be continuous. Consider a firm regulated by an emission standard which requires that its emissions, x_i, of some pollutant not exceed S. The firm's marginal cost of abatement is c_i and the expected penalty for non-compliance is described by $P(x_i, S)$ where $P_x > 0$, $P_S < 0$. In that case, the firm complies exactly ($x_i^* = S$) if c_i is less than some $\tilde{c}$, otherwise it violates, choosing a level of emission implicitly defined by

$$P_x(x_i^*, S) = c_i \qquad (1)$$

Importantly, whilst the level of penalties impacts the decision about whether or not to violate, once the decision to violate has been made (i.e. once we know we're talking about an interior solution), the decision about the extent of violation depends only upon the *marginal* properties of the penalty function. This is the 'principle of marginal deterrence' (Shavell, 1992; Friedman & Sjostrom, 1993; etc.). Failure to understand the principle underpins many of the apparent paradoxes of observed behavior – (e.g. that a higher level of penalties may worsen compliance). It is not, in fact, that surprising to an economist. It is analogous to the distinction between fixed and variable costs in first-year industrial economics. The magnitude of fixed costs effects the participation decision, but not the choice of level of output by those that are participant. It corresponds with the popular notion that – having decided to participate in the cattle rustling business – one 'might as well hang for a sheep not a lamb'.

Where mitigation options are available in the wake of an environmental event the case for being quite lenient in the handling of the initial wrongdoing is reinforced. If an oil tanker spills oil illegally, the system needs to be forgiving enough that the operator is not dissuaded from initiating prompt and thorough clean-up by the concern for 'giving away' their culpability for the initial spill (see, for example, the model presented in Heyes, 1996).

What of over-compliance and so called 'regulatory chill'. The discussion so far has been predicated on the assumption the enforcement process never mistakenly penalizes a compliant firm. Under monitoring uncertainty type II errors of this sort may, of course, occur and the standard analysis can straightforwardly be extended to take account of them (Segerson, 1988; Xepapedeas, 1997, are examples).[3] One of the most significant implications of such an extension is the possibility that low-cost firms will find it optimal to *over*-comply with the regulatory requirement to reduce the probability of wrongful penalty below what would be implied by exact compliance. Over Compliance has the potential to be as damaging, in welfare terms, as her Australian cousin Under.

1.3. CW 3: More Firms Would Comply With a Rule if
All Observed Violations Were Punished

This one seems blindingly obvious – if everyone who is caught wrong-doing gets punished, the incentive not to do wrong is increased. The usual compliance/ enforcement model assumes that the enforcement agency and firm: (a) interact only once; and (b) interact in only one context; and (c) there is no limit to how large a penalty the agency can levy. Neither of these is remotely realistic in almost any real-world context. It turns out that in a variety of models with much more plausible features this conventional wisdom is not sustained. Systematically letting known violators 'off' (perhaps with a slap on the wrist – but not with a substantive penalty) can make compliance more attractive.

Repeated playing of the enforcement-compliance game provides scope for the behavior of one or both players in any given play to be sensitive to previous actions and/or outcomes.

A variety of papers in the 'straight' law and economics literature model the treatment of repeat offenders. More sophisticated attempts have been made to use Markov models to characterize optimal state-dependent enforcement strategies when penalties are restricted, and these are likely to be particularly applicable in regulatory enforcement settings. Such regimes typically involve some degree of 'forgiveness' and are able to accommodate periodic type I monitoring errors.[4]

In a repeated, binary enforcement/compliance game with restricted penalties the EPA maximizes the rate of steady-state compliance, it can be shown, by operating a state-dependent enforcement regime. In the simplest case the agency groups sources according to recent inspection history – group 1 containing firms found to be compliant at last inspection, group two those found non-compliant – and levies no penalty upon a group 1 firm caught violating but a maximal penalty upon a group 2 firm caught likewise. In equilibrium a representative firm can be induced to comply a significant fraction of the time (i.e. whenever they find themselves resident in group) despite penalties never actually being levied. The model can thus be used to 'explain' the paradox with which Harrington opens his paper, namely that despite the fact that

(a) when the USEPA observes violations it often (almost always) chooses not to pursue the violator, and
(b) the expected penalty faced by a violator who is pursued is small compared to the cost of compliance,

it is still the case that

(c) most firms comply most of the time.[5]

Looking at the data on expected costs and benefits of violation presented by Harrington and others, the question that leaps out is not 'how come some firms violate?' but rather 'how come any firms comply?'.

Interestingly (and importantly, in some contexts where regulated firms tend to have small asset-bases) a firm can be induced to comply with requirements some (most) of the time even though the limit on penalties is such that if all violations were penalized with certainty it would never do so. The (crude) state-dependent regime described generates 'penalty leverage'. When in group 2 a source's incentive to comply is not just the maximal penalty which it avoids but also the present value of reinstatement to group 1 and the laxer treatment which that entails in the next period.

The optimal (compliance-maximizing) state-dependent policy can be characterized by refining the crude regime described here to allow for differential rates of random inspection amongst group 1 and group 2 firms, and by making reinstatement to group 1 less-than-automatic.

Heyes and Rickman (1998) provide a cross-sectional analogue to Harrington's model – consistent with the same set of stylized facts that motivated Harrington – in which an enforcement agency exploits issue-linkage opportunities. The underlying assumption driving their results is that the EPA typically interacts with a particular firm in more than one enforcement 'domain'. This is realistic. It may be that the agency enforces the same rule at more than one plant of a multi-plant firm, or in more than one geographical area in which the firm operates. It may, equally, enforce several different sets of regulations – those regarding airborne emissions, water-borne discharges, noise, etc. – at a single plant. In that case, when penalties do not permit full-compliance to be achieved, the EPA may be able to improve upon the population compliance rate achieved by a policy of full pursuit (penalizing all violations with certainty) by engaging in 'regulatory dealing'. A regulatory deal involves agreeing (perhaps tacitly) to tolerate non-compliance in some sub-set of domains in 'exchange' for compliance in others.

The scope for a compliance-enhancing trade can best be understood by thinking about a two domain example in which the cost of complying by firm i in domain j is c_{ij} per period, and the maximum penalty that the EPA can levy for violation is Λ. If, for illustration, $c_{i1} = c_{i2} = 15$ whilst $\Lambda = 10$, it is apparent that a regime which detects and penalizes every violation will induce a zero rate of compliance. The firm's decision problem is separable by domain, and in each domain it will violate since c_{i1}, $c_{i2} > \Lambda$. When offered a deal (which amounts to, in words, "comply in one domain in exchange for us turning a blind-eye to violation in the other"), the firm accepts (since $\min(c_{i1}, c_{i2}) < 2\Lambda$) – saving penalty in both regimes in exchange for compliance in one – increasing its global rate of compliance from zero to 50%.[6]

Of course the EPA will not, in general, know the values of c_{ij} and so cannot target the firms with which it offers to deal. Dealing will improve a firm's compliance rate from 0% to 50% if $\Lambda < \min(c_{i1}, c_{i2}) < 2.\Lambda$ (an example of this was provided in the last paragraph). It will, however, worsen it from 100% to 50% if $\max(c_{i1}, c_{i2}) < \Lambda$. The desirability of engaging in dealing – the type of 'horse trading' with polluters of which environment agencies are frequently accused – depends upon the distribution of costs. Interestingly, the gains from dealing are not necessarily increasing in the extent to which penalties are restricted – it's not necessarily the case that an agency has to offer industry more concessions when its (the agencies) 'stick' is less fearsome.[7]

So does the USEPA – or its counterparts agencies in other countries and other enforcement settings – exploit opportunities for penalty leverage or regulatory dealing? The two models are not, of course, mutually exclusive. It is realistic to suppose that the EPA in the U.S. interacts with most firms both across a variety of enforcement contexts and through time such that it could exploit both. In that case specification of optimal policy would permit the Agency to condition its enforcement stance in domain i in period t not only upon the firm's compliance-history in domain i but also its past and current performance in domain j. Heyes and Rickman (1998) argue that there is empirical evidence to suggest widespread use of both. More generally, the models can be seen as attempts to formalize the type of horse-trading and bargaining that routinely goes on between inspectors and sources, compelling institutional evidence for which is catalogued by Yaeger (1991), Hawkins (1983) and others who have presented accounts of what 'really happens' inside regulatory agencies.

1.4. CW 4: When Public Agencies are Over-Stretched, Involvement of Private Individuals and NGO's in Enforcing Environmental Laws is a Good Thing

Recent years have seen growing support amongst policy makers and pundits for increasing the role of individual citizens in regulation, and the regulatory enforcement process in particular (see Tietenberg, 1998).

There are a variety of channels through which citizens might be expected to participate in or influence the enforcement of environmental requirements:

Political behavior
Market behavior
Direct participation

The first channel is the most obvious – citizens impact environmental enforcement (as they do other aspects of policy) through their voting decisions – and we do not dwell on it here.

A private individual might also influence compliance behavior through market interaction with sources in his or her capacity as employee, investor or customer (see Grabosky, 1994). Voluntary (unenforced) compliance may, in many contexts, be a profit-maximizing strategy. It is the notion that firms can be 'shamed' into improving compliance that underlies the so-called 'third wave' instruments which involve making information about the environmental performance of firms more easily accessible (see Tietenberg, 1998, and citations therein). The first- and second-wave instruments are the command-and-control and market-based instruments respectively. The efficacy of one 'third wave' application (the Toxic Releases Inventory in the U.S.) has been investigated by Khanna and Quimio (1997).

Under some pieces of legislation there exist channels for citizens (either individually or in groups) to participate more directly and explicitly in the enforcement process. In the U.S. Naysnerski and Tietenberg (1992), Miller (1988) and others have noted that the USEPA increasingly relies upon private litigants to bring suit against polluters.

In the case of the U.S. Clean Water Act in particular, private litigation now constitutes a major component of the overall enforcement effort, whilst in its 1993 Green Paper EC93(47) the European Commission clearly signalled its desire to 'beef-up' the rights of individuals and environmental groups to pursue polluters to ensure compliance and restoration. Note that these sorts of suits are over and above the 'usual' cases in which specific individuals impacted by pollution sue for compensation for their private damages. Specific authorization of citizen suits is provided by a number of the major environmental Acts in the U.S., including the Clean Water Act (Section 505), Clean Air Act (Section 304), Endangered Species Act (Section 11(g)), Safe Drinking Water Act Section 1449) and Toxic Substances Control Act (Section 20).

There has not been much formal modelling of the likely role of private actors in the enforcement process. Conventional economic wisdom would suggest that because the environmental benefits of effective private enforcement are a public good there will be an *under*-provision of private enforcement effort for the standard reasons.

Heyes (1997), in contrast, provides a formal lottery-auction-based model of the interaction between private NGO's and regulated firms and establishes, in fact, the possibility of over-provision. Whilst in cases where there exists a welfare benefit from compliance, private enforcement effort generates a public good, in other cases where compliance implies a welfare loss (because of high firm-specific compliance costs, for example), that effort generates a public *bad*. The welfare effect of NGO intervention on balance is in theory, then, ambiguous. In practice the relative numbers of the two types of case will depend upon how discriminating the public Agency's program is, and the resulting extent to which those cases left unenforced tend to be the cases where enforcement would be welfare-reducing.

Naysnerski and Tietenberg (1992) provide a good statement of the conventional wisdom in the environmental setting. They also argue that public and private enforcement efforts are likely to be substitutable and additive – implying that in a world in which enforcement is incomplete, the addition of a private enforcement program to an existing public program must increase the compliance incentives: "Since the rise of private enforcement increases the likelihood that violations will be detected and prosecuted, it increases observed compliance with regulations" (Naysnerski & Tietenberg, 1992, p. 43)).

This is self-evidently true in a world in which the EPA operates a random enforcement program and is resource constrained. If the EPA does anything more subtle than this – exploit penalty leverage, or engage in regulatory dealing, for examples – the incentive impact may be perverse. In both Harrington (1988) and Heyes and Rickman (1998) the compliance-maximizing Agency lets known violators off without penalty because and only because it is compliance-enhancing to so do – this is the interesting feature of those papers, and the associated strands of literature. Private intervention in either of those cases could be expected to weaken population compliance rates and damage the environment.

1.5. CW 5: Penalties are Things that we Get to Choose

Much recent attention has been paid to the loss in reputation that firms face when found guilty of an environmental offence (or even when accused of such) and the scope for 'market enforcement'.

Badrinath and Bolster (1996) examine stock market reactions to EPA judicial actions on a sample of publicly traded firms between 1972 and 1991. They apply the type of event study analysis applied by Karpoff and Lott (1993) in the contest of prosecution for corporate fraud and Borenstein and Zimmerman (1988) in the context of airline crashes. They show that a firm's stock market valuation declines 0.43% in the week of settlement – which for anything but the smallest firm translates into a dollar amount far in excess of the nominal penalty, implying that investor responses to news of violation can substantially reinforce fiat penalties. Interestingly, this response is unrelated to violation size, more pronounced for citations under the CAA and greater for more recent citations. Other work on stock market reactions to environmental incidents includes Hamilton (1995) and Laplante and Lanoie (1994).

More effort should be applied to understanding the determinants of the size of the 'market penalty' – particularly if that knowledge then allows the enforcement agency to influence it.

1.6. CW 6: Carrot or Stick Can be Used to Supplement Self-Motivated 'Good Behavior'

As we have noted many individuals and firms will often engage in pro-social behavior (like not polluting) even in the absence of (say) monetary incentives, or of penalty for wrong-doing.

For example, many people will go to great personal efforts to recycle – so contributing to environmental protection. Many firms might voluntarily behave in a socially responsible way, report their emissions honestly, and so on.

The problem is that this sort of self-motivated good behavior ('intrinsic motivation') often doesn't get us far enough to achieve our aims. Conventional wisdom would then dictate that you reinforce that self-motivation by applying economic or legal incentives. These could take the form of sticks (penalties for wrong-doing) or carrots (reward for pro-environmental behavior). The motivation of regulatees may be more complex than suggested by simple expected cost minimization. Frey (1992, 1997) contends that the imposition of external motivation will crowd out intrinsic motivation – self-motivation may be crowded-out or diminished by the use of coercive instruments of enforcement. Examples abound. During a regional blood shortage in the 1970s some U.K. regions attempted to up the supply of blood (always previously collected on a voluntary basis) by paying for each unit supplied. They found that supply hemorrhaged (groan). Within a community service context, people were willing to behave pro-socially, once giving blood became a market transaction that voluntary behavior was pushed out.

Recent experimental research by Bohnet, Frey and Huck (2001) provides compelling support for the proposition that coercive attempts to enforce a contract (such as exists between a regulatory agency and a polluter) can crowd out voluntary compliance motivation. They find a non-monotonic relationship between economic incentives and behavior – with compliance rates being highest for either very high or very low intensities of enforcement, but lower at intermediate levels.

The relevance of this sort of motivational crowding out in the enforcement context is apparent (and the work of Bohnet et al. have been motivated by the enforcement setting). The fundamental ideas, however, have much wider application. Consider it, for example, in the context of planning law and planning decisions in settings where something noxious needs to be sited. Local opposition to many projects ("NIMBY'ism") makes it increasingly difficult to find local communities willing to host power stations, waste-sites, etc. A common policy response has been to induce communities to host by paying them compensation. Intriguingly, there is quite robust evidence from a number of studies that attaching compensation to projects – paying communities to host – doesn't increase local support for acceptance, and may even reduce it. The search for hazardous waste landfills and nuclear waste repositories in the United States is a good example – despite the use of hefty compensation only one small radioactive waste disposal facility has been sited since the mid-1970s (that being near the aptly named town of Last Chance, Colorado).

Bribery crowds out public spirit. This means that, as Frey et al. (1996) put it in the *Journal of Political Economy*: "... as siting issues are decided in the realm of politics, an economic theory of compensation must focus on the interplay between morals and markets" (p. 1297). They provide evidence of the crowding-out effect – particularly in the short run – from Switzerland.

Scholtz (1994) contends that people are more likely to comply with a requirement when those around are also compliant. Such motivation can serve to make compliant behavior 'infectious' and cause regulatory agencies to want to have intermittent bursts of very intensive enforcement activity rather than stationary programs (see Chu, 1993) who coins the phrase 'oscillatory enforcement').

1.7. CW 7: Judges (and Juries) Have Good Judgement

Four out of every five decisions I make are contested in court
William Reilly, EPA Administrator, 1994.

Different countries vary in the extent to which effective decision making on environmental matters is delegated to the courts. In the United States, in particular, they play a key role.

Even in contexts where an agency appears to exercise discretion – say in deciding whether a particular action is justified on a cost-benefit basis, or whether a particular activity should be covered by a particular piece of law – that discretion is exercised in the shadow of the courts (as the opening suggests). The same applies – to differing extents and in differing ways – in all EU countries.

Whilst all sorts of pros and cons can be listed for placing such discretion in the hands of judges and juries can be discussed in theoretical terms, it is instructive to ask how 'good' (given whatever interpretation we give that term) judges and juries are at making are the sorts of decisions that will crop up in the environmental policy sphere. In his recent work at Harvard, Kip Viscusi asked a panel of around 500 jury-eligible citizens – and a panel of almost 100 state judges – a series of questions (assessing liability, damages and making cost-benefit trade-off's) based hypothetical scenarios. He motivates the work thus:

Anomalies in individual behavior regarding risk are not restricted to private decisions. People's participation in juries also involves them in consideration of risk contexts. An important question for assessing the function of our judicial system is how juries in fact approach such decisions and whether their judgements are flawed in a systematic manner (Viscusi, 2000, p. 30).

Much of environmental policy making involves making decisions about risk (and uncertainty). It is a well-documented human frailty that we are simply not very good at evaluating, communicating, understanding, valuing or generally 'getting our heads around' risk.

Recognizing this, and given that these are the types of decisions that are going to predominate in this field (perhaps more so than in any other major policy area) it is reasonable to think that ceteris paribus we should: (a) want to put effective powers of decision into the hands of those institutions best equipped to make those decisions in a consistent and 'rational' way; and (b) want to look for ways to re-engineer those institutions in such a way as to take account of the types of decisions they are being expected to make.

Viscusi summarizes his experimental results:

> Jurors fall substantially short of what one might hope for in terms of a desired pattern of decisions . . . Several noteworthy discrepancies indicated comparatively better performance by judges (Viscusi, 2000, p. 34).

In particular he finds jurors susceptible to the zero-risk mentality, risk over-estimation, and undue emphasis on worst case outcomes. Anchoring effects were also influential with respect to corporate risk analyses, as was the role of hindsight bias.

Of course juries and (most) judges are also people, and the policy issue is how we design arbitration institutions that protect us from the systematic human frailties and shortcoming. Should juries be made up of 'experts' (the 'Science Court' model)? Be made up of lay-people but be given access to an impartial 'resident expert' to tutor them in the ways of science and risk? Is a jury-based approach suitable at all? Viscusi appears to advocate a transfer of discretion back from courts to regulatory agencies ("which are better equipped to deal with the complex scientific issues that often arise in these areas"), though a less sweeping proposal would be for courts to avail themselves of more scientific experts (and not just experts that *understand* the science, but those with an expertise in communicating it) as proposed by Justice Breyer (1993) amongst others. Sunstein et al. (1998) came up with broadly similar results, and advocated increasing the role played by judges (versus juries) in setting punitive damages.

Other question marks surround not just the powers given (or taken away) from juries, but the circumstances in which they operate. Does the process of deliberation, for example, enhance the decision making of juries (over, say, a simple vote taken amongst a panel of 12)?

Another strand of the literature aims to explain why differing enforcement actions may be taken in different circumstances in terms of political and bureaucratic incentives. Wood (1992), Seldon and Terrones (1993), Helland (1998), Kleit et al. (1998). Mixon (1993) provide empirical evidence of the public choice determinants of penalties assessed by the EPA for carbon emission violations, showing that whilst industry lobbying effort has only minor effect on the probability of EPA citation for a detected violation, it can impact the *degree* of that citation substantially.

This type of result provides empirical rationale for the 'contestable enforcement' type models of Kambhu (1989), Nowell and Shogren (1994) and others. The notion that enforcement outcomes are influentiable raises the issue of how to design enforcement institutions to contain (or efficiently manage) that influentiability – with the need to trade-off regulatory capture against capture of the legal system.

1.8. CW 8: Green Taxation Would Work Better (be Easier) if More Firms Reported Their Emissions Truthfully

Basic models of the interaction between an enforcement agency and would-be illegal polluter treat the inspection process as being something random – a benchmark that can straight-forwardly be extended to the case in which the agency conditions its inspection probabilities upon observable characteristics (size, location, outcome of previous inspections, etc.).

Many modern environmental programs, however, incorporate an element of self-reporting (Russell, 1990) found that on average 28% of air pollution sources and 84% of water pollution sources were required to self-report). The basic model of compliance has to be revised substantially if the incentive implications of this change – with the enforcer becoming *re*active to a report (or non-report) rather than first-mover in the enforcement relationship.

Self-reporting plays a particularly obvious role when the instrument chosen is a green tax – a tax based on reported discharges of some pollutant or other.

It is often taken for granted that if more firms were innately honest or ethical in the way they behaved this would be a good thing. Heyes (2001) uses the example of environmental regulation to show that such an claim cannot, in general, be sustained. If regulation is by pollution tax we show that – once optimal agency response is taken into account – social welfare is non-monotonic in the proportion of firms that report emissions honestly, with social welfare being highest when that proportion is either sufficiently close to zero or to one. The choice of policy instrument may itself be characterized by "reversals" with command-and-control methods being preferred for intermediate values of population honesty, a tax system being preferred at the extremes. This means that if – because of the spread of "ethical shareholding" or for whatever reason – the honesty of the corporate population increases through time, we should not be surprised to see at first a switch away from market-based instruments, and then a switch back.

The argument underlying the key results is a simple one, and relies on the ability of a tax-based system to 'handle' heterogeneity in the reporting tendencies of the taxed population of firms. Consider a simple model of a tax regime based on (imperfectly verified) self-reports in which tax plays an incentive role, as in the case of a tax designed to internalize an externality. It is not too surprising that such a regime may perform best – once optimal agency response is factored in – either when everyone is inherently honest or when everyone is inherently dishonest. When everyone is inherently honest, the agency can achieve first-best by setting the tax-rate equal to marginal damage (i.e. at its Pigouvian level). Supposing, instead, that everyone is inherently dishonest and is able to get away with under-reporting some fraction m of their true emissions, the agency can achieve the same outcome simply by raising the tax-rate to $(d/(1-m))$, thus maintaining the effective rate. The regime performs less well when the population is a mixture of some people who are routinely honest (i.e. do not exploit opportunities for evasion presented to them) and others who are not. Whilst the framework developed in this paper is somewhat more complex than this – incorporating differences in compliance costs, evasion opportunities etc. – the basic story is unchanged.

Heyes argues that the model is consistent with a number of the stylized features of American regulatory history. It may also provide a defence of the EPA's willingness to tolerate a 'culture of dishonesty' amongst those it is supposed to police.

Whilst the model is framed for tax as the instrument, the results can readily be extended to apply to alternative instruments involving a self-reporting component. Though environmental regulation is used as an example for the purposes of exposition, it is also apparent that the results are likely to be applicable in a variety of other enforcement/compliance settings.

The key advantage attributed to pollution taxation is the technological flexibility it offers to regulatees – empowering them with choice of technique – and its implied ability to "handle" technological heterogeneity. This is captured in the model presented here: In the absence of the variation in firms technological 'outside options' command-and-control would invariably be the preferred method of regulation.

The novelty in Heyes' analysis is to incorporate another type of heterogeneity – what we might refer to as 'motivational heterogeneity' – which a tax-based regime is comparatively bad at handling. When all (or a sufficient fraction – denoted β) – of firms are *either* honest or dishonest then a tax regime can be expected to perform well, and certainly no worse than the optimally-calibrated command-and-control alternative. It is for intermediate values of β that command-and-control – implying, as it does, the stipulation of a technology that for some firms is not cost-effective – may work better.

The significance of the analysis depends, to a large extent, upon the interpretation applied to β.

Suppose that β evolves exogenously through time, influenced by broader social trends, etc., then we should expect a responsive regulatory agency to adjust to take account of such evolution. This response might involve recalibrating a given instrument or – if one of the critical thresholds is traversed – switching *between* instruments. A long-term upward trend in the evolution of β – driven by, for example, the spreading influence of ethical investors – may generate 'reversals' in instrument choice, with an initial switch away from market-based instruments followed by a later switch back towards them.

What of regulatory attitudes towards changes in β? This depends upon the starting point and the size and direction of the change – certainly the simple view that more honesty is always good cannot be sustained. Starting from β = 0 (the 'default' assumption in conventional economic analysis) marginal increases in the parameter are an impediment to the agency in its work such that 'pioneer' ethical firms might legitimately be regarded as a nuisance, 'throwing a spanner' into the regulatory process.

Increases in population honesty are only welfare-improving at the margin once a certain critical prevalence has been achieved.

The recognition that this is the case has particularly interesting implications in contexts where β – rather than being exogenous – is something that the regulatory agency might have scope to influence, e.g. through 'good citizenship' campaigns. A policy aimed at encouraging increased honesty from a low base is only welfare-improving if it can work on a big enough scale to get 'over the hill' and far enough down the other side of the inverted-⌣.

The analysis provides a possible efficiency-interpretation for the USEPA's apparent willingness to tolerate widespread misreporting of environmental performance by those in its charge. This is alternative to the standard view that any such tolerance must necessarily indicate incompetence, sloth or regulatory capture on the part of the agency. Provided everyone dissembles and the EPA takes account of that in its own choice of strategy (in this case by increasing the unit tax appropriately) then such dissembling need not impact system-performance. What makes regulation difficult is heterogeneity coupled with asymmetric information.

2. CONCLUSIONS

None of this really helps in deciding what 'optimal' environmental law might be, nor what portfolio of institutions and practices might help us to implement it. What it might do is lead some of us to reflect on many of the preconceptions that we take for granted when thinking about enforcement issues. When thinking about regulation (of which law might be seen as an instrument) enforcement is where 'the rubber hits the road'. It is important, and it needs to be thought about energetically and creatively.

NOTES

1. There was no particular reason for picking '8' other than it seemed about right – 7 seemed like not enough, 9 like too many.

2. Russell (1990) points to examples of how untrustworthy estimates of compliance rates can be. In the context of hazardous waste regulation in the United States, for instance, he aserts that official documentation ". . . is largely a catalogue of speculations about the possible extent of illegal disposal" (p. 261).

3. See, for example, Segerson (1988), Fig. 1. Segerson's paper also makes a sharp distinction between single polluter and multiple polluter settings, and the problems that arise in the attribution of blame in the latter case. We do not pay much explicit attention to the multiple polluter in the current paper, but acknowledge its importance.

4. Whilst the contributions of Greenberg (1984) and Landsberger and Meilijson (1982) are motivated by the income tax setting they have more general application, and are adapted to the context of pollution enforcement by Harrington (1988).

5. Such (apparent) overcompliance has been observed in a variety of contexts by a number of authors. Harrington (1988) provides evidence of these and other stylised facts on pages 29–32, especially Table 1. To take a typical example – Connecticut – from that table, over the sampling period of 800 known violations (i.e. cases where Notices of Violations (NOVs) were issued) in an average year, penalties were assessed in only 21 cases, and the average penalty in those cases was a meagre $221.

6. Implying, note, the type of overcompliance consistent with the stylised facts. If every firm was like this firm then a compliance-maximising policy (characterised, as it would be, by dealing) would yield substantial compliance (50%) despite penalties never actually being levied. An external observer would calculate the expected benefit to compliance to be zero and so find the firms behaviour paradoxical in the sense of Harrington.

7. The notion that 'he who carries the big stick will choose not to give away his candy' is not borne out here. If, for example, $c_{ij} \sim U[0, 1]$, then the gains from dealing can be shown to be non-monotonic in Λ reaching an interior maximum of $(d/4)$ at $\Lambda = 1/4$ (i.e. if Λ happened to equal 1/4 the dealing would deliver a 25% improvement in population compliance over the becnhmark of a full-pursuit regime).

8. Haltiwanger and Waldman (1991, 1993) develop formal models in which individuals behave honestly or cooperatively until they come across another person who is dishonest. All of this type of contribution points to a need to provide a more considered behavioural foundation to our theory of compliance.

9. See Anderson and Kagan (2000) for an excellent survey of the available empirical and anecdotal evidence on the extent of that legalism – and the transactions costs burden that it imposes – on U.S. business. They provide evidence that the additional burden of legalism – over and above the direct cost of compliance – can significantly affect a firm's choice of the location of production. See also Kagan and Axelrad (1987).

10. There is a very long and distinguished literature (co-owned by economists, psychologists and others) which evidences this. The 'Monty Hall's Doors' problem has stumped not just the showcase-seeking finalists on *Let's Strike A Deal* but generations

of guinea-pigged students. Whilst historians of scientific thought may tell us that 'we are all Bayesians now', in day-to-day life most of us behave in ways that would have had the Reverend Bayes turning in his Tunbridge Wells grave.

11. An additional strand of the literature attempts to work backwards from observed patterns of enforcement to infer the underlying preferences of Agency's (Helland, 1998, is an excellent example).

12. Yaeger catalogues very widespread under-reporting of effluent discharges in the U.S. (e.g. Yaeger, 1991, pp. 276–283). Out of a sample of 251 reporting violations uncovered by EPA Region II, for example, either issuance of a warning letter or no action was the sanctioning response in 94.7% of cases. There were no civil or criminal referrals. Whilst these results relate to a licensing rather than tax-based regime the presence of a self-reporting component mean that the same principles apply.

13. The initial motivation for this paper came from comments made during part of an interview I conducted with a regulatory enforcement strategist (who shall remain nameless) in the context of another piece of research in 1996: ". . . there is an expectation that people will play the game, get away with what they can, and that's not necessarily such a bad thing. As long as you realise you're playing a game, then its not necessarily all such a big problem".

ACKNOWLEDGMENTS

This paper was prepared for the Law & Economics of Environmental Policy Symposium held at UCL, 5–7 September 2001. In writing it I have borrowed extensively from lots of my earlier work, including Heyes (1998, 1999, 2000, 2001), Heyes and Rickman (1999). I am grateful to Tim Swanson and Peth Tuppe for constructive comments.

REFERENCES

Anderson, L. C., & Kagan, R. A. (2000). Adversarial legalism and transactions costs: The industrial-flight hypothesis revisited. *International Review of Law & Economics, 20*(1), 1–19.

Badrinath, S. G., & Bolster, P. (1996). The Role of Market Forces in EPA Enforcement. *Journal of Regulatory Economics, 10*(2), 165–181.

Becker, G. (1968). Crime and Punishment: An Economic Approach. *Journal of Political Economy, 76*, 169–217.

Borenstein, S., & Zimmerman, M. B. (1988). Market Incentives for Safe Commercial Airline Operation. *American Economic Review, 78*(5), 913–935.

Bohnet, I., Huck, S., & Frey, B. (2001). More Order with Less Law: On Contract Enforcement, Trust and Crowding. *American Political Science Review, 95*(1), 131–143.

Breyer S. (1993). *Breaking the Vicious Circle.* Cambridge, MA.: Harvard University Press.

Chu, C. C. Y. (1993). Oscillatory vs Stationary Enforcement of Law. *International Review of Law and Economics, 13*, 303–315.

Cuccia, A. (1994). The Economics of Tax Compliance: What Do We Know and Where Do We Go? *Journal of Accounting Literature, 13*, 81–116.

Dimento, J. (1986). *They Treated Me Like a Criminal*. Pittsburgh: University of Pittsburgh Press.

Erard, B., & Feinstein, J. S. (1994). Honesty and Evasion in the Tax Compliance Game. *RAND Journal of Economics*, *25*(1), 1–19.

Feinstein, J. S. (1989). The Safety Regulation of US Nuclear Power Plants: Violations, Inspections and Abnormal Occurrences. *Journal of Political Economy*, *97*(1), 115–154.

Frey, B. S. (1992). Pricing and Regulating Affect Environmental Ethics. *Environmental & Resource Economics*, *2*, 399–414.

Frey, B., Oberholzer-Gee, F., & Eichenberger, R. (1996). The Old Lady Visits the Market: A Tale of Morals and Markets. *Journal of Political Economy*, *104*(6), 1297–1319.

Friedman, D., & Sjostrom, W. (1993). Hanging for a Sheep – The Economics of Marginal Deterrence. *Journal of Legal Studies*, *22*, 345–366.

Grabosky, P. N. (1994). Green Markets: Environmental Regulation by the Private Sector. *Law and Policy*, *16*(4), 419–448.

Greenberg, J. (1984). Avoiding Tax Avoidance. *Journal of Economic Theory*, *32*, 1–13.

Haltiwanger, J., & Waldman, M. (1991). Responders vs Nonresponders: A New Perspective on Heterogeneity. *Economic Journal*, *101*, 1085–1102.

Haltiwanger, J., & Waldman, M. (1993). The Role of Altruism in Economic Interaction. *Journal of Economic Behavior & Organisation*, *21*, 1–15.

Hamilton, J. T. (1995). Pollution as News: Media and Stock Market Reactions to the Toxics Release Data. *Journal of Environmental Economics and Management*, *28*(1), 31–43.

Harrington, W. (1988). Enforcement Leverage When Penalties are Restricted. *Journal of Public Economics*, *37*(1), 29–53.

Hawkins, K. (1983). Bargain and Bluff: Compliance Strategy and Deterrence in the Enforcement of Environmental Regulations. *Law and Policy Quarterly*, *5*(1), 35–73.

Helland, E. (1998). The Revealed Preferences of State EPAs: Stringency, Enforcement, and Substitution. *Journal of Environmental Economics and Management*, *35*(3), 242–261.

Heyes, A. G. (1996). Cutting Pollution Penalties to Protect the Environment. *Journal of Public Economics*, *60*(2), 251–265.

Heyes, A. G. (1996). Towards an Efficiency Interpretation of Regulatory Interpretation Lags. *Journal of Regulatory Economics*, *10*(2), 81–98.

Heyes, A. G. (1997). Environmental Regulation by Private Contest. *Journal of Public Economics*, *61*(2), 407–428.

Heyes, A. G. (2000). Implementing Environmental Regulation: Enforcement and Compliance. *Journal of Regulatory Economics*, *17*(2), 107–129.

Heyes, A. G. (2001). Honesty in a Regulatory Context – Good Thing or Bad? *European Economic Review*, *45*(1), 215–232.

Heyes, A. G., & Rickman, N. (1998). A Theory of Regulatory Dealing – Revisiting the Harrington Paradox. *Journal of Public Economics*, *72*(1), 361–378.

Kaplow, L., & Shavell, S. (1994). Optimal Law Enforcement with Self-Reporting of Behavior. *Journal of Political Economy*, *102*(3), 583–606.

Karpoff, J. M., & Lott, J. R. (1993). The Reputational Penalty Firms Bear From Committing Criminal Fraud. *Journal of Law and Economics*, *36*, 757–798.

Khanna, M., & Quimio, W. R. (1997). *Toxic Release Information: A Policy Tool for Environmental Protection*. Urbana-Champaign: Department of Agricultural and Consumer Economics, University of Illinois.

Khambu, J. (1989). Regulatory Standards, Compliance and Enforcement. *Journal of Regulatory Economics*, *1*(2), 103–114.

Khambu, J. (1990). Direct Controls and Incentives Systems of Regulation. *Journal of Environmental Economics and Management, 18*(2), 72–85.

Kleit, A., Pierce, M., & Hill, R. (1998). Environmental Protection, Agency Motivations, and Rent Extraction: The Regulation of Water Pollution. *Journal of Regulatory Economics, 13*(2), 121–137.

Laplante, B., & Lanoie, P. (1994). Market Response to Environmental Incidents in Canada. *Southern Economic Journal, 60*, 657–672.

Laplante, B., & Rilstone, P. (1996). Inspections and Emissions of the Pulp and Paper Industry in Quebec. *Journal of Environmental Economics and Management, 31*(1), 19–36.

Mixon, F. G. (1994). Public Choice and the EPA: Empirical Evidence on Carbon Emissions Violations. *Public Choice, 83*(1), 127–137.

Naysnerski, W., & Tietenberg, T. (1992). Private Enforcement of Federal Environmental Law. *Land Economics, 68*(1), 28–48.

Nowell, C., & Shogren, J. (1994). Challenging the Enforcement of Environmental Regulation. *Journal of Regulatory Economics, 6*, 265–282.

Russell, C. S. (1990). Monitoring and Enforcement. In: P. Portney (Ed.), *Public Policies for Environmental Protection*. Washington, D.C.: RFF.

Scholtz, J. T. (1984). Cooperation, Deterrence and the Ecology of Regulatory Enforcement. *Law and Society Review, 18*(2).

Scholtz, J. T. (1986). In Search of Regulatory Alternatives. *Journal of Policy Analysis and Management, 4*(1), 113–116.

Selden, T. M., & Terrones, M. (1993). Environmental Legislation and Enforcement. *Journal of Environmental Economics and Management, 24*(3), 212–228.

Segerson, K. (1988). Uncertainty and Incentives for Nonpoint Pollution Control. *Journal of Environmental Economics and Management, 15*, 87–98.

Shavell, S. (1992). A Note on Marginal Deterrence. *International Review of Law and Economics, 12*, 133–149.

Shavell, S. (1986). The Judgement Proof Problem. *International Review of Law and Economics, 6*(1), 45–58.

Sunstein, C., Schkade, D., & Kahneman, D. (2001). Do People Want Optimal Deterrence? *Journal of Legal Studies, 29*(1), 237–253.

Viscusi, W. K. (2000). Jurors, Judges and the Mistreatment of Risk by the Courts. Working Paper, Harvard Law School.

Wood, B. D. (1992). Modeling Federal Implementation as a System: The Clean Air Case. *American Journal of Political Science, 36*(1), 40–67.

Xepapadeas, A. (1997). *Advanced Principles in Environmental Policy*. Cheltenham, England: Edward Elgar.

Yaeger, P. (1991). *The Limits of the Law: The Public Regulation of Private Pollution*. Cambridge, England: Cambridge University Press.

COMMENT ON PAPER BY
ANTHONY G. HEYES

Timothy Swanson

Anthony Heyes has taken on several sacred cows in environmental compliance in his piece on conventional wisdoms. He has demonstrated that, in all 8 cases, it is not a given that the wisdom represented by the convention is true in each and every case. Such iconoclastic approaches are worthwhile in general, as they cause us all to question our assumptions. This is especially important in areas of applied policy making, and nothing could be much more applied than the area of enforcement in environmental policy. It is very important if our work is to have any of the desired impact, that we understand whether enforcement has the impact that we desire of it.

It is also important, however, to understand which of the conventional wisdoms are more or less wise than the others. It is not sufficient to show that there is a single case that is an exception, for it might be the one that proves the rule. So I proffer here my assessment of Heyes' assessment. Which of the cases he makes are strong enough to bring into question the wisdom of the convention, and which are not?

I believe that the cases he makes with respect to conventional wisdoms 2, 3, 6 and 8 are strong cases, and I think that they are correspondingly important points that he makes. 2, 3, and 8 are all variations on the same theme. They demonstrate that enforcement regimes can be constructed so as to take into consideration the absence of complete information in which they operate, and that they can even use this uncertainty to good effect. For this reason, the conventional wisdoms that derive from the assumed circumstances of complete information do not often transfer very well to the circumstances of reality. In short, policies that are developed with the complexity of the real world in mind

An Introduction to the Law and Economics of Environmental Policy: Issues in Institutional Design, Volume 20, pages 539–540.
© 2002 Published by Elsevier Science Ltd.
ISBN: 0-7623-0888-5

will work better there than those that are developed under more simplistic assumptions. This is an important point, and the conventional wisdoms should be thrown out and replaced by the convention of apposite complexity.

I am also impressed by statements 5 and 6 for similar reasons. These conventional wisdoms derive from the assumption that individuals are all motivated by similar and simplistic motivations. This is not so, and demonstrably not so in the case of public goods such as blood donations, etc. Since the environment is all about the regulation of public goods, it behooves us to consider these more complicated motives when we are designing environmental policies. These more complicated motives will both determine the public response to environmental regulatory frameworks and the public response to environmental enforcement actions. These complications justify more research than they are receiving in this area.

On the other hand, his assaults on conventional wisdoms 4 and 7 I find less convincing. It might be shown that there exist circumstances in which private actions under liability systems distort the regulatory system they are intended to substitute, but it is my observation that these are the exceptions rather than the rule. Similarly, it might be the case that judges and juries are poor environmental scientists, but this does not necessarily imply that environmental scientists are better representatives of society within the environmental decision making process. To contest the good judgement of judges and juries does not seem constructive criticism to me, unless there is some alternative decision-making device that is shown to perform more acceptably.

All in all, the paper surveys the state of the art in the enforcement literature, and in an interesting handful of pages. It demonstrates convincingly that complexity in motives and information makes complexity in compliance policies a requirement. An assault on simplistic thinking in this area is always welcome.

Printed in the United Kingdom
by Lightning Source UK Ltd.
120370UK00001B/55